# An Introduction to Number Theory with Cryptography

## Second Edition

James S. Kraft
Lawrence C. Washington

## CRC Press

Taylor & Francis Group
Boca Raton   London   New York

CRC Press is an imprint of the
Taylor & Francis Group, an **informa** business

A CHAPMAN & HALL BOOK

# TEXTBOOKS in MATHEMATICS

## Series Editors: Al Boggess and Ken Rosen

# PUBLISHED TITLES CONTINUED

DIFFERENTIAL EQUATIONS: THEORY, TECHNIQUE, AND PRACTICE, SECOND EDITION
Steven G. Krantz

DIFFERENTIAL EQUATIONS: THEORY, TECHNIQUE, AND PRACTICE WITH BOUNDARY VALUE PROBLEMS
Steven G. Krantz

DIFFERENTIAL EQUATIONS WITH APPLICATIONS AND HISTORICAL NOTES, THIRD EDITION
George F. Simmons

DIFFERENTIAL EQUATIONS WITH MATLAB®: EXPLORATION, APPLICATIONS, AND THEORY
Mark A. McKibben and Micah D. Webster

DISCOVERING GROUP THEORY: A TRANSITION TO ADVANCED MATHEMATICS
Tony Barnard and Hugh Neill

DISCRETE MATHEMATICS, SECOND EDITION
Kevin Ferland

ELEMENTARY DIFFERENTIAL EQUATIONS
Kenneth Kuttler

ELEMENTARY LINEAR ALGEBRA
James R. Kirkwood and Bessie H. Kirkwood

ELEMENTARY NUMBER THEORY
James S. Kraft and Lawrence C. Washington

THE ELEMENTS OF ADVANCED MATHEMATICS: FOURTH EDITION
Steven G. Krantz

ESSENTIALS OF MATHEMATICAL THINKING
Steven G. Krantz

EXPLORING CALCULUS: LABS AND PROJECTS WITH MATHEMATICA®
Crista Arangala and Karen A. Yokley

EXPLORING GEOMETRY, SECOND EDITION
Michael Hvidsten

EXPLORING LINEAR ALGEBRA: LABS AND PROJECTS WITH MATHEMATICA®
Crista Arangala

EXPLORING THE INFINITE: AN INTRODUCTION TO PROOF AND ANALYSIS
Jennifer Brooks

GRAPHS & DIGRAPHS, SIXTH EDITION
Gary Chartrand, Linda Lesniak, and Ping Zhang

INTRODUCTION TO ABSTRACT ALGEBRA, SECOND EDITION
Jonathan D. H. Smith

# PUBLISHED TITLES CONTINUED

# PUBLISHED TITLES CONTINUED

CRC Press
Taylor & Francis Group
6000 Broken Sound Parkway NW, Suite 300
Boca Raton, FL 33487-2742

© 2018 by Taylor & Francis Group, LLC
CRC Press is an imprint of Taylor & Francis Group, an Informa business

No claim to original U.S. Government works

Version Date: 20171221

International Standard Book Number-13: 978-1-1380-6347-1 (Hardback)

**Visit the Taylor & Francis Web site at**
**http://www.taylorandfrancis.com**

**and the CRC Press Web site at**
**http://www.crcpress.com**

MIX
Paper from
responsible sources
FSC
www.fsc.org    FSC® C013056

Printed and bound in Great Britain by
TJ International Ltd, Padstow, Cornwall

# Dedication

*To Kristi, Danny, and Aaron and*
*to Miriam Kraft and the memory of Norman Kraft and Steven Kraft*


*To Susan and Patrick and*
*to Ida H. Washington and Lawrence M. Washington*

# Contents

# Preface

Number theory has a rich history. For many years it was one of the purest areas of pure mathematics, studied because of the intellectual fascination with properties of integers. More recently, it has been an area that also has important applications to subjects such as cryptography. The goal of this book is to present both sides of the picture, giving a selection of topics that we find exciting.

The book is designed to be used at several levels. It should fit well with an undergraduate course in number theory, but the book has also been used in a course for advanced high school students. It could also be used for independent study. We have included several topics beyond the standard ones covered in classes in order to open up new vistas to interested students.

The main thing to remember is number theory is supposed to be fun. We hope you enjoy the book.

**The Chapters.** The flowchart (following this preface) gives the dependencies of the chapters. When a section number occurs with an arrow, it means that only that section is needed for the chapter at the end of the arrow. For example, only the statement of quadratic reciprocity (Section 13.1) from Chapter 13 is needed in Chapter 14.

The core material is Chapters 2, 3, 4, 6, 8, 11, along with Section 13.1. These should be covered if at all possible. At this point, there are several possibilities. It is highly recommended that some sections of Chapters 5, 7, 9, and 12 be covered. These present some of the exciting applications of number theory to various problems, especially in cryptography. If time permits, some of the more advanced topics from Chapters 14 through 20 can be covered. These chapters are mostly independent of one another, so the choices depend on the interests of the audience.

We have tried to keep the prerequisites to a minimum. Appendix A treats some topics such as methods of proof, induction, and the binomial theorem. Our experience is that many students have seen these topics but that a review is worthwhile. The appendix also treats Fibonacci numbers since they occur as examples in various places throughout the book.

**Notes to the reader.** At the end of each chapter, we have a short list of Chapter Highlights. We were tempted to use the label "If you don't know these, no one will believe you read the chapter." In other words, when you finish a chapter, make sure you thoroughly know the highlights. (Of course, there is more that is worth knowing.) At the end of several sections, there are problems labeled "CHECK YOUR UNDERSTANDING." These are problems that check whether you have learned some basic ideas. The solutions to these are given at the ends of the chapters. You should not leave a topic until you can do these problems easily.

**Problems.** At the end of every chapter, there are problems to solve. The *Exercises* are intended to give practice with the concepts and sometimes to introduce interesting ideas related to the chapter's topics. The *Projects* are more substantial problems. Often, they consist of several steps that develop ideas more extensively. Although there are exceptions, generally they should take much longer to complete. Several could be worked on in groups. Computations have had a great influence on number theory and the *Computer Explorations* introduce this type of experimentation. Sometimes they ask for specific data, sometimes they are more open-ended. But they represent the type of exploration that number theorists often do in their research.

Appendix B contains answers or hints for the odd-numbered problems. For the problems where the answer is a number, the answer is given. When the exercise asks for a proof, usually a sketch or a key step is given.

**Computers.** Many students are familiar with computers these days and many have access to software packages such as Sage, Pari, or various commercial products that perform number theoretical calculations with ease. Some of the exercises (the ones that use numbers of five or more digits) are intended to be used in conjunction with a computer. Many can probably be done with an advanced calculator. The Computer Explorations are definitely designed for students with computer skills.

**Acknowledgments.** Jim Kraft wants to thank the Gilman School for its generous support during the writing of this book and his students Rishi Bedi, John Chirikjian, Anthony Kim, and John Lee, whose comments helped make this a better book. Many thanks are also due to Manjit Bhatia, who made many very useful suggestions. We both want to thank our many students over the years who have taught us while we have taught them. This book would not have been possible without them.

We welcome comments, corrections, and suggestions. Corrections and related matter will be listed on the web site for the book (www.math.umd.edu/~lcw/numbertheory.html).

# What Is New in the 2nd Edition?

- Appendices on "What Is a Proof?" and on Matrices.

- Increased coverage of cryptography, including Vigenère, Stream, Transposition, and Block ciphers.

- Some of the basic (pre-RSA) cryptography has been moved to an earlier chapter so that it can be covered immediately after the basic material on congruences.

- Approximately 250 new exercises, so there are now over 800 exercises, projects, and computer explorations.

*James S. Kraft*
*Gilman School*
*jkraft@gilman.edu*

*Lawrence C. Washington*
*University of Maryland*
*lcw@math.umd.edu*

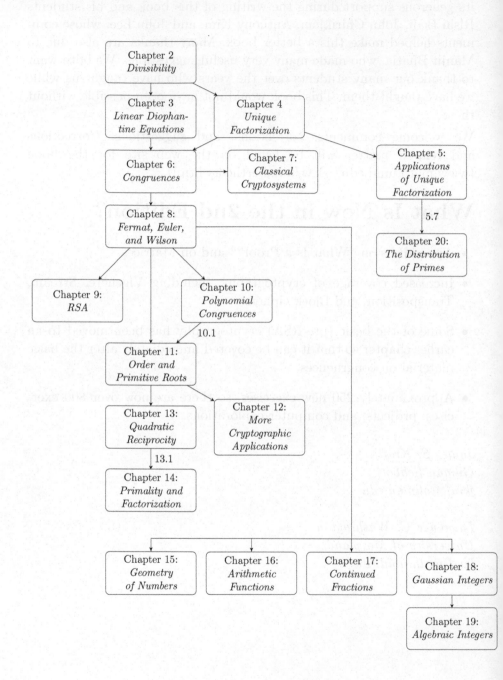

# Chapter 1

# Introduction

At Columbia University there is a Babylonian clay tablet called Plimpton 322 that is over 3800 years old and not much larger than a cell phone. Written in cuneiform script with four columns and 15 rows, it contains numbers written in base 60 (just as base 10 is standard today, base 60 was standard in Babylon). Each row gives a Pythagorean triple, that is, three whole numbers $x, y, z$ satisfying

$$x^2 + y^2 = z^2$$

(for example, $3^2 + 4^2 = 5^2$ and $12709^2 + 13500^2 = 18541^2$ are triples from the tablet). This is one of the earliest examples where integers are studied for their interesting properties, not just for counting objects.

Throughout history, there has been a fascination with whole numbers. For example, the Pythagorean school (ca. 500 BCE) believed strongly that every quantity could be expressed in terms of integers or ratios of integers, and they successfully applied the idea to the theory of musical scales. However, this overriding belief received a sharp setback when one of Pythagoreans, possibly Hippasus, proved that $\sqrt{2}$ is irrational. There is a story, which may be apocryphal, that he discovered this at sea and was promptly thrown overboard by his fellow Pythagoreans. Despite their attempt at suppressing the truth, the news of this discovery soon got out. Nevertheless, even though irrational numbers exist and are plentiful, properties of integers are still important.

Approximately 200 years after Pythagoras, Euclid's *Elements*, perhaps the most important mathematics book in history, was published. Although most people now think of the *Elements* as a book concerning geometry, a large portion of it is devoted to the theory of numbers. Euclid proves that there are infinitely many primes, demonstrates fundamental properties concerning divisibility of integers, and derives a formula that yields *all* possible Pythagorean triples, as well as many other seminal results. We will see and prove these results in the first five chapters of this book.

Number theory is a rich subject, with many aspects that are inextricably intertwined but which also retain their individual characters. In this introduction, we give a brief discussion of some of the ideas and some of the history of number theory as seen through the themes of Diophantine equations, modular arithmetic, the distribution of primes, and cryptography.

# 1.1   Diophantine Equations

Diophantus lived in Alexandria, Egypt, about 1800 years ago. His book *Arithmetica* gives methods for solving various algebraic equations and had a great influence on the development of algebra and number theory for many years. The part of number theory called **Diophantine equations**, which studies integer (and sometimes rational) solutions of equations, is named in his honor. However, the history of this subject goes back much before him. The Plimpton tablet shows that the Babylonians studied integer solutions of equations. Moreover, the Indian mathematician Baudhāyana ($\approx$ 800 BCE) looked at the equation $x^2 - 2y^2 = 1$ and found the solutions $(x, y) = (17, 12)$ and $(577, 408)$. The latter gives the approximation $577/408 \approx 1.4142157$ for $\sqrt{2}$, which is the diagonal of the unit square. This was a remarkable achievement considering that, at the time, a standardized system of algebraic notation did not yet exist.

The equation

$$x^2 - ny^2 = 1,$$

where $n$ is a positive integer not a square, was studied by Brahmagupta (598-668) and later mathematicians. In 1768, Joseph-Louis Lagrange (1736-1813) presented the first published proof that this equation always has a nontrivial solution (that is, with $y \neq 0$). Leonhard Euler (1707-1783) mistakenly attributed some work on this problem to the English mathematician John Pell (1611-1685), and ever since it has been known as **Pell's equation**, but there is little evidence that Pell did any work on it. In Chapters 15 and 17, we show how to solve Pell's equation, and in Chapter 19, we discuss its place in algebraic number theory.

Perhaps $x^2 + y^2 = z^2$, the equation for Pythagorean triples, is the most well-known Diophantine equation. Since sums of two nonzero squares can be a square, people began to wonder if this could be generalized. For example, Abu Mohammed Al-Khodjandi, who lived in the late 900s, claimed to have a proof that a sum of nonzero cubes cannot be a cube (that is, the equation $x^3 + y^3 = z^3$ has no nonzero solutions). Unfortunately, our only knowledge of this comes from another manuscript, which mentions that Al-Khodjandi's proof was defective, but gives no evidence to support this claim. The real excitement began when the great French mathematician Pierre de Fermat (1601-1665) penned a note in the margin of his copy of Diophantus's *Arithmetica* saying that it is impossible to solve $x^n + y^n = z^n$ in positive integers when $n \geq 3$ and that he had found a truly marvelous proof that the margin was too small to contain. After Fermat's son, Samuel Fermat, published an edition of Diophantus's book that included his father's comments, the claim became known as **Fermat's Last Theorem**. Today, it is believed that he actually had proofs only in the cases $n = 4$ (the only surviving proof by Fermat of any of his results) and possibly $n = 3$. But the statement acquired a life of its own and led to many developments in mathematics. Euler is usually credited with the first complete proof that Fermat's Last Theorem (abbreviated as FLT) is true for $n = 3$. Progress proceeded exponent by exponent, with Adrien-Marie Legendre (1752-1833) and Johann Peter Gustav Lejeune Dirichlet (1805-1859) each doing the case $n = 5$ around 1825 and Gabriel Lamé (1795-1870) treating $n = 7$ in 1839. Important general results were obtained by Sophie Germain (1776-1831), who showed that if $p < 100$ is prime and $xyz$ is not a multiple of $p$, then $x^p + y^p \neq z^p$.

The scene changed dramatically around 1850, when Ernst Eduard Kummer (1810-1893) developed his theory of *ideal numbers*, which are now known as *ideals* in ring theory. He used them to give general criteria that allowed him to prove FLT for all exponents up to 100, and many beyond that. His approach was a major step in the development of both algebraic number theory and abstract algebra, and it dominated the research on FLT until the 1980s. In the 1980s, new methods, based on work by Taniyama, Shimura, Weil, Serre, Langlands, Tunnell, Mazur, Frey, Ribet, and others, were brought to the problem, resulting in the proof of Fermat's Last Theorem by Andrew Wiles (with the help of Richard Taylor) in 1994. The techniques developed during this period have opened up new areas of research and have also proved useful in solving many classical mathematical problems.

## 1.2    Modular Arithmetic

Suppose you divide $1234^{25147}$ by 25147. What is the remainder? Why should you care? A theorem of Fermat tells us that the remainder is 1234. Moreover, as we'll see, results of this type are surprisingly vital in cryptographic applications (see Chapters 7, 9, and 12).

Questions about divisibility and remainders form the basis of modular arithmetic, which we introduce in Chapter 6. This is a very old topic and its development is implicit in the work of several early mathematicians. For example, the Chinese Remainder Theorem is a fundamental and essential result in modular arithmetic and was discussed by Sun Tzu around 1600 years ago.

Although early mathematicians discovered number theoretical results, the true beginnings of modern number theory began with the work of Fermat, whose contributions were both numerous and profound. We will discuss several of them in this book. For example, he proved that if $a$ is a whole number and $p$ is a prime, then $a^p - a$ is always a multiple of $p$. Results such as this are best understood in terms of modular arithmetic.

Euler and Karl Friedrich Gauss (1777-1855) greatly extended the work done by Fermat. Gauss's book *Disquisitiones Arithmeticae*, which was published in 1801, gives a treatment of modular arithmetic that is very close to the present-day version. Many of the original ideas in this book laid the groundwork for subsequent research in number theory.

One of Gauss's crowning achievements was the proof of Quadratic Reciprocity (see Chapter 13). Early progress towards this fundamental result, which gives a subtle relation between squares of integers and prime numbers, had been made by Euler and by Legendre. Efforts to generalize Quadratic Reciprocity to higher powers led to the development of algebraic number theory in the 1800s by Kummer, Richard Dedekind

(1831-1916), David Hilbert (1862-1943), and others. In the first half of the 1900s, this culminated in the development of *class field theory* by many mathematicians, including Hilbert, Weber, Takagi, and Artin. In the second half of the 1900s up to the present, the *Langlands Program*, which can be directly traced back to Quadratic Reciprocity, has been a driving force behind much number-theoretic research. Aspects of it played a crucial role in Wiles's proof of Fermat's Last Theorem in 1994.

## 1.3    Primes and the Distribution of Primes

> *There are two facts about the distribution of prime numbers of which I hope to convince you so overwhelmingly that they will be permanently engraved in your hearts. The first is that, despite their simple definitions and role as the building blocks of the natural numbers, the prime numbers belong to the most arbitrary and ornery objects studied by mathematicians: they grow like weeds among the natural numbers, seeming to obey no other law than that of chance, and nobody can predict where the next one will sprout. The second fact is even more astonishing, for it states just the opposite: that the prime numbers exhibit stunning regularity, that there are laws governing their behavior, and that they obey these laws with almost military precision.* - Don Zagier

Euclid proved that there are infinitely many primes, but we can ask for more precise information. Let $\pi(x)$ be the number of primes less than or equal to $x$. Legendre and Gauss used experimental data to conjecture that

$$\frac{\pi(x)}{x/\ln x} \approx 1,$$

and this approximation gets closer to equality as $x$ gets larger. For example,

$$\frac{\pi(10^4)}{10^4/\ln 10^4} = 1.132, \quad \text{and} \quad \frac{\pi(10^{10})}{10^{10}/\ln 10^{10}} = 1.048.$$

In 1852, Pafnuty Chebyshev (1821-1894) showed that the conjecture of

Legendre and Gauss is at least approximately true by showing that, for sufficiently large values of $x$,

$$0.921 \leq \frac{\pi(x)}{x/\ln x} \leq 1.106,$$

a result we'll discuss in Chapter 20. A few years later, Bernhard Riemann (1826-1866) introduced techniques from the theory of complex variables and showed how they could lead to more precise estimates for $\pi(x)$. Finally, in 1896, using Riemann's ideas, Jacques Hadamard (1865-1963) and Charles de la Vallée-Poussin (1866-1962) independently proved that

$$\lim_{x \to \infty} \frac{\pi(x)}{x/\ln x} = 1,$$

a result known as the **Prime Number Theorem**.

If we look at the list of *all* integers, we know that within that list there is an infinite number of primes. Suppose we look at a list like this:

$$1, 6, 11, 16, 21, 26, \ldots,$$

or like this:

$$3, 13, 23, 33, 43, 53, \ldots,$$

or like this:

$$1, 101, 201, 301, 401, \ldots.$$

Does each of the three lists contain an infinite number of primes as well? The answer is yes and we owe the proof of this remarkable fact to Dirichlet. In 1837, he proved that every arithmetic progression of the form $a, a + b, a + 2b, a + 3b, \ldots$ contains infinitely many primes if $a$ and $b$ are positive integers with no common factor greater than 1. We will not prove this result in this book; however, special cases are Projects and Exercises in Chapters 2, 8, and 13.

There are many other questions that can be asked about primes. One of the most famous is the **Goldbach Conjecture**. In 1742, Christian Goldbach (1690-1764) conjectured that every even integer greater than 2 is a sum of two primes (for example, $100 = 83 + 17$). Much progress has been made on this conjecture over the last century. In 1937, I. M. Vinogradov (1891-1983) proved that every sufficiently large odd integer is a sum of three primes, and in 2013, Harald Helfgott completed Vinogradov's work by showing that every odd number greater than 7 is a sum of three odd primes. In 1966, Jingrun Chen (1933-1996) proved that every sufficiently large even integer is either a sum of two primes or the sum of a prime and a number that is the product of two primes (for example, $100 = 23 + 7 \cdot 11$). These results require very delicate analytic techniques. Work on Goldbach's Conjecture and related questions remains a very active area of modern research in number theory.

# 1.4   Cryptography

For centuries, people have sent secret messages by various means. But in the 1970s, there was a dramatic change when Fermat's theorem and Euler's theorem (a generalization of Fermat's theorem), along with other results in modular arithmetic, became fundamental ingredients in many cryptographic systems. In fact, whenever you buy something over the Internet, it is likely that you are using Euler's theorem.

In 1976, Whitfield Diffie and Martin Hellman introduced the concept of public key cryptography and also gave a key establishment protocol (see Chapter 12) that uses large primes. A year later, Ron Rivest, Adi Shamir, and Len Adleman introduced the RSA cryptosystem (see Chapter 9), an implementation of the public key concept. It uses large prime numbers and its security is closely tied to the difficulty of factoring large integers.

Topics such as factorization and finding primes became very popular and soon there were several major advances in these subjects. For example, in the mid-1970s, factorization of 40-digit numbers was the limit of technology. As of 2016, the limit was 230 digits. Some of these factorization methods will be discussed in Chapter 14.

Cryptography brought about a fundamental change in how number theory is viewed. For many years, number theory was regarded as one of the purest areas of mathematics, with little or no application to real-world problems. In 1940, the famous British number theorist G. H. Hardy (1877-1947) declared, "No one has yet discovered any warlike purpose to be served by the theory of numbers or relativity, and it seems unlikely that anyone will do so for many years" (*A Mathematician's Apology*, Section 28). Clearly this statement is no longer true.

Although the basic purpose of cryptography is to protect communications, its ideas have inspired many related applications. In Chapters 9 and 12, we'll explain how to sign digital documents, along with more light-hearted topics such as playing mental poker and flipping coins over the telephone.

# Chapter 2

## Divisibility

## 2.1 Divisibility

A large portion of this book will be spent studying properties of the integers. You can add, subtract, and multiply integers and doing so always gives you another integer. Division is a little trickier. Sometimes when you divide one integer by another you get an integer (12 divided by 3) and sometimes you don't (12 divided by 5). Because of this, the first idea we have to make precise is that of divisibility.

**Definition 2.1.** *Given two integers $a$ and $d$ with $d$ non-zero, we say that $d$ **divides** $a$ (written $d \mid a$) if there is an integer $c$ with $a = cd$. If no such integer exists, so $d$ does **not** divide $a$, we write $d \nmid a$. If $d$ divides $a$, we say that $d$ is a **divisor** of $a$.*

**Examples.** $5 \mid 30$ since $30 = 5 \cdot 6$, and $3 \mid 102$ since $102 = 3 \cdot 34$, but $6 \nmid 23$ (since $23/6$ is not an integer) and $4 \nmid -3$ (since $4/(-3)$ is not an integer). Also, $-7 \mid 35$, $8 \mid 8$, $3 \mid 0$, $-2 \mid -10$, and $1 \mid 4$.

**Remark.** There are two technical points that need to be mentioned. First, we never consider 0 to be a divisor of anything. Of course, we could agree that $0 \mid 0$, but it's easiest to avoid this case completely since we never need it. Second, if $d$ is a divisor of $a$, then $-d$ is a divisor of $a$. However, whenever we talk about the *set of divisors* of a positive integer, we follow the convention that we mean the *positive* divisors. So we say that the divisors of 6 are 1, 2, 3, and 6 (and ignore $-1$, $-2$, $-3$, $-6$).

There are several basic results concerning divisibility that we will be using throughout this book.

**Proposition 2.2.**[1] *Assume that a, b, and c are integers. If $a \mid b$ and $b \mid c$, then $a \mid c$.*

*Proof.* Since $a \mid b$, we can write $b = ea$ and since $b \mid c$, we can write $c = fb$ with $e$ and $f$ integers. Then, $c = fb = f(ea) = (fe)a$. So, by definition, $a \mid c$.                                              □

**Example.** The proposition implies, for example, that a multiple of 6 is even: Let $a = 2$ and $b = 6$, and let $c$ be an arbitrary integer. Then $a \mid b$. If $6 \mid c$, the proposition says that $2 \mid c$, which says that $c$ is even.

**Proposition 2.3.** *Assume that $a, b, d, x$, and $y$ are integers. If $d \mid a$ and $d \mid b$ then $d \mid ax + by$.*

*Proof.* Write $a = md$ and $b = nd$. Then

$$ax + by = (md)x + (nd)y = d(mx + ny),$$

so $d \mid ax + by$ by definition.                                              □

Often, $ax + by$ is called a *linear combination* of $a$ and $b$, so Proposition 2.3 says that every divisor of both $a$ and $b$ is also a divisor of each linear combination of $a$ and $b$.

**Corollary 2.4.** *Assume that $a, b$, and $d$ are integers. If $d \mid a$ and $d \mid b$, then $d \mid a + b$ and $d \mid a - b$.*

*Proof.* To show that $d \mid a + b$, set $x = 1$ and $y = 1$ in the proposition and to show that $d \mid a - b$, set $x = 1$ and $y = -1$ in the proposition.   □

**Examples.** Since $3 \mid 9$ and $3 \mid 21$, the proposition tells us that

$$3 \mid 5 \cdot 9 + 4 \cdot 21 = 129.$$

Since $5 \mid 20$ and $5 \mid 30$, we have

$$5 \mid 20 + 30 = 50.$$

---

[1]There is a set of names for results: A *theorem* is an important result that is usually one of the highlights of the subject. A *proposition* is an important result, but not as important as a theorem. A *lemma* is a result that helps to prove a proposition or a theorem. It is often singled out because it is useful and interesting in its own right. A *corollary* is a result that is an easy consequence of a theorem or proposition.

Since $10 \mid 40$ and $10 \mid 60$, we have $10 \mid 40 - 60 = -20$.

---

## CHECK YOUR UNDERSTANDING[2]

1. Does 7 divide 1001?
2. Show that $7 \nmid 1005$.

---

# 2.2   Euclid's Theorem

Fundamental to the study of the integers is the idea of a prime number.

**Definition 2.5.** *A **prime number** is an integer $p \geq 2$ whose only divisors are 1 and $p$. A **composite number** is an integer $n \geq 2$ that is not prime.*

You may be wondering why 1 is not considered to be prime. After all, its only divisors are 1 and itself. Although there have been mathematicians in the past who have included 1 in the list of primes, nobody does so anymore. The reason for this is that mathematicians want to say there's *exactly one* way to factor an integer into a product of primes. If 1 were a prime, and we wanted to factor 6, for example, we'd have $6 = 2 \cdot 3 = 2 \cdot 3 \cdot 1 = 2 \cdot 3 \cdot 1 \cdot 1, \dots$ and we would have an infinite number of ways to factor an integer into primes. So, to avoid this, we simply declare that 1 is not prime.

The first ten prime numbers are

$$2, 3, 5, 7, 11, 13, 17, 19, 23, 29.$$

Notice that 2 is prime because its only divisors are 1 and 2, but no other even number can be prime because every other even number has 2 as a divisor.

It's natural to ask if the list of primes ever terminates. It turns out that it doesn't; that is, there are infinitely many primes. This fact is one of the most basic results in number theory. The first written record we

---

[2] Answers are at the end of the chapter.

have of it is in Euclid's *Elements*, which was written over 2300 years ago. In the next section, we'll discuss Euclid's original proof. Before we do that, here's a proof that is a variation of his idea. We begin with a lemma.

**Lemma 2.6.** *Every integer greater than 1 is either prime or is divisible by a prime.*

*Proof.* If an integer $n$ is not a prime, then it is divisible by some integer $a_1$, with $1 < a_1 < n$. If $a_1$ is prime, we've found a prime divisor of $n$. If $a_1$ is not prime, it must be divisible by some integer $a_2$ with $1 < a_2 < a_1$. If $a_2$ is prime, then since $a_2 \mid a_1$ and $a_1 \mid n$, we have $a_2 \mid n$, and $a_2$ is a prime divisor of $n$. If $a_2$ is not prime, we continue and get a decreasing sequence of positive integers

$$a_1 > a_2 > a_3 > a_4 > \cdots ,$$

all of which are divisors of $n$. Since you can't have a sequence of *positive* integers that decreases forever, this sequence must stop at some $a_m$ (this is an application of the Well-Ordering Principle; see Appendix A). The fact that the sequence stops means that $a_m$ must be prime, which means that $a_m$ is a prime divisor of $n$. □

**Example.** In the proof of the lemma, suppose $n = 72000 = 720 \times 100$. Take $a_1 = 720 = 10 \times 72$. Take $a_2 = 10 = 5 \times 2$. Finally, take $a_3 = 5$, which is prime. Working backwards, we see that $5 \mid 72000$.

**Euclid's Theorem.** *There are infinitely many primes.*

*Proof.* We assume that there is a finite number of primes and arrive at a contradiction. So, let

$$2, 3, 5, 7, 11, ..., p_n \tag{2.1}$$

be the list of all the prime numbers. Form the integer

$$N = 2 \cdot 3 \cdot 5 \cdot 7 \cdot 11 \cdots p_n + 1.$$

To begin, $N$ can't be prime since it's larger than $p_n$ and $p_n$ is assumed to be the largest prime. So, we can use the previous lemma to choose a prime divisor $p$ of $N$. Since Equation (2.1) is a list of every prime, $p$ is equal to one of the $p_i$ and therefore must divide $2 \cdot 3 \cdot 5 \cdot 7 \cdot 11 \cdots p_n$. But $p$ now divides both $N$ and $N - 1 = 2 \cdot 3 \cdot 5 \cdot 7 \cdot 11 \cdots p_n$. By Corollary 2.4, $p$ divides their difference, which is 1. This is a contradiction: $p \nmid 1$ because $p > 1$. This means that our initial assumption that there is a finite number of primes must be incorrect. $\qquad\square$

Since mathematicians like to prove the same result using different methods, we'll give several other proofs of this result throughout the book. As you'll see, each new proof will employ a different idea in number theory, reflecting the fact that Euclid's theorem is connected with many of its branches.

Here's one example of an alternative proof.

**Another Proof of Euclid's Theorem.** We'll show that for each $n > 0$, there is a prime number larger than $n$. Let $N = n! + 1$ and let $p$ be a prime divisor of $N$. Either $p > n$ or $p \leq n$. If $p > n$, we're done. If $p \leq n$, then $p$ is a factor of $n!$, so $p \mid N - 1$. Recall that $p$ was chosen so that $p \mid N$, so we now have $p \mid N$ and $p \mid N - 1$. Therefore, $p \mid N - (N - 1) = 1$, which is impossible. This means that $p \leq n$ is impossible, so we must have $p > n$.

In particular, if $n$ is prime, there is a prime $p$ larger than $n$, so there is no largest prime. This means that there are infinitely many primes. $\square$

---

**CHECK YOUR UNDERSTANDING**

3. Explain why $5 \nmid 2 \cdot 3 \cdot 5 \cdot 7 + 1$.

---

# 2.3  Euclid's Original Proof

Here is Euclid's proof that there is an infinite number of primes, using the standard translation of Sir Thomas Heath. Euclid's statements are written in italics. Since his terminology and notation may be unfamiliar, we have added comments in plaintext where appropriate. It will be helpful to know that when Euclid says *"A measures B"* or *"B is measured by A,"* he means that $A$ divides $B$ or, equivalently, that $B$ is a multiple of $A$.

What Euclid actually proved was that if we start with any finite set of primes, there is a prime not contained in this set. The modern interpretation is that there are infinitely many primes.

| **Euclid's Statements** | **Explanation** |
| --- | --- |
| *Let A, B, and C be the assigned prime numbers.* | This is the statement that we have some finite set of primes. Instead of assuming that there are $n$ of them as we did, Euclid assumes that there are only three. You can think of this as representing some arbitrary, unknown number of primes. |
| *I say that there are more prime numbers than A, B, and C.* | I will show that there is a prime not in our set. |
| *Take the least number DE measured by A, B, and C. Add the unit DF to DE.* | In this step, Euclid multiplies all the primes together and then adds 1. So, DE is the least common multiple of A, B, and C, and EF = DE + 1. |
| *Then EF is either prime or not. Let it be prime.* | Either EF is prime or it's not. First, assume that it's prime. |
| *Then the prime numbers A, B, C, and EF have been found which are more than A, B, and C.* | This yields a prime that is not in our set. |
| *Next, let EF not be prime. Therefore it is measured by some prime number. Let it be measured by the prime number G.* | Next, assume that EF is not prime. Then, EF is a multiple of some prime G. |
| *I say that G is not the same with any of the numbers A, B,* | We claim that G is not in our set. |

*and C.*

*For if possible, let it be so. Now A, B, and C measure DE, therefore G also will measure DE.*

Assume that G is in our set. Since DE is a multiple of A, and of B, and of C and since G is one of the listed primes, DE must also be a multiple of G.

*But it also measures EF.*

But EF is also a multiple of G.

*Therefore G, being a number, will measure the remainder, the unit DF, which is absurd.*

Since EF is a multiple of G and DE = EF + 1 is a multiple of G, their difference (EF + 1 − EF ), which equals 1, is also a multiple of G. This is a contradiction.

*Therefore G is not the same with any one of the numbers A, B, and C. And by hypothesis it is prime. Therefore the prime numbers A, B, C, and G have been found which are more than the assigned multitude of A, B, and C. Therefore, prime numbers are more than any assigned multitude of prime numbers. Q.E.D.*

So G is a prime number that is not in our set. And by hypothesis it is prime. Therefore the set consisting of the original set with G adjoined is a set of primes that is larger than the original finite set. Therefore, the set of all prime numbers is larger than any given finite set. □

## 2.4    The Sieve of Eratosthenes

Eratosthenes was born in Cyrene (in modern-day Libya) and lived in Alexandria, Egypt, around 2300 years ago. He made important contributions to many subjects, especially geography. In number theory, he is famous for a method of producing a list of prime numbers up to a given bound without using division. To see how this works, we'll find all the prime numbers up to 50.

List the integers from 1 to 50. Ignore 1 and put a circle around 2. Now cross out every second number after 2. This gives us (we display just the beginning of the list)

$$1 \quad ② \quad 3 \quad \cancel{4} \quad 5 \quad \cancel{6} \quad 7 \quad \cancel{8} \quad 9 \quad \cancel{10}$$
$$11 \quad \cancel{12} \quad 13 \quad \cancel{14} \quad 15 \quad \cancel{16} \quad 17 \quad \cancel{18} \quad 19 \quad \cancel{20}$$

Now look at the next number after 2 that is not crossed out. It's 3. Put a circle around 3 and cross out every third number after 3. This gives us

$$1 \quad ② \quad ③ \quad \cancel{4} \quad 5 \quad \cancel{6} \quad 7 \quad \cancel{8} \quad \cancel{9} \quad \cancel{10}$$
$$11 \quad \cancel{12} \quad 13 \quad \cancel{14} \quad \cancel{15} \quad \cancel{16} \quad 17 \quad \cancel{18} \quad 19 \quad \cancel{20}$$

The first number after 3 that is not crossed out is 5, so circle 5 and cross out every 5th number after 5. After we do this, the next number after 5 that is not crossed out is 7, so we cross out every 7th number after 7. Listing all numbers up to 50, we now have

$$1 \quad ② \quad ③ \quad \cancel{4} \quad ⑤ \quad \cancel{6} \quad ⑦ \quad \cancel{8} \quad \cancel{9} \quad \cancel{10}$$
$$11 \quad \cancel{12} \quad 13 \quad \cancel{14} \quad \cancel{15} \quad \cancel{16} \quad 17 \quad \cancel{18} \quad 19 \quad \cancel{20}$$
$$\cancel{21} \quad \cancel{22} \quad 23 \quad \cancel{24} \quad \cancel{25} \quad \cancel{26} \quad \cancel{27} \quad \cancel{28} \quad 29 \quad \cancel{30}$$
$$31 \quad \cancel{32} \quad \cancel{33} \quad \cancel{34} \quad \cancel{35} \quad \cancel{36} \quad 37 \quad \cancel{38} \quad \cancel{39} \quad \cancel{40}$$
$$41 \quad \cancel{42} \quad 43 \quad \cancel{44} \quad \cancel{45} \quad \cancel{46} \quad 47 \quad \cancel{48} \quad \cancel{49} \quad \cancel{50}.$$

The numbers that remain are 1 and the prime numbers up to 50. We can stop at 7 because of the following.

**Proposition 2.7.** *If n is composite, then n has a prime factor $p \le \sqrt{n}$.*

*Proof.* Since $n$ is composite, we can write $n = ab$ with $1 < a \le b < n$. Then
$$a^2 \le ab = n,$$
so $a \le \sqrt{n}$. Let $p$ be a prime number dividing $a$. Then $p \le a \le \sqrt{n}$. $\square$

The proposition says that the composite numbers up to 50 all have prime factors at most $\sqrt{50} \approx 7.07$, so we could stop after crossing out the multiples of 7. If we want to list the primes up to 1000, we need to cross out the multiples of only the primes through 31 (since $\sqrt{1000} \approx 31.6$).

Why is the process called a sieve? In our example, the multiples of the primes 2, 3, 5, 7 created a net. The numbers that fell through this net are the prime numbers.

---

## CHECK YOUR UNDERSTANDING

4. Use the Sieve of Eratosthenes to compute the prime numbers less than 20.

---

# 2.5    The Division Algorithm

If $a$ and $b$ are integers, when we divide $a$ by $b$ we get an integer if and only if $b \mid a$. What can we say when $b$ does not divide $a$? We can still make a statement using only integers by considering remainders. For example, we can say that 14 divided by 3 is 4 with a remainder of 2. We write this as

$$14 = 3 \cdot 4 + 2$$

to emphasize that our division statement can be written using addition and multiplication of integers. This is just the division with remainder that is taught in elementary school. Our next theorem says that this can always be done.

**The Division Algorithm.** *Let $a$ and $b$ be integers with $b > 0$. Then there exist unique integers $q$ (the quotient) and $r$ (the remainder) so that*

$$a = bq + r$$

*with $0 \leq r < b$.*

*Proof.* Let $q$ be the largest integer less than or equal to $a/b$, so

$$q \leq a/b < q + 1.$$

Multiplying by $b$ yields $bq \leq a < bq+b$, which implies that $0 \leq a-bq < b$. Let $r = a - bq$. Then

$$0 \leq r < b.$$

Since $a = bq + r$, we have proved that the desired $q$ and $r$ exist.

It remains to prove that $q$ and $r$ are unique. If

$$a = bq + r = bq_1 + r_1$$

with $0 \leq r, r_1 < b$ then

$$b(q - q_1) = r_1 - r.$$

Since the left-hand side of this equation is a multiple of $b$, so is $r_1 - r$. Because $0 \leq r, r_1 < b$, we must have

$$-b < r_1 - r < b. \tag{2.2}$$

The only multiple of $b$ that satisfies Equation (2.2) is 0, so $r_1 - r = 0$. Therefore, $r_1 = r$ and the choice of $r$ is unique. Since $b(q - q_1) = r_1 - r$, we now have $b(q - q_1) = 0$. Finally, because $b \neq 0$, we get that $q_1 - q = 0$, so $q_1 = q$ and $q$ is also unique. This completes the proof.    □

**Examples:** (a) Let $a = 27, b = 7$. Then $27 = 7 \cdot 3 + 6$, so $q = 3$ and $r = 6$.
(b) Let $a = -27, b = 7$. Then $-27 = 7 \cdot (-4) + 1$, so $q = -4$ and $r = 1$.
(c) Let $a = 24, b = 8$. Then $24 = 8 \cdot 3$, so $q = 3$ and $r = 0$.
(d) Let $a = 0$ and $b = 5$. Then $0 = 5 \cdot 0 + 0$, so $q = 0$ and $r = 0$.

If we write $a = bq + r$ with $0 \leq r < b$, then $b \mid a$ if and only if $r = 0$. This just says that if you divide $b$ into $a$ and the remainder is 0, then $b \mid a$, and if the remainder is not zero, then $b \nmid a$. Although this remark may seem obvious, it will be how we show one number divides another in various theoretical contexts. See, for example, the proofs of Theorem 2.12 and Theorem 11.1.

---

## CHECK YOUR UNDERSTANDING

5. Let $a = 200, b = 7$. Compute $q$ and $r$ such that $a = bq + r$ and $0 \leq r < b$.
6. Let $a = -200, b = 7$. Compute $q$ and $r$ such that $a = bq + r$ and $0 \leq r < b$.

---

## 2.5.1    A Cryptographic Application

Here's an amusing cryptographic application of the Division Algorithm. Let's say there is a 16-person committee that has to vote to approve a budget. The members prefer to keep their votes anonymous. Here's a mathematical way to have every person vote Yes, vote No, or Abstain, while ensuring that all votes are kept secret.

We'll call the chair $A_1$ and the other 15 members $A_2, A_3, ..., A_{16}$. The chair takes a blank piece of paper, writes a large number, say 7923, on it, and passes this to $A_2$. Then $A_2$ adds 17 for Yes, 1 for No, or 0 for Abstain. $A_2$ writes this sum on a new piece of paper, hands the new number to $A_3$, and returns the paper with 7923 written on it back to the chair. $A_3$ now has a piece of paper with either 7940 (if $A_2$ voted Yes), 7924 (if $A_2$ voted No), or 7923 (if $A_2$ abstained). Because $A_3$ does not know the original number, there is no way to know how $A_2$ voted. This process continues with $A_3$ adding 17 for Yes, 1 for No, or 0 for an abstention, and then passing the result to $A_4$. They continue until $A_{16}$ gives a number to $A_1$, who adds a number for $A_1$'s vote. Let's say the final sum is 8050. The chair subtracts the secret number 7923 from 8050 and gets 127. Then 127 is divided by 17 using the Division Algorithm:

$$127 = 7 \cdot 17 + 8$$

The chair announces that 7 people voted Yes, 8 people voted No, and there was 1 abstention (since $7 + 8$ is one less than 16, one person must have abstained).

Why do we count a Yes vote as 17 in this example? It's one more than the number of voters. If we used 16 for a Yes vote, we couldn't tell the difference between 16 No votes, and one Yes plus 15 abstentions since both give a total of 16.

Let's do another example with 23 people voting. Let's say the chair's random number is 27938. Now, committee members add 24 if they vote Yes and 1 if they vote No. We'll tell you what the votes were so that you can see why the method works. Let's say there are 16 Yes votes, 5 No votes, and 2 abstentions. Then the chair receives the number

$$27938 + 16 \cdot 24 + 5 + 2 \cdot 0 = 27938 + 389 = 28327.$$

Of course, when the chair subtracts 27938 from 28327 the answer is

389, and the Division Algorithm says that

$$389 = 16 \cdot 24 + 5.$$

The voting scheme does have a security flaw. If $A_2$ and $A_4$ compare notes, they can figure out how $A_3$ voted. Therefore, this method should be used only with a friendly committee.

## 2.6    The Greatest Common Divisor

The divisors of 12 are 1, 2, 3, 4, 6, and 12. The divisors of 18 are 1, 2, 3, 6, 9, and 18. Then $\{1, 2, 3, 6\}$ is the set of common divisors of 12 and 18. Notice that this set has a largest element, 6. If you have any two non-zero integers $a$ and $b$, you can always form the set of their common divisors. Since 1 is a divisor of every integer, this set is nonempty. Because this set is finite, it must have a largest element. This idea is so basic, we make special note of it:

**Definition 2.8.** *Assume that $a$ and $b$ are integers and they are not both zero. Then the set of their common divisors has a largest element $d$, called the* **greatest common divisor** *of $a$ and $b$. We write $d = \gcd(a, b)$.*

**Examples.** $\gcd(24, 52) = 4$, $\gcd(9, 27) = 9$, $\gcd(14, 35) = 7$, $\gcd(15, 28) = 1$.

**Definition 2.9.** *Two integers $a$ and $b$ are said to be* **relatively prime** *if $\gcd(a, b) = 1$.*

**Examples.** 14 and 15 are relatively prime. So are 21 and 40.

**Remark.** If $a \neq 0$, then $\gcd(a, 0) = a$. However, we do not define $\gcd(0, 0)$. Since arbitrarily large integers divide 0, there is no largest divisor. This is the reason we often explicitly write that at least one of $a$ and $b$ is nonzero when we are going to make a statement about $\gcd(a, b)$. In any case, whenever we write $\gcd(a, b)$, it is implicitly assumed that at least one of $a$ and $b$ is nonzero.

We saw that $\gcd(24, 52) = 4$, so 24 and 52 are not relatively prime. If we divide both 24 and 52 by 4, we get 6 and 13, which are relatively prime. This makes sense, since we've divided these numbers by their gcd, which is the largest possible common divisor. We now prove in the following that dividing two integers by their gcd always results in two relatively prime integers.

**Proposition 2.10.** *If $a$ and $b$ are integers with $d = \gcd(a, b)$, then*

$$\gcd\left(\frac{a}{d}, \frac{b}{d}\right) = 1.$$

*Proof.* If $c = \gcd(a/d, b/d)$, then $c \mid (a/d)$ and $c \mid (b/d)$. This means that there are integers $k_1$ and $k_2$ with

$$\frac{a}{d} = ck_1 \text{ and } \frac{b}{d} = ck_2,$$

which tells us that $a = cdk_1$ and $b = cdk_2$. So, $cd$ is a common divisor of $a$ and $b$. Since $d$ is the greatest common divisor and $cd \geq d$, we must have $c = 1$. □

The following is closely related to Proposition 2.10, but we will have to wait until Subsection 2.7.1 to give a complete proof.

**Proposition 2.11.** *If $a$ and $b$ are integers, not both 0, and $e$ is a positive integer, then*

$$\gcd(ea, eb) = e \cdot \gcd(a, b).$$

*Proof.* Let's try the "obvious" proof. Let $d = \gcd(ea, eb)$. Then clearly we can take $d$ to be a multiple of $e$, and the rest of $d$ is obtained by taking the largest possible divisor of $a$ and $b$. So $d = e \cdot \gcd(a, b)$. What's wrong with this? Well, can we obtain an even bigger divisor by not starting with $d$ being a multiple of $e$ and instead using some other method? The answer is "no," but this needs to be justified.

To satisfy anyone who is impatient, we'll give the proof now, with the understanding that it is not complete until we finish Subsection 2.7.1. We will prove in Theorem 2.12 in Subsection 2.7.1 that there exist integers $x, y$ such that $\gcd(a, b) = ax + by$. Multiply by $e$ to obtain $e \cdot \gcd(a, b) = eax + eby$. If $d$ is a common divisor of $ea$ and $eb$, then

$d$ divides $e \cdot \gcd(a, b)$, by Proposition 2.3. Therefore, $d \leq e \cdot \gcd(a, b)$. Since $e \cdot \gcd(a, b)$ is a common divisor of $ea$ and $eb$, it must be the gcd, as claimed.                                                                                    □

We'll see later that calculating the greatest common divisor has important applications. So, it's natural to ask, how do we go about finding the gcd when the answer is not immediately obvious? One way would be to factor each integer into primes and then take the product of all the primes that they have in common, including repetitions. For example, to find $\gcd(84, 264)$, we write

$$84 = 2 \cdot 2 \cdot 3 \cdot 7 \text{ and } 264 = 2 \cdot 2 \cdot 2 \cdot 3 \cdot 11,$$

so their common primes are 2, 2, and 3. We see that $\gcd(84, 264) = 2 \cdot 2 \cdot 3 = 12$. This may seem to be quite efficient but, as we'll see later on, for numbers of the size (i.e., hundreds of digits) that we'll be interested in for cryptography, factoring is so slow as to be completely impractical. It's much easier to calculate $\gcd(a, b)$ by the method of the next section.

For reasonably small numbers, Proposition 2.3 can useful for calculating gcd's. For example, suppose we want to calculate $d = \gcd(1005, 500)$. Then $d \mid 1005$ and $d \mid 500$, so $d \mid 1005 - 2 \cdot 500$. Therefore, $d \mid 5$, which means that $d = 1$ or 5. Since $5 \mid 1005$ and $5 \mid 500$, we see that $5 = \gcd(1005, 500)$.

As another example, suppose $n$ is an integer and we want to find all possibilities for $d = \gcd(2n + 3, 3n - 6)$. By Proposition 2.3,

$$d \mid 2n + 3, \quad d \mid 3n - 6 \Longrightarrow d \mid 3(2n + 3) - 2(3n - 6) = 21,$$

so $d = 1, 3, 7$, or 21. In fact, all possibilities occur: when $n = 1$ we have $d = \gcd(5, -3) = 1$, when $n = 3$ we have $d = \gcd(9, 3) = 3$, when $n = 2$ we have $d = \gcd(7, 0) = 7$, and when $n = 9$ we have $d = \gcd(21, 21) = 21$.

---

## CHECK YOUR UNDERSTANDING

7. Evaluate $\gcd(24, 42)$.

8. Find an $n$ with $1 < n < 10$ such that $\gcd(n, 60) = 1$.

9. Let $n$ be an integer. Show that $\gcd(n, n + 3) = 1$ or 3, and show that both possibilities occur.

# 2.7    The Euclidean Algorithm

The Euclidean Algorithm is one of the oldest and most useful algorithms in all of number theory. It is found as Proposition 2 in Book VII of Euclid's *Elements*. One of its features is that it allows us to compute gcd's without factoring. In cryptographic situations, where the numbers often have several hundred digits and are hard to factor, this is very important.

Suppose that we want to compute gcd(123, 456). Consider the following calculation:

$$456 = 3 \cdot 123 + 87$$
$$123 = 1 \cdot 87 + 36$$
$$87 = 2 \cdot 36 + 15$$
$$36 = 2 \cdot 15 + 6$$
$$15 = 2 \cdot 6 + 3$$
$$6 = 2 \cdot 3 + 0.$$

By looking at the prime factorizations of 456 and 123 we see that the last non-zero remainder, namely 3, is the gcd. Let's look at what we did. We divided the smaller of the original two numbers into the larger and got the remainder 87. Then we shifted the 123 and the 87 to the left, did the division, and got a remainder of 36. We continued the "shift left and divide" procedure until we got a remainder of 0.

Let's try another example. Compute gcd(119, 259):

$$259 = 2 \cdot 119 + 21$$
$$119 = 5 \cdot 21 + 14$$
$$21 = 1 \cdot 14 + 7$$
$$14 = 2 \cdot 7 + 0.$$

Again, the last non-zero remainder is the gcd. Why does this work? Let's start by showing why 7 is a common divisor in the second example. The fact that the remainder on the last line is 0 says that $7 \mid 14$. Since $7 \mid 7$ and $7 \mid 14$, the next-to-last line says that $7 \mid 21$, since 21 is a linear combination of 7 and 14. Now move up one line. We have just shown that $7 \mid 14$ and $7 \mid 21$. Since 119 is a linear combination of 21 and 14, we deduce that $7 \mid 119$. Finally, moving to the top line, we see that $7 \mid 259$ because 259 is a linear combination of 119 and 21, both of which are multiples of 7. Since $7 \mid 119$ and $7 \mid 259$, we have proved that 7 is a common divisor of 119 and 259.

We now want to show that 7 is the largest common divisor. Let $d$ be any divisor of 119 and 259. The top line implies that 21, which is a linear combination of 259 and 119 (namely, $259 - 2 \cdot 119$), is a multiple of $d$. Next, go to the second line. Both 119 and 21 are multiples of $d$, so 14 must be a multiple of $d$. The third line tells us that since $d \mid 21$ and $d \mid 14$, we must have $d \mid 7$. In particular, $d \leq 7$, so 7 is the greatest common divisor, as claimed. We also have proved the additional fact that any common divisor must divide 7.

All of this generalizes to the following:

**Euclidean Algorithm.** *Let $a$ and $b$ be non-negative integers and assume that $b \neq 0$. Do the following computation:*

$$a = q_1 b + r_1, \ \text{with } 0 \leq r_1 < b$$
$$b = q_2 r_1 + r_2, \ \text{with } 0 \leq r_2 < r_1$$
$$r_1 = q_3 r_2 + r_3, \ \text{with } 0 \leq r_3 < r_2$$
$$\vdots$$
$$r_{n-3} = q_{n-1} r_{n-2} + r_{n-1}, \ \text{with } 0 \leq r_{n-1} < r_{n-2}$$
$$r_{n-2} = q_n r_{n-1} + 0.$$

*The last non-zero remainder, namely $r_{n-1}$, equals $\gcd(a, b)$.*

The proof that $r_{n-1} = \gcd(a, b)$ follows exactly the steps used in the example of $7 = \gcd(259, 119)$. Since the last remainder is 0, $r_{n-1}$ divides $r_{n-2}$. The next-to-last line yields $r_{n-1} \mid r_{n-3}$. Moving up, line by line, we eventually find that $r_{n-1}$ is a common divisor of $a$ and $b$.

Now suppose that $d$ is a common divisor of $a$ and $b$. The first line yields $d \mid r_1$. Since $d \mid b$ and $d \mid r_1$, the second line yields $d \mid r_2$. Continuing downwards, line by line, we eventually find that $d \mid r_{n-1}$. Therefore, $d \leq r_{n-1}$, so $r_{n-1}$ is the largest divisor, which means that $r_{n-1} = \gcd(a, b)$. We also obtain the extra fact that each common divisor of $a$ and $b$ divides $\gcd(a, b)$.

There is a geometrical way to view the Euclidean Algorithm. For example, suppose we want to compute $\gcd(48, 21)$. Start at the point $(48, 21)$

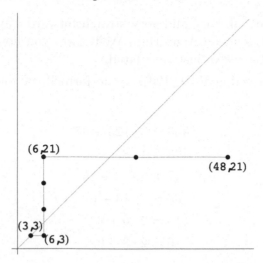

**FIGURE 2.1: Computation of gcd(48, 21).**

in the plane. Move to the left in steps of size 21 until you land on or
cross the line $y = x$. In this case, we take two steps of size 21 and move
to $(6, 21)$. Now move downward in steps of size 6 (the smaller of the
two coordinates) until you land on or cross the line $y = x$. In this case,
we take three steps of size 6 and move to $(6, 3)$. Now move to the left
in steps of size 3. In one step we end up at $(3, 3)$ on the line $y = x$. The
$x$-coordinate (also the $y$-coordinate) is the gcd.

In each set of moves, the number of steps is the quotient in the Euclidean Algorithm and the remainder is the amount that the last step
overshoots the line $y = x$.

## 2.7.1    The Extended Euclidean Algorithm

The Euclidean Algorithm yields an amazing and very useful fact:
$\gcd(a, b)$ can be expressed as a linear combination of $a$ and $b$. That
is, there exist integers $x$ and $y$ such that $\gcd(a, b) = ax + by$. For example,

$$3 = \gcd(456, 123) = 456 \cdot 17 - 123 \cdot 63$$
$$7 = \gcd(259, 119) = 259 \cdot 6 - 119 \cdot 13.$$

The method for obtaining $x$ and $y$ is called the **Extended Euclidean
Algorithm**. Once you've used the Euclidean Algorithm to arrive at

gcd($a$, $b$), there's an easy and very straightforward way to implement the Extended Euclidean Algorithm. We'll show you how it works with the two examples we've just calculated.

When we computed gcd($456, 123$), we performed the following calculation:

$$456 = 3 \cdot 123 + 87$$
$$123 = 1 \cdot 87 + 36$$
$$87 = 2 \cdot 36 + 15$$
$$36 = 2 \cdot 15 + 6$$
$$15 = 2 \cdot 6 + 3$$
$$6 = 2 \cdot 3 + 0.$$

We'll form a table with three columns and explain how they arise as we compute them.

We begin by forming two rows and three columns. The first entries in the rows are the numbers we started with. In this case these numbers are 456 and 123. The columns tell us how to form each of these numbers as a linear combination of 456 and 123. In other words, we will always have

$$\text{entry in first column } = 456x + 123y,$$

where $x$ and $y$ are integers. Initially, this is trivial: $456 = 1 \cdot 456 + 0 \cdot 123$ and $123 = 0 \cdot 456 + 1 \cdot 123$:

$$
\begin{array}{ccl}
 & x & y \\
456 & 1 & 0 \quad (456 = 1 \cdot 456 + 0 \cdot 123) \\
123 & 0 & 1 \quad (123 = 0 \cdot 456 + 1 \cdot 123).
\end{array}
$$

Now things get more interesting. If we look at the first line in our gcd($456$, $123$) calculation, we see $456 = 3 \cdot 123 + 87$. We rewrite this as $87 = 456 - 3 \cdot 123$. Using this as a guide, we compute

$$(\text{1st row}) \; - 3 \cdot (\text{2nd row}),$$

yielding the following

$$
\begin{array}{ccc}
 & x & y \\
456 & 1 & 0 \\
123 & 0 & 1 \\
87 & 1 & -3 \quad \text{(1st row)} - 3 \cdot \text{(2nd row).}
\end{array}
$$

The last line tells us that $87 = 456 \cdot 1 + 123 \cdot (-3)$.

We now move to the second row of our gcd calculation. This says that $123 = 1 \cdot 87 + 36$, which we rewrite as $36 = 123 - 1 \cdot 87$. Again, in the column and row language, this tells us to compute (2nd row) − (3rd row). We write this as

$$
\begin{array}{ccc}
 & x & y \\
456 & 1 & 0 \\
123 & 0 & 1 \\
87 & 1 & -3 \\
36 & -1 & 4 \quad \text{(2nd row)} - \text{(3rd row).}
\end{array}
$$

The last line tells us that $36 = 456 \cdot (-1) + 123 \cdot 4$.

Moving to the third row of our gcd calculation, we see that $15 = 87 - 2 \cdot 36 =$(3rd row) $-2 \cdot$(4th row) in our row and column language. This becomes

$$
\begin{array}{ccc}
 & x & y \\
456 & 1 & 0 \\
123 & 0 & 1 \\
87 & 1 & -3 \\
36 & -1 & 4 \\
15 & 3 & -11 \quad \text{(3rd row)} - 2 \cdot \text{(4th row).}
\end{array}
$$

We continue in this way and end when we have $3 = \gcd(456,\ 123)$ in the first column:

$$
\begin{array}{ccc}
 & x & y \\
456 & 1 & 0 \\
123 & 0 & 1 \\
87 & 1 & -3 \\
36 & -1 & 4 \\
15 & 3 & -11 \\
6 & -7 & 26 \quad \text{(4th row)} - 2 \cdot \text{(5th row)} \\
3 & 17 & -63 \quad \text{(5th row)} - 2 \cdot \text{(6th row).}
\end{array}
$$

This tells us that $3 = 456 \cdot 17 + 123 \cdot (-63)$.

Notice that as we proceeded, we were doing the Euclidean Algorithm in the first column. The first entry of each row is a remainder from the gcd calculation and the second and third entries allow us to express the number in the first column as a linear combination of 456 and 123. The quotients in the Euclidean Algorithm told us what to multiply a row by before subtracting it from the previous row.

Here's another example, where we calculate gcd(259, 119). You should go step-by-step to make sure that you understand how we're arriving at the numbers in each row.

|     | $x$ | $y$ |                                  |
|-----|-----|-----|----------------------------------|
| 259 | 1   | 0   |                                  |
| 119 | 0   | 1   |                                  |
| 21  | 1   | $-2$ | (1st row) $-$ 2·(2nd row)       |
| 14  | $-5$ | 11  | (2nd row) $-$ 5·(3rd row)       |
| 7   | 6   | $-13$ | (3rd row) $-$ (4th row).        |

The end result is $7 = 259 \cdot 6 - 119 \cdot 13$.

To summarize, we state the following.

**Theorem 2.12.** *Let $a$ and $b$ be integers with at least one of $a, b$ nonzero. There exist integers $x$ and $y$, which can be found by the Extended Euclidean Algorithm, such that*

$$\gcd(a, b) = ax + by.$$

*Proof.* Although it would be fairly straightforward to write a detailed proof that follows the reasoning of the examples, the numerous indices and variables would make the proof rather unenlightening. Therefore, we spare the reader. Instead, we give the following non-constructive proof that $\gcd(a, b)$ is a linear combination of $a$ and $b$.

Let $S$ be the set of integers that can be written in the form $ax + by$ with integers $x$ and $y$. Since $a$, $b$, $-a$, and $-b$ are in $S$, we see that $S$ contains at least one positive integer. Let $d$ be the smallest positive integer in $S$ (this is an application of the Well-Ordering Principle; see Appendix A). Since $d \in S$, we know that $d = ax_0 + by_0$ for some integers $x_0$ and $y_0$. We claim that $a$ and $b$ are multiples of $d$, so $d$ is a *common divisor* of both $a$ and $b$. To see this, write $a = dq + r$ with integers $q$ and $r$ such that $0 \le r < d$. Since

$$r = a - dq = a - (ax_0 + by_0)q = a(1 - x_0 q) + b(-y_0 q),$$

we have that $r \in S$. Since $d$ is the smallest positive element of $S$ and $0 \le r < d$, we must have $r = 0$. This means that $d \mid a$. Similarly, $d \mid b$, so $d$ is a common divisor of $a$ and $b$.

Now suppose that $e$ is any common divisor of $a$ and $b$. Proposition 2.3 implies that $e$ divides $ax_0 + by_0 = d$, so $e \le d$. Therefore, $d$ is the greatest common divisor. By construction, $d$ is a linear combination of $a$ and $b$. □

Finally, we give a version of Theorem 2.12 that applies to more than two numbers.

**Theorem 2.13.** *Let $n \ge 2$ and let $a_1, a_2, \ldots, a_n$ be integers (at least one of them must be nonzero). Then there exist integers $x_1, x_2, \ldots, x_n$ such that*

$$\gcd(a_1, a_2, \ldots, a_n) = a_1 x_1 + a_2 x_2 + \cdots + a_n x_n.$$

*Proof.* We'll use mathematical induction (see Appendix A). By Theorem 2.12, the result is true for $n = 2$. Assume that it is true for $n = k$. Then

$$\gcd(a_1, a_2, \cdots, a_k) = a_1 y_1 + a_2 y_2 + \cdots + a_k y_k \qquad (2.3)$$

for some integers $y_1, y_2, \ldots, y_k$. But

$$\gcd(a_1, a_2, \ldots, a_{k+1}) = \gcd(\gcd(a_1, a_2, \ldots, a_k), a_{k+1})$$
$$= \gcd(a_1, a_2, \ldots, a_k)x + a_{k+1}y$$

for some integers $x$ and $y$, again by Theorem 2.12. Substituting (2.3) into this equation yields

$$\gcd(a_1, a_2, \ldots, a_{k+1}) = (a_1 y_1 + a_2 y_2 + \cdots + a_k y_k)x + a_{k+1}y$$
$$= a_1(xy_1) + a_2(xy_2) + \cdots + a_k(xy_k) + a_{k+1}y_{k+1},$$

which is the desired result, with $x_i = xy_i$ for $1 \le i \le k$ and $x_{k+1} = y$. Therefore, the result is true for $n = k + 1$. By induction, the result holds for all positive integers $n \ge 2$. □

Theorem 2.12 (and its generalization 2.13) are among the most important tools in number theory and they'll be used to deduce many fundamental properties of the integers.

One immediate consequence is the following, which we previously saw was true when we proved that the Euclidean Algorithm gives us the greatest common divisor. However, it is now easy to give another proof.

**Corollary 2.14.** *If e is a common divisor of a and b, then e divides* $\gcd(a, b)$.

*Proof.* Both $e \mid a$ and $e \mid b$, and $\gcd(a, b)$ is a linear combination of $a$ and $b$, so Proposition 2.3 implies that $e \mid \gcd(a, b)$, as desired. □

Here is another consequence.

**Proposition 2.15.** *If a, b, c are integers with* $\gcd(a, c) = \gcd(b, c) = 1$, *then* $\gcd(ab, c) = 1$.

*Proof.* After we prove the uniqueness of factorizations into primes, the result will be easy to prove: $\gcd(a, c) = 1$ says that the factorizations of $a$ and $c$ have no primes in common. Similarly, $b$ and $c$ have no primes in common. Since the factorization of $ab$ is constructed from the factorizations of $a$ and $b$, and this is the only way to do this, $ab$ and $c$ have no primes in common, so $\gcd(ab, c) = 1$. But we can prove the result now without using this uniqueness.

Since $\gcd(a, c) = 1$, there are integers $x_1, y_1$ with $ax_1 + cy_1 = 1$. There are also integers $x_2$ and $y_2$ such that $bx_2 + cy_2 = 1$. Multiplying these two relations yields

$$1 = (ax_1 + cy_1)(bx_2 + cy_2) = (ab)(x_1x_2) + (c)(by_1x_2 + ax_1y_2 + cy_1y_2).$$

Proposition 2.3 implies that a common divisor of $ab$ and $c$ must divide 1, so $\gcd(ab, c) = 1$. □

The following two propositions will be used several times in later chapters. In the first, the condition $\gcd(a, b) = 1$ can be interpreted as saying that $b$ is not helping at all with $a \mid bc$, and the consequence is that the divisibility of $bc$ by $a$ requires that $a \mid c$. In the second proposition, the main thing that could keep $ab$ from dividing $c$ is, for example, a prime $p$ that occurs to only the first power in the factorizations of $a, b, c$. Then

$ab$ has $p^2$ while $c$ has only $p$ in its factorization. The condition that $\gcd(a, b) = 1$ keeps this situation from happening, since the prime $p$ can occur in at most one of $a, b$.

**Proposition 2.16.** *Let $a, b, c$ be integers with $a \neq 0$ and $\gcd(a, b) = 1$. If $a \mid bc$ then $a \mid c$.*

*Proof.* Theorem 2.12 says that we can write $1 = ax + by$ for some integers $x$ and $y$. Multiply by $c$ to obtain $c = acx + bcy$. Since $a \mid a$ and $a \mid bc$, Proposition 2.3 implies that $a \mid c$.                                $\square$

**Proposition 2.17.** *Let $a, b, c$ be integers with $a, b$ nonzero and $\gcd(a, b) = 1$. If $a \mid c$ and $b \mid c$, then $ab \mid c$.*

*Proof.* Theorem 2.12 says there exist integers $x$ and $y$ such that $ax + by = 1$. Multiply by $c$ to obtain $acx + bcy = c$. Since $b \mid c$, we have $ab \mid ac$. Since $a \mid c$, we have $ba \mid bc$. Since $c$ is a linear combination of $ac$ and $bc$, we have $ab \mid c$.                                $\square$

Proposition 2.17 is false if we don't assume that $\gcd(a, b) = 1$. For example, $6 \mid 30$ and $10 \mid 30$, but $6 \cdot 10 \nmid 30$. The problem is that both 6 and 10 have 2 in their factorizations, and $6 \cdot 10$ contains this 2 twice, which is too much for 30.

---

## CHECK YOUR UNDERSTANDING

10. Compute $\gcd(654, 321)$ without factoring.
11. Find $x$ and $y$ such that $17x + 12y = 1$.

---

# 2.8    Other Bases

The numbers that we use in our everyday life are written using base 10 notation. For example, 783 means $7 \cdot 10^2 + 8 \cdot 10^1 + 3 \cdot 10^0$. The position of each digit tells us what power of 10 that digit will be multiplied by to give us our number, so 58 and 85 represent different numbers because of the positions of the 5 and 8. In the past there have been other ways to represent integers. When Abraham Lincoln wrote the Gettysburg Address, he didn't begin with "Eighty-seven years ago," but with "Four score and seven years ago" using the word *score* (which comes from the Norse *skar*, meaning *mark* or *tally*) for the number 20. In Britain, people still say they weigh 10 stone 7 pounds instead of 147 pounds, using the word stone for 14 from an old unit of measurement.

Our reliance on base 10 is most likely an accident of evolution, and is a reflection of the ten fingers that we use to count. The Babylonians used a base 60 for their number system, and the Mayans used base 20. (Perhaps they also used their toes.) Computers are based on a binary system and often use base 16 $(= 2^4)$ to represent numbers.

If we have a number in a different base, let's say base 7, then it's easy to rewrite it as a base 10 number. Let's say we had $3524_7$ where the subscript 7 means we are working in base 7. Then,

$$3524_7 = 3 \cdot 7^3 + 5 \cdot 7^2 + 2 \cdot 7^1 + 4 \cdot 7^0 = 3 \cdot 343 + 5 \cdot 49 + 2 \cdot 7 + 4 \cdot 1 = 1292_{10}.$$

We can also convert a number from base 10 to any other base with the use of the Division Algorithm.

We give three examples to show how this works.

**Example.** Convert the base 10 number 21963 to a base 8 number.

We proceed by dividing 21963 by 8, then dividing the quotient by 8, and continuing until the quotient is 0. At the end of the example, we'll show why the process works.

$$
\begin{aligned}
21963 &= 2745 \cdot 8 &+ 3 \\
2745 &= 343 \cdot 8 &+ 1 \\
343 &= 42 \cdot 8 &+ 7 \\
42 &= 5 \cdot 8 &+ 2 \\
5 &= 0 \cdot 8 &+ 5.
\end{aligned}
$$

This tells us that $21963_{10} = 52713_8$. Observe that we read the remainders upward to get the base 8 expansion. To see why this works, we start from the beginning, making sure to group our factors of 8 together.

$$21963 = 2745 \cdot 8 + 3 = (343 \cdot 8 + 1)8 + 3 =$$

$$343 \cdot 8^2 + 1 \cdot 8 + 3 = (42 \cdot 8 + 7)8^2 + 1 \cdot 8 + 3 =$$

$$42 \cdot 8^3 + 7 \cdot 8^2 + 1 \cdot 8 + 3 = (5 \cdot 8 + 2)8^3 + 7 \cdot 8^2 + 1 \cdot 8 + 3 =$$

$$5 \cdot 8^4 + 2 \cdot 8^3 + 7 \cdot 8^2 + 1 \cdot 8 + 3 =$$

$$52713_8.$$

**Example.** Convert the base 10 number 1671 to base 2.

$$
\begin{array}{rclcl}
1671 & = & 835 \cdot 2 & + & 1 \\
835 & = & 417 \cdot 2 & + & 1 \\
417 & = & 208 \cdot 2 & + & 1 \\
208 & = & 104 \cdot 2 & + & 0 \\
104 & = & 52 \cdot 2 & + & 0 \\
52 & = & 26 \cdot 2 & + & 0 \\
26 & = & 13 \cdot 2 & + & 0 \\
13 & = & 6 \cdot 2 & + & 1 \\
6 & = & 3 \cdot 2 & + & 0 \\
3 & = & 1 \cdot 2 & + & 1 \\
1 & = & 0 \cdot 2 & + & 1.
\end{array}
$$

So, $1671_{10} = 11010000111_2$.

**Example.** It's always a good idea to make sure that any mathematical method works for an example where you already know the answer. This serves as a type of "reality check." So, let's take a base 10 number, say 314159, and use the above algorithm to "convert" it to base 10:

$$
\begin{array}{rclcl}
314159 & = & 31415 \cdot 10 & + & 9 \\
31415 & = & 3141 \cdot 10 & + & 5 \\
3141 & = & 314 \cdot 10 & + & 1 \\
314 & = & 31 \cdot 10 & + & 4 \\
31 & = & 3 \cdot 10 & + & 1 \\
3 & = & 0 \cdot 10 & + & 3.
\end{array}
$$

It should be reassuring that this gives back the original 314159.

---

## CHECK YOUR UNDERSTANDING

12. Convert $1234_{10}$ to base 7.
13. Convert $321_5$ to base 10.

# 2.9 Fermat and Mersenne Numbers

When Euclid proved that there are infinitely many primes, he did so by showing that there cannot be a largest prime number. Nevertheless, there is a long history, going back at least 2000 years ago to the Sieve of Eratosthenes, of finding primes, especially large ones. It is remarkable that this area of mathematics, which was long considered a somewhat recreational academic exercise, has now turned out to have important cryptographic uses. We'll discuss these applications in Chapter 9.

Most of the large primes that are found are too large (and not random enough) for cryptographic applications, but they are often a way to test algorithmic advances. Since exponential functions grow very rapidly, it seems reasonable to look for exponential expressions that take on prime values in order to find large prime numbers. This was the motivation behind the work of some seventeenth century mathematicians, whose ideas on generating large primes still spark our interest today.

The most common function to look at was $2^n - 1$. Primes of this form appear in Euclid's *Elements* in his discussion of perfect numbers (see Chapter 16). Many people have thought that this expression is always prime if $n$ is prime, but Hudalricus Regius, in 1536, seems to have been the first person to realize that $2^{11} - 1 = 23 \cdot 89$. (This was before calculators, computers, and the widespread knowledge of algebra. Factoring was very hard and very slow.) Today, numbers of the form

$$M_n = 2^n - 1 \qquad (2.4)$$

are called **Mersenne numbers**, after the French monk Marin Mersenne who lived from 1588 to 1648. Mersenne, and others before him, were most likely aware that in order for $M_n$ to be prime, $n$ must be prime. This is the content of the following result.

**Proposition 2.18.** *If $n$ is composite then $2^n - 1$ is composite.*

*Proof.* Recall that for every $k \geq 1$,

$$x^k - 1 = (x - 1)\left(x^{k-1} + x^{k-2} + x^{k-3} + \cdots + x + 1\right) \qquad (2.5)$$

(if we divide both sides of Equation (2.5) by $x-1$, we get the well known formula for a geometric sum; see Appendix A). Since $n$ is composite, $n = ab$ with $1 < a < n$ and $1 < b < n$. Substitute $x = 2^a$ and $k = b$ to get

$$2^{ab} - 1 = (2^a - 1)\left(2^{a(b-1)} + 2^{a(b-2)} + \cdots + 2^a + 1\right).$$

Since $1 < a < n$, we have $1 < 2^a - 1 < 2^n - 1$, so the factor $2^a - 1$ is nontrivial. This means that $2^{ab} - 1$ is composite.        □

The proposition tells us that if we want to find $n$ such that $M_n$ is prime, we should look only at prime values of $n$. This works well at the beginning:

$$2^2 - 1 = 3, \quad 2^3 - 1 = 7, \quad 2^5 - 1 = 31, \quad 2^7 - 1 = 127.$$

All of these are primes. However, Hudalricus Regius's example with $n = 11$ shows that if $n$ is prime, $M_n$ may still be composite. Continuing past the case $n = 11$, we find that

$$2^{13} - 1, \quad 2^{17} - 1, \quad 2^{19} - 1$$

are prime, but $2^{23} - 1$ is composite. As of 2016, there are 49 values of $n$ for which $M_n$ is known to be prime.

In honor of Mersenne, who made some early (and partly incorrect) lists of values of $n$ that make $M_n$ prime, a prime number of the form $M_n$ is called a *Mersenne prime.*

Because of the special form of Mersenne primes, there are very fast tests (see the Lucas-Lehmer Test in Chapters 14 and 19) to determine whether $M_n$ is a prime number. Whenever you read that a new "largest" prime has been discovered, you'll likely see that it's a Mersenne prime. In fact, for much of recent history, the largest known prime has been a Mersenne prime. In 1876, Eduoard Lucas proved that $M_{127}$ is prime, and that remained the largest known prime until 1951, when the non-Mersenne $180(M_{127})^2 + 1$ became the champion. In 1952, Mersenne primes took back the lead, with $M_{521}$ being proved prime. Subsequently, larger and larger Mersenne primes were discovered through 1985, when $M_{216091}$ became the leader. Then, in 1989 the non-Mersenne $391581 \times 2^{216193} - 1$ took over and reigned until 1992, when $M_{756839}$ regained the title for the Mersennes. Since then, the largest known prime has always been Mersenne. As of March 2017, the largest known Mersenne prime was $2^{74207281} - 1$. The base 10 expansion of this number has more than 22 million digits. The Great Internet Mersenne Prime Search (GIMPS) is an Internet-based distributed computing project with numerous volunteers using spare time on their computers to carry out the search for Mersenne primes. The 15 most recent Mersenne primes were found by GIMPS.

Instead of looking at $2^n - 1$, the great French mathematician Pierre de Fermat (1601-1665) examined $2^m + 1$. Fermat realized that the only way for $2^m + 1$ to be prime is for the exponent $m$ to be a power of 2.

**Proposition 2.19.** *If $m > 1$ is not a power of 2, then $2^m + 1$ is composite.*

*Proof.* If $k$ is odd then

$$x^k + 1 = (x + 1)\left(x^{k-1} - x^{k-2} + x^{k-3} - \cdots - x + 1\right).$$

Since $m$ is not a power of 2, it has a nontrivial odd factor: $m = ab$ with $a$ an odd integer and $a \geq 3$. Let $k = a$ and let $x = 2^b$. Then

$$2^{ab} + 1 = (2^b + 1)\left(2^{b(a-1)} - 2^{b(a-2)} + 2^{b(a-3)} - \cdots - 2^b + 1\right).$$

Since $1 \leq b < m$, we have $1 < 2^b + 1 < 2^m + 1$, so the factor $2^b + 1$ is nontrivial (a "non-trivial" factor of an integer $n$ is a factor $d$ with $1 < d < n$). Therefore, $2^{ab} + 1$ is composite. $\qquad\square$

The proposition tells us that if we want to find $m$ such that $2^m + 1$ is prime, we should take $m$ to be a power of 2, leading to the definition of the $n$th Fermat number:

$$F_n = 2^{2^n} + 1.$$

Fermat incorrectly believed that $F_n$ is prime for every integer $n \geq 0$. Although this is true for small values of $n$:

$$F_0 = 2^1 + 1 = 3, \quad F_1 = 2^2 + 1 = 5, \quad F_2 = 2^4 + 1 = 17,$$
$$F_3 = 2^8 + 1 = 257, \quad F_4 = 2^{16} + 1 = 65537,$$

Euler showed in 1732 that $F_5 = 4294967297 = 641 \cdot 6700417$. More recently, $F_n$ has been factored for $6 \leq n \leq 11$, and many more have been proved to be composite, although they are yet to be factored. Many people now believe that $F_n$ is *never* prime if $n \geq 5$.

Fermat primes occur in compass and straightedge constructions in geometry. Using only a compass and a straightedge, it is easy to make an equilateral triangle or a square. It's a little harder to make a regular pentagon, but it's possible. The constructions of equilateral triangles and regular pentagons can be combined to produce a regular 15-gon. Moreover, by bisecting angles, it's easy to double the number of sides of a polygon that is already constructed, so 30-sided, 60-sided, and 120-sided regular polygons can be constructed. What else can be done? In 1796, just before he turned 19, Gauss discovered how to construct a regular 17-gon. Moreover, his methods (completed by Wantzel) yield the following:

*Let $n \geq 3$. A regular $n$-gon can be constructed by compass and straightedge if and only if $n$ is a power of 2 times a product of distinct Fermat primes:*

$$n = 2^a F_{n_1} \cdot F_{n_2} \cdots F_{n_r}$$

*with $a \geq 0$ and $r \geq 0$.*

As mentioned above, the largest known Fermat prime is 65537. Johann Hermes spent 10 years writing down the explicit construction of a 65537-gon, finishing in 1894. The result is stored in a box in Göttingen.

There are heuristic arguments that predict that the number of Mersenne primes should be infinite while the number of Fermat primes should be finite. Arguments such as the ones we'll give are common in number theory when people want to get a rough idea of what is happening. They are not proofs; instead, they are arguments that point in the direction of what is hoped to be the truth.

Here are the heuristic arguments. The Prime Number Theorem (Theorem 20.7) says that the probability that a randomly chosen integer of size $x$ is prime is approximately $1/\ln x$. Therefore, the probability that $2^p - 1$ is prime is approximately

$$\frac{1}{\ln(2^p - 1)} \geq \frac{1}{\ln(2^p)} = \frac{1}{p \ln 2}.$$

The number of Mersenne primes $2^p - 1$ as $p$ ranges through all primes should be approximately

$$\frac{1}{\ln 2} \sum_p \frac{1}{p} = \infty$$

(the fact that the sum diverges is proved in Chapter 5).

On the other hand, the probability that $2^{2^n} + 1$ is prime should be approximately

$$\frac{1}{\ln(2^{2^n} + 1)} \leq \frac{1}{\ln(2^{2^n})} = \frac{1}{2^n \ln 2}.$$

Since

$$\sum_{n=0}^{\infty} \frac{1}{2^n \ln 2} = \frac{2}{\ln 2} \approx 2.885,$$

the expectation is that the number of Fermat primes is finite. You might have noticed that there are five Fermat primes, while the prediction is for only 2.885. This can be explained by the fact that all the Fermat numbers are odd, and the chance that a random odd number of size $x$ is prime is $2/\ln x$. When we double 2.885, we get a prediction of 5.77, which is closer to the actual count.

## 2.10   Chapter Highlights

1. Euclid's theorem: There are infinitely many prime numbers.

2. The Euclidean Algorithm for computing gcd's.

3. If $\gcd(a, b) = 1$, then there exist integers $x, y$ with $ax + by = 1$.

4. Finding solutions of $ax + by = \gcd(a, b)$ using the Extended Euclidean Algorithm.

## 2.11   Problems

### 2.11.1   Exercises

**Section 2.1: Divisibility**

1. Show that $5 \mid 120$,   $11 \mid 165$,   and $14 \mid 98$.
2. Show that $7 \mid 7$,   $10 \nmid 25$,   and $32 \mid -160$.
3. Show that $7 \mid 21$,   $3 \mid 21$,   $14 \mid 42$,   and $6 \mid 42$.
4. Show that $3 \mid 63$,   $9 \mid 27$,   $18 \mid 54$,   and $10 \mid 150$.

5.  Show that $8 \nmid 15$,   $3 \nmid 20$,   $5 \nmid 72$,   and $4 \nmid 22$.

6.  Show that $11 \nmid 27$,   $15 \nmid 9$,   $12 \nmid 44$,   and $7 \nmid 90$.

7.  Find all positive divisors of the following integers:
    (a) 20      (b) 52      (c) 195      (d) 203

8.  Find all positive divisors of the following integers:
    (a) 12      (b) 13      (c) 15      (c) 16

9.  Prove or give a counterexample for the following statements:
    (a) If $6 \nmid a$ and $6 \nmid b$, then $6 \nmid ab$.
    (b) If $6 \nmid a$ and $6 \nmid b$, then $36 \nmid ab$.
    (c) If $6 \nmid a$ and $6 \nmid b$, then $6 \nmid (a+b)$.

10. Prove or give a counterexample for the following statements:
    (a) If $6 \mid a$ and $6 \mid b$, then $6 \mid ab$.
    (b) If $6 \mid a$ and $6 \mid b$, then $36 \mid ab$.

11. Prove or give a counterexample for the following statements:
    (a) If $c \mid a$ and $c \mid b$, then $c \mid ab$.
    (b) If $c \mid a$ and $c \mid b$, then $c^2 \mid ab$.
    (c) If $c \nmid a$ and $c \nmid b$, then $c \nmid (a+b)$.

12. Recall that a number $n$ is even if $n = 2k$ and is odd if $n = 2k + 1$. Prove the following:
    (a) The sum of two even numbers is even.
    (b) The sum of two odd numbers is even.
    (c) The product of two even numbers is even *and* is divisible by 4.
    (d) The product of two odd numbers is odd.

13. Find all integers $n$ (positive or negative) such that $n^2 - n$ is prime.

14. Show that if $n$ is an integer then 2 divides $n^2 - n$

15. Show that if $n$ is an integer, then $n^2 - 3n$ is even.

16. Show that if $n$ is an integer, then $n^2 + n$ is even.

17. Show that if $a, m$, and $n$ are positive integers, then $a^m - a^n$ is even.

18. Show that if $a, m$, and $n$ are positive integers, then $a^m + a^n$ is even.

19. Suppose that $a$ and $b$ are positive integers such that $a \mid b$ and $b \mid a$. Show that $a = b$.

20. Suppose that $a$ and $b$ are nonzero integers such that $a \mid b$ and $b \mid a$. Show that $a = \pm b$.

21. Suppose that $a, b, c$ are nonzero integers.
    (a) Show that if $a \mid b$ then $ac \mid bc$.
    (b) Show that if $ac \mid bc$ then $a \mid b$.

22. Show that if $a$ is an integer then 3 divides $a^3 - a$.

23. Find all positive integers $n$ such that $n^3 + 12n$ is prime.

24. (a) If $n \geq 2$, show that each of

$$n! + 2, \ n! + 3, \ n! + 4, \ \dots, n! + n$$

is composite.
(b) Find a list of 100 consecutive composite integers.

25.  (a) Find all prime numbers that can be written as a difference of two squares.
(b) Find all prime numbers that can be written as a difference of two fourth powers.

## Section 2.2: Euclid's Theorem

26.  Let $p_k$ denote the $k$th prime. Show that

$$p_{n+1} \leq p_1 p_2 p_3 \cdots p_n + 1$$

for all $n \geq 1$.

27.  Consider the numbers

$$2 + 1, \quad 2 \cdot 3 + 1, \quad 2 \cdot 3 \cdot 5 + 1, \quad 2 \cdot 3 \cdot 5 \cdot 7 + 1, \quad \cdots .$$

Show, by computing several values, that there are composite numbers in this sequence. (This shows that in the proof of Euclid's theorem, these numbers are not necessarily prime, so it is necessary to look at prime factors of these numbers.)

28.  A friend claims that the difference of cubes of successive positive integers is always a prime (for example, $2^3 - 1^3 = 7$ and $4^3 - 3^3 = 37$). Is this claim correct? Prove it or find a counterexample.

29.  (a) Let $p$ be a prime larger than 10. Show that the last digit of $p$ (in decimal notation) must be 1, 3, 7, or 9.
(b) The numbers 11, 13, 15, 17, 19 are four odd integers ending in 1, 3, 7, 9 that lie between two multiples of 10. Find the next such set of prime numbers ending in 1, 3, 7, 9.

30.  Goldbach's Conjecture says that every even integer larger than 4 can be written as the sum of two odd primes. Find all ways to write each of 66, 68, and 128 as a sum of two primes (one has two ways, one has three, and one has six).

31.  It was proved in 2013 that every odd integer greater than equal to 9 can be written as a sum of three odd primes. Verify that 37, 59, and 2013 are sums of three primes.

32.  The famous Goldbach Conjecture says that every even integer $n \geq 4$ is a sum of two primes. This conjecture is not yet proved. Prove the weaker statement that there are infinitely many even integers that are sums of two primes.

33.  It is conjectured, but not proved, that there is always a prime between every two consecutive squares $n^2$ and $(n+1)^2$ for $n \geq 1$. Prove the weaker statement that there are infinitely many $n$ such that there is a prime between $n^2$ and $(n+1)^2$.

34.  Here is a variant of Euclid's proof that there are infinitely many primes. Suppose that $\{p_1, p_2, \ldots, p_n\}$ is the set of all primes. Let $N = p_1 p_2 \cdots p_n - 1$.
(a) Use the fact that $p_1 = 2$ and $p_2 = 3$ to show that $N > 1$, hence has a prime factor $p$.
(b) Show that $p \notin \{p_1, p_2, \ldots, p_n\}$, so no finite set can contain all the primes.

## Section 2.5: The Division Algorithm

35.  For each of the following, find the quotient and remainder when $a$ is divided by $b$ in the Division Algorithm.
     (a) $a = 29, b = 2$    (b) $a = 53, b = 6$,
     (c) $a = 205, b = 13$    (d) $a = 45, b = 5$.

36.  For each of the following, find the quotient and remainder when $a$ is divided by $b$ in the Division Algorithm.
     (a) $a = 43, b = 7$    (b) $a = 96, b = 11$,
     (c) $a = 140, b = 12$    (d) $a = -200, b = 21$.

37.  For each of the following, find the quotient and remainder when $a$ is divided by $b$ in the Division Algorithm.
     (a) $a = 7, b = 7$    (b) $a = 9, b = 11$,
     (c) $a = 246, b = 10$    (d) $a = 18, b = 3$.

38.  For each of the following, find the quotient and remainder when $a$ is divided by $b$ in the Division Algorithm.
     (a) $a = 11, b = 11$    (b) $a = 47, b = 58$,
     (c) $a = -87, b = 3$    (d) $a = -87, b = 5$.

39.  For each of the following, find the quotient and remainder when $a$ is divided by $b$ in the Division Algorithm.
     (a) $a = 12, b = 3$    (b) $a = 45, b = 97$,
     (c) $a = 138, b = 111$    (d) $a = 170, b = 20$.

40.  When dividing $a$ by $b$ in the Division Algorithm, find the quotient $q$ and remainder $r$ when
     (a) $1 \leq a = b$.
     (b) $a = kb$, $k$ an integer.
     (c) $0 \leq a < b$.
     (d) $a < 0$ and $0 < |a| < b$.

41.  Suppose that today is Monday. What day of the week will it be in 100 days? (*Hint:* What is the remainder when 100 is divided by 7?)

42.  Suppose that this month is July. What month will it be 100 months from now? (*Hint:* What is the remainder when 100 is divided by 12?)

43.  (a) Divide $1! + 2! + 3! + \cdots + 100!$ by 7. What is the remainder? (*Hint:* Do not evaluate the sum before dividing.)
     (b) Divide $1! + 2! + 3! + \cdots + 100!$ by 8. What is the remainder?

44.  Divide 11111111111 (eleven ones) by 11. What are the quotient and remainder?

45.  Find all integers $n$ (positive, negative, or zero) so that $n^2 + 1$ is divisible by $n + 1$.

46.  Find all integers $n$ (positive, negative, or zero) so that $n^3 - 1$ is divisible by $n + 1$.

47.  Consider the voting scheme described in Section 2.5.1. Suppose there are 16 people who vote, and a Yes counts as 17, a No counts as 1, and Abstain counts as 0. The initial number chosen by the Chair is 12345. The number after the voting is 12440. How many Yes, No, and Abstain votes were there?

48. Five sailors and a monkey are shipwrecked on a desert island. During the day, they gather many coconuts and put them in a pile. Since they are tired, they decide to split them among the five sailors the next morning. However, the first sailor wakes up in the middle of the night and decides to take his share then. He divides the coconuts into five equal piles. There is one coconut left over (that is, he does the division algorithm and has a remainder $r = 1$), which he gives to the monkey. He takes his share and leaves the other 4 shares in a pile. Soon thereafter, another sailor wakes up and does a similar procedure, unaware of what the first sailor has done. He divides the pile into 5 equal piles. There is one coconut left over, which he gives to the monkey. He takes his share and leaves the other four shares in a pile. Soon, a third sailor wakes up and does the same procedure, and so do the fourth and fifth sailors, each unaware of what has been done previously. In the morning, there is a (somewhat smaller) pile of coconuts. Not wanting to admit to their actions of the previous night, the sailors divide the remaining coconuts into 5 equal piles, which they take. There is one left over, which they give to the monkey. The classic question is how many coconuts were in the original pile.

    (a) Show that if $n$ is a solution to the problem, then $n + 15625$ is also a solution (*Note:* $15625 = 5^6$).

    (b) Show that $n = -4$ is a "solution" (note that taking away $-1$ coconuts is the same as adding a coconut to the pile).

    (c) Find a positive integer $n$ that solves the puzzle.

## Section 2.6: The Greatest Common Divisor

49. Evaluate
    (a) $\gcd(12, 36)$
    (b) $\gcd(3, 0)$
    (c) $\gcd(6, 7)$

50. Evaluate
    (a) $\gcd(6, 9)$
    (b) $\gcd(10, 14)$
    (c) $\gcd(8, -9)$

51. Show that if $n$ is an integer then $\gcd(n, n + 1) = 1$.

52. If $\gcd(a, b) = 1$, prove that $\gcd(a + b, a - b) = 1$ or 2. (*Hint:* Show that a common divisor divides $\gcd(2a, 2b)$.)

53. Show that if $n$ is an integer then $\gcd(2n - 1, 2n + 1) = 1$.

54. If $n$ is a positive integer, what is $\gcd(n, n^2)$?

55. If $m$ is a positive integer, prove that $\gcd(6m + 5, 7m + 6) = 1$.

56. If $k$ is a positive integer, prove that $\gcd(2k + 5, 3k + 7) = 1$.

57. Show that if $k$ is odd, then $\gcd(k, k - 2) = 1$.

58. Show that if $k$ is even, then $\gcd(k, k - 2) = 2$.

59. Show that if $n$ is an integer then $\gcd(2n^2 - 1, n + 1) = 1$.

60. Show that if $n$ is an integer then $\gcd(n^2 + n + 1, 4) = 1$.

61. Show that if $\gcd(a, b) = 1$ and $c \mid a$, then $\gcd(b, c) = 1$.

62. Let $a$ be an integer. Find all possibilities for $\gcd(5a + 1, 12a + 9)$ (there are four of them). For each possibility, find an integer $a$ such that the gcd has that value. (*Hint:* If $d$ is the gcd, then $d$ divides any combination of the two numbers. Find a combination that gets rid of the $a$.)

63. Show that $n! + 1$ and $(n + 1)! + 1$ are relatively prime. (*Hint:* Try multiplying $n! + 1$ by $n + 1$ and then using various linear combinations.)

64. Let $n$ be an integer.
    (a) Show that $\gcd(n^2 + n + 6, n^2 + n + 4) = 2$. (*Hint:* $n^2 + n$ is always even.)
    (b) Show that $\gcd(n^2 + n + 5, n^2 + n + 3) = 1$.

65. Let $a, b$ be integers and let $d = \gcd(a, b)$. Write $a = da_1$ and $b = db_1$. Show that $a_1/b_1$ is a reduced fraction.

66. Let $a, b, c, d$ be nonzero integers with $ad - bc = \pm 1$. Show that $\gcd(a + b, c + d) = 1$.

## Section 2.7: The Euclidean Algorithm

67. Evaluate each of the following and use the Extended Euclidean Algorithm to express each $\gcd(a, b)$ as a linear combination of $a$ and $b$.
    (a) $\gcd(14, 100)$
    (b) $\gcd(6, 84)$
    (c) $\gcd(182, 630)$
    (d) $\gcd(1776, 1848)$

68. Evaluate each of the following and use the Extended Euclidean Algorithm to express each $\gcd(a, b)$ as a linear combination of $a$ and $b$.
    (a) $\gcd(13, 203)$
    (b) $\gcd(57, 209)$
    (c) $\gcd(465, 2205)$
    (d) $\gcd(1066, 42)$

69. Let $n$ be an integer. Show that $\gcd(n^2, n^2 + n + 1) = 1$.

70. (a) Compute $\gcd(89, 55)$.
    (b) If $F_n$ denotes the $n$th Fibonacci number, describe the quotients and remainders in the Euclidean Algorithm for $\gcd(F_{n+1}, F_n)$.

71. (a) Use the Euclidean Algorithm to compute $\gcd(13, 5)$.
    (b) Let $n = 1111111111111$ (that's thirteen 1's) and $m = 11111$. Use the Euclidean Algorithm to compute $\gcd(n, m)$.
    (c) Observe the relation between the calculations in (a) and (b). Generalize this to show that if $c = (10^a - 1)/9$ and $d = (10^b - 1)/9$ (so $c$ has $a$ 1's and $d$ has $b$ 1's), then $\gcd(c, d)$ is the number with $\gcd(a, b)$ 1's.

72. Let $a_1, a_2, \ldots, a_n$ be integers (with at least one of them nonzero). Let $T$ be the set of integers of the form

$$a_1 x_1 + a_2 x_2 + \cdots + a_n x_n,$$

with integers $x_1, x_2, \ldots, x_n$. Use the proof of Theorem 2.12 with appropriate changes to show that $\gcd(a_1, a_2, \ldots, a_n)$ is a linear combination of $a_1, a_2, \ldots, a_n$. (This gives another proof of Theorem 2.13.)

73.  Suppose we want to compute the gcd of three numbers, for example, $\gcd(210, 294, 490)$. Do the following computations:
     (a) Compute $d' = \gcd(210, 294)$.
     (b) Compute $d = \gcd(d', 490)$.
     (c) It was pointed out in the discussion following the Euclidean Algorithm that every common divisor of $a$ and $b$ divides $\gcd(a, b)$. Use this fact (first applied to $d'$, then to $d$) to show that every common divisor of 210, 294, and 490 divides $d$, so $d$ is the largest common divisor of these numbers.

74.  Using the steps of the previous exercise as a guide, calculate that $1 = \gcd(147, 45, 175)$ and then find integers $x, y, z$ such that $1 = 147x + 45y + 175z$.

## Section 2.8: Other Bases

75.  Express the following in base 10:
     (a) $1234_5$
     (b) $10101_2$
     (c) $111_{11}$

76.  (a) Convert the base 10 number 54321 to base 6.
     (b) Convert the base 10 number 1000000 to base 2.
     (c) Convert the base 10 number 31416 to base 7.

77.  (a) Write $1011_2, 10101_2$, and $111110_2$ in base 4.
     (b) How would you convert from base 2 to base 4?

78.  (a) Write $302_4, 2131_4$, and $2032_4$ in base 2.
     (b) How would you convert from base 4 to base 2?

79.  (a) Write $2104_9, 1375_9$, and $561_9$ in base 3.
     (b) How would you convert from base 9 to base 3?

80.  (a) Write $2011_3, 1121_3$, and $21022_3$ in base 9.
     (b) How would you convert from base 3 to base 9?

81.  The numbers 6, 28, and 496 are examples of what are called *perfect numbers* (see Section 16.1). Express there 3 numbers in base 2.

82.  Compute $123_8 \times 321_8$:
     (a) by working completely in base 8
     (b) by changing to base 10, multiplying, and then changing back to base 8.

83.  Suppose that a group of 18 billionaires want to see how many in their group are worth 1 billion, how many are worth 2 billion, etc., but they do not want to reveal anyone's individual wealth. The first person chooses a large integer $N$ (around 100 digits) and computes $N + 20^i$, where $i$ is an integer indicating a net worth of $i$ billions (they round off to the nearest billion). This sum, call it $N_1$, is given to the second person. This second person computes $N_2 = N_1 + 20^j$, where $j$ is an integer indicating a net worth of $j$ billions. The sum $N_2$ is given to the third person. This continues until all 18 have added on their numbers. The first person then subtracts $N$. Show how to determine how many people are worth 0 billion, how many are worth 1 billion, etc.

### Section 2.9: Fermat and Mersenne Numbers

84.  Show that if $n > 1$ and $a$ is a positive integer with $a^n - 1$ prime, then $a = 2$ and $n$ is prime. (*Hint:* Look at the proof of Proposition 2.18.)

85.  Show that if $n > 1$ and $a \geq 2$ is an integer with $a^n + 1$ prime, then $n = 2^k$ for some $k \geq 0$. (*Hint:* Look at the proof of Proposition 2.19.)

86.  (a) Show that if $n$ is a positive integer such that $(3^n - 1)/2$ is prime, then $n$ is prime.
     (b) Find the first two primes of the form in part (a) (the second such prime is greater than 1000).

87.  A *repunit* is an integer of the form $(10^n - 1)/9$, whose decimal expansion consists of $n$ 1's. Show that if $(10^n - 1)/9$ is prime then $n$ is prime. (The first two repunit primes have $n = 2$ and $n = 19$.)

## 2.11.2   Projects

1.  (a) Use the Division Algorithm to show that every odd number is of the form
    $$4k + 1 \text{ or } 4k + 3$$
    for some integer $k$.
    (b) From (a), every odd prime falls into one of the two sets:
    $$S_1 = \{4k + 1 \mid k \text{ is an integer, and } 4k + 1 \text{ is prime}\}$$
    $$S_3 = \{4k + 3 \mid k \text{ is an integer, and } 4k + 3 \text{ is prime}\}.$$

    It turns out that both of these sets are infinite. Explain why you already know that at least one of them is infinite (without specifying which one).
    (c) We'll now show that $S_3$ is infinite. (In Exercise 27 in Chapter 8, we'll show that $S_1$ is also infinite.) To do this, we proceed in steps.
    (i) Show that the product of elements of $S_1$ has the form $4k + 1$ for some integer $k$.
    (ii) Suppose that $S_3$ is finite, and $3, 7, \ldots, p_n$ are *all* the elements in $S_3$. Show that $N = 4p_1p_2p_3 \cdots p_n - 1$ cannot be divisible by a prime in $S_3$.
    (iii) Use (i) to show that $N$ cannot be a product only of elements of $S_1$.
    (iv) Show that we have a contradiction, and therefore $S_3$ is infinite.
    For more on $S_1$ and $S_3$, see the Computer Explorations.
    (d) We will see in Chapter 8 (Exercise 24) that a prime divisor of a number of the form $4N^2 + 1$ must be in $S_1$. Assume this for the moment. Show that $S_1$ must be infinite. (*Hint:* What should you use as $N$?)
    (e) Let
    $$T_1 = \{6k + 1 \mid k \text{ is an integer, and } 6k + 1 \text{ is prime}\}$$
    $$T_5 = \{6k + 5 \mid k \text{ is an integer, and } 6k + 5 \text{ is prime}\}.$$

    Use the method of (c) to show that $T_5$ is infinite.
    (f) We'll see later (Exercise 27 of Chapter 13) that every prime divisor

of a number of the form $N^2 + N + 1$ is 3 or is in $T_1$. Use this fact to show that $T_1$ is infinite.

Dirichlet proved in 1837 that if $\gcd(a, b) = 1$ then there are infinitely many primes of the form $ak + b$. His proof is one of the first examples of analytic number theory, where techniques from calculus (in this case, infinite series) are used to prove results about the distribution of prime numbers. One of the key ingredients of his proof is the fact that $\sum 1/p$ (the sum is over all primes) diverges. See Chapter 5 for a proof of this last fact.

2. The Euclidean Algorithm for the greatest common divisor can be visualized in terms of a tiling analogy. Assume that we wish to cover an $a \times b$ rectangle with square tiles exactly, where $a$ is the larger of the two numbers. We first attempt to tile the rectangle using $b \times b$ square tiles; however, this leaves an $r_0 \times b$ rectangle untiled, where $0 \leq r_0 < b$. We then attempt to tile the residual rectangle with $r_0 \times r_0$ square tiles. This leaves a second rectangle untiled of size $r_1 \times r_0$, which we attempt to tile using $r_1 \times r_1$ square tiles with $0 \leq r_1 < r_0$. We continue with this process, ending when there is no residual rectangle, that is, when the square tiles cover the previous residual rectangle exactly. The length of the sides of the smallest square tile is the gcd of the dimensions of the original rectangle.
   (a) Draw a picture for each step in this process when $a = 3$ and $b = 2$.
   (b) Draw a picture for each step in this process when $a = 11$ and $b = 8$.
   (c) Draw a picture for each step in this process when $a = 8$ and $b = 6$.
   (d) Draw a picture for what happens when $a = F_n$ and $b = F_{n-1}$ are successive Fibonacci numbers.

3. Let $a \geq b > 0$. In the Euclidean Algorithm for computing $\gcd(a, b)$, let $r_{n-1}$ be the last nonzero remainder (we may assume that $b \nmid a$, so $r_{n-1}$ exists, since otherwise the conclusion of (e) is trivial).
   (a) Let $F_k$ be the $k$th Fibonacci number (see Appendix A). Use induction to show that $r_{n-k} \geq F_{k+1}$ for $k = 0, 1, 2, 3, \ldots, n - 1$. (*Hint:* We must have $r_{n-2} \geq 2 = F_3$; otherwise, the division by $r_{n-2}$ would leave no remainder.)
   (b) Show that $b \geq F_{n+1}$.
   (c) Let $\phi = (1 + \sqrt{5})/2$ be the Golden Ratio. Use induction to show that $F_k > \phi^{k-2}$ for all $k \geq 1$. (*Hint:* Use the relations $\phi^2 = \phi + 1$ and $F_{k+1} = F_k + F_{k-1}$.)
   (d) Show that $\log_{10}(b) > (n - 1)/5$.
   (e) Suppose that $b$ has $m$ decimal digits. Show that the number of divisions (which is just $n$) in the Euclidean Algorithm for $\gcd(a, b)$ is at most $5m$.
   This result, namely that the number of divisions in the Euclidean Algorithm for two numbers is at most 5 times the number of decimal digits in the smaller number, is known as *Lamé's theorem*. It is generally regarded as the first theorem in the complexity theory of algorithms.

## 2.11.3   Computer Explorations

1.  Define a function on positive integers by

$$f(n) = \begin{cases} n/2 \text{ if } n \text{ is even} \\ 3n+1 \text{ if } n \text{ is odd.} \end{cases}$$

For example, $f(5) = 16$ and $f(6) = 3$. If we start with a positive integer
$m$, we can form the sequence $m_1 = f(m)$, $m_2 = f(m_1)$, $m_3 = f(m_2)$, etc.
The *Collatz Conjecture* predicts that we eventually get $m_k = 1$ for some
$k$. For example, if we start with $m = 7$, we get $m_1 = 22$, $m_2 = 11$, $m_3 = 34$, $m_4 = 17$, $m_5 = 52$, $m_6 = 26$, $m_7 = 13$, $m_8 = 40$, $m_9 = 20$, $m_{10} = 10$, $m_{11} = 5, m_{12} = 16$, $m_{13} = 8$, $m_{14} = 4$, $m_{15} = 2$, $m_{16} = 1$.
(a) Show that the Collatz Conjecture is true for all $m \leq 60$. Which
starting value of $m$ required the most steps?
(b) Suppose you change $3n+1$ to $n+1$ in the definition of $f(n)$. What
happens? Can you prove this?
(c) Suppose you change $3n+1$ to $5n+1$ in the definition of $f(n)$. Try a
few examples and see what happens. Do you see a different behavior for
starting values $m = 5$, $m = 6$, and $m = 7$?

2.  Let $\pi_{n,a}(x)$ be the number of primes of the form $nk+a$ that are less than
or equal to $x$. For example, $\pi_{4,3}(20) = 4$ since it is counting the primes
$3, 7, 11, 19$.
(a) Let $\pi(x)$ be the number of primes less than or equal to $x$. Compute
the ratios

$$\frac{\pi_{4,1}(x)}{\pi(x)} \text{ and } \frac{\pi_{4,3}(x)}{\pi(x)}$$

for $x = 1000$, $x = 100000$, and $x = 1000000$. Make a guess as to what
each ratio approaches as $x$ gets larger and larger.
(b) Show that $\pi_{4,1}(x) \leq \pi_{4,3}(x)$ for all integers $x < 30000$ except for two
(consecutive) integers $x$. This comparison is sometimes called a "prime
number race." Chebyshev observed in 1853 that usually $\pi_{4,3}(x)$ is ahead
of $\pi_{4,1}(x)$, but Littlewood showed in 1914 that the lead changes infinitely
often. In 1994, Rubinstein and Sarnak showed (under some yet-to-be-
proved assumptions) that $\pi_{4,3}(x)$ is ahead most of the time.

3.  Find 20 examples of numbers that are sixth powers. What is true about
the remainders when these sixth powers are divided by 7? Can you make
a conjecture and then prove it?

4.  Write a program that generates 10000 "random" pairs $(a, b)$ of integers,
and then use the Euclidean Algorithm (or your software gcd routine)
to decide whether $\gcd(a, b) = 1$. If $m$ is the number of pairs that are
relatively prime, calculate

$$\frac{m}{10000},$$

which is the fraction of these pairs that are relatively prime. Follow the
same procedure for $10^5$ pairs and then $10^6$ pairs. Can you make a guess
as to what number this ratio approaches? (*Hint:* It is an integer divided
by $\pi^2$.)

5.  By Exercise 24 above, there are arbitrarily long sequences of consecutive composite numbers. Find the first string of (at least) 10 consecutive composites, of (at least) 50 consecutive composites, and of (at least) 100 consecutive composites. How do these numbers compare in size to the numbers used in Exercise 24?

6.  (a) Show that $n^4 + 4$ is composite for several values of $n \geq 2$. There is a reason for this. Let's find it.
    (b) Show that in each example in (a), $n^4 + 4$ always can be factored as $a \times b$, where $a - b = 4n$.
    (c) Write $a = x + 2n$ and $b = x - 2n$ for some $x$. Use $n^4 + 4 = ab$ to find $x$ and then find the numbers $a$ and $b$.

7.  Start with a 4-digit integer (0 is allowed as any of the 4 digits) other than 0000, 1111, 2222, 3333, ..., 9999. Write the digits in decreasing order and in increasing order, and subtract, yielding a new 4-digit number. Repeat the process with this new number to obtain another number, and then repeat the process again, etc. For example, if you start with 6876, you compute $8766 - 6678 = 2088$. Then $8820 - 0288 = 8532$, then $8532 - 2358 = 6174$, and then $7641 - 1467 = 6174$. Show that the process always yields 6174. (*Remark:* The integer 6174 is called the *Kaprekar constant.*)

## 2.11.4   Answers to "CHECK YOUR UNDERSTANDING"

1.  Since $1001 = 7 \times 143$, we have $7 \mid 1001$.

2.  Since $1005 = 7 \times 143 + 4$, 1005 is not a multiple of 7.

3.  Suppose $5 \mid 2 \cdot 3 \cdot 5 \cdot 7 + 1$. Since $5 \mid 2 \cdot 3 \cdot 5 \cdot 7$, Corollary 2.4 says that $5 \mid (2 \cdot 3 \cdot 5 \cdot 7 + 1) - (2 \cdot 3 \cdot 5 \cdot 7) = 1$, which is not true. Therefore, $5 \nmid 2 \cdot 3 \cdot 5 \cdot 7 + 1$.

4.  We have to use only the primes less than $\sqrt{20} \approx 4.5$, so 2 and 3 suffice. After crossing out 1 and the multiples of 2, the numbers that remain are 2, 3, 5, 7, 9, 11, 13, 15, 17, 19. Of these, 3 crosses out 9 and 15. The numbers 2, 3, 5, 7, 11, 13, 17, 19 remain and are the primes less than 20.

5.  Divide 7 into 200: $q = 28$ and $r = 4$, so $200 = 7 \cdot 28 + 4$.

6.  We get $-200 = 7 \cdot (-28) - 4$, but we want $0 \leq r < 7$. Therefore, we write $-200 = 7 \cdot (-29) + 3$, so $q = -29$ and $r = 3$.

7.  The divisors of 24 are 1, 2, 3, 4, 6, 8, 12, 24. The largest of these that divides 42 is 6, so $\gcd(24, 42) = 6$.

8.  Since $60 = 2^2 \cdot 3 \cdot 5$, we need $n$ to be odd and not a multiple of 3 or 5. The only $n$ between 1 and 10 that satisfies these requirements is $n = 7$.

9.  If $d \mid n$ and $d \mid n + 3$, then $d \mid (n + 3) - n = 3$. Therefore, $d = 1$ or 3, so any common divisor is either 1 or 3. If $n = 1$, then $\gcd(1, 4) = 1$, so $d = 1$ occurs. If $n = 3$, then $\gcd(3, 6) = 3$, so $d = 3$ occurs.

10.    Use the Euclidean Algorithm:

$$654 = 2 \cdot 321 + 12$$
$$321 = 26 \cdot 12 + 9$$
$$12 = 1 \cdot 9 + 3$$
$$9 = 3 \cdot 3 + 0.$$

Therefore, $3 = \gcd(654, 321)$.

11.    Use the Extended Euclidean Algorithm:

| | $x$ | $y$ | |
|---|---|---|---|
| 17 | 1 | 0 | |
| 12 | 0 | 1 | |
| 5 | 1 | $-1$ | (1st row) $-$(2nd row) |
| 2 | $-2$ | 3 | (2nd row) $-$ 2$\cdot$(3rd row) |
| 1 | 5 | $-7$ | (3rd row) $-$ 2$\cdot$(4th row). |

The end result is $1 = 17 \cdot 5 - 12 \cdot 7$, so $x = 5$ and $y = -7$.

12.    Use the Division Algorithm:

$$1234 = 176 \cdot 7 + 2$$
$$176 = 25 \cdot 7 + 1$$
$$25 = 3 \cdot 7 + 4$$
$$3 = 0 \cdot 7 + 3.$$

Therefore, $1234_{10} = 3412_7$.

13.    $321_5 = 3 \cdot 5^2 + 2 \cdot 5 + 1 = 75 + 10 + 1 = 86.$

# Chapter 3

# Linear Diophantine Equations

## 3.1   $ax + by = c$

As we mentioned in the introduction, Diophantus lived in Alexandria, Egypt, about 1800 years ago. His book *Arithmetica* gave methods for solving various algebraic equations and had a great influence on the development of algebra and number theory for many years. The part of number theory called Diophantine equations, which studies integer (and sometimes rational) solutions of equations, is named in his honor.

In this section we study the equation

$$ax + by = c$$

where $a$, $b$, and $c$ are integers. Our goal is to find out when **integer** solutions to this equation exist, and when they do exist, to find all of them.

Equations of this form can arise in real life. For example, how many dimes and quarters are needed to pay someone \$1.05? This means we have to solve $10x + 25y = 105$. One solution is $x = 3$, $y = 3$. Another solution is $x = 8$, $y = 1$. There are also solutions such as $x = -2$, $y = 5$, which means you pay 5 quarters and get back 2 dimes.

Before we get to the main result of this section, we look at two more examples that will help us understand the general situation. First, consider $6x - 9y = 20$. Notice that 3 must divide the left-hand side but 3 is not a divisor of the right-hand side. This tells us that this equation can never have an integer solution. To make things notationally simpler, let $d = \gcd(a, b)$. We then see that in order for $ax + by = c$ to have a solution, we must have $d \mid c$. Now let's look at an example where this does occur, say $6x + 9y = 21$. We can divide both sides by 3, giving us $2x + 3y = 7$. After a brief inspection, we see that $x = 2$ and $y = 1$ is a solution. Are there others? It's easy to see that if $t$ is any integer, then $x = 2 + 3t$ and $y = 1 - 2t$ is also a solution. Let's verify this by substituting these expressions for $x$ and $y$ into the original equation, $6x + 9y = 21$:

$$6(2 + 3t) + 9(1 - 2t) = 12 + 18t + 9 - 18t = 21,$$

so our single solution gives rise to an infinite number of them. This can be generalized in the following theorem:

**Theorem 3.1.** *Assume that $a, b$, and $c$ are integers where at least one of $a, b$ is non-zero. Then the equation*

$$ax + by = c \tag{3.1}$$

*has a solution if and only if $\gcd(a, b) \mid c$. If it has one solution, then it has an infinite number. If $(x_0, y_0)$ is any particular solution, then all solutions are of the form*

$$x = x_0 + \frac{b}{\gcd(a, b)}t, \qquad y = y_0 - \frac{a}{\gcd(a, b)}t \tag{3.2}$$

*with $t$ an integer.*

*Proof.* We begin by setting $\gcd(a, b) = d$. We have already seen that if $d \nmid c$, then there are no solutions. Now, assume $d \mid c$. From Theorem 2.12 we know that there are integers $r$ and $s$ so that $ar + bs = d$. Since $d \mid c$, we have that $df = c$ for some integer $f$. Therefore,

$$a(rf) + b(sf) = df = c.$$

So, $x_0 = rf$ and $y_0 = sf$ is a solution to $ax + by = c$.

Now let

$$x = x_0 + \frac{b}{d}t \text{ and } y = y_0 - \frac{a}{d}t,$$

where $t$ is an integer. Since $a/d$ and $b/d$ are integers, $x$ and $y$ are integers. Then

$$ax + by = a(x_0 + \frac{b}{d}t) + b(y_0 - \frac{a}{d}t) = ax_0 + by_0 + \frac{ab}{d}t - \frac{ba}{d}t = c.$$

This shows that a solution to (3.1) exists (assuming that $\gcd(a, b)$ divides $c$) and that once we have one solution, we have an infinite number of a specific form.

Next, we need to prove that every solution of Equation (3.1) is of the stated form. Fix one solution $x_0, y_0$ and let $u, v$ be *any* solution of Equation (3.1). (Any solution continues to mean any integer solution.) Then

$$au + bv = c \tag{3.3}$$

and

$$ax_0 + by_0 = c. \tag{3.4}$$

Subtracting Equation (3.4) from Equation (3.3) gives us

$$a(u - x_0) + b(v - y_0) = 0,$$

so

$$a(u - x_0) = -b(v - y_0) = b(y_0 - v). \tag{3.5}$$

After dividing both sides of Equation (3.5) by $d$, we get

$$\frac{a}{d}(u - x_0) = \frac{b}{d}(y_0 - v). \tag{3.6}$$

There is a small technicality that needs to be dealt with. If $a = 0$, then we can't say that $a/d$ divides the right-hand side, because we don't allow 0 to divide anything. But if $a = 0$ then our original equation is $by = c$. This means that $v = y_0 = c/b$ and $x$ can be arbitrary, since there is no restriction on $x$. This is exactly the conclusion of the theorem, which says that all solutions have the form $y = y_0$ and $x = x_0 + t$ (since $\gcd(a, b) = \gcd(0, b) = b$). For the rest of the proof, we now assume that $a \neq 0$.

Equation (3.6) implies that

$$(a/d) \mid (b/d)(y_0 - v).$$

Since $\gcd(a/d, b/d) = 1$ from Proposition 2.10, Proposition 2.16 implies that $(a/d)$ divides $(y_0 - v)$. By definition, this means that there is an integer $t$ with

$$y_0 - v = t\frac{a}{d}. \tag{3.7}$$

Substituting the value for $y_0 - v$ from (3.7) into (3.6), we get

$$\frac{a}{d}(u - x_0) = \frac{b}{d}\left(\frac{a}{d}t\right). \tag{3.8}$$

Multiplying both sides by $\frac{d}{a}$, we have

$$u - x_0 = \frac{b}{d}t \quad \text{or} \quad u = x_0 + \frac{b}{d}t. \tag{3.9}$$

Combining (3.7) and (3.9), we have

$$u = x_0 + \frac{b}{d}t \quad \text{and} \quad v = y_0 - \frac{a}{d}t. \tag{3.10}$$

Since $u$ and $v$ were arbitrary solutions of (3.1), we have completed the proof.                                                                $\square$

In practice, if we want to solve Equation (3.1), we first verify that $d \mid c$. If it doesn't, we're done since there are no solutions. If it does, we divide both sides by $d$ to get a new equation

$$a'x + b'y = c'$$

and in this equation, $\gcd(a', b') = 1$. For example, if we want to solve $6x + 15y = 30$, we divide by 3 and instead solve $2x + 5y = 10$. This means that we will usually be using the following:

**Corollary 3.2.** *Assume that $a, b$, and $c$ are integers with at least one of $a, b$ non-zero. If $\gcd(a, b) = 1$, then the equation*

$$ax + by = c$$

*always has an infinite number of solutions. If $(x_0, y_0)$ is any particular solution, then all solutions are of the form*

$$x = x_0 + bt, \quad y = y_0 - at$$

*with $t$ an integer.*

It may seem that we've ignored the problem of actually finding a solution to a linear Diophantine equation; however, the Extended Euclidean Algorithm from the previous section provides an efficient method. For example, to solve $13x + 7y = 5$, we write $\gcd(7, 13) = 1$ as a linear combination of 7 and 13 and then multiply our solution by 5. Here's how it works.

We begin by calculating $\gcd(7, 13)$ using the Euclidean Algorithm.

$$13 = 1 \cdot 7 + 6$$
$$7 = 1 \cdot 6 + 1$$
$$6 = 6 \cdot 1 + 0.$$

Now, we use the Extended Euclidean Algorithm to express 1 as a linear combination of 7 and 13:

| | $x$ | $y$ | |
|---|---|---|---|
| 13 | 1 | 0 | |
| 7 | 0 | 1 | |
| 6 | 1 | −1 | (1st row) − (2nd row) |
| 1 | −1 | 2 | (2nd row) − (3rd row). |

We see that $1 = -1 \cdot 13 + 2 \cdot 7$, so that $x = -1$, $y = 2$ is a solution to $13x + 7y = 1$:

$$13(-1) + 7(2) = 1.$$

Multiplying both $x$ and $y$ by 5 gives us $x = -5$, $y = 10$ is the desired solution to the original equation, $13x + 7y = 5$:

$$13(-5) + 7(10) = 5.$$

Theorem 3.1 tells us that all solutions have the form

$$x = -5 + 7t, \quad y = 10 - 13t,$$

where $t$ is an integer.

Here is another example. Let's find all solutions of $10x + 25y = 105$, the equation for paying \$1.05 in dimes and quarters. First, divide by $5 = \gcd(10, 25)$ to get

$$2x + 5y = 21.$$

At this point, you can find a solution by any method. For example you can try values until something works or use the Extended Euclidean

Algorithm. In any case, one solution is $x_0 = 8$, $y_0 = 1$. The set of all solutions is

$$x = 8 + 5t, \quad y = 1 - 2t.$$

The solution $x = 3$, $y = 3$ given at the beginning of this section is obtained by letting $t = -1$. The solution with $x = -2$, $y = 5$ is obtained by letting $t = -2$.

This example points out that you can find the particular solution by any method that works. Often, it is easier to find a solution by trial and error rather than by the Extended Euclidean Algorithm. For example, if you need a particular solution of $12x + 11y = 131$, you could use the Extended Euclidean Algorithm to find that $12(1) + 11(-1) = 1$, then multiply by 131 to get $x = 131$, $y = -131$ as a particular solution. However, you could also notice that $x = 10$, $y = 1$ also gives a particular solution and has the advantage of smaller numbers.

Now, a warning. It's quite possible that two people working on the same problem may get correct answers that look different. If a problem says find all solutions to $5x - 3y = 1$, you may notice that $(2, 3)$ is a particular solution, so all solutions look like $x = 2 - 3t, \quad y = 3 - 5t$. A friend may choose a particular solution to be $(-1, -2)$ and say that all solutions are of the form $x = -1 - 3t, y = -2 - 5t$. These two apparently different sets of solutions are in fact the same, as the following shows:

Solutions of the form $x = 2 - 3t, y = 3 - 5t$:

$$\ldots, (-4, -7), (-1, -2), (\mathbf{2}, \mathbf{3}), (5, 8), (8, 13), (11, 18), \ldots$$

Solutions of the form $x = -1 - 3t, y = -2 - 5t$:

$$\ldots, (-4, -7), (\mathbf{-1}, \mathbf{-2}), (2, 3), (5, 8), (8, 13), (11, 18), \ldots$$

Finally, suppose we want to find the solutions to

$$-17x + 14y = 30.$$

The Extended Euclidean Algorithm could be carried out with negative numbers, but instead we use it to solve $17x + 14y = 30$ and then change the sign of $x$. For example, the algorithm with 17 and 14 yields $17(5) + 14(-6) = 1$, so $17(150) + 14(-180) = 30$. A particular solution of $-17x + 14y = 30$ is therefore $x_0 = -150$, $y_0 = -180$. The general solution is given by Corollary 3.2:

$$x = -150 + 14t, \quad y = -180 + 17t.$$

---

**CHECK YOUR UNDERSTANDING**

1. Find all integer solutions to $6x + 8y = 4$.

---

# 3.2    The Postage Stamp Problem

If you went to the post office to mail a letter and discovered that they had only three-cent and five-cent stamps, what postage values would you be able to put on your mail? What values are unobtainable from these two stamps? These questions are special cases of what is called the Postage Stamp Problem.

**The Postage Stamp Problem:** If $a$ and $b$ are positive integers, what positive integers can be written as $ax + by$ with both $x$ and $y$ non-negative?

To begin, notice that we want to consider only the case where $a$ and $b$ are relatively prime. If, for example, they were both even, then no odd numbers would ever be expressible as a linear combination of them, and the problem becomes less interesting.

We'll call numbers that can be written as $ax + by$ with both $x$ and $y$ non-negative **feasible**. For example, $a, b$, and $ab$ are always feasible since

$$a = 1 \cdot a + 0 \cdot b, \quad b = 0 \cdot a + 1 \cdot b,$$
$$ab = b \cdot a + 0 \cdot b = 0 \cdot a + a \cdot b.$$

The requirement that $x$ and $y$ both be non-negative is what makes this an interesting problem. For example, our initial question had three-cent and five-cent stamps, so $a = 3$ and $b = 5$. Since 3 and 5 are relatively prime, if negative coefficients were allowed, then *every* integer could be expressed as a linear combination of them from Theorem 2.12.

Let's try to understand which numbers are feasible and which are not by making a chart to see if any patterns occur:

Postage Stamp Problem with $a = 3$ and $b = 5$

| Number | 1 | 2 | 3 | 4 | 5 | 6 | 7 | 8 | 9 | 10 | 11 | 12 | 13 | 14 |
|--------|---|---|---|---|---|---|---|---|---|----|----|----|----|----|
| Feasible | | | ✓ | | ✓ | ✓ | | ✓ | ✓ | ✓ | ✓ | ✓ | ✓ | ✓ |

An empty space means that the number above it cannot be written as a permissible linear combination of 3 and 5, while a ✓ means that it can be. For example, 13 has a ✓ underneath it because $13 = 1 \cdot 3 + 2 \cdot 5$.

Notice that every number in our list that's greater than 7 is feasible. In fact, once we notice that the three consecutive integers 8, 9, and 10 are feasible, we can show that every number greater than 10 is also feasible. Here's how the argument goes. Every number greater than 10 differs from 8, 9, or 10 by a multiple of 3. (You can use the Division Algorithm to formally prove this.) So, to get any number greater than 10, just add enough three-cent stamps to get the desired amount. For example, suppose you want 92 cents. Since $92 - 8 = 84 = 3 \cdot 28$, take the three-cent and five-cent stamps needed to get eight cents, and add 28 more 3-cent stamps to get 92 cents.

One more thing to point out before we do another example. The last number in our list that is not feasible is 7, which can be written as $3 \cdot 5 - 3 - 5$. Let's see if this pattern holds.

Postage Stamp Problem with $a = 2$ and $b = 7$

| Number | 1 | **2** | 3 | 4 | 5 | 6 | **7** | 8 | 9 | 10 | 11 | 12 | 13 | **14** |
|--------|---|---|---|---|---|---|---|---|---|----|----|----|----|----|
| Feasible | | ✓ | | ✓ | | ✓ | ✓ | ✓ | ✓ | ✓ | ✓ | ✓ | ✓ | ✓ |

Again, $5 = 2 \cdot 7 - 2 - 7$ is the last number that is not feasible. If you try another small example (say with $a = 4$ and $b = 9$) you will see the same thing occurs, namely every integer greater than $ab - a - b$ can be written as a linear combination of $a$ and $b$ while $ab - a - b$ cannot be. We'll now use two steps to show that this is always the case.

- $ab - a - b$ is not feasible.

- If $n > ab - a - b$, then $n$ is feasible.

We begin by verifying the first step.

**Proposition 3.3.** *Assume that a and b are positive, relatively prime integers. Then there are no non-negative integers x and y with $ax + by = ab - a - b$.*

*Proof.* We start by observing that

$$a(-1) + b(a - 1) = ab - a - b,$$

so $x = -1, y = a - 1$ is a solution to $ax + by = ab - a - b$.

Since $\gcd(a, b) = 1$, we see from Corollary 3.2 that every solution to $ax + by = ab - a - b$ has the form

$$x = -1 + bt, \quad y = a - 1 - at$$

for some integer $t$. In order to get an $x$ value that's non-negative, we have to choose a value of $t$ that's positive. But if $t > 0$, then $1 - t \leq 0$ so $y = a - 1 - at = a(1 - t) - 1 \leq -1$. Forcing $x$ to be non-negative forces $y$ to be negative. This means it is impossible to find a non-negative solution to $ax + by = ab - a - b$. ☐

We've now shown that $ab - a - b$ can never be feasible. We next show that every integer larger than $ab - a - b$ is feasible.

**Proposition 3.4.** *Assume that a and b are positive with $\gcd(a, b) = 1$. If $n \geq ab - a - b + 1$, there are non-negative integers x and y with $ax + by = n$.*

*Proof.* We again appeal to Corollary 3.2. We find a pair of integers $(x_0, y_0)$ with

$$ax_0 + by_0 = n \geq ab - a - b + 1, \tag{3.11}$$

which enables us to express every solution in the form

$$x = x_0 + bt, \quad y = y_0 - at.$$

It is possible that $x_0$ or $y_0$ is negative. What we'll do is find the solution to $ax + by = n$ with the smallest possible non-negative $y$, and then show that the corresponding $x$ must be non-negative. After all, this is our best chance, because making $y$ smaller makes $x$ bigger, so if anything works, it must be the smallest non-negative value of $y$. Using the Division Algorithm, we can divide $y_0$ by $a$ and write $y_0 = at + y_1$, with $0 \leq y_1 \leq a - 1$, for some integer $t$. This $y_1$ is our choice of $y$. Since

$y_1 = y_0 - at$, we take $x_1 = x_0 + bt$ as our choice of $x$. We'll show that $x_1 \geq 0$. If $x_1 \leq -1$, then, because $y_1 \leq a - 1$,

$$
\begin{aligned}
n &= ax_0 + by_0 \\
&= a(x_1 - bt) + b(y_1 + at) \\
&= ax_1 + by_1 \\
&\leq a(-1) + b(a - 1) \\
&= ab - a - b,
\end{aligned}
$$

and this contradicts our assumption that $n \geq ab - a - b + 1$. So, $(x_1, y_1)$ is a non-negative solution.                                                                 $\square$

Somewhat surprisingly, if there are three or more stamps, there are no known formulas analogous to those we have when there are two stamps.

---

## 3.3   Chapter Highlights

1.  $ax + by = c$ has a solution in integers $x$, $y$ if and only if $\gcd(a, b) \mid c$.

2.  The solutions to the linear Diophantine equation $ax + by = c$ can be found using the Extended Euclidean Algorithm.

---

## 3.4   Problems

### 3.4.1   Exercises

**Section 3.1: Linear Diophantine Equations**

1.  For each of the following equations, find all integral solutions or show that it has none.
    (a) $3x + 4y = 10$
    (b) $5x - 7y = 9$
    (c) $9x + 23y = 1$
    (d) $4x + 6y = 11$

2.  For each of the following equations, find all integral solutions or show that it has none.
    (a) $8x + 2y = 26$
    (b) $44x - 17y = 9$
    (c) $60x + 9y = 31$
    (d) $60x + 9y = 51$

3.  For the following equations, find all *positive* solutions or show that it has none.
    (a) $10x + 2y = 100$
    (b) $5x + 3y = 21$
    (c) $9x - y = 44$
    (d) $2x + 7y = 3$

4.  For the following equations, find all *positive* solutions or show that it has none.
    (a) $15x + y = 150$
    (b) $4x + 6y = 20$
    (c) $11x - 5y = 60$
    (d) $7x + 17y = 81$

5.  A farmer pays 1770 crowns in purchasing horses and oxen. He pays 31 crowns for each horse and 21 crowns for each ox. How many horses and oxen did he buy? (This problem is from Euler's *Algebra*, which was published in 1770.)

6.  You cash a check at the bank, but the teller accidentally switches the dollars and the cents. The amount you are given is 47 cents more than twice the correct amount. What was the amount on the original check? Assume that the number of dollars and the number of cents are less than 100. (*Hint:* If the original check is for $x$ dollars and $y$ cents, then the check was for $100x + y$. You received $100y + x$.)

7.  Show that $10x + 11y = 880$ has *exactly* 7 solutions in positive integers $x, y$.

8.  Show that 880 is the smallest integer value of $n$ for which $10x + 11y$ has *exactly* 7 positive solutions.

9.  Find the smallest integer $x \geq 100$ such that there is an integer $y$ with $19x + 13y = 20$.

10. Find all solutions of $43x + 23y = 1$ in integers $x, y$ such that $0 < y < 100$.

11. Let $a, b, c$ be integers with $\gcd(a, b) = 1$ and with $a > 0$ and $b > 0$. Show that
$$ax - by = c$$
has infinitely many solutions in positive integers $x, y$.

12. You have nickels, dimes, and quarters, for a total of 20 coins. If the total value of the coins is $2.00, what are the possibilities for the numbers of nickels, dimes, and quarters? (Assume that you have at least one of each type of coin.)

13. An all-you-can-eat restaurant charges $15 for teenagers and $13 for adults. At the end of a day, the restaurant had collected $500. What is the largest number of people who could have eaten there that day? What is the smallest number of people?

14.  (a) What is the smallest nonzero value of $\left|\frac{x}{36} - \frac{y}{31}\right|$, where $x$ and $y$ are integers?
     (b) A yardstick has a blue mark at each inch. Suppose someone miscounts and also divides the yardstick into 31 equally spaced parts with a red mark at each division. What is the smallest distance between a blue mark and a red mark, and where does this smallest distance occur? (There are no marks at the 0-inch and 36-inch ends.)

15.  Find all integer solutions to the equation $3x + 5y + 7z = 1$.

16.  Find all integer solutions to the equation $2x + 4y + 3z = 5$.

17.  Find all integer solutions to the equation $3x + 6y + 5z = 2$.

18.  Find all integer solutions to the equation $x + 2y + 3z = 9$.

### Section 3.2: The Postage Stamp Problem

19.  The customer claims that she paid $50 with a set of 7-dollar and 11-dollar bills. The store owner claims she paid only $48. Exactly one of them is right. Which one? Explain why.

20.  (a) Suppose you have only 13-dollar and 7-dollar bills. You need to pay someone 71 dollars. Is this possible without receiving change? If so, show how to do it. If not, explain why it is impossible.
     (b) Suppose you need to pay someone 75 dollars. Is this possible? If so, show how to do it. If not, explain why it is impossible.

21.  (a) In the Postage Stamp Problem with stamps worth 3 cents and 5 cents, use the fact that 8, 9, and 10 are feasible to show that
         (i) 100 is feasible,   (ii) 260 is feasible,   (iii) 302 is feasible.
     (b) If we have stamps of denomination $a$ and $b$ and we can show that the $a$ consecutive numbers $k, k+1, k+2, ..., k+a-1$ are feasible, show that every number greater than $k + a - 1$ must also be feasible.

22.  Suppose you have piles of 5-cent and 8-cent stamps. You want to put 95 cents postage on a letter.
     (a) What combination of stamps uses the smallest number of stamps?
     (b) What combination of stamps uses the largest number of stamps?

23.  Some children are making a tower out of blocks. They have 12-inch blocks and 7-inch blocks, and they want to make a tower that is 71 inches tall. What combination of blocks achieves this?

## 3.4.2    Answers to "CHECK YOUR UNDERSTANDING"

1.  Use Theorem 3.1. Trial and error, for example, yields the solution $(x_0, y_0) = (-2, 2)$. Since $b/\gcd(a, b) = 8/2 = 4$ and $a/\gcd(a, b) = 6/2 = 3$, all solutions are given by

$$x = -2 + 4t, \qquad y = 2 - 3t,$$

where $t$ is an integer (there are similar solutions corresponding to other choices of $(x_0, y_0)$).

# Chapter 4

# Unique Factorization

We already know from Lemma 2.6 that every integer greater than 1 has a prime factor, and we'll see that this easily implies that each such integer is prime or a product of primes. The main purpose of this chapter is to prove that this factorization into primes can be done in only one way.

## 4.1 The Starting Point

The key to everything we do in this chapter is a consequence of the Division Algorithm. Recall Theorem 2.12: *Let $a$ and $b$ be integers with at least one of $a, b$ non-zero. There exist integers $x$ and $y$, which can be found by the Extended Euclidean Algorithm, such that*

$$\gcd(a, b) = ax + by.$$

The following result is by far the most important property of primes. The theorem makes the transition from a property that can be verified numerically ($p$ doesn't factor) to this important divisibility property for $p$.

**Theorem 4.1.** *Let $p$ be prime and let $a$ and $b$ be integers with $p \mid ab$. Then $p \mid a$ or $p \mid b$.*

*Proof.* Let $d = \gcd(a, p)$. Then $d \mid p$, so $d = 1$ or $p$. If $d = p$, then the fact that $d$ is a divisor of $a$ implies that $p \mid a$, so we're done in this case. If $d = 1$, we use Theorem 2.12 to write $1 = ax + py$ for some integers $x$ and $y$. Multiply by $b$ to get

$$b = abx + pby.$$

63

Since $p \mid ab$ and $p \mid p$, and $b$ is a linear combination of $ab$ and $p$, Proposition 2.3 implies that $p \mid b$, so we're done in this case, too.    □

**Remark.** If $n$ is composite and $n \mid ab$, we cannot conclude that $n \mid a$ or $n \mid b$. For example, $6 \mid 9 \cdot 8$, but $6 \nmid 9$ and $6 \nmid 8$. In fact, when $n$ is composite, there is always an example: write $n = ab$ with $1 < a, b < n$. Then $n \mid ab$ but $n \nmid a$ and $n \nmid b$.

**Corollary 4.2.** *Let $p$ be prime and let $a_1, a_2, \ldots, a_r$ be integers such that*

$$p \mid a_1 a_2 \cdots a_r.$$

*Then $p \mid a_i$ for some $i$.*

*Proof.* We'll prove this by induction on $r$ (see Appendix A). The result is clearly true for $r = 1$ since if $p \mid a_1$, then $p \mid a_1$. Suppose it is true for $r = k$ with $k \geq 2$. We'll now prove it must be true for $r = k + 1$. Let $a = a_1 a_2 \cdots a_k$ and let $b = a_{k+1}$. We are assuming that

$$p \mid a_1 a_2 \cdots a_r = a_1 a_2 \cdots a_{k+1} = ab.$$

Theorem 4.1 says that $p \mid a$ or $p \mid b$. If $p \mid a = a_1 a_2 \cdots a_k$, we have $p \mid a_i$ for some $i$, by the induction assumption. If $p \mid b$, then $p \mid a_{k+1}$ since $a_{k+1} = b$. Therefore, the corollary is true for $r = k + 1$. By induction, it is true for all $r \geq 1$.    □

---

### CHECK YOUR UNDERSTANDING

1. If $p$ is prime and $p^2 \mid ab$, is it true that $p^2 \mid a$ or $p^2 \mid b$? Give a proof or a counterexample.

---

# 4.2   The Fundamental Theorem of Arithmetic

The Fundamental Theorem of Arithmetic says that any positive integer greater than 1 is either prime or can be factored in *exactly one way* as a product of primes. Put more colloquially, this says that every integer greater than 1 is built up out of primes in exactly one way. This type of theorem is called an existence-uniqueness theorem because it makes two assertions. The first, the existence part, is that every integer can be written as a product of prime numbers. This will be relatively easy to show. The second, the uniqueness part, says that this can be done in only one way. This is harder to show and its proof relies on the Euclidean Algorithm. Later on, when we work in number systems other than the integers, we'll see that the uniqueness part of this theorem is not always true. See Chapter 19.

Although people often say that the Fundamental Theorem of Arithmetic (FTA) was first proved by Euclid, that is not entirely true. There is enough information in Euclid's *Elements* to deduce the FTA, but it's never explicitly stated there. We will have a more detailed discussion of this at the end of the section.

**The Fundamental Theorem of Arithmetic.** *Every integer $n > 1$ can be uniquely written as a prime or a product of primes.*

*Proof.* We begin our proof with the existence part.

**Lemma 4.3.** *Every integer can be written as a product of prime numbers.*

*Proof.* We prove this lemma by assuming there are non-prime integers that cannot be written as products of primes and arriving at a contradiction. Let

$$S = \{\text{integers } n > 1 |\ n \text{ is not prime and}$$
$$n \text{ cannot be written as a product of primes}\}.$$

Assume that S is nonempty. Since all the elements in $S$ are positive integers, $S$ has a smallest element; call it $s$. (This is an example of the Well-Ordering Principle. See Appendix A.) Now, $s$ cannot be prime, so $s$ must be composite. This means that $s = ab$ where $a$ and $b$ are integers that are greater than 1, so $1 < a < s$ and $1 < b < s$. Since $s$ was the smallest integer that could not be written as a product of primes, we know that $a$ is a product of primes and so is $b$. This means that $s = ab$ is a product of primes, contradicting that $s \in S$. The only other possibility is that $S$ must be empty, so every integer greater than 1 is either prime or is a product of primes. $\qquad\square$

Having shown that every integer $n > 1$ is either prime or a product of primes, we now need to show uniqueness: an integer can be written as a product of primes in only one way. The proof of this relies on Corollary 4.2. We suppose that there is at least one integer that can be written as a product of primes in more than one way. Let $n$ be the smallest of these, so

$$n = p_1 p_2 \cdots p_r = q_1 q_2 \cdots q_s, \tag{4.1}$$

where the $p_i$ and $q_j$ are primes. Since $p_1$ divides the left-hand side, it has to divide the right-hand side as well. From Corollary 4.2, we see that $p_1 \mid q_i$ for some $i$. After rearranging terms, we can assume that $q_i = q_1$. Since $q_1$ is prime, and $p_1 \mid q_1$, we must have that $p_1 = q_1$. So, we write Equation (4.1) as

$$n = p_1 p_2 \cdots p_r = p_1 q_2 \cdots q_s$$

and then divide both sides by $p_1$. This gives us

$$m = p_2 p_3 \cdots p_r = q_2 q_3 \cdots q_s$$

for some integer $m$. Since the two factorizations of $n$ were different, these must be two different factorizations of $m$.

Because $m = n/p_1$, we know that $m < n$. This contradicts our assumption that $n$ was the smallest integer that could be factored as a product of primes in more than one way. Therefore, our initial assumption on non-unique factorization was incorrect and every integer can be factored into a product of primes in exactly one way.                      □

Throughout the book, we'll see applications of the Fundamental Theorem of Arithmetic. We start by rephrasing divisibility in terms of prime factorizations. Often, we'll write the factorization of an integer as follows:

$$a = 2^{a_2} 3^{a_3} 5^{a_5} \cdots .$$

Each of the exponents is a nonnegative integer, and all but finitely many of the exponents are 0. For example, if $a = 20 = 2^2 5$, then $a_2 = 2, a_3 = 0, a_5 = 1, a_7 = 0, a_{11} = 0, \ldots$.

**Proposition 4.4.** *Let $a$ and $b$ be positive integers with*

$$a = 2^{a_2} 3^{a_3} 5^{a_5} \cdots \quad and \quad b = 2^{b_2} 3^{b_3} 5^{b_5} \cdots$$

*the prime factorizations of $a$ and $b$. Then $a \mid b$ if and only if $a_p \leq b_p$ for all $p$.*

*Proof.* If $a \mid b$, then there exists $c$ with $ac = b$. Write

$$c = 2^{c_2} 3^{c_3} 5^{c_5} \cdots .$$

Then

$$2^{a_2 + c_2} 3^{a_3 + c_3} 5^{a_5 + c_5} \cdots = ac = b = 2^{b_2} 3^{b_3} 5^{b_5} \cdots .$$

By uniqueness of factorization into primes, we must have

$$a_p + c_p = b_p$$

for each $p$. In particular, $a_p \leq b_p$ for each $p$.

Conversely, suppose $a_p \leq b_p$ for all $p$. Let $c_p = b_p - a_p$, so $c_p \geq 0$. Let

$$c = 2^{c_2} 3^{c_3} 5^{c_5} \cdots .$$

Then $ac = b$ so $a \mid b$.

There is one technical point that needs to be mentioned. Only finitely many $a_p$ and $b_p$ are nonzero, so only finitely many $c_p$ are nonzero. Therefore, there are only finitely many factors in the expression for $c$ so $c$ is an integer. (If, for some reason, infinitely many $c_p$ were nonzero, $c$ would not be an integer since we would have an infinite product.)  $\square$

The **least common multiple** of numbers $a$ and $b$, written $[a, b]$ or $\mathrm{lcm}(a, b)$, is defined to the smallest positive integer that is divisible by both $a$ and $b$. Proposition 4.4 immediately yields the following method to compute gcd's and lcm's when the factorizations of $a$ and $b$ are known.

**Proposition 4.5.** *Let $a$ and $b$ be two positive integers. Let*

$$a = 2^{a_2} 3^{a_3} 5^{a_5} \cdots \text{ and } b = 2^{b_2} 3^{b_3} 5^{b_5} \cdots$$

*be the prime factorizations of $a$ and $b$. Let $d_p = \min(a_p, b_p)$ and $e_p = \max(a_p . b_p)$. Then,*

$$\gcd(a, b) = 2^{d_2} 3^{d_3} 5^{d_5} \cdots ,$$

*and*

$$\mathrm{lcm}(a, b) = 2^{e_2} 3^{e_p} 5^{e_5} \cdots .$$

**Example.** Let $a = 720 = 2^4 3^2 5$ and $b = 6300 = 2^2 3^2 5^2 7$. Then $d_2 = 2, d_3 = 2, d_5 = 1, d_7 = 0$ (note that the exponent for 7 in $a$ is 0). Therefore, $\gcd(720, 6300) = 2^2 3^2 5 = 180$. This method of finding gcd's is the one many people use before learning the Euclidean Algorithm. It works well for small numbers, but it has a drawback because it requires that you know the factorizations of at least one of $a$ and $b$. (If you can factor only $a$, you can test each of the prime factors of $a$ and see if they divide $b$. The prime factors of $a$ are the only ones that can occur in the gcd, so it doesn't matter if you can't factor $b$ completely and find its other prime factors.)

*Proof.* Let $d$ be any common divisor and let

$$d = 2^{d_2} 3^{d_3} 5^{d_5} \cdots$$

be the prime factorization of $d$. Since $d \mid a$, we have $d_p \le a_p$ for each $p$. Similarly, $d_p \le b_p$ for each $p$. The largest common divisor occurs when each exponent $d_p$ is as large as possible. This is when $d_p = \min(a_p, b_p)$ for each $p$.

Similarly, let $e$ be any common multiple of $a$ and $b$ and let

$$e = 2^{e_2} 3^{e_p} 5^{e_5} \cdots$$

be the prime factorization of $e$. Since $a \mid e$, we have $a_{\le e_p}$ for each $p$. Similarly, we have $b_p \le e_p$ for each $p$. The least common multiple occurs when $e_p$ is as small as possible. This is when $e_p = \max(a_p, b_p)$ for each $p$. $\qquad\qquad\square$

Here is another application of factorization into primes. First, we need a definition.

**Definition 4.6.** *A positive integer $n$ is **squarefree** if all the factors that occur in its prime factorization are distinct. In other words, $n$ is not divisible by a square larger than 1.*

**Examples.** 165 is squarefree since $165 = 3 \cdot 5 \cdot 11$, but 585 is *not* squarefree since $585 = 3^2 \cdot 5 \cdot 13$. Also, $250 = 2 \cdot 5^3$ is not squarefree since it is divisible by $5^2$.

**Proposition 4.7.** *Let $n$ be a positive integer. Then there is an integer $r \geq 1$ and a squarefree integer $s \geq 1$ such that $n = r^2 s$.*

*Proof.* Let

$$n = p_1^{a_1} p_2^{a_2} \cdots p_k^{a_k}$$

be the prime factorization of $n$. If $a_i$ is even, write $a_i = 2b_i$ and if $a_i$ is odd, write $a_i = 2b_i + 1$. (Notice that $b_i = \lfloor a_i/2 \rfloor$, where $\lfloor x \rfloor$ denotes the largest integer less than or equal to the real number $x$.) Let

$$r = p_1^{b_1} p_2^{b_2} \cdots p_k^{b_k},$$

and let $s$ be the product of the primes $p_i$ for which $a_i$ is odd (if no $a_i$ is odd, let $s = 1$). Then

$$r^2 s = p_1^{2b_1} p_2^{2b_2} \cdots p_k^{2b_k} s = p_1^{a_1} p_2^{a_2} \cdots p_k^{a_k} = n.$$

$\square$

**Example.** Let $n = 2^5 \cdot 3^2 \cdot 7^3 \cdot 17$. Then $r = 2^2 \cdot 3 \cdot 7$ and $s = 2 \cdot 7 \cdot 17$.

---

**CHECK YOUR UNDERSTANDING**

2. Let $m = 2^6 5^3 7^2 17$ and let $n = 2^4 3^4 5^3 11$. Find $\gcd(m, n)$.
3. Write 1440 in the form $r^2 s$, where $s$ is squarefree.

---

## 4.3  Euclid and the Fundamental Theorem of Arithmetic

As we said, although Euclid never proved the Fundamental Theorem of Arithmetic, he did come very close to doing so. Here are the relevant propositions from his *Elements*.

**Book VII, Proposition 30:** If two numbers by multiplying one another make some number, and any prime number measures the product, it will also measure one of the original numbers.

This is Theorem 4.1: If $p$ divides the product of two numbers, then it divides at least one of them. This theorem played an essential role in the uniqueness piece of the Fundamental Theorem of Arithmetic.

**Book VII, Proposition 31:** Any composite number is measured by some prime number.

This says that every composite number is divisible by a prime. This is what Lemma 2.6 says.

Finally, Euclid does state a unique factorization type of result.

**Book IX, Proposition 14:** If a number be the least that is measured by prime numbers, it will not be measured by any other prime number except those originally measuring it.

In essence, this says that the least common multiple of prime numbers can be divisible only by those prime numbers. It appears that Euclid was only considering the least common multiple of *distinct* primes, so the least common multiple had to be squarefree.

As you can see, Euclid had all the ingredients necessary to prove the Fundamental Theorem of Arithmetic. Why he never did so explicitly is a mystery.

---

# 4.4   Chapter Highlights

1. If $p$ is prime and $p \mid ab$, then $p \mid a$ or $p \mid b$ (Theorem 4.1).

2. The Fundamental Theorem of Arithmetic: Every integer greater than 1 is either prime or a product of primes in exactly one way.

# 4.5    Problems

## 4.5.1    Exercises

1. (a) Find the prime factorization of 5625.
   (b) All of the exponents in the factorization in (a) are even. What does this tell you about 5625?

2. (a) Find the prime factorization of 1728.
   (b) All of the exponents in the factorization in (a) are multiples of 3. What does this tell you about 1728?

3. Show that if $p$ is prime and $p \mid a^2$, then $p \mid a$.

4. Suppose that $p$ is prime and $p^2 \mid ab$. Show that if $\gcd(a,b) = 1$, then $p^2 \mid a$ or $p^2 \mid b$.

5. Find integers $a, b$ such that $25 \nmid a$ and $25 \nmid b$ but $25 \mid ab$.

6. Let $n = p_1^{m_1} p_2^{m_2} \cdots p_k^{m_k}$ be the prime factorization of $n$. Show that $n$ is a perfect square (that is, a square of an integer) if and only if every $m_i$ is even.

7. (a) Let $a, b, c$ be integers. Show that if $a+b$, $b+c$, and $a+b$ are multiples of 3, then each of $a$, $b$, $c$ is a multiple of 3.
   (b) Let $r, s$, and $t$ be positive integers. Show that if $rs, rt$, and $st$ are perfect cubes, then $r, s$, and $t$ have to be perfect cubes. (*Hint:* For each prime $p$, look at the exponents of $p$ in the factorizations of $r, s, t$. Then use part (a).)

8. Let $p$ be prime. Show that $\log_{10}(p)$ is irrational.

9. Find all positive integers $n$ such that $\log_{10}(n)$ is rational.

10. Find all positive integers $n$ such that $\log_n(2)$ is rational.

11. Let $n > 0$. Prove
    (a) If $a^n \mid b^n$, then $a \mid b$.
    (b) If $a^m \mid b^n$ and $m \geq n$, then $a \mid b$.
    (c) Give an example where $a^m \mid b^n$ with $0 < m < n$, and $a \nmid b$.

12. (a) Show that $\gcd(a^2, b^2) = \gcd(a, b)^2$.
    (b) Show that if $n \geq 1$, then $\gcd(a^n, b^n) = \gcd(a, b)^n$.

13. If $\gcd(a, b) = 1$ and $c \mid (a+b)$, show that $\gcd(a, c) = \gcd(b, c) = 1$. (*Hint:* Let $d = \gcd(a, c)$ and show that $d \mid \gcd(a, b)$.)

14. (a) Let $a, b, x, y$ be integers such that $ax + by = 1$. Show that $\gcd(a, b) = 1$.
    (b) Use part (a) to show that $\gcd(101, 50) = 1$.
    (c) Use part (a) to show that $\gcd(101, 51) = 1$.

15. What is the largest number of consecutive squarefree positive integers?

16. What is the largest number of consecutive odd squarefree positive integers?

17. (a) Find all primes $p$ such that $3p + 1$ is a square.
    (b) Find all primes such that $5p + 1$ is a square.
    (c) Find all primes such that $29p + 1$ is a square.

(d) If we fix a prime $p$ and ask to find all primes $q$ so that $qp + 1$ is a square, what relationship must $p$ and $q$ have if we find a $q$ that works?

18. Let $p$ be a prime, and let $a, b$ be nonzero integers. Suppose $p^c \mid a$ but $p^{c+1} \nmid a$, and $p^d \mid b$ but $p^{d+1} \nmid b$.
    (a) Show that if $c < d$ then $p^c \mid (a + b)$ but $p^{c+1} \nmid (a + b)$.
    (b) Give an example to show that if $c = d$ then it is possible that $p^{c+1} \mid (a + b)$.

19. Show that every positive integer $n$ can be uniquely written as $2^r m$ with $r \geq 0$ and $m$ a positive odd integer.

20. If $p$ is a prime and $\gcd(a, b) = p$, what are the possible values of
    (a) $\gcd(a^2, b)$?
    (b) $\gcd(a^3, b)$?

21. Assume that $a$ and $b$ are integers and $p$ is a prime. If $\gcd(a, p^2) = p$ and $\gcd(b, p^3) = p^2$, find
    (a) $\gcd(ab, p^4)$,
    (b) $\gcd(a + b, p^4)$.

22. Here is a variant of Euclid's proof that there are infinitely many primes. Suppose that $\{p_1, p_2, \ldots, p_n\}$ is the set of all primes. Let

$$N = (p_2 p_3 \cdots p_n) + (p_1 p_3 p_4 \cdots p_n) + (p_1 p_2 p_4 \cdots p_n) + \cdots + (p_1 p_2 \cdots p_{n-1}).$$

Show that none of the primes $p_i$ can divide $N$, so $N$ must have a prime factor $p \notin \{p_1, p_2, \ldots, p_n\}$, which means that no finite set can contain all the primes. Explain where you used Corollary 4.2.

23. The following is a classic problem and is one of the few facts we have about Diophantus (but we don't know that it is actually true): Diophantus passed one sixth of his life in childhood, one twelfth in youth, and one seventh as a bachelor. Five years after his marriage his son was born, who died four years before his father, at half his father's age. How many years did Diophantus live?
    (a) Set up an equation and solve the problem using elementary algebra.
    (b) Assuming that one twelfth of his life is an integer, and so is one seventh, use Proposition 2.17 to guess his age without solving the equation in part (a).

24. Find $\operatorname{lcm}(15, 21)$, $\operatorname{lcm}(30, 40)$, and $\operatorname{lcm}(5, 47)$.

25. Find two integers with greatest common divisor 15 and least common multiple 60.

26. Let $a$ and $b$ be positive integers. Show that $\gcd(a, b) \mid \operatorname{lcm}(a, b)$.

27. When is $\gcd(a, b) = \operatorname{lcm}(a, b)$?

28. If $p$, $q$, and $r$ are distinct odd primes, find
    (i) $\operatorname{lcm}(p, q)$, (ii) $\operatorname{lcm}(pq, p^2 r)$, (iii) $\operatorname{lcm}(pq, 2q^2 r^3)$.

29. (a) By looking at the prime factorizations of $a$ and $b$, show that

$$\operatorname{lcm}(a, b) \times \gcd(a, b) = ab.$$

(b) Give an example of integers $a, b, c$ such that $\operatorname{lcm}(a, b, c) \times \gcd(a, b, c) \neq abc$. (*Note:* $\operatorname{lcm}(a, b, c)$ is the least common multiple of $a, b, c$. That is, it is the smallest positive integer that is a multiple of $a$, $b$, and $c$. Similarly, $\gcd(a, b, c)$ is the largest integer that divides $a$, $b$, and $c$.)

30. Let $a, b, c$ be positive integers and suppose $a \mid c$ and $b \mid c$. Show that $\operatorname{lcm}(a, b) \mid c$.

31. Let $p$ be prime. Show that the equation $\operatorname{lcm}(a, b) = p^2$ has five solutions in positive integers $a, b$.

32. Let $p$ and $q$ be distinct primes. Show that the equation $\operatorname{lcm}(a, b) = p^2 q$ has fifteen solutions in positive integers $a, b$.

33. Let $n = 2^{n_2} 3^{n_3} 5^{n_5} \cdots$. Show that the equation $\operatorname{lcm}(a, b) = n$ has

$$(2n_2 + 1)(2n_3 + 1)(2n_5 + 1) \cdots$$

solutions in positive integers $a, b$.

34. Let $a$ and $b$ be positive integers. Prove that the following are equivalent:
    (1) $a \mid b$
    (2) $\gcd(a, b) = a$
    (3) $\operatorname{lcm}(a, b) = b$.

## 4.5.2   Projects

1. Let $\mathcal{H} = \{n \mid n = 1 + 4k, k \text{ a non-negative integer}\} = \{1, 5, 9, 13, \dots\}$. Then $\mathcal{H}$ is called the set of Hilbert numbers.

   (a) Show that $\mathcal{H}$ is closed under multiplication. (This means that if we multiply two numbers in $\mathcal{H}$ by each other, the result is still in $\mathcal{H}$.) Would this still be true if we looked instead at the set $\mathcal{H}' = \{n \mid n = 3 + 4k, k \text{ a non-negative integer}\}$?

   (b) Write out the first 10 Hilbert numbers.

   (c) Define a Hilbert prime to be a Hilbert number larger than 1 that cannot be written as the product of two smaller Hilbert numbers. Write out the first 10 Hilbert primes. What is the first Hilbert prime that is *not* a prime number?

   (d) Show that if $p$ and $q$ are primes of the form $4k+3$, then $pq$ is a Hilbert prime (for example, $77 = 7 \times 11$ is a Hilbert prime).

   (e) Show that each Hilbert prime is either a prime of the form $4k + 1$ or a product of two (not necessarily distinct) primes of the form $4k + 3$.

   (f) Show that 441 has two different factorizations into Hilbert primes.

   (g) Find two examples of Hilbert numbers that have at least three different factorizations into Hilbert primes.

   (h) Adapt the Sieve of Eratosthenes to the Hilbert numbers to find the Hilbert primes.

## 4.5.3   Answers to "CHECK YOUR UNDERSTANDING"

1. The statement is false. If $p^2 \mid ab$, we could have one $p$ divide $a$ and one $p$ divide $b$. For example, $3^2 \mid 6 \cdot 12$, but $3^2 \nmid 6$ and $3^2 \nmid 12$.

2. We take the minimum of the exponents for each prime, so $\gcd(m, n) = 2^4 3^0 5^3 7^0 11^0 17^0 = 2^4 5^3$.

3.   Factor 1440 as $2^5 3^2 5$. This equals $(2^2 3)^2 2 \cdot 5$, so $r = 2^2 3 = 12$ and $s = 2 \cdot 5 = 10$.

# Chapter 5

# Applications of Unique Factorization

## 5.1 A Puzzle

> There was a young lady named Chris
> Who, when asked her age, answered this.
> Two-thirds of its square
> Is a cube, I declare.
> Now what was the age of the miss?

(This limerick appeared in the *San Francisco Examiner* in the mid-1970s.)

Let $A$ be her age and let

$$A = 2^a 3^b 5^c 7^d \cdots$$

be the prime factorization of $A$ (for convenience, we include all primes and note that only finitely many of the exponents $a, b, c, \ldots$ are nonzero). Then two-thirds of its square is

$$(2/3)A^2 = (2/3)2^{2a} 3^{2b} 5^{2c} 7^{2d} \cdots = 2^{2a+1} 3^{2b-1} 5^{2c} 7^{2d} \cdots.$$

We need the following useful fact.

**Proposition 5.1.** *Let $k \geq 2$ be an integer and let $m$ be a positive integer. Then $m$ is a $k$th power if and only if all the exponents in the prime factorization of $m$ are multiples of $k$.*

**Examples.** We know that $8000 = 2^6 5^3$ is a cube (it equals $20^3$) since the exponents 6 and 3 are multiples of 3, but is not a sixth power since the exponent 3 is not a multiple of 6. Similarly, $2000 = 2^4 5^3$ is not a cube since the exponent 4 is not a multiple of 3.

*Proof.* Suppose
$$m = 2^{y_2} 3^{y_3} 5^{y_5} \cdots .$$

If each exponent $y_p$ is a multiple of $k$, then we can write $y_p = k z_p$ for each prime $p$, so
$$m = \left( 2^{z_2} 3^{z_3} 5^{z_5} \cdots \right)^k ,$$

which is a $k$th power. Conversely, if $m = n^k$, let

$$n = 2^{w_2} 3^{w_3} 5^{w_5} \cdots$$

be the prime factorization of $n$. Then

$$2^{y_2} 3^{y_3} 5^{y_5} \cdots = m = n^k = 2^{k w_2} 3^{k w_3} 5^{k w_5} \cdots .$$

By the uniqueness of the factorization, we must have $y_p = k w_p$ for each $p$. Therefore, every exponent for $m$ is a multiple of $k$. This proves the proposition. $\qquad\qquad\square$

We now return to the limerick. Since $(2/3)A^2$ is a cube, the proposition implies that
$$2a + 1, \quad 2b - 1, \quad 2c, \quad 2d, \ldots$$
are multiples of 3. An easy way to accomplish this is to take $a = 1$, $b = 2$, $c = 0$, $d = 0$, ..., which yields $A = 18$. But let's ignore the word "young" and find the general solution.

Since $3 \mid 2c$ and $\gcd(3, 2) = 1$, Proposition 2.16 says that $c$ must be a multiple of 3, say $c = 3c'$. Similarly, $d$ and all later exponents are multiples of 3. Since $2a + 1$ is odd and a multiple of 3, we can write $2a + 1 = 3(2j + 1)$, which implies $a = 3j + 1$ for some $j$. For $b$, we have $2b - 1 = 3(2k + 1)$, hence $b = 3k + 2$ for some $k$. We conclude that

$$A = 2^a 3^b 5^c \cdots = 2 \cdot 3^2 \cdot \left( 2^j 3^k 5^{c'} \cdots \right)^3 = 18B^3$$

for some $B$. If $B = 1$, we obtain Chris's age as $A = 18$. The next possibility (with $B = 2$) is $A = 18 \cdot 2^3 = 144$.

## CHECK YOUR UNDERSTANDING

1. How does the prime factorization of 5400 tell you that it's not a perfect cube?

2. Find a positive integer $n$ such that $n/3$ is a square and $n/5$ is a cube.

## 5.2    Irrationality Proofs

A real number is **rational** if it can be expressed as the ratio of two integers. For example,

$$3 = 3/1, \quad 3.7 = 37/10, \quad 0 = 0/1, \quad -.001 = -1/1000$$

are rational numbers. A real number is **irrational** if it is not rational. Later, in Chapter 11, we will see that rational numbers are exactly those whose decimal expansions eventually become periodic (that is, repeating the same string of digits forever). For example,

$$\frac{6187}{3300} = 1.874848484848\ldots$$

is rational and the decimal expansion eventually becomes the string 48 repeating forever. Sometimes, irrational numbers are defined as being those whose decimal expansions never become periodic. This is correct, but it is hard to use this definition to prove that a number is irrational. If we compute a billion digits of the decimal expansion and don't find a pattern, how do we know that the repetition doesn't start at the trillionth digit?

In Chapter 17, we'll show that $e = 2.71828\ldots$ and $\pi = 3.14159\ldots$ are irrational. For the present, we prove the famous fact that $\sqrt{2}$ is irrational.

**Theorem 5.2.** $\sqrt{2}$ *is irrational.*

*Proof.* Suppose instead that $\sqrt{2}$ is rational. Then

$$\sqrt{2} = \frac{a}{b}$$

with some integers $a, b$. We assume that $a/b$ is a reduced fraction; that is, $\gcd(a, b) = 1$. In particular, we may assume that at least one of $a, b$ is odd.

Squaring the equation $\sqrt{2} = a/b$ yields $2 = a^2/b^2$ and therefore $2b^2 = a^2$. Since $a^2$ is even (it equals $2b^2$), $a$ must be even. Write $a = 2a_1$ for some integer $a_1$. Then

$$2b^2 = a^2 = (2a_1)^2 = 4a_1^2.$$

Dividing by 2 yields $b^2 = 2a_1^2$, so $b^2$ is even. This implies that $b$ is even. We have now shown that both $a$ and $b$ are even. This contradicts the fact that at least one of $a, b$ is odd. Therefore, we conclude that we cannot write $\sqrt{2} = a/b$, so $\sqrt{2}$ is irrational.          □

Mathematicians often like to look at things and give proofs from different points of view. In this spirit, we'll give four more proofs at the end of this section and one more in the next section.

Using unique factorization, we can generalize the theorem.

**Theorem 5.3.** *Let $k \geq 2$ be an integer. Let $n$ be a positive integer that is not a perfect $k$th power. Then $\sqrt[k]{n}$ is irrational.*

**Examples.** $\sqrt{3}$,   $\sqrt[7]{16}$, and $\sqrt[3]{5}$ are irrational.

*Proof.* We show the equivalent statement that if $\sqrt[k]{n}$ is rational, then $n$ is a perfect $k$th power. Suppose $\sqrt[k]{n} = a/b$, where $a, b$ are positive integers. Then

$$nb^k = a^k.$$

Write

$$n = 2^{x_2} 3^{x_3} 5^{x_5} \cdots$$
$$b = 2^{z_2} 3^{z_3} 5^{z_5} \cdots$$

with exponents $x_p \geq 0$ and $z_p \geq 0$. Then

$$nb^k = 2^{x_2 + kz_2} 3^{x_3 + kz_3} \cdots .$$

Since $nb^k = a^k$ is a perfect $k$th power, Proposition 5.1 implies that each exponent must be a multiple of $k$:

$$x_p + kz_p = ky_p,$$

for some $y_p$. This implies that $x_p = k(y_p - z_p)$, which is a multiple of $k$. It follows that $n$ is a perfect $k$th power.          □

## 5.2.1   Four More Proofs That $\sqrt{2}$ Is Irrational

**Proof 1:** This proof follows immediately from Theorem 5.3 by choosing both $k$ and $n$ to be 2. Since 2 is not a perfect square, $\sqrt{2}$ ($= \sqrt[2]{2}$) cannot be a rational number.

**Proof 2:** Here is a proof that does not use "even" and "odd." Suppose $\sqrt{2} = a/b$ with positive integers $a, b$. Then $a^2 - 2b^2 = 0$. Let's assume that $b$ is chosen to be as small as possible. (The fact that a smallest $b$ exists is an example of the Well-Ordering Principle. See Appendix A.) Note that $1 < \sqrt{2} < 2$ implies that $b < \sqrt{2}b < 2b$. Since $a = \sqrt{2}b$ by assumption,

$$b < a < 2b,$$

so $0 < a - b < b$. But

$$(2b - a)^2 - 2(a - b)^2 = 4b^2 - 4ab + a^2 - 2(a^2 - 2ab + b^2) = 2b^2 - a^2 = 0.$$

Let $a_1 = 2b - a$ and $b_1 = a - b$. Then $a_1^2 - 2b_1^2 = 0$ and $0 < b_1 < b$, which contradicts the fact that $b$ was chosen to be smallest. Therefore, there is no smallest positive integer $b$ such that $a^2 - 2b^2 = 0$ for some $a$. If there is any such integer $b$, then there must be a smallest one, so we are forced to conclude that there is no such $b$. Therefore, $\sqrt{2}$ is irrational.

**Remark.** Where did $2b - a$ and $a - b$ come from? Note that

$$(a + b\sqrt{2})(\sqrt{2} - 1) = (2b - a) + (a - b)\sqrt{2}.$$

Multiplying by $\sqrt{2}-1$ does not change the norm, in the sense of Chapter 19. Since $0 < \sqrt{2} - 1 < 1$, we expect the product might have a smaller coefficient of $\sqrt{2}$, which it does.

**Proof 3:** Here is a geometric proof. Suppose that $\sqrt{2} = c/a$ with positive integers $a$ and $c$, where $a$ and $c$ are chosen as small as possible. This gives a 45-45-90 isosceles right triangle $ABC$ with legs $BC = AC = a$ and hypotenuse $AB = \sqrt{2}a = c$ (see Figure 5.1). Our assumption is that this is the smallest 45-45-90 triangle with integer sides. Put point $P$ on $AB$ such that $AP = a$. This is possible since $a < c$. Then $PB = c - a$. Draw a line perpendicular to $AB$ at $P$. Since $a > c/2$, the point $P$ is between $B$ and the midpoint of $AB$, so this line lies to one side of the altitude from $C$ to $AB$. Therefore, it intersects $BC$ at a point $Q$ between $B$ and $C$. Since $\angle BPQ = 90$ and $\angle PBQ = 45$, we must

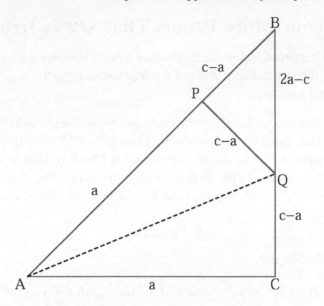

**FIGURE 5.1: Irrationality of $\sqrt{2}$.**

have $\angle PQB = 45$. Therefore, triangle $PBQ$ is a 45-45-90 triangle. In particular, $PQ = PB = c - a$.

Note that triangles $ACQ$ and $APQ$ are right triangles sharing a common hypotenuse and each having a leg of length $a$. By the Pythagorean Theorem (or by theorems on congruence of right triangles), the legs $PQ$ and $CQ$ are congruent. Therefore, $CQ = c - a$, so $QB = a - (c - a) = 2a - c$.

Summarizing, we see that triangle $PBQ$ is a smaller 45-45-90 triangle and has legs of length $c - a$ and hypotenuse $2a - c$. Since $a$ and $c$ are integers, the new triangle has integer sides. Therefore, the original triangle was not the smallest one with integer sides. This contradiction shows that $\sqrt{2}$ is irrational.

**Remarks.** Proof 3 is a geometric version of Proof 2. It is reported that the Greeks had geometric proofs for the irrationality of $\sqrt{n}$ for nonsquares $n \leq 15$.

**Proof 4:** Suppose $\sqrt{2} = a/b$ with integers $a, b$. We may assume that $\gcd(a, b) = 1$. Divide $a$ and $b$ by 3 using the division algorithm:

$$a = 3q_1 + r_1, \quad b = 3q_2 + r_2,$$

with $r_1, r_2 \in \{0, 1, 2\}$. Then

$$0 = a^2 - 2b^2 = (3q_1 + r_1)^2 - 2(3q_2 + r_2)^2$$
$$= 3\left(3q_1^2 + 2q_1 r_1 - 6q_2^2 - 4q_2 r_2\right) + (r_1^2 - 2r_2^2).$$

Since the first term in the last expression is a multiple of 3, and the sum of the two terms is 0, we conclude that

$$r_1^2 - 2r_2^2$$

is a multiple of 3. If we let $r_1 = 0, 1, 2$ and $r_2 = 0, 1, 2$, we obtain

$$r_1^2 - 2r_2^2 = 0, -2, -8, 1, -1, -7, 4, 2, -4.$$

The only choice that gives a multiple of 3 is $r_1 = r_2 = 0$. But this means that both $a$ and $b$ are multiples of 3, contradicting the assumption that $\gcd(a, b) = 1$. Therefore, $a, b$ cannot exist, and $\sqrt{2}$ is irrational.

**Remarks.** The proof becomes slightly shorter if we take remainders $r_1, r_2 \in \{0, 1, -1\}$. In the language of congruences (see Chapter 6), we have proved that $a^2 \equiv 2b^2 \pmod{3}$ has only the solution $a \equiv b \equiv 0 \pmod{3}$. In other words, 2 is not a quadratic residue mod 3 (see Chapter 13).

---

## CHECK YOUR UNDERSTANDING

3. How do you know that both $\sqrt[3]{28}$ and $\sqrt[9]{50}$ are irrational?

---

# 5.3    The Rational Root Theorem

The results of the previous section are all part of a more general result. Recall that a *root* of a polynomial $P(x)$ is a number $r$ satisfying $P(r) = 0$.

**Theorem 5.4. (Rational Root Theorem)** *Let*

$$P(X) = a_n X^n + a_{n-1} X^{n-1} + \cdots + a_1 X + a_0$$

*be a polynomial with integer coefficients (that is, $a_0, a_1, \ldots, a_n$ are integers) such that $a_0 \neq 0$ and $a_n \neq 0$. If $r = u/v$ is a rational number (with $\gcd(u, v) = 1$) and $P(u/v) = 0$, then $u$ divides $a_0$ and $v$ divides $a_n$.*

*Proof.* Since $P(u/v) = 0$, we have

$$a_n(u/v)^n + a_{n-1}(u/v)^{n-1} + \cdots + a_0 = 0.$$

Multiply by $v^n$:

$$a_n u^n + a_{n-1} u^{n-1} v + \cdots + a_1 u v^{n-1} + a_0 v^n = 0.$$

All of the terms except the first are multiples of $v$, so $v$ must divide $a_n u^n$, too. Since $\gcd(u, v) = 1$, Proposition 2.16 implies that $v$ divides $a_n$.

Similarly, all the terms except the last are multiples of $u$, so $u$ divides $a_0 v^n$. Since $\gcd(u, v) = 1$, it follows that $u$ divides $a_0$.          □

Note that, although the proof of the theorem looks easy, it relies heavily on Proposition 2.16, which was proved as a consequence of Theorem 2.12, which was a key step in proving that factorization into primes is unique.

**Examples.** Let $P(X) = 2X^3 - 5X^2 + X + 3$. The possible (rational) roots are

$$\pm 3, \quad \pm 1, \quad \pm \frac{3}{2}, \quad \pm \frac{1}{2}.$$

If we try these, we find that $P(3/2) = 0$.

The Rational Root Theorem works well when there are only a few factors of the highest and lowest coefficients. However, it is still possible to use it to advantage when there are many possibilities. Consider the polynomial

$$78X^3 - 547X^2 + 310X - 45.$$

The factors of 45 are

$$\pm 1, \quad \pm 3, \quad \pm 5, \quad \pm 9, \quad \pm 15, \quad \pm 45.$$

The factors of 78 are

$$\pm 1, \quad \pm 2, \quad \pm 3, \quad \pm 6, \quad \pm 13, \quad \pm 26, \quad \pm 39, \quad \pm 78.$$

There are $96 = 12 \cdot 8$ combinations:

$$\pm\frac{1}{1}, \quad \pm\frac{1}{2}, \quad \cdots, \quad \pm\frac{45}{39}, \quad \pm\frac{45}{78};$$

however, some fractions appear more than once. For example, $\frac{45}{39} = \frac{15}{13}$ and $-1/1 = -3/3$. Removing the repeats leaves 64 possible roots. These can be substituted into the polynomial to see if any of them are roots. But it is easier to proceed as follows. Numerical methods (for example, the Newton–Raphson Method or simply using a graphing calculator) yield the roots

$$0.26015\cdots, \quad .34615\cdots, \quad 6.40651\cdots.$$

If a root is rational, the Rational Root Theorem tells us that its denominator divides 78. Therefore, multiply these roots by 78:

$$20.2918\cdots, \quad 27.0000\cdots, \quad 499.7082\cdots.$$

Consequently, the first and third roots are not rational and the second one, if rational, is $27/78 = 9/26$. In fact, substituting $9/26$ into the polynomial gives 0, so we have found a rational root.

The Rational Root Theorem immediately implies that $\sqrt{2}$ is irrational: The only possible rational roots of $X^2 - 2$ are $\pm 1$ and $\pm 2$. Since these are not roots, we see that $X^2 - 2$ has no rational roots. In particular, $\sqrt{2}$, which is a root of this polynomial, is irrational.

Finally, we can use the Rational Root Theorem to give another proof of Theorem 5.3. Let $k \geq 2$ be an integer and let $n$ be a positive integer. Consider the polynomial $X^k - n$. By the Rational Root Theorem, the only possible roots have denominator 1. That is, the only possible rational roots are integers. If $m$ is an integer that is a root, then $m^k - n = 0$, so $n = m^k$ is a perfect $k$th power. Therefore, if $n$ is not a perfect $k$th power, then $X^k - n$ has no rational roots, which means that $\sqrt[k]{n}$ is irrational. This is Theorem 5.3.

---

## CHECK YOUR UNDERSTANDING

4. List all the possible rational roots of $2x^3 - 3x^2 + 5x - 10$. Are any of the possibilities actually roots?

## 5.4   Pythagorean Triples

We all have seen the relation

$$3^2 + 4^2 = 5^2,$$

corresponding to the 3-4-5 right triangle, and maybe also the relations

$$5^2 + 12^2 = 13^2, \qquad 8^2 + 15^2 = 17^2.$$

A triple $(a, b, c)$ of positive integers satisfying

$$a^2 + b^2 = c^2$$

is called a **Pythagorean triple**. If $\gcd(a, b, c) = 1$, then the Pythagorean triple $(a, b, c)$ is called a **primitive Pythagorean triple**. Of course, if $\gcd(a, b, c) = d$, then we may divide $a, b, c$ by $d$ and obtain a primitive triple. For example, $(6, 8, 10)$ is a Pythagorean triple with $\gcd(6, 8, 10) = 2$. Dividing by 2 yields the primitive triple $(3, 4, 5)$.

**Questions:** Are there infinitely many primitive Pythagorean triples? Is there a way of generating all of them?

There is an easy way of producing many Pythagorean triples. Take an odd number, for example, 9. Square it: $9^2 = 81$. Split the result into two pieces that differ by 1: $81 = 40 + 41$. The result produces a Pythagorean triple: $9^2 + 40^2 = 41^2$. This yields many examples:

$$(3, 4, 5), \quad (5, 12, 13), \quad (7, 24, 25), \ldots.$$

Since the last two entries in each triple differ by 1, their gcd is 1, so the triples are primitive.

Algebraically, we have done the following: Let the odd number be $2n+1$. Its square is

$$(2n + 1)^2 = 4n^2 + 4n + 1.$$

Split this into two parts:

$$4n^2 + 4n + 1 = (2n^2 + 2n) + (2n^2 + 2n + 1).$$

Then $(2n + 1,\ 2n^2 + 2n,\ 2n^2 + 2n + 1)$ is a Pythagorean triple. This can be verified using the identity

$$(2n + 1)^2 + (2n^2 + 2n)^2 = (2n^2 + 2n + 1)^2.$$

There is a similar construction that uses a multiple of 4. For example, start with 8. Square it: $8^2 = 64$. Take one-fourth of this square and subtract 1 and add 1:

$$(1/4)\,8^2 - 1 = 15 \quad \text{and} \quad (1/4)\,8^2 + 1 = 17.$$

This gives the triple $(8, 15, 17)$. Here are some more examples:

$$(12, 35, 37), \quad (16, 63, 65), \quad \dots .$$

Since the last two entries in each triple are odd and differ by 2, their gcd is 1, so the triples are primitive. We could do this process starting with any even integer, but the ones that are not multiples of 4 yield triples all of whose entries are even, so they are imprimitive.

We could continue to find series of triples similar to these examples, but instead, we state a general result that gives all primitive triples.

**Theorem 5.5.** *Let $(a, b, c)$ be a primitive Pythagorean triple. Then $c$ is odd and exactly one of $a, b$ is even. Assume that $b$ is even. Then there exist relatively prime integers $m < n$, one even and one odd, such that*

$$a = n^2 - m^2, \quad b = 2mn, \quad c = m^2 + n^2.$$

**Examples.** If $n = 2$ and $m = 1$, we obtain $(a, b, c) = (3, 4, 5)$. If $n = 7$ and $m = 2$, we obtain the less familiar triple $(a, b, c) = (45, 28, 53)$.

*Proof.* We start with a very important lemma.

**Lemma 5.6.** *Let $k \geq 2$, let $a, b$ be relatively prime positive integers such that $ab = n^k$ for some integer $n$. Then $a$ and $b$ are each $k$th powers of integers.*

*Proof.* Factor $n$ into prime powers:

$$n = 2^{x_2} 3^{x_3} 5^{x_5} \cdots$$

with integers $x_p \geq 0$. Then

$$ab = n^k = 2^{kx_2}3^{kx_3}5^{kx_5}\cdots.$$

Let $p$ be a prime occurring in the prime factorization of $a$; let's say $p^c$ is the exact power of $p$ in $a$. Since $\gcd(a, b) = 1$, the prime $p$ does not occur in the factorization of $b$. Therefore, $p^c$ is the power of $p$ occurring in $ab$. But $n^k$ has $p^{kx_p}$ as the power of $p$ in its prime factorization. Since the prime factorization of $ab = n^k$ is unique, we must have $p^c = p^{kx_p}$, so $c = kx_p$. This means that every prime in the factorization of $a$ occurs with exponent that is a multiple of $k$, and this implies that $a$ is a $k$th power. By exactly the same reasoning, $b$ is a $k$th power.    $\square$

We need one more lemma.

**Lemma 5.7.** *The square of an odd number is 1 more than a multiple of 8. The square of an even number is a multiple of 4.*

*Proof.* Let $n$ be even. Then $n = 2j$ for some integer $j$, and $n^2 = 4j^2$, which is a multiple of 4.

Now, let $n$ be odd, so $n = 2k + 1$ for some $k$. Then

$$n^2 = (2k + 1)^2 = 4k(k + 1) + 1.$$

Since either $k$ or $k + 1$ is even, $k(k + 1)$ is a multiple of 2, so $4k(k + 1)$ is a multiple of 8. Therefore, $n^2$ is 1 more than a multiple of 8.    $\square$

**Remark.** Arguments such as the one in the proof of Lemma 5.7 will be much more routine after we develop the machinery of congruences in Chapter 6.

We now can prove the theorem. Suppose

$$a^2 + b^2 = c^2$$

and $\gcd(a, b, c) = 1$.

First, there are some technical details involving parity and gcd's that must be treated — not very exciting, but necessary.

Suppose that both $a$ and $b$ are odd. Then $a^2$ and $b^2$ are each 1 more than multiples of 8, so $a^2 + b^2$ is 2 more than a multiple of 8. In particular, $a^2 + b^2$ is even but not a multiple of 4. By Lemma 5.7, even squares are multiples of 4, so $a^2 + b^2$ cannot be a square. Therefore, at least one of $a, b$ is even.

If both $a$ and $b$ are even, then $c^2 = a^2 + b^2$ is even, so $c$ is even. But then 2 is a common divisor of $a, b, c$, so $\gcd(a, b, c) \neq 1$, which is contrary to our assumption that the triple is primitive. It follows that exactly one of $a, b$ is even and the other is odd.

Switching the names of $a$ and $b$ if necessary, we may assume that $a$ is odd and $b$ is even. Therefore, $a^2 + b^2$ is an odd plus an even, hence is odd, so $c$ is odd.

Write $b = 2b_1$ so $(c + a)(c - a) = c^2 - a^2 = b^2 = 4b_1^2$. Dividing by 4 yields

$$\left(\frac{c+a}{2}\right)\left(\frac{c-a}{2}\right) = b_1^2.$$

Since $c$ and $a$ are odd, $c + a$ is even, so $(c+a)/2$ is an integer. Similarly, $(c - a)/2$ is an integer.

Let $d = \gcd((c + a)/2, (c - a)/2)$. Suppose $d > 1$. Let $p$ be a prime dividing $d$. Then

$$c = \frac{c+a}{2} + \frac{c-a}{2} \text{ and } a = \frac{c+a}{2} - \frac{c-a}{2}$$

are multiples of $p$ since these are sums and differences of multiples of $p$. Therefore, $b^2 = c^2 - a^2$ is a difference of multiples of $p$, hence $b^2$ is a multiple of $p$. Since $p$ is prime and divides $b^2$, it must divide $b$, so $p$ is a common factor of $a, b, c$. This contradicts the primitivity of $(a, b, c)$. Therefore, there is no prime dividing $d$, so $d = 1$.

We have now arrived at the heart of the proof. We have two relatively prime positive integers, namely $(c + a)/2$ and $(c - a)/2$, whose product is a square. By Lemma 5.6, each factor is a square:

$$\frac{c-a}{2} = m^2, \quad \frac{c+a}{2} = n^2,$$

for some positive integers $m, n$. Then

$$c = \frac{c+a}{2} + \frac{c-a}{2} = n^2 + m^2 \text{ and } a = \frac{c+a}{2} - \frac{c-a}{2} = n^2 - m^2.$$

Moreover,

$$b^2 = c^2 - a^2 = (n^2 + m^2)^2 - (n^2 - m^2)^2 = 4m^2n^2,$$

so $b = 2mn$.

Because $(c - a)/2 = m^2$ and $(c + a)/2 = n^2$ are relatively prime, we must have $\gcd(m, n) = 1$. Finally, since $m^2 + n^2 = c$, which is odd, one of $m$ and $n$ is odd and the other is even.                    □

There is a geometric way of obtaining the formula for primitive Pythagorean triples. The method goes back at least to Diophantus, around 1800 years ago.

Suppose $a^2 + b^2 = c^2$. Then $(a/c)^2 + (b/c)^2 = 1$, so $(a/c, b/c)$ is a point with rational coordinates on the circle $x^2 + y^2 = 1$. Conversely, suppose $x$ and $y$ are rational numbers satisfying $x^2 + y^2 = 1$. Let $d$ be a common denominator for $x$ and $y$, so $xd$ and $yd$ are integers. Then

$$x^2 + y^2 = 1 \implies (xd)^2 + (yd)^2 = d^2,$$

so $(xd, yd, d)$ is a Pythagorean triple. Therefore, finding Pythagorean triples corresponds to finding points with rational coordinates on the unit circle.

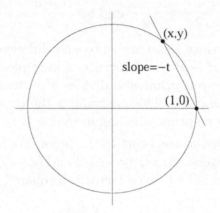

**FIGURE 5.2: Finding Pythagorean triples.**

We already know one point: $(x, y) = (1, 0)$. Suppose we have another point $(x, y) = (a/c, b/c)$ with positive rational coordinates. The slope of the line through $(a/c, b/c)$ is a negative rational number; call it $-t$. The line through $(1, 0)$ with slope $-t$ has equation

$$y = -t(x - 1).$$

To find the intersection of this line with the circle, substitute $y = -t(x - 1)$ into $x^2 + y^2 - 1 = 0$ to obtain

$$
\begin{aligned}
0 &= x^2 + (-t(x - 1))^2 - 1 \\
&= (1 + t^2)x^2 - 2t^2x + t^2 - 1 \\
&= (x - 1)\left((1 + t^2)x - (t^2 - 1)\right).
\end{aligned}
$$

The solution $x = 1$ corresponds to the starting point $(1, 0)$. The other solution is

$$
x = \frac{t^2 - 1}{t^2 + 1}.
$$

This yields

$$
y = -t(x - 1) = -t\left(\frac{t^2 - 1}{t^2 + 1} - 1\right) = \frac{2t}{t^2 + 1}.
$$

Since $t$ is a positive rational number, we write $t = n/m$ with integers $n, m$. Then

$$
x = \frac{(n/m)^2 - 1}{(n/m)^2 + 1} = \frac{n^2 - m^2}{n^2 + m^2}, \qquad y = \frac{2(n/m)}{(n/m)^2 + 1} = \frac{2nm}{n^2 + m^2}.
$$

At this point, it might look like we're done: We have

$$
a/c = x = (n^2 - m^2)/(n^2 + m^2).
$$

Doesn't this imply that $a = n^2 - m^2$? Yes, if $\gcd(n^2 - m^2, n^2 + m^2) = 1$, since then both are reduced fractions. But suppose both $n$ and $m$ are odd. Then $n^2 - m^2$ and $n^2 + m^2$ are even, so we don't have equality. It is possible to show that $d = \gcd(n^2 - m^2, n^2 + m^2) = 1$ or 2. If $d = 1$, then

$$
a = n^2 - m^2, \quad c = n^2 + m^2, \quad b = 2mn.
$$

If $d = 2$, then

$$
a = 2m_1n_1, \quad b = n_1^2 - m_1^2, \quad c = m_1^2 + n_1^2,
$$

where $n_1 = (n + m)/2$ and $m_1 = (n - m)/2$. See Project 8.

---

## CHECK YOUR UNDERSTANDING

5. Using Theorem 5.5, find the Pythagorean triple with $n = 8$ and $m = 3$.

6. Find the values of $m$ and $n$ in Theorem 5.5 that yield the Pythagorean triple $(5, 12, 13)$.

## 5.5   Differences of Squares

Perhaps you have noticed that odd numbers appear as the differences between successive squares:

$$1^2 - 0^2 = 1, \quad 2^2 - 1^2 = 3, \quad 3^2 - 2^2 = 5, \quad 4^2 - 3^2 = 7, \dots.$$

This is a consequence of the identity

$$(n+1)^2 - n^2 = 2n + 1.$$

The identity

$$(n+1)^2 - (n-1)^2 = 4n$$

shows that every multiple of 4 is the difference of two squares. Is that all?

**Theorem 5.8.** *Let $m$ be a positive integer. Then $m$ is the difference of two squares if and only if either $m$ is odd or $m$ is a multiple of 4.*

*Proof.* We have shown that odd $m$ and multiples of 4 are differences of squares. We need to show the converse. Suppose

$$m = x^2 - y^2 = (x+y)(x-y).$$

Since $x + y$ and $x - y$ differ by the even number $2y$, they are both even or both odd. If they are both even, then $m = (x+y)(x-y)$ is a product of two even numbers, hence a multiple of 4. If they are both odd, then $m$ is a product of two odd numbers, so $m$ is odd. Therefore, either $m$ is odd or $m$ is a multiple of 4.                                      ☐

A natural question is how many ways can $m$ be written as the difference of two squares? If $m = x^2 - y^2 = (x+y)(x-y)$, then $m$ is the product of two factors $(x + y)$ and $(x - y)$. As we just saw, both factors have the same parity; that is, both are even or both are odd. Conversely, suppose $m = uv$ where $u, v$ have the same parity. Assume $u \geq v$. Let $x = (u+v)/2$ and $y = (u-v)/2$. Then $x, y$ are integers (because $u, v$ have the same parity) and

$$x^2 - y^2 = \frac{(u+v)^2}{4} - \frac{(u-v)^2}{4} = uv = m.$$

Therefore, the ways of writing $m$ as the difference of two squares correspond to the factorizations of $m$ into two factors of the same parity. Both Euclid and Diophantus were aware of this result. Euclid's method appears in his *Elements* (Book II, Propositions 9 and 10) and is geometric. Diophantus's method appears in his *Arithmetica* (Book III, Proposition 19).

**Examples.**

Let $m = 15$. Then

$$15 \times 1 \longleftrightarrow 8^2 - 7^2 = 15 \quad (\text{since } 8 + 7 = 15 \text{ and } 8 - 7 = 1)$$
$$5 \times 3 \longleftrightarrow 4^2 - 1^2 = 15 \quad (\text{since } 4 + 1 = 5 \text{ and } 4 - 1 = 3).$$

Let $m = 60$. Then

$$30 \times 2 \longleftrightarrow 16^2 - 14^2 = 60$$
$$10 \times 6 \longleftrightarrow \phantom{1}8^2 - \phantom{1}2^2 = 60.$$

Let $m = 144$. Then

$$72 \times 2 \longleftrightarrow 37^2 - 35^2 = 144$$
$$36 \times 4 \longleftrightarrow 20^2 - 16^2 = 144$$
$$24 \times 6 \longleftrightarrow 15^2 - \phantom{1}9^2 = 144$$
$$18 \times 8 \longleftrightarrow 13^2 - \phantom{1}5^2 = 144$$
$$12 \times 12 \longleftrightarrow 12^2 - \phantom{1}0^2 = 144.$$

Note that the second example ($m = 60$) can be obtained from the first ($m = 15$) by multiplying each factor of 15 by 2. As in the above discussion, the factorization $m = uv$ corresponds to $x^2 - y^2 = m$ with $x = (u + v)/2$ and $y = (u - v)/2$.

---

**CHECK YOUR UNDERSTANDING**

7. Can 50 be written as the difference of two squares? Explain.

8. How do you know that 12 can be written as the difference of two squares? Find all possible ways to do so.

# 5.6    Prime Factorization of Factorials

How many zeros does 30! end with? We can write

$$30! = 265252859812191058636308480000000$$

and count the seven zeros, but fortunately, there is a better way. A zero at the end of a number means that the number has factor of $10 = 2 \cdot 5$. Two zeros means that there's a factor of $100 = 2^2 \cdot 5^2$. So, we count up how many factors of 2 and 5 there are. There are six multiples of 5 up to 30 (namely, 5, 10, 15, 20, 25, 30), and the 25 supplies an extra factor of 5, so $5^7$ divides 30!. There are more than seven even numbers up to 30, so seven factors of 2 can combine with $5^7$ to yield $10^7$. This accounts for the seven zeros we see.

All of this can be done much more generally. We need a useful notation. If $x$ is a real number, let

$$\lfloor x \rfloor = \text{ the largest integer } \leq x.$$

This is called the *floor* of $x$. For example,

$$\lfloor 3.14 \rfloor = 3, \quad \lfloor -3.14 \rfloor = -4, \quad \lfloor 3 \rfloor = 3$$

(note that $\lfloor x \rfloor = x$ if and only if $x$ is an integer).

**Theorem 5.9.** *Let $n \geq 1$ and let $p$ be a prime. Write $n! = p^b c$, with $p \nmid c$. Then*

$$b = \left\lfloor \frac{n}{p} \right\rfloor + \left\lfloor \frac{n}{p^2} \right\rfloor + \left\lfloor \frac{n}{p^3} \right\rfloor + \cdots \tag{5.1}$$

*(note that the terms become zero when $p^j > n$, so this is a finite sum).*

*Proof.* Write $n = qp + r$, with $0 \leq r < p$. The multiples of $p$ up to $n$ are $p, 2p, 3p, \ldots, qp$, so there are $q$ multiples of $p$ up to $n$. But

$$\left\lfloor \frac{n}{p} \right\rfloor = \lfloor q + (r/p) \rfloor = q.$$

Therefore, there are $\lfloor n/p \rfloor$ multiples of $p$ up to $n$.

Similarly, there are $\lfloor n/p^j \rfloor$ multiples of $p^j$ up to $n$ for each $j \geq 1$. We have proved that

$$\left\lfloor \frac{n}{p} \right\rfloor + \left\lfloor \frac{n}{p^2} \right\rfloor + \left\lfloor \frac{n}{p^3} \right\rfloor + \cdots \tag{5.2}$$

$$\tag{5.3}$$

$$= \#\{\text{multiples of } p \text{ up to } n\} + \#\{\text{multiples of } p^2 \text{ up to } n\} + \cdots .$$

Let $1 \leq m \leq n$ and let $m = p^k m_1$ with $p \nmid m_1$. Then $m$ contributes $p^k$ to $n!$, so $m$ contributes $k$ to the exponent $b$. Since $m$ is a multiple of $p^j$ for each $j \leq k$, it gets counted $k$ times by the right-hand side of Equation (5.2). Therefore, $m$ contributes $k$ to both sides of (5.1). Since this is true for all $m \leq n$, the two sides are equal, as desired.          $\square$

**Examples.** Let $n = 30$ and $p = 5$. Then

$$\left\lfloor \frac{30}{5} \right\rfloor + \left\lfloor \frac{30}{25} \right\rfloor + \left\lfloor \frac{30}{125} \right\rfloor + \cdots = 6 + 1 + 0 = 7,$$

so $5^7$ is the power of 5 in 30!.

Now let $n = 30$ and $p = 2$. Then

$$\left\lfloor \frac{30}{2} \right\rfloor + \left\lfloor \frac{30}{4} \right\rfloor + \left\lfloor \frac{30}{8} \right\rfloor + \left\lfloor \frac{30}{16} \right\rfloor + \cdots = 15 + 7 + 3 + 1 = 26,$$

so $2^{26}$ is the power of 2 in 30!. Since $2^{26}5^7 = 2^{19}10^7$, we obtain seven zeros at the end of 30!, as we saw before.

**Remark.** There is a simple formula for the right-hand side of Equation (5.1). See Project 4.

---

## CHECK YOUR UNDERSTANDING

9. Find the power of 3 in the prime factorization of 10!.

## 5.7   The Riemann Zeta Function

Let $s > 1$ be a real number. The **Riemann zeta function** is defined as

$$\zeta(s) = \sum_{n=1}^{\infty} \frac{1}{n^s}.$$

It is a standard application of the integral test to show that this infinite series converges for each $s > 1$. In the 1700s, Euler discovered a striking connection between the zeta function and prime numbers that gives an analytic expression of the fact that every positive integer is uniquely a product of primes.

**Theorem 5.10.** *Let $s > 1$. Then*

$$\zeta(s) = \prod_{p}(1 - p^{-s})^{-1},$$

*where the product is over all prime numbers. (The symbol $\prod$ denotes a product in the same way that $\Sigma$ denotes a sum.)*

*Proof.* We ignore technicalities about convergence except to say that the absolute convergence (not just conditional convergence) of the sum and product can be used to justify our manipulations. Recall the formula for the sum of a geometric series:

$$1 + r + r^2 + r^3 + \cdots = \frac{1}{1-r} = (1-r)^{-1}$$

when $|r| < 1$. Letting $r = p^{-s}$, we obtain

$$1 + \frac{1}{p^s} + \frac{1}{p^{2s}} + \frac{1}{p^{3s}} + \cdots = (1 - p^{-s})^{-1}.$$

Consider the product

$$(1 - 2^{-s})^{-1}(1 - 3^{-s})^{-1}$$

$$= \left(1 + \frac{1}{2^s} + \frac{1}{4^s} + \frac{1}{8^s} + \cdots\right)\left(1 + \frac{1}{3^s} + \frac{1}{9^s} + \cdots\right).$$

When we expand this, we obtain

$$1 + \frac{1}{2^s} + \frac{1}{4^s} + \frac{1}{8^s} + \cdots + \frac{1}{3^s} + \frac{1}{2^s 3^s}$$
$$+ \frac{1}{4^s 3^s} + \cdots + \frac{1}{9^s} + \frac{1}{2^s 9^s} + \frac{1}{4^s 9^s} + \cdots$$
$$= 1 + \frac{1}{2^s} + \frac{1}{3^s} + \frac{1}{4^s} + \frac{1}{6^s} + \frac{1}{8^s} + \frac{1}{9^s} + \frac{1}{12^s} + \cdots$$
$$= \sum_{n \in S(2,3)} \frac{1}{n^s},$$

where $S(p, q, \dots)$ denotes those integers (including $1 = p^0 q^0 \cdots$) whose prime factorizations involve only primes in $\{p, q, \dots\}$. Note that each integer in $S(2, 3)$ occurs exactly once in the sum. For example, $12 \in S(2, 3)$, and the term $1/12^s$ comes from $(1/4^s) \times (1/3)^s$. The uniqueness of factorization says that this is the only way it can occur.

When we multiply by $(1 - 5^{-s})^{-1}$, we obtain

$$(1 - 2^{-s})^{-1}(1 - 3^{-s})^{-1}(1 - 5^{-s})^{-1} = \sum_{n \in S(2,3,5)} \frac{1}{n^s}.$$

In general, if $p_m$ is the $m$th prime, we have

$$(1 - 2^{-s})^{-1}(1 - 3^{-s})^{-1}(1 - 5^{-s})^{-1} \cdots (1 - p_m^{-s})^{-1} = \sum_{n \in S(2,3,5,\dots,p_m)} \frac{1}{n^s}.$$

Let $m \to \infty$. The left side converges to the product over all primes. Since every positive integer has a prime factorization, each positive integer $n$ lies in $S(2, 3, 5, \cdots, p_m)$ for a sufficiently large integer $m$. Therefore, the right side converges to the sum over all positive integers $n$. This gives the identity in the theorem.                                                      □

Euler used his identity to give a completely new proof of the classical theorem of Euclid.

**Euler's proof that there are infinitely many primes.**

*Proof.* Euler's proof relies on analyzing $\zeta(s)$. We need to see what happens to $\zeta(s)$ as $s \to 1^+$ (that is, as $s \to 1$ through values greater than 1).

**Lemma 5.11.** *If $s > 1$, then $\zeta(s) \geq 1/(s-1)$.*

*Proof.* Let $n \geq 1$. If $n \leq x$, then

$$\frac{1}{n^s} \geq \frac{1}{x^s},$$

so

$$\frac{1}{n^s} = \int_{x=n}^{n+1} \frac{1}{n^s}\, dx \geq \int_{x=n}^{n+1} \frac{1}{x^s}\, dx$$

(the first integral is just the integral of a constant over an interval of length 1). Therefore,

$$\zeta(s) = 1 + \frac{1}{2^s} + \cdots + \frac{1}{n^s} + \cdots$$

$$\geq \int_1^2 \frac{1}{x^s}\, dx + \int_2^3 \frac{1}{x^s}\, dx + \cdots + \int_n^{n+1} \frac{1}{x^s}\, dx + \cdots$$

$$= \int_1^\infty \frac{1}{x^s}\, dx = \frac{x^{1-s}}{1-s}\Big|_{x=1}^\infty = 0 - \frac{1}{1-s} = \frac{1}{s-1}.$$

$\square$

Since $1/(s-1) \to \infty$ as $s \to 1^+$ and $\zeta(s) \geq 1/(s-1)$, we must have $\zeta(s) \to \infty$ as $s \to 1^+$.

Suppose that there are finitely many primes. Then the product

$$\prod_p (1 - p^{-s})^{-1}$$

over all primes $p$ is a finite product. Each factor is continuous at $s = 1$, so

$$\lim_{s \to 1^+} \sum_{n=1}^\infty \frac{1}{n^s} = \lim_{s \to 1^+} \prod_p (1 - p^{-s})^{-1} = \prod_p (1 - p^{-1})^{-1} < \infty.$$

As we just proved, this is not the case. Therefore, there must be infinitely many primes. $\square$

**Remark.** We know that $\sum 1/n$ diverges, so we expect that $\lim_{s \to 1^+} \sum 1/n^s = \infty$, but we had to be a little careful since we could not automatically claim that

$$\lim_{s \to 1^+} \sum 1/n^s = \sum \lim_{s \to 1^+} 1/n^s = \sum 1/n = \infty$$

without some justification (switching the order of limits is a delicate procedure). The integral giving the lower bound for $\zeta(s)$ provided the justification.

Euler initiated the study of $\zeta(s)$ and showed, for example, that

$$\zeta(2) = \sum_{n=1}^{\infty} \frac{1}{n^2} = \frac{\pi^2}{6} \text{ and } \zeta(4) = \frac{\pi^4}{90}.$$

In general, when $m \geq 1$ is an integer,

$$\zeta(2m) = \frac{(-1)^{m+1}(2\pi)^{2m} B_{2m}}{2 \cdot (2m)!},$$

where $B_k$ is the $k$th *Bernoulli number*, defined as the coefficient of $t^k/k!$ in the Taylor series

$$\frac{t}{e^t - 1} = 1 - \frac{1}{2}t + \frac{1}{12}t^2 - \frac{1}{720}t^4 + \frac{1}{30240}t^6 + \cdots = \sum_{k=0}^{\infty} B_k \frac{t^k}{k!}.$$

This implies that $\zeta(2m)$ is $\pi^{2m}$ times a rational number. On the other hand, little is known about $\zeta(2m+1)$ where $2m+1 \geq 3$ is an odd integer. Numerical experiments indicate that it is probably not a power of $\pi$ times a rational number. In 1978, Roger Apéry proved that $\zeta(3)$ is irrational. In 2000, Tanguy Rivoal and Wadim Zudilin proved, for example, that at least one of $\zeta(5), \zeta(7), \zeta(9), \zeta(11)$ is irrational (it is conjectured that all of these are irrational).

The fact that $\zeta(2) = \pi^2/6$, combined with Theorem 5.10, has the following interesting consequence.

**Proposition 5.12.** *The probability that two randomly chosen integers are relatively prime is $6/\pi^2$. More precisely,*

$$\frac{\#\{pairs\ (m,n)\ with\ 1 \leq m,n \leq x\ and\ \gcd(m,n) = 1\}}{\#\{pairs\ (m,n)\ with\ 1 \leq m,n \leq x\}} \to \frac{6}{\pi^2}$$

*as $x \to \infty$.*

*Sketch of proof:* Observe that $\gcd(m,n) = 1$ is equivalent to saying that there is no prime $p$ that divides $\gcd(m,n)$.

There are approximately $x/p$ values of $m \leq x$ and $x/p$ values of $n \leq x$ that are multiples of $p$. Therefore, there are approximately $x^2/p^2$ pairs

$(m, n)$ with $m, n \leq x$ and $p \mid \gcd(m, n)$. The number of pairs with $p \nmid \gcd(m, n)$ is approximately

$$x^2 - (x^2/p^2) = x^2(1 - p^{-2}).$$

Start with the pairs $(m, n)$ with $1 \leq m, n \leq x$. The number of pairs with $2 \nmid \gcd(m, n)$ is approximately $x^2(1 - 2^{-2})$. Among these remaining pairs, approximately $(1 - 3^{-2}) = 8/9$ of them have $3 \nmid \gcd(m, n)$. Therefore, we have approximately

$$x^2(1 - 2^{-2})(1 - 3^{-2})$$

pairs with $2, 3 \nmid \gcd(m, n)$. As we go through the list of *all* primes, each new prime $p$ introduces a factor of $(1 - p^{-2})$ so that approximately

$$x^2(1 - 2^{-2})(1 - 3^{-2})(1 - 5^{-2}) \cdots$$

pairs have no prime dividing $\gcd(m, n)$. Since the total number of pairs is $x^2$, the ratio in the proposition is

$$(1 - 2^{-2})(1 - 3^{-2})(1 - 5^{-2}) \cdots = \frac{1}{\zeta(2)} = \frac{6}{\pi^2}.$$

(A complete proof would use the Chinese Remainder Theorem to justify the independence of the primes and the convergence of the product to show that the many uses of "approximately" do not affect the result.) $\square$

**Technical point:** It is impossible to choose a random integer in such a way that all integers have the same probability. If the probability of choosing any given integer is $\beta$, then the sum of the probabilities for all integers, which must be 1, is

$$1 = \mathrm{Prob}(1) + \mathrm{Prob}(2) + \mathrm{Prob}(3) + \cdots = \beta + \beta + \beta + \cdots.$$

This is either 0 (if $\beta = 0$) or infinite. In particular, it cannot equal 1. This is why we needed to be careful about saying "randomly chosen integers" in Proposition 5.12.

The sum defining the Riemann zeta function $\zeta(s)$ also converges for complex values of $s$ with $Re(s) > 1$ (if $s = x + iy$ is a complex number, then $Re(s) = x$) is called the *real part* of $s$). There is a natural extension of $\zeta(s)$ to all complex numbers $s \neq 1$ (for those who have studied complex analysis, the technical term is "analytic continuation"). In 1859, Bernhard Riemann showed that there is a remarkable connection between the distribution of prime numbers and the location of the solutions of $\zeta(s) = 0$ with $0 \leq Re(s) \leq 1$. He conjectured the **Riemann Hypothesis**, which predicts that the only complex solutions to $\zeta(s) = 0$ with $0 \leq Re(s) \leq 1$ have $Re(s) = 1/2$. This problem is perhaps the most famous unsolved problem in mathematics. In 1900, David Hilbert listed the Riemann Hypothesis as one of 23 important unsolved problems in a list that had a lot of influence on 20th century mathematics. In 2000, the Clay Mathematics Institute included the Riemann Hypothesis as one of its seven Millennium Prize Problems and offered \$1 million for its solution. We make no prediction as to whether or not it will be on 2100's list.

Theorem 5.13, which we prove in Subsection 5.7.1 below and which says that the sum

$$\frac{1}{2} + \frac{1}{3} + \frac{1}{5} + \frac{1}{7} + \frac{1}{11} + \cdots = \infty,$$

and Euler's proof that there are infinitely many primes are the first examples of how properties of $\zeta(s)$ be used to deduce results about prime numbers.

In 1896, Jacques Hadamard and Charles Jean de la Vallée-Poussin independently proved that $\zeta(s) \neq 0$ when $Re(s) = 1$ and they used this to deduce the Prime Number Theorem, which states that the number of primes less than $x$ is asymptotic to $x/\ln(x)$ as $x \to \infty$ (see Chapter 20). If the full Riemann Hypothesis is proved, it will yield much more precise estimates on the number of primes less than $x$ and also supply technical steps needed in many situations in number theory.

In 2000, Jeffrey Lagarias, building on work of Guy Robin, gave an elementary problem that is equivalent to the Riemann Hypothesis: For each positive integer $n$, let

$$H_n = \sum_{j=1}^{n} \frac{1}{j}.$$

The Riemann Hypothesis is equivalent to the statement that

$$\sum_{d|n} d \leq H_n + e^{H_n} \ln(H_n)$$

for all $n \geq 1$, with equality only for $n = 1$ (the sum is over the divisors $d$ of $n$ with $1 \leq d \leq n$).

For example, let $n = 6$. Then

$$H_6 = 1 + (1/2) + (1/3) + (1/4) + (1/5) + (1/6) = 49/20.$$

We have

$$H_6 + e^{H_6} \ln(H_6) = 12.834 \cdots .$$

Since

$$\sum_{d|6} d = 1 + 2 + 3 + 6 = 12 < 12.834,$$

the inequality is verified for $n = 6$. If we could verify it for all $n \geq 1$, then we would win \$1 million.

Admittedly, if you tried to tell someone that proving that $\sum_{d|n} d \leq H_n + e^{H_n} \ln(H_n)$ is the most famous problem in mathematics, then you would probably get a weird look. But it's interesting that such a famous problem has such an elementary equivalent formulation.

## 5.7.1    $\sum 1/p$ Diverges

**Theorem 5.13.** *Let $n \geq 1$ be an integer. Then*

$$\sum_{p \leq n} \frac{1}{p} \geq \ln(\ln n) - .48,$$

*where the sum is over all primes $p \leq n$. In particular,*

$$\sum_{all\ primes\ p} \frac{1}{p}$$

*diverges.*

Let's think about what this says. We know (from the Integral Test for infinite series) that $\sum_{n=1}^{\infty} 1/n$ diverges and that $\sum_{n=1}^{\infty} 1/n^2$ converges. One of the reasons is that squares are rare in the positive integers, and the second sum is missing so many terms that it is able to converge. You can regard the fact that $\sum 1/p$ diverges as saying that the primes occur rather frequently.

*Proof.* Before getting to the heart of the proof, we need the following three lemmas.

The first lemma uses the method of comparing sums to integrals. You might recall this technique from the Integral Test in calculus.

**Lemma 5.14.** *Let $n \geq 1$ be an integer. Then*

$$\ln n \leq \sum_{j=1}^{n} \frac{1}{j}.$$

*Proof.* When $j \leq x \leq j+1$, we have $1/x \leq 1/j$, so

$$\int_{x=j}^{j+1} \frac{1}{x}\, dx \leq \int_{x=j}^{j+1} \frac{1}{j}\, dx = \frac{1}{j}$$

(the last integral is the integral of a constant over an interval of length 1). Therefore,

$$\ln n \leq \ln(n+1) = \int_{1}^{n+1} \frac{1}{x}\, dx = \int_{1}^{2} \frac{1}{x}\, dx + \int_{2}^{3} \frac{1}{x}\, dx + \cdots + \int_{n}^{n+1} \frac{1}{x}\, dx$$

$$\leq \frac{1}{1} + \frac{1}{2} + \cdots + \frac{1}{n}.$$

This proves the lemma.                                                                                        □

Recall the Taylor series

$$-\ln(1-x) = x + \frac{1}{2}x^2 + \frac{1}{3}x^3 + \cdots,$$

which converges for $|x| < 1$. This tells us that $-\ln(1-x) \approx x$ when $x$ is small. We need an explicit upper bound, so we make the $x^2$ term a little bigger to obtain the upper bound given in the next lemma.

**Lemma 5.15.** *If $0 \leq x \leq 1/2$ then*

$$-\ln(1-x) \leq x + x^2.$$

*Proof.*

$$0 \leq t \leq 1/2 \Longrightarrow 0 \leq 2t \leq 1 \Longrightarrow 2t^2 \leq t$$
$$\Longrightarrow 0 \leq t - 2t^2$$
$$\Longrightarrow 1 \leq 1 + t - 2t^2 = (1-t)(1+2t)$$
$$\Longrightarrow \frac{1}{1-t} \leq 1 + 2t.$$

This implies that if $0 \le x \le 1/2$ then

$$-\ln(1-x) = \int_0^x \frac{1}{1-t}\, dt \le \int_0^x (1+2t)\, dt = x + x^2,$$

which proves the lemma.                                      □

The final lemma estimates an error term that occurs later.

**Lemma 5.16.**

$$\sum_p \frac{1}{p^2} < .48$$

*(the sum is over all primes $p$).*

*Proof.* For $0 < x \le j$, we have $1/j^2 \le 1/x^2$, so

$$\frac{1}{j^2} = \int_{x=j-1}^j \frac{1}{j^2}\, dx \le \int_{x=j-1}^j \frac{1}{x^2}\, dx$$

(the first integral is the integral of the constant function $1/j^2$ over an interval of length 1). Therefore,

$$\sum_{p \ge 29} \frac{1}{p^2} \le \sum_{j=29}^\infty \frac{1}{j^2}$$

$$\le \int_{28}^{29} \frac{1}{x^2}\, dx + \int_{29}^{30} \frac{1}{x^2}\, dx + \int_{30}^{31} \frac{1}{x^2}\, dx + \cdots$$

$$= \int_{28}^\infty \frac{1}{x^2}\, dx = \frac{1}{28} = .0357 \cdots < .036.$$

A calculation shows that

$$\sum_{p \le 23} \frac{1}{p^2} = 0.4438 \cdots < .444.$$

Therefore,

$$\sum_p \frac{1}{p^2} < .444 + .036 = .48,$$

as desired.                                                  □

We can now prove the theorem. The proof of Theorem 5.10 shows that

$$\sum_{j \in S_n} \frac{1}{j} = \prod_{p \leq n} (1 - p^{-1})^{-1}, \tag{5.4}$$

where $S_n$ is the set of positive integers whose prime factorizations use only the primes $p \leq n$. Since the factorization of each $j \leq n$ uses only the primes $p \leq n$, the set $S_n$ contains all $j \leq n$. Therefore,

$$\sum_{j=1}^{n} \frac{1}{j} \leq \sum_{j \in S_n} \frac{1}{j}. \tag{5.5}$$

Putting everything together, we obtain

$$\ln(\ln n) \leq \ln \left( \sum_{j=1}^{n} \frac{1}{j} \right) \quad \text{(by Lemma 5.14)}$$

$$\leq \ln \left( \sum_{j \in S_n} \frac{1}{j} \right) \quad \text{(by Equation (5.5))}$$

$$= \ln \left( \prod_{p \leq n} (1 - p^{-1})^{-1} \right) \quad \text{(by Equation (5.4))}$$

$$= \sum_{p \leq n} -\ln \left( 1 - p^{-1} \right) \leq \sum_{p \leq n} \left( \frac{1}{p} + \frac{1}{p^2} \right) \quad \text{(by Lemma 5.15)}$$

$$\leq \sum_{p \leq n} \frac{1}{p} + \sum_{p \leq n} \frac{1}{p^2}$$

$$\leq \sum_{p \leq n} \frac{1}{p} + .48 \quad \text{(by Lemma 5.16)},$$

which is the desired result.

In particular, since $\ln(\ln n) \to \infty$ as $n \to \infty$, we see that $\sum 1/p$ diverges. $\square$

It can be shown that

$$0.251 + \ln(\ln n) \leq \sum_{p \leq n} 1/p \leq 0.272 + \ln(\ln n)$$

for all $n \geq 1000$ (see J. Rosser and L. Schoenfeld, "Approximate Formulas for Some Functions of Prime Numbers," *Illinois Journal of Mathematics* 6 (1962), 64-94, for more precise estimates).

A computer calculation shows that

$$\sum_{p \leq 10} 1/p = 1.176, \qquad \sum_{p \leq 10^6} 1/p = 2.887, \qquad \sum_{p \leq 10^9} 1/p = 3.293.$$

Of course, these agree with the estimate we just gave. The estimate shows also that

$$\sum_{p \leq 10^{100}} 1/p \approx 5.7.$$

Let's make some observations:

1. There are estimated to be around $10^{85}$ particles in the universe (this could easily be wrong by a few powers of 10, but you get the idea).

2. It is therefore unlikely that more than $4 \times 10^{97}$ primes will ever be written down (this even includes primes getting erased and replaced by others).

3. The Prime Number Theorem (Theorem 20.7) tells us that there are approximately $4 \times 10^{97}$ primes less than $10^{100}$.

4. The sum of $1/p$ for any set of $4 \times 10^{97}$ primes is less than or equal to the sum of $1/p$ for the first $4 \times 10^{97}$ primes.

We conclude the following:

*The sum of $1/p$ over all primes that will ever be written down is less than 6.*

(We thank Don Zagier for this observation.)

In particular, it is unlikely that a computer calculation would convince anyone that the sum of $1/p$ over all primes $p$ diverges.

The divergence of $\sum 1/p$ implies that there are infinitely many primes. In contrast, in 1919 Viggo Brun proved the *convergence* of the sum

$$\left(\frac{1}{3} + \frac{1}{5}\right) + \left(\frac{1}{5} + \frac{1}{7}\right) + \left(\frac{1}{11} + \frac{1}{13}\right) + \left(\frac{1}{17} + \frac{1}{19}\right) + \cdots,$$

which is the sum over all pairs $(p, p + 2)$ where both $p$ and $p + 2$ are prime (we call these **twin primes**). The value is approximately 1.902. It is conjectured, but not proved, that there are infinitely many twin primes, but the convergence of the sum is evidence that they are much rarer than primes.

Let $p_1 = 2$, $p_2 = 3$, $p_3 = 5, \ldots$ be the primes. One way to state the twin prime conjecture is to say that $p_{n+1} - p_n \leq 2$ for infinitely many $n$. In 2013, Yitang Zhang proved that $p_{n+1} - p_n \leq 7 \times 10^7$ for infinitely many $n$. This is an amazing result since previously it was not known whether there is any constant $C$ such that $p_{n+1} - p_n \leq C$ for infinitely many $n$. We know by Exercise 24 in Chapter 2 that there are arbitrarily long strings of composite numbers, which means that $p_{n+1} - p_n$ can be arbitrarily large. Zhang's result shows that large spacing between primes is not the general rule; often the primes are close together. The result has been improved by the Terence Tao, James Maynard, and others: $p_{n+1} - p_n \leq 246$ for infinitely many $n$.

## 5.8   Chapter Highlights

1.  An integer $m$ is a $k$th power if and only if all the exponents in its prime factorization are multiples of $k$.

2.  If an integer $n$ is not a $k$th power, then $\sqrt[k]{n}$ is irrational.

3.  All primitive Pythagorean triples $(a, b, c)$ with $b$ even are given by the formula

$$a = n^2 - m^2, \quad b = 2mn, \quad c = m^2 + n^2.$$

    for some relatively prime integers $m < n$ with one even and one odd.

4.  $\sum \dfrac{1}{p}$ diverges.

# 5.9   Problems

## 5.8.1   Exercises

### Section 5.1: A Puzzle

1. Find a positive integer $n$ such that
   (a) $n/2$ is a square and $n/3$ is a cube.
   (b) $n/2$ is a square, $n/3$ is a cube, and $n/5$ is a fifth power.

2. Find a positive integer $n$ such that
   (a) $2n$ is a square and $3n$ is a cube.
   (b) $2n$ is a square, $3n$ is a cube, and $5n$ is a fifth power.

3. Let $a$ and $b$ be positive integers with $\gcd(a, b) = 1$ and let $n \geq ab-a-b+1$. Show that every $n$th power of an integer can be written as the product of an $a$th power and a $b$th power. (*Hint:* Look at Section 3.2.)

4. A positive integer $n$ is called *powerful* if every prime factor of $n$ occurs with exponent larger than 1. Show that if $n$ is powerful then there are integers $x$ and $y$ such that $n = x^2 y^3$. (*Hint:* Use Proposition 3.4 with $a = 2$ and $b = 3$ and work separately with each prime dividing $n$.)

5. Let $n$ be an integer such that $12n$ is a cube. Show that $18 \mid n$.

6. Let $n$ be an integer such that $5n$ is a square. Show that $5 \mid n$.

7. Let $n$ be an integer such that $10n$ is a cube. Show that $100 \mid n$.

8. (a) Five people, call them $A, B, C, D$, and $E$, have a pancake eating contest, and they eat $a, b, c, d$, and $e$ pancakes, respectively. They each eat a different nonzero number of pancakes and all of the ratios $b/a, c/b, d/c$, and $e/d$ are integers. If they eat a total of 47 pancakes, how many pancakes did each eat? Justify that the solution you find is the only solution.
   (b) Find a sequence of primes $p_1, p_2, p_3, p_4, p_5$ such that $p_{i+1} = 2p_i + 1$ for $i = 1, 2, 3, 4$.
   (c) What is the relation between (a) and (b)?

### Section 5.2: Irrationality Proofs

9. For which positive integers $n$ is $\sqrt[n]{64}$ rational?

10. For which positive integers $n$ is $\sqrt[n]{81}$ rational?

11. (a) Show that if $x$ is rational then $x^2$ is rational.
    (b) Show that $\sqrt{2} + \sqrt{3}$ is irrational.

12. Use the technique of the previous problem to prove that if $a$ and $b$ are positive integers and $ab$ is not a square, then $\sqrt{a} + \sqrt{b}$ is irrational.

### Section 5.3: The Rational Root Theorem

13. Find all rational solutions of $x^4 - 2x^3 - 3x^2 + x + 10 = 0$.

14. Find all rational solutions of $x^3 + 3x^2 - x - 3 = 0$.

15. Find all rational solutions of $2x^3 + 5x^2 - x - 1 = 0$.

16. Show that $x^7 + 3x^3 + 1 = 0$ has no rational solutions.

17. The roots of the polynomial $36x^3 - 81x^2 - 10x + 75$ are $-0.8637854$, $1.4471187$, and $1.6666667$ (rounded to seven decimal places). Use the fact that the rational roots of this polynomial have denominators dividing 36 to find the rational roots of the polynomial.

18. The roots of the polynomial $1092x^3 - 745x^2 - 57x + 70$ are $-0.2857143$, $0.3846154$, and $0.5833333$ (rounded to 7 decimal places). Use the fact that the rational roots of this polynomial have denominators dividing 1092 to find the rational roots of the polynomial.

19. (a) Show that $\sqrt{2} + \sqrt{3}$ is a root of $x^4 - 10x^2 + 1$.
    (b) Use the Rational Root Theorem to show that $\sqrt{2} + \sqrt{3}$ is irrational. (Compare with Exercise 11.)

## Section 5.4: Pythagorean Triples

20. Find the values of $m$ and $n$ in Theorem 5.5 corresponding to the Pythagorean triples $(7, 24, 25)$ and $(8, 15, 17)$.

21. Find a Pythagorean triple $(a, b, c)$ with $a = 37$.

22. Find all primitive Pythagorean triples $(a, b, c)$ with $b = 44$.

23. Find two primitive Pythagorean triples $(a, b, c)$ with $c = 65$.

24. Find two imprimitive Pythagorean triples $(a, b, c)$ with $c = 65$.

25. Find all (not necessarily primitive) Pythagorean triples $(a, b, c)$ with $b = 18$.

26. Find a primitive Pythagorean triple $(a, b, c)$ with $c = 85$ and find an imprimitive Pythagorean triple $(a, b, c)$ with $c = 85$.

27. Find two primitive Pythagorean triples $(a, b, c)$ with $a = 85$ and find two imprimitive Pythagorean triple $(a, b, c)$ with $a = 85$.

28. Show that $(3, 4, 5)$ is the only primitive Pythagorean triple whose terms are in an arithmetic progression. Then, find *all* Pythagorean triples whose terms are in an arithmetic progression.

29. Show that if $n \geq 3$ is an integer then there is a (possibly imprimitive) Pythagorean triple that has $n$ as one of its numbers. (*Hint:* If $n$ is odd, $n = (k + 1)^2 - k^2$ for some even $k$. If $n$ is even, write $n = 2^r m$ with $m$ odd and separately examine the cases $m = 1$ and $m \geq 3$.)

30. (a) Let $n = 2^j$, where $j$ is an odd positive integer, and let $m$ be an odd positive integer such that $n + m$ is a square. Show that $2n(n + m)$ is a square.
    (b) Find a primitive Pythagorean triangle whose perimeter has length equal to the square of an integer.

31. At the beginning of the section on Pythagorean triples, we saw two easy ways of producing Pythagorean triples, one starting with squaring an odd number and the other starting with squaring a multiple of 4. Find the values of $m$ and $n$ in Theorem 5.5 that correspond to these two families.

## Section 5.5: Differences of Squares

32. Show that 7 is the difference of two squares in exactly one way.

33. Let $p$ be an odd prime. Show that $p$ is the difference of two squares in exactly one way.

34. Find all solutions of $x^2 - y^2 = 20$ in positive integers $x, y$.

35. Find all solutions of $x^2 - y^2 = 15$ in positive integers $x, y$.

36. Show that if $x$ is an integer then $x^3 + y^2 = z^2$ has a solution in integers $y, z$. (*Hint:* Rewrite as $x \cdot x^2 = z^2 - y^2$.)

37. Show that if $x \geq 3$ and $n \geq 2$ then $x^n = z^2 - y^2$ has a solution in positive integers $y, z$. (*Hint:* Rewrite $x^n$ as $x \cdot x^{n-1}$ when $n \geq 3$, or as $1 \cdot x^2$ or $2 \cdot 2x_1^2$ for some $x_1$ when $n = 2$.)

38. (a) Use Lemma 5.6 to show that if $a$ and $b$ are positive integers with $a(a + 1) = b^2$ then both $a$ and $a + 1$ are squares, which is impossible.
    (b) Show that the only solution of $x(x + 11) = y^2$ in positive integers is $x = 25$, $y = 30$. (*Hint:* If $\gcd(x, x + 11) = 11$, reduce to part (a).)

## Section 5.6: Prime Factorization of Factorials

39. How many 0's are at the end of 1000! (in its usual base 10 expression)?

40. How many 0's are at the end of 123! (in its usual base 10 expression)?

41. What is the largest power of 2 dividing 50!?

42. What is the largest power of 3 dividing 100!?

43. Does there exist $n$ such that $n!$ ends in exactly 30 zeros?

44. Find all values of $n$ such that $n!$ ends in exactly 100 zeros.

45. Let $p$ be a prime. Show that $\binom{p^2}{p}$ is a multiple of $p$ but not a multiple of $p^2$.

## Section 5.7: The Riemann Zeta Function

46. Use an argument similar to the one for Proposition 5.12 to show that the probability that a randomly chosen integer is squarefree is $6/\pi^2$.

47. (a) Show that

$$\prod_{p \text{ odd prime}} \left(1 - \frac{1}{p^2}\right)^{-1} = 1 + \frac{1}{9} + \frac{1}{25} + \frac{1}{49} + \frac{1}{81} + \cdots$$

(the sum of the reciprocals of the odd squares).
(b) Using the fact that $\zeta(2) = \pi^2/6$, show that the sum of the series in part (a) is $\pi^2/8$.

## 5.9.2    Projects

1. Diophantus appears to have been aware of the following identities.

$$(x^2 + y^2)(z^2 + t^2) = (xz + yt)^2 + (xt - yz)^2$$

$$(x^2 + y^2)(z^2 + t^2) = (xz - yt)^2 + (xt + yz)^2$$

(a) Show that these identities are correct.

(b) By factoring $n = 65$ as $5 \cdot 13 = (1^2 + 2^2)(2^2 + 3^2)$ use these identities to find two Pythagorean triples $(a, b, c)$ with $a < b$ and $c = 65$.

(c) Find two Pythagorean triples $(a, b, c)$ with $a < b$ and $c = 221$.

(d) Find two Pythagorean triples $(a, b, c)$ with $a < b$ and $c = 493(=17 \cdot 29)$.

(e) What happens when you use this idea to find all Pythagorean triples $(a, b, c)$ with $a < b$ and $c = 91$?

2.  Let $k \geq 2$ be an integer.

(a) Let $x$ and $y$ be positive integers. Show that $x^k - y^k > 2$.

(b) Show that the product of two consecutive positive integers cannot be a $k$th power. (*Hint:* $\gcd(n, n+1) = 1$.)

(c) Show that the product of three consecutive positive integers cannot be a $k$th power. (*Hint:* Show that $\gcd(n, n^2 - 1) = 1$, and note that the square of a $k$th power is a $k$th power.)

(d) Find an algebraic identity that shows that the product of four consecutive positive integers is always 1 less than a square.

(e) Show that the product of four consecutive positive integers is never a square.

*Note:* Erdős and Selfridge proved in 1975 that a product of two or more consecutive positive integers is never a $k$th power.

3.  We want to find all solutions of $a^2 + 2b^2 = c^2$ with $\gcd(a, c) = 1$ and $a, b, c \geq 0$.

(a) Show that $a$ and $c$ are odd.

(b) Show that $(c + a)/2$ and $(c - a)/2$ are relatively prime.

(c) Show that either $(c + a)/2 = k^2$ or $(c + a)/2 = 2k^2$ for an integer $k$.

(d) Show that there are positive integers $m$ and $n$ such that

$$a = \pm(n^2 - 2m^2), \quad b = 2mn, \quad c = n^2 + 2m^2.$$

(e) Give 5 solutions of $a^2 + 2b^2 = c^2$.

(f) Consider the ellipse $x^2 + 2y^2 = 1$. There is an "obvious" point $x = 1$, $y = 0$. Starting with this point, use Diophantus's method of drawing lines with various slopes to obtain the 5 solutions from part (e).

4.  Let $p$ be prime and let $n \geq 1$. Write $n$ in base $p$ as

$$n = a_0 + a_1 p + a_2 p^2 + \cdots + a_m p^m,$$

where $0 \leq a_i \leq p - 1$ for each $i$. Let $s = a_0 + a_1 + \cdots + a_m$ and let $p^b$ be the power of $p$ dividing $n!$.

(a) Show that

$$\left\lfloor \frac{n}{p^j} \right\rfloor = a_j + a_{j+1} p + \cdots + a_m p^{m-j}.$$

(b) Show that $b =$

$$a_1(1) + a_2(1 + p) + a_3(1 + p + p^2) + \cdots + a_m(1 + p + \cdots + p^{m-1}).$$

(c) Show that

$$b = \frac{n-s}{p-1}.$$

(d) Let $k \geq 1$. Show that the binomial coefficient

$$\binom{2^k}{2^{k-1}}$$

is a multiple of 2 but is not a multiple of 4.

5.   (a) Let $k \geq 1$. Show that

$$x^{2^k} - 1 = (x+1)(x^{2^k-1} - x^{2^k-2} + x^{2^k-3} \cdots - 1).$$

(b) Let $F_j = 2^{2^j} + 1$ denote the $j$th Fermat number. Let $1 \leq m < n$. Show that $F_m \mid F_n - 2$. (*Hint:* Let $k = n - m$ in part (a).)

(c) Let $0 \leq m < n$. Show that $\gcd(F_m, F_n) = 1$.

(d) Use part (c) to show that, for $n \geq 1$, the $n$th prime is less than $2^{2^n}$. Since (d) says that there are at least $n$ primes less than $2^{2^n}$, it implies that there are infinitely many primes.

(e) Use Euclid's proof to show by induction (that is, without Fermat numbers) that, for $n \geq 1$, the $n$th prime is less than $2^{2^n}$.

(It follows from the Prime Number Theorem that the $n$th prime is approximately $n \ln n$ (see Project 2 in Chapter 20), so the upper bound $2^{2^n}$ is not very accurate.)

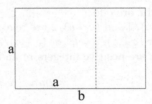

**FIGURE 5.3: Irrationality of $\phi$.**

6.   The **Golden Ratio** is defined to be

$$\phi = \frac{1 + \sqrt{5}}{2} \approx 1.618.$$

It is a root of the polynomial $x^2 - x - 1$.

(a) Suppose we have an $a \times b$ rectangle, where $b/a = \phi$. Suppose we cut off an $a \times a$ square (see Figure 5.3). Show that the remaining rectangle, which is $a \times (b-a)$, satisfies $a/(b-a) = \phi$.

(b) Use part(a) to show that $\phi$ is irrational.

(c) Show that $\sqrt{5}$ is irrational.

7.   Here is another proof that $\sqrt{2}$ is irrational. Suppose that $a$ and $b$ are positive integers with $b^2 = 2a^2$. Form a $b \times b$ square and cut an $a \times a$ square out of each of two diagonally opposite corners, say the northeast

and the southwest corners, as in Figure 5.4.

(a) Use the relation $b^2 = 2a^2$ to deduce that the area of the center square (the overlap in the middle) equals the sum of the areas of the squares in the northwest and southeast corners. (This can be done algebraically, but it is easier to realize that the overlap must equal the amount not covered.)

(b) Let the center square be $r \times r$ and the northwest and southeast corner squares be $s \times s$. Show that $r^2 = 2s^2$.

(c) Use the above steps to prove that $\sqrt{2}$ is irrational. (This proof is due to Tennenbaum.)

(d) Write out an algebraic expression in terms of $a$ and $b$ for the relation $r^2 = 2s^2$. Which other proof(s) does this relate to?

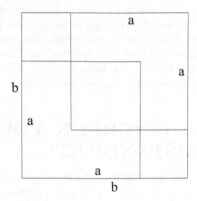

**FIGURE 5.4: Irrationality of $\sqrt{2}$.**

8.  This project finishes the geometric method at the end of Section 5.4 for finding all primitive Pythagorean triples. Let $m, n$ be integers with $\gcd(m, n) = 1$. Write

$$\frac{n^2 - m^2}{n^2 + m^2} = \frac{a}{c}$$

with $c > 0$ and $\gcd(a, c) = 1$.

(a) Show that $\gcd(n^2 - m^2, n^2 + m^2) = 1$ or 2. Let $d$ denote this gcd.

(b) Show that if $d = 1$ then $a = n^2 - m^2$ and $c = n^2 + m^2$.

(c) Suppose $d = 2$. Let $n_1 = (n+m)/2$ and $m_1 = (n-m)/2$. Show that $n$ and $m$ are odd, so $n_1$ and $n_2$ are integers, and show that $\gcd(m_1, n_1) = 1$.

(d) Show that one of $m_1, n_1$ is even and one is odd.

(e) Show that $\gcd(2m_1n_1, m_1^2 + n_1^2) = 1$.

(f) Show that $a/c = 2m_1n_1/(m_1^2 + n_1^2)$.

(g) Conclude that $a = 2m_1n_1$ and $c = m_1^2 + n_1^2$.

## 5.9.3    Computer Explorations

1.  Let $H_n$ be as in Section 5.7. Compute

$$\frac{\sum_{d \mid n} d}{H_n + e^{H_n} \ln(H_n)}$$

for several values of $n$. Are there any $n$ for which this ratio is close to
1? What can you say about the values of $n$ that make the ratio large?
Which values of $n$ make the ratio small?

2.  The six primes

$$7, \quad 37, \quad 67, \quad 97, \quad 127, \quad 157$$

form an arithmetic progression with common difference 30. Can you find
an arithmetic progression of ten primes with common difference 210?
There is one whose first prime is less than 300.

*Remark:* In 2004, Ben Green and Terry Tao proved that there are arbitrarily large arithmetic progressions in the primes. For example, there is
a sequence of one million very large primes that form an arithmetic progression. Note, however, that this does not say that there are arithmetic
progressions of primes that are infinitely long, just that there exists an
arithmetic progression of any desired finite length.

## 5.9.4    Answers to "CHECK YOUR UNDERSTANDING"

1.  The prime factorization of 5400 is $2^3 3^3 5^2$. Since the exponent for 5 is not
    a multiple of 3, we know that 5400 is not a perfect cube.

2.  Write $n = 2^a 3^b 5^c \cdots$. Then $n/3 = 2^a 3^{b-1} 5^c \cdots$ and $n/5 = 2^a 3^b 5^{c-1} \cdots$.
    Since $n/3$ is a square, $a, b-1, c$ are even. Since $n/5$ is a cube, $a, b, c-1$
    are multiples of 3. We can take $a = 0$, $b = 3$, $c = 4$ and obtain $n = 3^3 5^4$.

3.  Since 28 is not a perfect cube, $\sqrt[3]{28}$ is irrational. Since 50 is not a perfect
    ninth power, $\sqrt[9]{50}$ is irrational.

4.  The possible rational roots of $2x^3 - 3x^2 + 5x - 10$ are

$$\pm 10, \quad \pm 5, \quad \pm \frac{5}{2}, \quad \pm 2, \quad \pm \frac{1}{2}, \quad \pm 1.$$

    None of these are actually roots.

5.  $a = 8^2 - 3^2 = 55, b = 2 \cdot 3 \cdot 8 = 48$ and $c = 8^2 + 3^2 = 73$.

6.  We need $m < n$ and $m^2 + n^2 = 13$. The only solution to $m^2 + n^2 = 13$
    with $n > m$ is $n = 3, m = 2$. This produces the triple $(5, 12, 13)$.

7.  No, because 50 is neither odd nor a multiple of 4.

8.  $12 = 4^2 - 2^2$. This is the only way to write 12 as the difference of two
    squares since $12 = 6 \cdot 2$ is the only way to write 12 as the product of two
    numbers with the same parity.

9.  The power of 3 in 10! is $3^4$ since

$$\left\lfloor \frac{10}{3} \right\rfloor + \left\lfloor \frac{10}{9} \right\rfloor + \cdots = 3 + 1 + 0 = 4.$$

# Chapter 6

# Congruences

## 6.1 Definitions and Examples

The Division Algorithm tells us that an even number is a number that leaves a remainder of 0 when divided by 2 and an odd number leaves a remainder of 1 when divided by 2. Although nobody would say that 14 and 96 are the same number, they do share the property of having the same remainder when divided by 2. Similarly, when you look at a clock and you notice that it's 10:00 now and think to yourself that in 5 hours it will be 3:00, you're not saying that $10 + 5 = 3$. What you're really doing, perhaps without realizing it, is saying $10 + 5 = 15$, and when I divide 15 by 12, the remainder is 3. Congruences generalize these ideas to all positive integers $n$ by saying (as we'll see in a bit) that two numbers are "congruent for the number $n$" if they leave the same remainder when divided by $n$. So, to return to our clock example, 3 and 15 "are congruent for the number 12" because they both leave a remainder of 3 when divided by 12. It will take us a few paragraphs to make this precise and we'll have to develop some notation along the way, so let's begin.

**Definition 6.1.** *Two integers $a$ and $b$ are* **congruent mod $m$** *(written $a \equiv b \pmod{m}$) if $a - b$ is a multiple of $m$. The integer $m$ is called the* **modulus** *of the congruence and is assumed to be positive.*

**Examples.**

1. $7 \equiv 1 \pmod 2$, $14 \equiv 0 \pmod 2$, $19 \equiv 7 \pmod 6$,
   $-8 \equiv 12 \pmod 5$, $4 \equiv 4 \pmod{11}$, $6 \equiv 16 \pmod 5$,
   $34 \equiv 12 \pmod{11}$, $7 \equiv -2 \pmod 3$, $-11 \equiv 3 \pmod 7$

2. We can see how the notion of congruence applies to numbers being even or odd. A number is even if it's congruent to 0 mod 2 and it's odd if it's congruent to 1 mod 2. Using congruence notation, $n$ is even if and only if $n \equiv 0 \pmod 2$ and $n$ is odd if and only if $n \equiv 1 \pmod 2$.

3. When we look at a number mod 10, we get its last digit and if we look at it mod 100, we get its last two digits. So, $2347 \equiv 7 \pmod{10}$, $65931 \equiv 1 \pmod{10}$, $2347 \equiv 47 \pmod{100}$, and $65931 \equiv 31 \pmod{100}$. More generally, two positive integers are congruent mod 10 if their last digits are the same, and they are congruent mod 100 if the last two digits of one are the same as the last two digits of the other.

4. If we consider a modulus of 5 we see that $11 \equiv 1 \pmod 5$, $27 \equiv 2 \pmod 5$ and $114 \equiv 4 \pmod 5$.

   As we'll see, congruences behave a lot like equalities but there are some important differences. For example, $3 = 3$ and 3 is the only number equal to 3. (Yes, we do feel silly writing that.) But, notice that $12 \equiv 4 \pmod 8$, $20 \equiv 4 \pmod 8$, and $28 \equiv 4 \pmod 8$. In fact, there is an infinite number of integers that are congruent to 4 mod 8. That's because of the following.

**Proposition 6.2.** $a \equiv b \pmod m$ *if and only if $a = b + km$ for some integer $k$.*

*Proof.* $a \equiv b \pmod m$ if and only if $a - b$ is a multiple of $m$. This means that $a - b = km$ for some integer $k$, or equivalently that $a = b + km$.  $\square$

When we look at the integers mod $m$, we get $m$ different sets. One set consists of all the integers that are congruent to 0 mod $m$, another of the integers congruent to 1 mod $m$, then the integers congruent to 2 mod $m$ until we get to the set which is made of the integers that are $m - 1$ mod $m$. (Notice that the set made up of the integers congruent to $m$ mod $m$ is the same as the set of integers congruent to 0 mod $m$). These $m$ sets are called **congruence classes** mod $m$. Given a positive integer $m$, every integer is in one and only one congruence class modulo $m$. For example, if $m = 5$, the five congruence classes are

$$\{\cdots - 10, -5, 0, 5, 10, 15, \cdots\}$$
$$\{\cdots - 9, -4, 1, 6, 11, 16, \cdots\}$$
$$\{\cdots - 8, -3, 2, 7, 12, 17, \cdots\}$$
$$\{\cdots - 7, -2, 3, 8, 13, 18, \cdots\}$$
$$\{\cdots - 6, -1, 4, 9, 14, 19, \cdots\}$$

The first line has all integers congruent to 0 mod 5, the second all integers congruent to 1 mod 5, etc.

Sometimes, instead of saying that two numbers are congruent mod $m$, people say that these numbers are in the same congruence class mod $m$. For example, $-3$ and $17$ are in the same congruence class mod 5 and $-3 \equiv 17 \pmod 5$.

Proposition 6.2 gives us an embarrassment of riches. If we're given an integer $a$ and a modulus $m$, there are infinitely many integers that are congruent to $a$ $\pmod m$ (the integers in the same congruence class as $a$). Nevertheless, the following proposition tells us that if we're given an $a$ and $m$ as above, we can always single out one.

**Proposition 6.3.** *If $a$ is an integer and $m$ is a positive integer, then there is a unique integer $r$ with $0 \le r \le m - 1$ so that $a \equiv r \pmod m$. This integer $r$ is called the* **least non-negative residue** *of $a$ mod $m$.*

*Proof.* We know from the Division Algorithm that we can write $a = mq + r$ with a unique $r$ satisfying $0 \le r \le m - 1$. By the previous proposition, this means that $a \equiv r \pmod m$. $\qquad\square$

**Examples.** The least non-negative residue of $7 \pmod 4$ is 3, and the least non-negative residue of $-3 \pmod 9$ is 6.

Many computer programs interpret $7 \pmod 4$ as equal to 3. In general, they replace a number mod $n$ by its least non-negative residue.

**Remark.** Congruences occurred, at least implicitly, in ancient number theory. Euler, around 1750, was responsible for introducing them into modern number theory. The terminology and notation $a \equiv b \pmod m$ was introduced by Gauss in his book *Disquisitiones Arithmeticae* in 1801. The "mod" is short for "modulo," which means "with respect to the modulus" in Latin. The use of parentheses around "mod $m$" is traditional, but not necessary.

The following is another useful property of congruences. You may recognize this as saying that congruence mod $m$ forms an equivalence relation.

**Proposition 6.4.** *If $a$, $b$, $c$, and $m$ are integers with $m > 0$, then*
*(a) $a \equiv a \pmod{m}$*
*(b) If $a \equiv b \pmod{m}$, then $b \equiv a \pmod{m}$*
*(c) If $a \equiv b \pmod{m}$ and $b \equiv c \pmod{m}$ then $a \equiv c \pmod{m}$.*

*Proof.* (a) Since $a = a + 0 \cdot m$, Proposition 6.2 says $a \equiv a \pmod{m}$.
(b) If $a \equiv b \pmod{m}$, then $a = b + km$, for some $k$. Therefore,

$$b = a + (-k)m,$$

so $b \equiv a \pmod{m}$.
(c) If $a \equiv b \pmod{m}$, then there is an integer $k_1$ with $a - b = k_1 m$. Similarly, if $b \equiv c \pmod{m}$ then there is an integer $k_2$ with $b - c = k_2 m$. Then,

$$a - c = (a - b) + (b - c) = k_1 m + k_2 m = (k_1 + k_2)m$$

so $a - c$ is a multiple of $m$, which means that $a \equiv c \pmod{m}$.            □

As we said before, congruences behave a lot like equalities. In Section 6.4, we'll learn how to solve linear congruences using techniques similar to those used to solve linear equations. But before we do that, we have to make sure that the relevant rules still apply. For example, if we want to solve the equation $x - 4 = 10$, we add 4 to both sides and get that $x = 14$. We're adding equal numbers (in this case 4) to both sides of an equality and the result is still equality. For the integers, it's pretty obvious that "equals added to equals are equal," because after all there's only one number equal to 4. But, as we pointed out, if we fix a modulus, say 15, there are infinitely many integers that are congruent to 4 mod 15, for example (4, 19, 34, 49, ...). Since 49 and 4 are the same mod 15, we need to know that adding 4 to one side and 49 to the other side of a congruence maintains the congruence. In this case, if we want to solve the congruence $x - 4 \equiv 10 \pmod{15}$, if we add 4 to the left side and 49 to the right side, do we get a correct solution? We see that we do: we have $x \equiv 59$ (mod 15). Plugging 59 into $x - 4 \equiv 10$ (mod 15), we get the valid congruence $55 \equiv 10$ (mod 15). The following proposition tells us that congruences obey the same fundamental rules as the integers do when adding, subtracting, and multiplying.

**Proposition 6.5.** *Assume that $a$, $b$, $c$, $d$, and $m$ are integers with $m$ positive. If $a \equiv b$ (mod $m$) and $c \equiv d$ (mod $m$), then*
*(a) $a + c \equiv b + d$ (mod $m$).*
*(b) $a - c \equiv b - d$ (mod $m$).*
*(c) $ac \equiv bd$ (mod $m$).*

*Proof.* Since $a \equiv b$ (mod $m$) and $c \equiv d$ (mod $m$), we know that $a = b + k_1 m$ and $c = d + k_2 m$ for some integers $k_1$ and $k_2$. To prove (a), we see that

$$a + c = (b + k_1 m) + (d + k_2 m) = (b + d) + (k_1 + k_2)m \equiv b + d \ (\text{mod } m).$$

The proof of (b) follows in the same way:

$$a - c = (b + k_1 m) - (d + k_2 m) = (b - d) + (k_1 - k_2)m \equiv b - d \ (\text{mod } m).$$

Although the algebra is a little messier, the proof of (c) is very similar to that of (a) and (b):

$$ac = (b + k_1 m)(d + k_2 m) = bd + bk_2 m + dk_1 m + k_1 k_2 m^2$$
$$= bd + (bk_2 + dk_1 + k_1 k_2 m)m \equiv bd \ (\text{mod } m). \qquad \square$$

So, just as we say "equals added to equals are equal" and "equals multiplied by equals are equal," we can say "congruents added to congruents are congruent" and "congruents multiplied by congruents are congruent."

Proposition 6.5 was easy to prove. Nevertheless, it is an extremely powerful and useful result since it allows us to calculate with congruences very rapidly. For example, if we want to compute $7^2$ mod 6, we just have to notice that

$$7^2 \equiv 7 \cdot 7 \ (\text{mod } 6)$$

and since $7 \equiv 1$ (mod 6), Proposition 6.5 allows us to conclude that

$$7^2 \equiv 7 \cdot 7 \equiv 1 \cdot 1 \equiv 1 \ (\text{mod } 6)$$

Now, it's true that you could have said that $7^2 = 49$ and quickly seen that $49 \equiv 1 \pmod 6$. But the exact same technique works to show that $7^{200} \equiv 1 \pmod 6$, and calculating $7^{200}$ is much harder. In fact, here is a corollary of Proposition 6.5.

**Corollary 6.6.** *If $a \equiv b \pmod m$, then $a^n \equiv b^n \pmod m$ for any positive integer $n$.*

*Proof.* Since $a \equiv b \pmod m$, we can multiply this congruence by itself to get $a^2 \equiv b^2 \pmod m$. Continuing to multiply, we get $a^n \equiv b^n \pmod m$ for any positive integer $n$. $\qquad\square$

As we'll see later, being able to compute $a^n \pmod m$ quickly is extremely important in cryptographic applications.

**Example.** Here is another example of calculating with congruences. Suppose someone claims that $3987^{12} + 4365^{12} = 4472^{12}$. We'll show that this is incorrect by using the fact that if two quantities are equal, then they are congruent to each other mod $m$ for every positive integer $m$. First, we'll look at this mod 10, which is equivalent to looking at the last digits. We have $3987 \equiv 7 \pmod{10}$, $4365 \equiv 5 \pmod{10}$, and $4472 \equiv 2 \pmod{10}$. A quick calculation shows that $7^4 = 2401 \equiv 1 \pmod{10}$. Cubing this yields $7^{12} = (7^4)^3 \equiv 1^3 \equiv 1 \pmod{10}$. Therefore, $3987^{12} \equiv 7^{12} \equiv 1 \pmod{10}$. Also, since $5^n$ has last digit 5 for each $n \geq 1$, we have $4365^{12} \equiv 5^{12} \equiv 5 \pmod{10}$. Finally, $4472^{12} \equiv 2^{12} \equiv 4096 \equiv 6 \pmod{10}$. Therefore,

$$3987^{12} + 4365^{12} \equiv 1 + 5 = 6 \equiv 4472^{12} \pmod{10}.$$

In other words, the last digits are correct. Suppose we have a calculator that displays 10 decimal digits. We calculate

$$3987^{12} + 4365^{12} = 6.397665635 \times 10^{43} = 4472^{12}.$$

So it looks like we might have equality. But if you look ahead in this book and in Chapter 21 you will see that Fermat's Last Theorem says that the sum of two $n$th powers of integers cannot be an $n$th power if $n \geq 3$. In particular, the sum of two twelfth powers cannot be a twelfth power. So you return to the problem and try looking at it mod 9. We have $3987 \equiv 0 \pmod 9$, $4365 \equiv 0 \pmod 9$, and $4472 \equiv 8 \pmod 9$. Since $8 \equiv -1 \pmod 9$, we have

$$4472^{12} \equiv 8^{12} \equiv (-1)^{12} \equiv 1 \pmod 9.$$

Therefore,

$$3987^{12} + 4365^{12} \equiv 0^{12} + 0^{12} = 0 \not\equiv 1 \equiv 4472^{12} \pmod 9.$$

Since $3987^{12} + 4365^{12}$ and $4472^{12}$ are not congruent mod 9, they cannot be equal. (This "near miss" for Fermat's Last Theorem once appeared in an episode of the television show *The Simpsons*.)

In the preceding example, note that when we computed $8^{12} \pmod 9$, we first changed 8 to $-1$, which is easier to work with. We're allowed to do this because $8 \equiv -1 \pmod 9$, and Proposition 6.5 and Corollary 6.6 tell us that working with 8 (mod 9) is the same as working with $-1$ (mod 9).

Notice that Proposition 6.5 doesn't say anything about dividing in congruences. That's because it is *not* always true that if $ac \equiv bc \pmod m$ then $a \equiv b \pmod m$: you can't always divide a congruence and get a true statement. For example, $4 \cdot 5 \equiv 4 \cdot 2 \pmod{12}$, but if we divide by 4 we see that $5 \not\equiv 2 \pmod{12}$. The problem is that we're dividing by 4, which is a divisor of the modulus of 12. However, you can divide when the number you're dividing by is relatively prime to the modulus. To see this, we recast our congruence in terms of divisibility. If $ac \equiv bc \pmod m$, then by definition, $m \mid (ac - bc)$ so that $m \mid c(a - b)$. We recall from Proposition 2.16 that if $c$ and $m$ are relatively prime, then $m \mid c(a-b)$ implies that $m \mid (a-b)$ so $a \equiv b \pmod m$. We have proved the following:

**Proposition 6.7.** *If $ac \equiv bc \pmod m$ and $\gcd(c, m) = 1$, then $a \equiv b$ (mod $m$).*

Not all is lost if the number you're dividing by and the modulus do have a common factor. We can still get a true statement, but with a smaller modulus.

**Proposition 6.8.** *If $ac \equiv bc \pmod m$ and $\gcd(c, m) = d$, then*

$$a \equiv b \pmod{(m/d)}$$

*and*

$$a \equiv b + (m/d)k \pmod m \text{ with } 0 \le k \le d - 1.$$

*Proof.* If $ac \equiv bc \pmod m$, then $m \mid (ac - bc)$ so $m \mid c(a - b)$ and

$$(m/d) \mid (c/d)(a - b).$$

But $m/d$ and $c/d$ are relatively prime from Proposition 2.10. Using Proposition 2.16, we see that $(m/d) \mid a - b$, hence

$$a \equiv b \pmod{m/d}.$$

This means that $a - b = (m/d)k = m(d/k)$ for some $k$. Therefore, $a - b$ is a multiple of $m$ exactly when $k/d$ is an integer, so we get all possibilities mod $m$ by taking $0 \leq k \leq d - 1$. $\qquad\square$

Here are some examples of the previous two propositions.

1. If $3c \equiv 12 \pmod 5$, we can divide both sides by 3 because 3 and 5 are relatively prime. (We're using Proposition 6.7.) After dividing, we get $c \equiv 4 \pmod 5$.

**Remark.** There's a subtlety that we need to discuss about this example. The congruence $3c \equiv 12 \pmod 5$ is often thought of as an equation in the "mod 5 world." When we solve this, we usually think of having a solution ($c = 4$) in that "mod 5 world." If we changed our perspective to the "integer world," there would be an infinite number of solutions, namely $c = 4, 9, 14, 19, \dots$. When working with congruences, we will stay in the world we started with unless our situation dictates otherwise.

2. If $10c \equiv 30 \pmod 7$, then because 10 and 7 are relatively prime, we can divide both sides by 10 and get $c \equiv 3 \pmod 7$.

3. If $2c \equiv 7 \pmod 9$, we could use guess and check to find that $c = 8$ is a solution. A faster way is to exploit the fact that $9 \equiv 0 \pmod 9$ means that adding multiples of 9 to the right-hand side of this congruence is the same as adding 0. By adding 9 to 7 just once, we get $16 \pmod 9$, so that $2c \equiv 16 \pmod 9$. Because 2 and 9 are relatively prime, we can divide both sides by 2 to get $c \equiv 8 \pmod 9$, so $c = 8$ is a solution.

4. If $6c \equiv 18 \pmod{21}$, we cannot simply divide both sides by 6 because $\gcd(6, 21) = 3$ so 3 and 21 are *not* relatively prime. In this situation, we use Proposition 6.8, which tells us that after dividing by 6 our modulus changes to $21/3 = 7$. So, we get that $c \equiv 3 \pmod 7$ and $c \equiv 3, 10, 17$ is the list of solutions mod 21.

We also could have solved this in two steps. First, divide everything by the gcd, namely 3, to get $2c \equiv 6 \pmod 7$. Then solve the new congruence to get $c \equiv 3 \pmod 7$. This gives us the list 3, 10, 17 of solutions mod 21.

Notice that in this example, our equation began in the "mod 21" world. We left this world when we divided by 3, but our final answer was written as a mod 21 result. We see the same situation in the next example as well.

5. If $10c \equiv 80 \pmod{45}$, when we divide by 10 the modulus changes to $45/\gcd(10,45) = 45/5 = 9$ and we have $c \equiv 8 \pmod 9$. Then $c \equiv 8, 17, 26, 35, 44$ is our list of solutions $\pmod{45}$. Notice that if we divided by 10 and forgot to divide the modulus by 5, we would have gotten $c \equiv 8 \pmod{45}$. Then we would have lost 17, 26, 35, 44 as solutions.

6. Suppose we want to solve $15x \equiv 25 \pmod{55}$. Divide everything by 5 to get $3x \equiv 5 \pmod{11}$. Trying a few numbers (or changing $5 \pmod{11}$ to $27 \pmod{11}$) gives us $x \equiv 9 \pmod{11}$, which is the same as $x \equiv 9, 20, 31, 42, 53 \pmod{55}$.

We conclude this section with some useful results about congruences. The first says that the greatest common divisor depends only on a congruence, which allows us to talk about congruence classes being relatively prime to the modulus.

**Proposition 6.9.** *Let $n$ be a positive integer, and let $a$ and $b$ be integers. If $a \equiv b \pmod n$, then $\gcd(a,n) = \gcd(b,n)$.*

*Proof.* Write $a = b + nk$ for some $k$. Let $d$ be a common divisor of $b$ and $n$. By Proposition 2.3, $d \mid a$. Rewriting our relation as $b = a - nk$, we see that any common divisor of $a$ and $n$ is also a divisor of $b$. Therefore, the set of common divisors of $a$ and $n$ is the same as the set of common divisors of $b$ and $n$. This implies that $\gcd(a,n) = \gcd(b,n)$. □

For example, $\gcd(1234, 10) = \gcd(4 \ 10) = 2$ since $1234 \equiv 4 \pmod{10}$.

The next proposition and its corollary are examples of a general phenomenon that congruences mod a prime behave in many ways like equalities in real numbers (if the product of two real numbers is 0 then one of the factors is 0, and if $x$ is a real number with $x^2 = 1$, then $x = \pm 1$).

**Proposition 6.10.** *If $p$ is a prime and $ab \equiv 0 \pmod{p}$, then $a \equiv 0 \pmod{p}$ or $b \equiv 0 \pmod{p}$.*

*Proof.* If $ab \equiv 0 \pmod{p}$, then $p \mid ab$. By Theorem 4.1, this implies that $p \mid a$ or $p \mid b$, so $a \equiv 0 \pmod{p}$ or $b \equiv 0 \pmod{p}$.            $\square$

A useful corollary of this proposition is

**Corollary 6.11.** *Let $p$ be prime. The congruence*

$$x^2 \equiv 1 \pmod{p}$$

*has only the solutions $x \equiv \pm 1 \pmod{p}$.*

*Proof.*

$$x^2 \equiv 1 \pmod{p} \Longleftrightarrow x^2 - 1 \equiv 0 \pmod{p}$$
$$\Longleftrightarrow (x-1)(x+1) \equiv 0 \pmod{p}. \qquad (6.1)$$

From the previous proposition, congruence (6.1) has a solution if and only if $x - 1 \equiv 0 \pmod{p}$ or $x + 1 \equiv 0 \pmod{p}$, which we can rewrite as $x \equiv \pm 1 \pmod{p}$.            $\square$

---

**CHECK YOUR UNDERSTANDING**

1. Which of the following are true? (a) $17 \equiv 29 \pmod{4}$, (b) $321 \equiv 123 \pmod{10}$, (c) $321 \equiv 123 \pmod{9}$.
2. Solve $2x \equiv 3 \pmod{11}$.
3. Solve $4x \equiv 6 \pmod{22}$.
4. Solve $(x-1)(x-2) \equiv 0 \pmod{11}$.

---

# 6.2   Modular Exponentiation

As we said, being able to calculate $a^n \pmod{m}$ for large values of $n$ and $m$ is extremely important in cryptographic applications. Let's say that we need to calculate

$$3^{385} \pmod{479}. \tag{6.2}$$

We could, of course, compute $3^{385}$, divide it by 479, and then obtain the remainder as the answer. Or, we could repeatedly multiply 3 by itself and then reduce mod 479 every time our product exceeds 479. Fortunately, there's a more practical and faster method for modular exponentiation that's based on successive squaring. Here's how it works in Equation (6.2). We begin by squaring 3:

$$3^2 \equiv 9 \pmod{479}.$$

We now repeatedly square and reduce mod 479. Doing so gives us the following list of congruences, all of which are taken mod 479.

$$3^4 \equiv 9^2 \equiv 81$$
$$3^8 \equiv 81^2 \equiv 6561 \equiv 334$$
$$3^{16} \equiv 334^2 \equiv 111556 \equiv 428$$
$$3^{32} \equiv 428^2 \equiv 183184 \equiv 206$$
$$3^{64} \equiv 206^2 \equiv 284$$
$$3^{128} \equiv 284^2 \equiv 184$$
$$3^{256} \equiv 184^2 \equiv 326.$$

How does this help? Since the exponent $385 = 256 + 128 + 1$, we have

$$3^{385} \equiv 3^{256} \cdot 3^{128} \cdot 3^1 \equiv 326 \cdot 184 \cdot 3 \equiv 327.$$

So, $3^{385} \equiv 327 \pmod{479}$.

In order to understand how to use this to calculate $a^m \pmod{n}$ for any $a, m$, and $n$, let's look more closely at what we've done. Essentially, there were three steps:

1. Precalculate successive squarings of $a$, stopping when you exceed the exponent $m$.

2. Write the exponent $m$ in binary form.

3. Multiply together the powers $a^{2^i}$ corresponding to the 1's in the binary expansion of $m$.

Notice that we did Step 2 when we wrote $385 = 256 + 128 + 1$. This says that the binary form of 385 is 110000001. We saw how to calculate this using successive divisions by 2 in Section 2.8:

$$385 = 2 \cdot 192 + 1$$
$$192 = 2 \cdot 96 + 0$$
$$96 = 2 \cdot 48 + 0$$
$$48 = 2 \cdot 24 + 0$$
$$24 = 2 \cdot 12 + 0$$
$$12 = 2 \cdot 6 + 0$$
$$6 = 2 \cdot 3 + 0$$
$$3 = 2 \cdot 1 + 1$$
$$1 = 2 \cdot 0 + 1,$$

which shows that the base 2, or binary, representation of 385 is 110000001. In practice, the process of modular exponentiation is remarkably fast.

In Project 4, we describe this algorithm in detail.

---

## 6.3   Divisibility Tests

When you look at a number, you immediately know whether it's divisible by 2 by looking at the last digit. If the last digit of the number is even, it's divisible by 2, and if the last digit is odd, it's not. In this section, we use the language of congruences to develop divisibility tests for 2, 3, 4, 5, 6, 8, 9, 10, and 11. These tests have a long history, first appearing over 1200 years ago in the work of the Persian mathematician al-Khwarizini (whose name when Latinized was *Algoritmi*, giving us the word *algorithm*).

The divisibility tests all exploit the fact that we use a base 10 number system. So, for example, the number 37569 is a shorthand expression of $3 \cdot 10^4 + 7 \cdot 10^3 + 5 \cdot 10^2 + 6 \cdot 10^1 + 9 \cdot 10^0$. In general, every positive integer can be written in an expanded form:

$$n = a_m 10^m + a_{m-1} 10^{m-1} + \cdots + a_1 10 + a_0 \qquad (6.3)$$

Written this way, our observation about divisibility by 2 can be expressed as "A number is divisible by 2 if and only if its last digit, $a_0$, is divisible by 2."

We will present the other divisibility tests as a series of propositions, referring back to Equation (6.3) when necessary.

**Proposition 6.12.** *The number $n$ is divisible by 10 if and only if its last digit is 0, and is divisible by 5 if and only if its last digit is 0 or 5.*

*Proof.* Since $10 \equiv 0 \pmod{10}$, we have $10^k \equiv 0 \pmod{10}$ for every positive integer $k$, so

$$n = a_m 10^m + a_{m-1} 10^{m-1} + \cdots + a_1 10 + a_0$$
$$\equiv a_m \cdot 0 + a_{m-1} \cdot 0 + \cdots + a_1 \cdot 0 + a_0$$
$$\equiv a_0 \pmod{10}.$$

This means that $n$ is divisible by 10 if and only if $a_0 \equiv 0 \pmod{10}$, which happens exactly when $a_0 = 0$. The proof for divisibility for 5 is similar and rests on the observation that $10 \equiv 0 \pmod{5}$. So, we again have $n \equiv a_0 \pmod 5$. This means that 5 divides $n$ if and only if $a_0 \equiv 0 \pmod 5$, which happens if and only if $a_0 = 0$ or 5. $\qquad\square$

**Examples.** Since the last digit of $n = 681094136$ is neither 0 nor 5, $n$ is divisible by neither 5 nor 10. On the other hand, 2107195 is divisible by 5 but not by 10 since the last digit is 5, and 94979420 is divisible by both 5 and 10 since the last digit is 0.

**Proposition 6.13.** *The integer $n$ is divisible by 4 if the number formed by its last two digits is divisible by 4, and $n$ is divisible by 8 if the number formed by its last three digits is divisible by 8.*

*Proof.* We begin by observing that if $k \geq 2$ then

$$10^k = 10^2 10^{k-2} = 4 \cdot 25 \cdot 10^{k-2} \equiv 0 \pmod 4.$$

Therefore

$$n = a_m 10^m + a_{m-1} 10^{m-1} + \cdots + a_1 10 + a_0$$
$$\equiv a_m \cdot 0 + a_{m-1} \cdot 0 + \cdots + a_1 \cdot 10 + a_0$$
$$\equiv a_1 10 + a_0 \pmod 4.$$

So, a number is congruent mod 4 to the number formed by its last two digits and it's divisible by 4 if and only if the number formed by its last two digits is divisible by 4. The proof for divisibility by 8 is almost identical: if $k \geq 3$ then

$$10^k = 10^3 10^{k-3} = 8 \cdot 125 \cdot 10^{k-3} \equiv 0 \pmod{8}.$$

This means that a number mod 8 reduces down to its last three digits and is therefore divisible by 8 if and only if the number formed by its last three digits is divisible by 8.                                      □

**Examples.** 5675972 is divisible by 4 because 72 is, but 5675972 is not divisible by 8 because 972 is not divisible by 8. On the other hand, 32589104 is divisible by both 4 and 8, while 16932594 is divisible by neither 8 nor 4.

**Proposition 6.14.** *An integer mod 3 (respectively, mod 9) is congruent to the sum of its digits mod 3 (respectively, mod 9).*

*Proof.* We begin with the observation that $10 \equiv 1 \pmod{3}$. Since $1^k = 1$ for every integer $k$, we have

$$10^k \equiv 1^k \equiv 1 \pmod{3} \text{ for } k \geq 0.$$

This means that when we look at $n$ expanded in its base 10 form mod 3 we get

$$n = a_m 10^m + a_{m-1} 10^{m-1} + \cdots + a_1 10 + a_0$$
$$\equiv a_m + a_{m-1} + \cdots + a_1 + a_0 \pmod{3},$$

where we get the right-hand side of the congruence because each power of 10 is 1 mod 3. We now see that $n \pmod 3$ is the same as the sum of the digits of $n \pmod 3$ and that $n$ is divisible by 3 if and only if the sum of its digits is divisible by 3.

The proof of divisibility by 9 is exactly the same after we note that

$$10 \equiv 1 \pmod{9} \implies 10^k \equiv 1 \pmod{9} \text{ for every integer } k \geq 0.$$

□

**Corollary 6.15.** *The integer n is divisible by 3 if and only if the sum of its digits is divisible by 3. It is divisible by 9 if and only if the sum of its digits is divisible by 9.*

**Example.** The sum of the digits of 9789042 is $9+7+8+9+0+4+2 = 39$. Since 39 is divisible by 3, so is 9789042. Note that 39 is not divisible by 9, since $39 \equiv 3 \pmod 9$, and when we divide 9789042 by 9, we get a remainder of 3. We can also see this by adding the digits three times:

$$9789042 \equiv 9 + 7 + 8 + 9 + 0 + 4 + 2$$
$$\equiv 39 \equiv 3 + 9 = 12$$
$$\equiv 1 + 2 = 3 \pmod 9.$$

We could also do this calculation by discarding 9s as they occur:

$$9789042 \equiv 9 + 7 + 8 + 9 + 0 + 4 + 2$$
$$\equiv 7 + 8 + 4 + 2 \equiv 8 + 4 = 12 \text{ (since } 7 + 2 = 9)$$
$$\equiv 1 + 2 = 3 \pmod 9.$$

**Remark.** We can also prove Proposition 6.14 without congruences in the following elementary way. If we write

$$n = a_0 + a_1(9 + 1) + a_2(99 + 1) + a_3(999 + 1) + \cdots$$
$$= a_0 + a_1 + a_2 + \cdots + a_n + 9(a_1 + 11a_2 + 111a_3 + \cdots),$$

we see that
$$n = a_0 + a_1 + a_2 + \cdots + a_n + 9k,$$

for some integer $k$. Therefore, $n$ is congruent to the sum of its digits mod 9.

This also shows that $n$ is congruent to the sum of its digits mod 3 since

$$n = a_0 + a_1 + a_2 + \cdots + a_n + 9k = a_0 + a_1 + a_2 + \cdots + a_n + 3 \cdot (3k).$$

**Example.** Here is a way to impress your friends with your powers of mental calculation. Ask them to give a random 5-digit number, for example 31415. Tell them to square it (on their calculator), and then give you all but one of the digits of the answer in random order. In this case, suppose they say 2, 2, 8, 0, 9, 6, 5, 9. You immediately say that the missing digit is 2. How do you do this? You know that $31415 \equiv 3 + 1 + 1 \equiv 5 \pmod 9$ (omit the $4 + 5$ because they add to 9). Therefore, $31415^2 \equiv 5^2 = 25 \equiv 7 \pmod 9$. Add up the digits of the answer, letting $x$ be the missing digit. Leave out the 0 and the two 9s since we are working mod 9:

$$x + 2 + 2 + 8 + 6 + 5 = x + 23 \equiv x + 2 + 3 \equiv x + 5 \pmod 9.$$

Therefore,
$$7 \equiv (31415)^2 \equiv x + 5 \pmod 9.$$

This tells you that $x = 2$ is the missing digit. In fact, this is correct: $31415^2 = 986902225$.

The only problem that could arise is when the missing digit is 0 or 9, since they are congruent mod 9. If you find that the missing digit is 0 mod 9, ask one more question: is the missing digit is even or odd?

The final divisibility test that we'll do is for 11.

**Proposition 6.16.** *An integer mod* 11 *is congruent to the alternating sum of its digits starting with the ones digit* $(a_0)$, *subtracting the tens digit* $(a_1)$, *etc. So,*

$$n = a_m 10^m + a_{m-1} 10^{m-1} + \cdots + a_1 10 + a_0$$

*is congruent mod* 11 *to*

$$a_0 - a_1 + a_2 - a_3 + \cdots + (-1)^m a_m.$$

*Proof.* We begin with the observation that

$$10 \equiv -1 \pmod{11} \implies 10^k \equiv (-1)^k \pmod{11}.$$

This means that when we look at the base 10 expansion of $n$ (mod 11), we get

$$n \equiv a_0 - a_1 + a_2 - a_3 + \cdots + (-1)^m a_m \pmod{11}.$$

$\square$

**Corollary 6.17.** *An integer* $n$ *is divisible by* 11 *if and only if the alternating sum of its digits is divisible by* 11.

**Examples.** If $n = 8563981$ then the alternating sum of the digits of $n$ is $1 - 8 + 9 - 3 + 6 - 5 + 8 = 8$, so $n \equiv 8 \pmod{11}$. But $n = 639162513$ has alternating sum $3 - 1 + 5 - 2 + 6 - 1 + 9 - 3 + 6 = 22 \equiv 0 \pmod{11}$ so $639162513$ is divisible by 11. Three-digit numbers such as 264 and 385, where the middle digit is the sum of the first and third digits, are immediately recognized as multiples of 11 because the alternating sum is 0.

---

## CHECK YOUR UNDERSTANDING

5. Show that $987654321 \equiv 0 \pmod{9}$.
6. Show that $343434 \equiv 58 \pmod{11}$.

---

# 6.4   Linear Congruences

Let's say that we want to solve $3x \equiv 4 \pmod{7}$. We can try different values of $x$ until we discover a solution. In this case, the smallest positive one is $x = 6$. Now, suppose we want to solve $3x \equiv 6 \pmod{12}$. We find that $x \equiv 2, 6, 10 \pmod{12}$ are the solutions. On the other hand, the equation $2x \equiv 1 \pmod{4}$ has no solutions because $2x$ is always congruent to 0 or 2 mod 4. In general, we have the following.

**Theorem 6.18.** *Let $m$ be a positive integer and let $a \neq 0$. The congruence*

$$ax \equiv b \pmod{m} \tag{6.4}$$

*has a solution if and only if $d = \gcd(a, m)$ is a divisor of $b$. If $d \mid b$, then there are exactly $d$ solutions that are distinct mod $m$. In this case, if $x_0$ is a solution, the solutions mod $m$ are*

$$x = x_0 + (m/d)k \text{ for } 0 \leq k < d.$$

*The solution $x_0$ can be taken to be any integer satisfying*

$$(a/d)x_0 \equiv (b/d) \pmod{m/d}.$$

*Proof.* We begin by rewriting Equation (6.4) without congruences:

$$ax \equiv b \pmod{m}$$

is equivalent to

$$ax = b + my, \quad \text{or} \quad ax - my = b \tag{6.5}$$

for some integer $y$. We now need to use Theorem 3.1. Note that this theorem talked about the equation $ax + by = c$, while we're looking at the equation $ax - my = b$, so we replace $(a, b, c)$ in Theorem 3.1 with $(a, -m, b)$.

If $d \nmid b$, Theorem 3.1 tells us that Equation (6.5) has no solutions.

We now assume that $d \mid b$. Theorem 3.1 tells us that the solutions of (6.5) have the form

$$x = x_0 + (m/d)k, \quad y = y_0 + (a/d)k,$$

where $k$ is an integer and $(x_0, y_0)$ is any particular solution of Equation (6.5). Therefore, all solutions of $ax \equiv b \pmod{m}$ have the form $x = x_0 + (m/d)k$. This means that the solutions $x$ are exactly the integers congruent to $x_0 \bmod (m/d)$. We need to see when these solutions are different $\bmod m$.

Let $x_1 = x_0 + (m/d)k_1$ and $x_2 = x_0 + (m/d)k_2$ be two solutions and suppose that $x_1 \equiv x_2 \pmod{m}$. Then $x_1 - x_2 = mk_3$ for some $k_3$. But

$$\begin{aligned}
x_1 - x_2 = mk_3 &\iff \left(x_0 + (m/d)k_1\right) - \left(x_0 + (m/d)k_2\right) = mk_3 \\
&\iff (m/d)\left(k_1 - k_2\right) = mk_3 \\
&\iff k_1 - k_2 = dk_3,
\end{aligned}$$

which is the same as saying that $k_1 \equiv k_2 \pmod{d}$. Therefore, we get the distinct solutions $\bmod m$ by letting $x = x_0 + k(m/d)$ with $0 \le k < d$ (or with $k$ running through any complete set of residue classes $\bmod d$).

Finally, we need to show that $x_0$ arises from solving $(a/d)x_0 \equiv (b/d) \pmod{m/d}$. Rewrite this congruence as

$$(a/d)x_0 = (b/d) + (m/d)z,$$

where $z$ is an integer. Multiplying by $d$ yields

$$ax_0 = b + mz,$$

which says that $x_0$ is a solution of (6.4), and therefore can be used as our particular solution. $\qquad \square$

By far, the most common case of Theorem 6.18 is when $d = 1$. Since it is so important, we single it out.

**Corollary 6.19.** *If* $\gcd(a, m) = 1$, *the congruence* $ax \equiv b \pmod{m}$ *has exactly one solution mod* $m$.

*Proof.* Set $d = 1$ in the theorem. Then $d \mid b$, so there is a solution. The distinct solutions mod $m$ are $x_0 + mk$ with $0 \le k < 1$, so only $k = 0$ is allowed, and therefore there is exactly one solution mod $m$. $\qquad\square$

**Examples.**

- Theorem 6.18 says that the congruence $6x \equiv 7 \pmod{15}$ has no solutions because $\gcd(6, 15) = 3$ and $3 \nmid 7$.

- The congruence $-10x \equiv 12 \pmod{15}$ has no solutions because we have $\gcd(-10, 15) = 5$ and $5 \nmid 12$.

- The congruence $5x \equiv 6 \pmod{11}$ has $x = 10$ as a solution. Since $\gcd(5, 11) = 1$, this is the only solution mod 11.

- The congruence $3x \equiv 8 \pmod{10}$ has $x = 6$ as a solution and it is the only solution mod 10.

- In the congruence $9x \equiv 6 \pmod{15}$, we divide by $3 = \gcd(9, 15)$ and get the new congruence $3x \equiv 2 \pmod{5}$. This has $x_0 = 4$ as a solution. The theorem tells us that the solutions to $9x \equiv 6 \pmod{15}$ are obtained by adding multiples of $5 = 15/3$ to 4:

$$x \equiv 4,\ 9,\ 14 \pmod{15}.$$

- The congruence $10x \equiv 15 \pmod{45}$ has $5 = \gcd(10, 45)$ solutions. To find them, we divide by 5 and get $2x \equiv 3 \pmod{9}$. This has $x_0 = 6$ as a solution. The solutions are obtained by adding multiples of $9 = 45/5$ to 6:

$$x \equiv 6,\ 15,\ 24,\ 33,\ 42 \pmod{45}.$$

There is a subtlety here that you should be aware of. When we had the linear equation $ax - my = b$, we could divide both sides by $\gcd(a, m)$. This made the computations involved simpler and nothing was lost. But if we try the same thing with a linear congruence, we get solutions to a congruence mod $m/d$, where $d = \gcd(a, m)$. These yield $d$ solutions to the original congruence.

For example, $4x \equiv 8 \pmod{20}$ has 4 solutions mod 20 (namely, $x = 2$, 7, 12, 17). Dividing by $\gcd(4, 20) = 4$ gives us just one solution mod 5, namely $x \equiv 2 \pmod 5$. So we take 2, $2 + 5$, $2 + 10$, $2 + 15$, and stop at $2 + 20 = 22$ since 22 is the same as 2 mod 20. Just remember that if you divide by the gcd, then you need to remember this extra step to get back all solutions for the original modulus.

A linear congruence can be solved using the Extended Euclidean Algorithm by transforming the congruence to an equation. Here's an example.

**Example.** We'll show how to solve the linear congruence

$$183x \equiv 15 \pmod{31}. \tag{6.6}$$

We begin by reducing 183 mod 31 to get $28x \equiv 15 \pmod{31}$, and then we change this linear congruence to a linear Diophantine equation:

$$28x - 31y = 15.$$

The next step is to use the Euclidean Algorithm to find $\gcd(28, 31)$:

$$31 = 1 \cdot 28 + 3$$
$$28 = 9 \cdot 3 + 1$$
$$3 = 3 \cdot 1 + 0.$$

So $\gcd(28, 31) = 1$. We now use the Extended Euclidean Algorithm:

| | $x$ | $y$ | |
|---|---|---|---|
| 31 | 1 | 0 | |
| 28 | 0 | 1 | |
| 3 | 1 | $-1$ | (1st row) $-$ (2nd row) |
| 1 | $-9$ | 10 | (2nd row)$- 9 \cdot$(3rd row). |

This tells us that $28 \cdot (10) + 31 \cdot (-9) = 1$ or $28 \cdot 10 - 31 \cdot 9 = 1$. After multiplying both sides by 15, we find

$$28 \cdot 150 - 31 \cdot 135 = 15.$$

Since mod 31 means that we can discard multiples of 31, we obtain

$28 \cdot 150 \equiv 15 \pmod{31}$. This tells us that $x = 150$ is a solution to (6.6), and after reducing $150 \bmod 31$, we get that $x = 26$ is a solution.

An alternative approach uses inverses, which we now discuss.

**Definition 6.20.** *We say that integers $a$ and $b$ are* **multiplicative inverses mod $m$** *if $ab \equiv 1 \pmod{m}$.*

If the context is clear, we drop the word multiplicative and simply say that $a$ and $b$ are inverses mod m.

**Examples.** Since $2 \cdot 3 \equiv 1 \pmod{5}$, we see that 2 and 3 are inverses mod 5. Similarly, 5 and 7 are inverses mod 34, and 8 and 7 are inverses mod 11 (since $8 \cdot 7 = 56 \equiv 1 \pmod{11}$).

To find the inverse mod $m$ of an integer $a$, we need to solve the congruence

$$ax \equiv 1 \pmod{m}. \tag{6.7}$$

We immediately see that Theorem 6.18 has the following as a consequence.

**Corollary 6.21.** *The integer $a$ has an inverse mod $m$ if and only if $\gcd(a, m) = 1$.*

*Proof.* We know from Theorem 6.18 that $ax \equiv 1 \pmod{m}$ has a solution if and only if $\gcd(a, m) \mid 1$. This is true if and only if $\gcd(a, m) = 1$. $\qquad\qquad\square$

**Example.** In order to find the inverse of 13 (mod 100), we need to solve the congruence

$$13x \equiv 1 \pmod{100}. \tag{6.8}$$

To solve this, we use the same technique used to solve an arbitrary linear congruence. First, we change the congruence into a linear Diophantine equation:

$$13x - 100y = 1.$$

Next we use the Euclidean Algorithm to find $\gcd(13, 100)$:

$$100 = 7 \cdot 13 + 9$$
$$13 = 1 \cdot 9 + 4$$
$$9 = 2 \cdot 4 + 1$$
$$4 = 4 \cdot 1 + 0.$$

So $\gcd(13, 100) = 1$. We now use the Extended Euclidean Algorithm:

$$
\begin{array}{rrrl}
 & x & y & \\
100 & 1 & 0 & \\
13 & 0 & 1 & \\
9 & 1 & -7 & \text{(1st row)} - 7 \cdot \text{(2nd row)} \\
4 & -1 & 8 & \text{(2nd row)} - \text{(3rd row)} \\
1 & 3 & -23 & \text{(3rd row)} - 2 \cdot \text{(4th row)}.
\end{array}
$$

Therefore, $1 = 100 \cdot 3 - 13 \cdot 23$, so $13(-23) \equiv 1 \pmod{100}$. The inverse of $13 \pmod{100}$ is $-23 \equiv 77 \pmod{100}$.

**Example.** To find the inverse of $7 \pmod{19}$, we need to solve the congruence $7x \equiv 1 \pmod{19}$. We can do this by writing this as the linear equation $7x - 19y = 1$ and then solving by using the Extended Euclidean Algorithm as in the previous example. An alternative that is often quicker for small moduli is to manipulate the equation in order to get a coefficient of 1 for $x$. Here is an example.

$$
\begin{array}{ll}
7x \equiv 1 \pmod{19} & \text{Now multiply by 3} \\
21x \equiv 3 \pmod{19} & \text{Now reduce mod 19} \\
2x \equiv 3 \pmod{19} & \text{Now add 19 to 3} \\
2x \equiv 22 \pmod{19} & \text{Now divide by 2} \\
x \equiv 11 \pmod{19} & \text{We're done!}
\end{array}
$$

This says that a solution to $7x \equiv 1 \pmod{19}$ is $x = 11$, so $11 \pmod{19}$ is the inverse to $7 \pmod{19}$. We can write this as $7^{-1} \equiv 11 \pmod{19}$.

**Example.** Here's how we can solve $7x \equiv 4 \pmod{19}$ by using the (multiplicative) inverse of 7. Multiplying both sides by 11, which we just saw is the inverse of $7 \pmod{19}$, yields the following:

$$7x \equiv 4 \pmod{19} \iff 77x \equiv 44 \pmod{19} \iff x \equiv 6 \pmod{19}$$

since $77 \equiv 1 \pmod{19}$ and $44 \equiv 6 \pmod{19}$. Therefore, $x \equiv 6 \pmod{19}$ is the solution.

In effect, we've solved this equation by multiplying by 11, instead of dividing by 7. This may seem strange, but we do things like this all the time. For example, with real numbers, instead of dividing by 2 we can multiply by 0.50. What we've just done is the modular analogue of this.

**Example.** Let's try this technique to solve $3x \equiv 4 \pmod{26}$. Since $3 \cdot 9 \equiv 1 \pmod{26}$, we have $3^{-1} \equiv 9 \pmod{26}$. Therefore,

$$3x \equiv 4 \pmod{26} \Longleftrightarrow 27x \equiv 36 \pmod{26} \Longleftrightarrow x \equiv 10 \pmod{26},$$

so $x \equiv 10 \pmod{26}$ is the solution.

The following are summaries of two important algorithms. Note that the second is a special case of the first and occurs as an intermediate step in the first.

---

**Solve $ax \equiv b \pmod{m}$ when $\gcd(a, m) = 1$**

1. Form a linear Diophantine equation: $ax - my = b$.

2. Use the Extended Euclidean Algorithm to find $x_0, y_0$ with $ax_0 - my_0 = 1$.

3. Compute $x \equiv bx_0 \pmod{m}$.

---

**Find the inverse of $a \pmod{m}$ when $\gcd(a, m) = 1$**

1. Form the linear Diophantine equation $ax - my = 1$.

2. Use the Extended Euclidean Algorithm to find $x_0, y_0$ with $ax_0 - my_0 = 1$.

3. $x_0 \pmod{m}$ is the inverse of $a \pmod{m}$.

---

## CHECK YOUR UNDERSTANDING

7. Find the multiplicative inverse of 15 $\pmod{73}$.
8. Solve $15x \equiv 11 \pmod{73}$.

# 6.5   The Chinese Remainder Theorem

Around 1700 years ago, the Chinese mathematician Sun Tzu asked for a number that leaves a remainder of 2 when divided by 3, leaves a remainder of 3 when divided by 5, and leaves a remainder of 2 when divided by 7. In terms of congruences, this means

$$x \equiv 2 \pmod 3, \quad x \equiv 3 \pmod 5, \quad x \equiv 2 \pmod 7.$$

Similar problems subsequently appeared in various writings in various countries, along with algorithms for solving them. Eventually, the theorem that guarantees (under certain hypotheses) that a solution exists became known as the Chinese Remainder Theorem.

If you try to find a solution to Sun Tzu's problem, you'll probably devise methods that are close to some that we discuss in this section.

In the previous section we learned how to solve a linear congruence. Let's start with asking what happens if we have two (or more) linear congruences with different moduli. For example,

$$x \equiv 2 \pmod 3, \quad \text{and } x \equiv 4 \pmod 5. \tag{6.9}$$

We'll begin by solving this by using a somewhat ad hoc method: Start by looking at the congruence that has the larger modulus, in this case $x \equiv 4 \pmod 5$. Make a list of positive integers that are congruent to 4 mod 5:

$$4, \ 9, \ 14, \ 19, \ 24$$

and then make a list of these numbers mod 3, stopping when you get a solution to $x \equiv 2 \pmod 3$:

$$\begin{aligned} x \equiv 4 \pmod 5 : \quad & 4 \quad 9 \quad 14 \quad 19 \quad 24 \\ x \pmod 3 : \quad & \ \ 1 \quad 0 \quad \ \, \mathbf{2}. \end{aligned}$$

This tells us that 14 is a solution to both congruences. If we had started our list with the smaller modulus 3, we would have arrived at the answer of 14, but it would have taken us longer to get there since our initial list would be 2, 5, 8, 11, 14.

Here's another example of this method, this time with three congruences.

$$x \equiv 4 \pmod 5, \quad x \equiv 5 \pmod 7, \quad \text{and} \quad x \equiv 3 \pmod{11}. \tag{6.10}$$

Since 11 is the largest of the moduli, we begin with a list of integers that are 3 (mod 11):

$$3, \ 14, \ 25, \ 36, \ 47, \ 58, \ 69, \ 80, \ldots.$$

Next, put each number mod 7 underneath this:

$$x \equiv 3 \ (\text{mod } 11): \quad 3 \quad 14 \quad 25 \quad 36 \quad 47 \quad 58$$
$$x \ (\text{mod } 7): \qquad \ 3 \quad \ 0 \quad \ 4 \quad \ 1 \quad \ \mathbf{5} \quad \ 2.$$

This tells us that $x \equiv 47$ (mod 77) is the solution to $x \equiv 3$ (mod 11) and $x \equiv 5$ (mod 7). Now make a list of the numbers that are 47 (mod 77) and look at them mod 5:

$$x \equiv 47 \ (\text{mod } 77): \quad 47 \quad 124 \quad 201 \quad \cdots$$
$$x \ (\text{mod } 5): \qquad \quad 2 \quad \ \ \mathbf{4}$$

So, $x = 124$ is a solution to the three congruences.

Although this method always works, if the moduli are large the lists can become very long, which makes this impractical. A more systematic technique is given by the second proofs of Theorems 6.22 and 6.23.

Before we get to the general situation of $r$ linear congruences with $r$ different moduli, we consider two congruences.

**Theorem 6.22.** *If $m$ and $n$ are relatively prime, then the system of congruences*

$$x \equiv a \ (\text{mod } m) \ and \ x \equiv b \ (\text{mod } n) \tag{6.11}$$

*has a unique solution mod $mn$.*

*Proof.* We give two proofs for the existence of $x$.

*First Proof:* Our first proof closely follows our solution to (6.9). Because $x \equiv a$ (mod $m$),

$$x = a + mt \tag{6.12}$$

for some integer $t$. But $x \equiv b$ (mod $n$) as well, so we can substitute and get

$$a + mt \equiv b \ (\text{mod } n).$$

After subtracting, we have

$$mt \equiv b - a \ (\text{mod } n),$$

which we know has a unique solution $t_0$ (mod $n$), by Corollary 6.19, because $\gcd(m, n) = 1$.

Then $x = a + mt_0$ is a solution to both congruences: Clearly $x \equiv a$ (mod $m$). Moreover, $x = a + mt_0 \equiv a + (b - a) = b$ (mod $n$).

*Second Proof:* Since $\gcd(m, n) = 1$, there exist $u, v$ with $mu + nv = 1$. Let

$$x = bmu + anv.$$

Note that $mu \equiv 0$ (mod $m$) and $nv = 1 - mu \equiv 1$ (mod $m$), so we have $x \equiv b \cdot 0 + a \cdot 1 \equiv a$ (mod $m$). Similarly, $x \equiv b$ (mod $n$). Therefore, $x$ gives a solution.

To see that the solution $x$ is unique, assume that there are two solutions, $x_1$ and $x_2$. Then

$$x_1 \equiv a \ (\text{mod } m), \quad x_1 \equiv b \ (\text{mod } n),$$
$$x_2 \equiv a \ (\text{mod } m), \quad x_2 \equiv b \ (\text{mod } n).$$

This means that $x_1 \equiv x_2$ (mod $m$) and $x_1 \equiv x_2$ (mod $n$), so

$$m \mid (x_1 - x_2) \text{ and } n \mid (x_1 - x_2).$$

Therefore, $x_1 - x_2$ is a multiple of both $m$ and $n$. Because $\gcd(m, n) = 1$, we have that $x_1 - x_2$ is a multiple of $mn$ by Proposition 2.17. By definition, this is the same as

$$x_1 \equiv x_2 \ (\text{mod } mn).$$

Therefore, any two solutions are congruent mod $mn$, which is the same as saying that the solution is unique mod $mn$. □

**Example.** Let's use the second proof to solve Equation 6.9. That is, we want $x$ satisfying $x \equiv 2$ (mod 3), $x \equiv 4$ (mod 5). We have $\gcd(3, 5) = 1$ and we can write $3(2) + 5(-1) = 1$. Then

$$x \equiv bmu + anv \equiv (4)(3)(2) + (2)(5)(-1) = 14 \ (\text{mod } 15).$$

In the example with three congruences, we first solved two of the congruences and reduced our system from three congruences to two. This is how we prove the main result of this section. We have $r$ congruences with $r$ moduli. We pair off the first two and get $r - 1$ congruences with $r - 1$ moduli. We continue in this manner until we are left with a single congruence and this will yield our solution.

**Theorem 6.23. (Chinese Remainder Theorem)** *Assume that* $m_1, m_2, \ldots, m_r$ *are positive integers that are pairwise relatively prime (that is,* $\gcd(m_i, m_j) = 1$ *if* $i \neq j$*). Then the system of congruences*

$$x \equiv a_1 \pmod{m_1}$$
$$x \equiv a_2 \pmod{m_2}$$
$$\vdots$$
$$x \equiv a_r \pmod{m_r}$$

*has a unique solution mod* $m_1 m_2 \cdots m_r$.

*Proof.* Again, we give two proofs for the existence of a solution.

*First Proof:* We pair up our first two equations and use Theorem 6.22 to find a solution, say

$$x \equiv b_1 \pmod{m_1 m_2}. \tag{6.13}$$

Next, we pair off (6.13) with the third congruence,

$$x \equiv a_3 \pmod{m_3}.$$

Because all the moduli are pairwise relatively prime, $m_3$ and $m_1 m_2$ have no common divisors, which allows us to use Theorem 6.22 again and get a solution to the first three congruences, say

$$x \equiv b_2 \pmod{m_1 m_2 m_3}.$$

We can continue in this manner $r-1$ times to finally arrive at a solution

$$x \equiv b_{r-1} \pmod{m_1 m_2 \cdots m_r}. \tag{6.14}$$

*Second Proof:* Let $m = m_1 m_2 \cdots m_r$ and let $n_i = m/m_i$, which is the product of all the $m_j$ except $m_i$. We claim that $\gcd(n_i, m_i) = 1$. Suppose $p$ is a prime dividing the gcd. Then $p \mid n_i$. Corollary 4.2 implies that $p \mid m_j$ for some $j \neq i$. Since $p \mid m_i$, too, we have $p \mid \gcd(m_j, m_i)$, which contradicts the assumption that $\gcd(m_j, m_i) = 1$. Therefore, $\gcd(n_i, m_i) = 1$. By Corollary 6.21, there exists $u_i$ such that $n_i u_i \equiv 1 \pmod{m_i}$. Let

$$x = a_1 n_1 u_1 + \cdots + a_r n_r u_r.$$

Since $m_i$ is a factor of $n_j$ for $j \neq i$, we find that all terms in the sum $a_1 n_1 u_1 + \cdots + a_r n_r u_r$ are 0 mod $m_i$ except for $a_i n_i u_i$. Therefore, $x \equiv a_i n_i u_i \equiv a_i \pmod{m_i}$. Since this is true for each $i$, the number $x$ is a solution.

The only thing left to prove is the uniqueness of $x$. Assume that there are an $x_1$ and an $x_2$ with

$$x_1 \equiv a_1 \pmod{m_1} \quad \text{and} \quad x_2 \equiv a_1 \pmod{m_1}$$
$$x_1 \equiv a_2 \pmod{m_2} \quad \text{and} \quad x_2 \equiv a_2 \pmod{m_2}$$
$$\vdots$$
$$x_1 \equiv a_r \pmod{m_r} \quad \text{and} \quad x_2 \equiv a_r \pmod{m_r}$$

with the $m_i$ pairwise relatively prime. Since

$$m_i \mid (x_1 - x_2) \text{ for } 1 \leq i \leq r,$$

the assumption that the $m_i$ are pairwise relatively prime implies that

$$m_1 m_2 \cdots m_r \mid x_1 - x_2$$

(see Exercise 91). So, $x_1 \equiv x_2 \pmod{m_1 m_2 \cdots m_r}$, which shows the uniqueness of our solution.                                                    $\square$

**Example.** Let's use the second proof to solve Sun Tzu's problem:

$$x \equiv 2 \pmod{3}, \quad x \equiv 3 \pmod{5}, \quad x \equiv 2 \pmod{7}.$$

We have $n_1 = 35, n_2 = 21, n_3 = 15$. Solve $35u_1 \equiv 1 \pmod{3}$ to get $u_1 = 2$. Solve $21u_2 \equiv 1 \pmod{5}$ to get $u_2 = 1$. Solve $15u_3 \equiv 1 \pmod{7}$ to get $u_3 = 1$. Therefore,

$$x = a_1 n_1 u_1 + a_2 u_2 u_2 + a_3 n_3 u_3$$
$$= (2)(35)(2) + (3)(21)(1) + (2)(15)(1) = 233 \equiv 23 \pmod{105}.$$

The Chinese Remainder Theorem gives us a new way of looking at congruences for composite moduli. If we factor the modulus $m$ into distinct prime powers, then either we can look at a congruence mod $m$, or we can look at a system of congruences mod these prime powers. Sometimes the latter are easier to work with. For example, suppose we want to solve

$$x^2 \equiv 1 \pmod{275}.$$

Since $275 = 5^2 \cdot 11$, this is the same as the system

$$x^2 \equiv 1 \pmod{25}. \quad x^2 \equiv 1 \pmod{11}.$$

The congruence $x^2 \equiv 1 \pmod{25}$ has the solutions $x \equiv 1$ or $24$ (mod 25). The congruence $x^2 \equiv 1 \pmod{11}$ has the solutions $x \equiv 1$ or $10 \pmod{11}$. There are four ways to combine these:

$$x \equiv 1 \pmod{25}, \quad x \equiv 1 \pmod{11} \Longrightarrow x \equiv 1 \pmod{275}$$
$$x \equiv 1 \pmod{25}, \quad x \equiv 10 \pmod{11} \Longrightarrow x \equiv 76 \pmod{275}$$
$$x \equiv 24 \pmod{25}, \quad x \equiv 1 \pmod{11} \Longrightarrow x \equiv 199 \pmod{275}$$
$$x \equiv 24 \pmod{25}, \quad x \equiv 10 \pmod{11} \Longrightarrow x \equiv 274 \pmod{275}.$$

Therefore, the solutions are 1, 76, 199, and 274 mod 275.

---

**CHECK YOUR UNDERSTANDING**

9. Solve $x \equiv 3 \pmod{11}$, $x \equiv 2 \pmod 7$.

---

# 6.6   Fractions mod $m$

Let's look at two calculations:

1. *Problem:* Solve $7x = 2$. *Solution:* Multiply by 1/7 to get $x = 2/7$.

2. *Problem:* Solve $7x \equiv 2 \pmod{19}$. *Solution:* Multiply by 11 to get $77x \equiv 22 \pmod{19}$, which becomes $x \equiv 22 \pmod{19}$ because $77 \equiv 1 \pmod{19}$.

In the second calculation, the role of 1/7 was played by 11, which is the multiplicative inverse 7 in the mod 19 world.

What is 1/7, anyway? It's a symbol with exactly one property: when you multiply it by 7 you get 1 as the answer. In the mod 19 world, if you multiply 11 by 7, you get 77, which is the same as 1, since $77 \equiv 1 \pmod{19}$. Therefore, if you had grown up in the mod 19 world, you would have learned that you could write 11 in place of 1/7. This might seem strange, but you already know a similar situation. You probably grew up in a decimal world, so you don't get upset when someone writes $1/7 = 0.142857\cdots$, even though the second expression does not look much like dividing by 7. In different worlds, $1/7$ can be expressed in different ways.

In many situations in number theory, it is useful to be able to work with fractions mod $m$ during intermediate steps in calculations. In this section, we show how this can be done.

Here are two warnings:

1. When working mod $m$, work only with fractions $a/b$ with $\gcd(b, m) = 1$. This is because such $b$ are the only ones that have multiplicative inverses mod $m$.

2. At the end of the calculation, change the result back to an integer. This is particularly important when the numbers involved are exponents, since fractional exponents cause problems and you should never have them when you are working mod $m$.

The fraction $a/b$ is the same as $a(1/b)$. When a fraction $a/b$ occurs mod $m$, interpret it as $a \cdot b^{-1}$, where $b \cdot b^{-1} \equiv 1 \pmod{m}$; that is, $b^{-1}$ is the multiplicative inverse of $b$ mod $m$. This is why we need $\gcd(b, m) = 1$.

**Example.** Since $1/7 \equiv 11 \pmod{19}$, we see that $2/7 \pmod{19}$ is the same as $2 \cdot 11 \pmod{19}$, which is the same as $3 \pmod{19}$.

Here is another example. Suppose we want to compute the sum $x_1 + x_2 + x_3 \pmod{13}$, and we know that $2x_1 \equiv 1 \pmod{13}$, $3x_2 \equiv 1 \pmod{13}$, and $6x_3 \equiv 1 \pmod{13}$. In terms of fractions, we have

$$x_1 \equiv 1/2, \quad x_2 \equiv 1/3, \quad x_3 \equiv 1/6 \pmod{13}.$$

Since

$$\frac{1}{2} + \frac{1}{3} + \frac{1}{6} = 1,$$

$x_1 + x_2 + x_3 \equiv 1 \pmod{13}$. To check this, we compute $x_1 \equiv 7$, $x_2 \equiv 9$, and $x_3 \equiv 11$. Therefore,

$$x_1 + x_2 + x_3 \equiv 7 + 9 + 11 \equiv 1 \pmod{13}.$$

Fractions are convenient when writing certain formulas. For example, the quadratic formula mod an odd prime $p$ (see Section 13.3) says that the solutions to $x^2 + bx + c \equiv 0 \pmod{p}$ are given by

$$x \equiv \frac{-b \pm \sqrt{b^2 - 4c}}{2} \pmod{p}.$$

Of course, we could write this as $(-b \pm \sqrt{b^2 - 4c})(2^{-1})$, where $2^{-1}$ is the inverse of 2 mod $p$, but many people find it more convenient to write the formula as a fraction.

As another example, in intermediate steps of some calculations, we start with congruences like $2^x \equiv 2^y \cdot b \pmod{m}$ and then write $2^{x-y} \equiv b \pmod{m}$. We don't want to worry about whether $x - y$ is positive or negative. If it's negative, $2^{x-y}$ is a fraction, which means that $2^{x-y}$ is interpreted as the multiplicative inverse of $2^{-(x-y)}$.

---

## 6.7   Queens on a Chessboard

The *eight queens problem* is to place eight queens on a chessboard so that no queen can attack another. This means that there is exactly one queen in each row and exactly one in each column, and no two can lie on the same diagonal (that is, any line with slope $\pm 1$). One solution is the following:

More generally, the $n$ queens problem is to place $n$ queens on an $n \times n$ chessboard with the same requirements. It has been proved that there is a solution for every positive integer $n$ except for $n = 2$ and $n = 3$. In the following, we give a simple solution for all $n$ with $\gcd(n, 6) = 1$.

Label the rows and columns from $0$ to $n - 1$, so $(x, y)$ denotes the square in column $x$ and row $y$. Regard the coordinates as numbers mod $n$. Diagonal lines are given by equations $x + y = c$ or $x - y = c$, where $c$ is a constant.

There are $n$ queens, which we'll number from $0$ to $n - 1$. Place the $j$th queen in square $(j, 2j)$, where $2j$ means $2j \pmod{n}$. Since the $x$-coordinates take on all $n$ values, there is one queen in each column. If queen $i$ and queen $j$ are in the same row, their $y$-coordinates are the same, so $2i \equiv 2j \pmod{n}$. Since $\gcd(n, 6) = 1$, we have $\gcd(2, n) = 1$, so we can divide by 2 and obtain $i \equiv j \pmod{n}$. Therefore, there is only one queen in each row.

Now suppose that the $i$th queen and the $j$th queen lie on the diagonal $x + y = c$. Since their coordinates are $(i, 2i)$ and $(j, 2j)$ mod $n$, we have

$$i + 2i \equiv c \pmod{n} \quad \text{and} \quad j + 2j \equiv c \pmod{n}.$$

Therefore, $3i \equiv c \equiv 3j \pmod{n}$. Since $\gcd(3, n) = 1$, we can divide by 3 and obtain $i \equiv j \pmod{n}$. This means that $i = j$, so at most one queen can lie on this diagonal.

Similarly, suppose that the $i$th queen and the $j$th queen lie on the diagonal $x - y = c$. Then

$$i - 2i \equiv c \pmod{n} \quad \text{and} \quad j - 2j \equiv c \pmod{n}.$$

Therefore, $-i \equiv c \equiv -j \pmod{n}$. This means that $i = j$, so at most one queen can lie on this diagonal.

Therefore, this placement of the queens solves the problem. Here is an example for $n = 5$. We take the lower left corner as the square $(0, 0)$.

|   |   | q |   |   |
|---|---|---|---|---|
|   |   |   | q |   |
|   | q |   |   |   |
|   |   |   | q |   |
| q |   |   |   |   |

It is interesting to observe that our solution works "on a torus." Imagine that the chess board is a torus (that is, the surface of a donut), so the top edge is attached to the bottom edge and the left edge is attached to the right edge. Then the diagonals wrap around. For example, on the $5 \times 5$ board, the diagonal that starts at square $(0, 1)$ and continues through $(1, 2)$, $(2, 3)$, and $(3, 4)$ is allowed to exit at the top edge and then re-enter at the bottom at $(4, 0)$ and continue back to $(0, 1)$. In this situation, the diagonal lines are given by congruence $x + y \equiv c \pmod{n}$ and $x - y \equiv c \pmod{n}$ rather than equalities. Our solution shows that the queens cannot attack one another, even if they are allowed to move along these extended diagonals.

In contrast, in the solution for the $8 \times 8$ board at the beginning of this section, the queen in the third row (counting from the top) can attack the queen in the eighth row along an extended diagonal in the northeast direction.

---

## 6.8    Chapter Highlights

1. If $\gcd(a, m) = 1$, then $a$ has a multiplicative inverse mod $m$ and $ax \equiv b \pmod{m}$ has a unique solution mod $m$.

2. The Chinese Remainder Theorem.

---

## 6.9    Problems

### 6.9.1    Exercises

**Section 6.1: Definitions and Examples**

1. Find the least non-negative residue mod 14 for each of the following:
   (a) 27    (b) 16    (c) $-20$    (d) 311    (e) $-91$    (f) 42

2. Find the least non-negative residue mod 8 for each of the following:
   (a) 11    (b) $-5$    (c) 121    (d) 2014    (e) $-83$    (f) 57

3. Verify that each of the following congruences is true:
   (a) $96 \equiv 6 \pmod{10}$    (b) $101 \equiv -9 \pmod{10}$
   (c) $-5 \equiv 13 \pmod{9}$

4.  Find the least non-negative residue mod 17 for each of the following:
    (a) 30    (b) 41    (c) 130    (d) −43    (e) −70

5.  Find the least non-negative residue mod 6 for each of the following:
    (a) 25    (b) 91    (c) 200    (d) −25    (e) −111

6.  Verify that each of the following congruences is true:
    (a) $77 \equiv 12 \pmod 5$    (b) $136 \equiv 31 \pmod 5$
    (c) $-11 \equiv -60 \pmod 7$

7.  Verify that each of the following congruences is true:
    (a) $95 \equiv 5 \pmod{10}$    (b) $213 \equiv -7 \pmod{10}$
    (c) $4 \equiv -20 \pmod 6$

8.  Verify that each of the following congruences is true:
    (a) $93 \equiv 13 \pmod{20}$    (b) $106 \equiv 2 \pmod 4$
    (c) $-8 \equiv -64 \pmod 7$

9.  Find all positive integers $n$ for which the congruence
    $123 \equiv 234 \pmod n$ holds.

10. Find all positive integers $n$ for which the congruence
    $1855 \equiv 1777 \pmod n$ holds.

11. Find all positive integers $n$ for which the congruence
    $11 \equiv 4 \pmod n$ holds.

12. Find all positive integers $n$ for which the congruence
    $38 \equiv -2 \pmod n$ holds.

13. A bug starts at the 12 on a (non-digital) 12-hour clock. It jumps to the 1, then to the 2, then to the 3, etc. After 12345 jumps, what number is it on? Explain your answer in terms of a congruence.

14. Suppose today is Monday. What day of the week will it be in 150 days? Explain your answer in terms of a congruence.

15. Let $n$ be an integer.
    (a) Use congruences to show that $n(n + 1)$ is always even.
    (b) Use congruences to show that $n(n + 1)(n + 2)$ is always a multiple of 6.
    (c) Use the fact that binomial coefficients are integers to prove (a) and to prove (b). (See Appendix A for binomial coefficients.)

16. Show that if $n \equiv 2 \pmod 4$, then $n$ cannot be written as $a^m$ with $m > 1$.

17. Show that if you sum the cubes of three consecutive integers, the result is a number that's congruent to 0 (mod 9).

18. Prove that the square of any integer must be a number ending in 0, 1, 4, 5, 6, or 9.

19. If $n$ is an odd integer, what are the possibilities for the last digit of $n^4$?

20. If $n$ is an integer, what possibilities are there for $n^3 \pmod 9$?

21. Let $a, b, d, k, m$ be positive integers. Decide whether each of the following is true or false. If true, give a proof; if false, give a counterexample.
    (a) If $d \mid m$ and $a \equiv b \pmod m$, then $a \equiv b \pmod d$.
    (b) If $d \mid m$ and $a \equiv b \pmod d$, then $a \equiv b \pmod m$.
    (c) If $a = b$, then $a \equiv b \pmod n$ for every positive integer $n$.
    (d) If $a \equiv b \pmod n$ for every positive integer $n$, then $a = b$.
    (e) If $\gcd(k, m) = 1$ and $ka \equiv kb \pmod m$, then $a \equiv b \pmod m$.

22.  If $(a, b, c)$ is a Pythagorean triple, prove that
     (a) at least one of $a, b$, or $c$ is divisible by 3.
     (b) at least one of $a, b$, or $c$ is divisible by 5.
     (c) we always have $abc$ divisible by 60.
     (*Hint:* Squares are congruent to $0, 1$ (mod 3) and to $0, 1, 4$ (mod 5).)

23.  Show that if $n \equiv 2$ (mod 3), then $n$ cannot be written in the form $x^2 + 3y^2$ (originally due to Fermat).

24.  Show that the fourth power of any odd integer is of the form $16k + 1$.

25.  Suppose that $p$ and $q$ are twin primes, so that $p$ and $q$ are primes whose difference is 2. (Examples are 3 and 5, 5 and 7, 11 and 13, ....)
     (a) Show that if $p, q \geq 5$ then $p + q$ is divisible by 3.
     (b) Show that $p + q$ is a multiple of 4.
     (c) Use (a) and (b) to show that $p + q$ is divisible by 12.

26.  (a) Show that a square is congruent to 0, 1, or 4 mod 8.
     (b) Show that a sum of two squares is not 3 mod 4.
     (c) Show that a sum of three squares is not 7 mod 8.

27.  Use induction to show that if $n \geq 0$ then $5^n \equiv 1 + 4n$ (mod 16).

28.  Use induction to show that if $n \geq 0$ then $7^n \equiv 1 + 6n$ (mod 36).

29.  (a) Show that if $x$ is an integer then $x^3 \equiv 0, \pm 1$ (mod 9).
     (b) Show that $x^3 + y^3 \equiv 3$ (mod 9) has no solutions.

30.  Find all prime numbers $p$ such that $p^2 + 2$ is also prime.

31.  Show that if $9 \mid (a^3 + b^3 + c^3)$ then 3 divides at least one of the integers $a, b$, or $c$.

32.  (a) Show that if $1 \leq x \leq y \leq z$ and $x! + y! = z!$, then $x = y = 1$ and $z = 2$. (*Hint:* Use congruences mod $y!$.)
     (b) Show that $x! \, y! = z!$ has infinitely many solutions. (*Hint:* For example, $5! \, 119! = 120!$.)

33.  Alice and Bob have a pile of $n$ markers. They fix an integer $m$ and play the following game. A move consists of removing at least one and at most $m$ markers from the pile. They take turns making moves. Alice plays first.
     (a) Suppose the rules state that the person who removes the last marker wins. Show that Alice has a winning strategy if and only if $n \not\equiv 0$ (mod $m + 1$).
     (b) Suppose the rules state that the person who removes the last marker loses. Show that Alice has a winning strategy if and only if $n \not\equiv 1$ (mod $m + 1$).

34.  Let $p$ be an odd prime.
     (a) Let $a \not\equiv 0$ (mod $p$). Show that $x^2 \equiv a^2$ (mod $p$) has exactly two solutions.
     (b) Show that there are exactly $(p - 1)/2$ nonzero squares mod $p$.

35.  Let $F_n$ be the $n$th Fibonacci number (see the Appendix). Show that if $n \geq 1$, then $F_n \equiv F_{n+2} \equiv F_{n+3}$ (mod $F_{n+1}$).

36.  Let $F_n$ be the $n$th Fibonacci number (see the Appendix).
     (a) Use strong induction to show that if $n \geq 1$, then $F_{n+3} \equiv F_n$ (mod 2).
     (b) Show that if $n \geq 1$, then $F_{n+6} \equiv 5F_n$ (mod 8).
     (c) Show that if $n \geq 1$, then $F_{n+12} \equiv F_n$ (mod 8).

37.  Show that there are infinitely many primes $p \equiv 3 \pmod{4}$. (*Hint:* Let $N = 4p_1 p_2 \cdots p_r - 1$ and show that $N$ must have a prime factor $p \equiv 3 \pmod{4}$.)

38.  Let $m \geq 3$.
     (a) Let $n$ be a positive integer. Show that if $n \not\equiv 1 \pmod{m}$, then $n$ has a prime factor $p \not\equiv 1 \pmod{m}$.
     (b) Show that there are infinitely many primes $p \not\equiv 1 \pmod{m}$. (*Hint:* Let $N = mp_1 p_2 \cdots p_r - 1$.)

## Section 6.2: Modular Exponentiation

39.  Use Modular Exponentiation to calculate (a) $2^{17} \pmod{13}$, (b) $5^{25} \pmod{11}$, (c) $2^{16} \pmod{10}$.

40.  Use Modular Exponentiation to calculate (a) $3^{40} \pmod{100}$, (b) $7^9 \pmod{11}$, (c) $6^{10} \pmod{17}$.

41.  Use Modular Exponentiation to calculate
     (a) $2^{20} \pmod{97}$, (b) $3^{20} \pmod{97}$, (c) $5^{20} \pmod{97}$, (d) $4^{20} \pmod{97}$

42.  Use Modular Exponentiation to calculate
     (a) $2^{80} \pmod{255}$    (b) $2^{70} \pmod{257}$ (*Hint:* You do not need to square 256. Instead, use the fact that $256 \equiv -1 \pmod{257}$.)

43.  Use Modular Exponentiation to calculate and then make a conjecture:
     (a) $2^{96} \pmod{97}$, (b) $3^{96} \pmod{97}$, (c) $5^{96} \pmod{97}$, (d) $4^{96} \pmod{97}$

44.  Use Modular Exponentiation to calculate and then make a conjecture:
     (a) $2^{36} \pmod{37}$, (b) $3^{36} \pmod{37}$, (c) $5^{36} \pmod{37}$, (d) $4^{36} \pmod{37}$

## Section 6.3: Divisibility Tests

45.  Find the least non-negative residue mod 9 for each of the following:
     (a) 1453   (b) 1927   (c) 1066   (d) 1855   (e) $-4004$   (f) $-753$

46.  Find the least non-negative residue mod 11 for each of the following:
     (a) 1777   (b) $-43$   (c) 275   (d) 1234567   (e) $-83$   (f) 235711

47.  You calculated

     $$123456789 \times 987654321 = 121932a31112635269,$$

     but your handwriting was so bad that you cannot figure out what digit $a$ represents. Use congruences mod 9 to get the answer.

48.  You receive a message from an extraterrestrial alien, who is calculating $4343434343^2$. The answer is $18865ab151841649$, where the two digits represented as $a$ and $b$ were lost in transmission. Use congruences mod 9 and mod 11 to determine the answer to this fundamental problem.

49.  Suppose you know that $12^{10} = 61917ab4224$, where $a$ and $b$ are digits. Use congruences mod 9 and mod 11 to determine $a$ and $b$.

50.  Divide $123456789^{1223456789}$ by 9. What is the remainder?

51.  Suppose $123456789^2 = 15241578750a90521$, where $a$ represents a digit. Find $a$. (*Hint:* $1+5+2+4+1+5+7+8+7+5+0+9+0+5+2+1 = 62$.)

52.  Let $n$ be a positive integer. Write the digits of $n$ in decreasing order, then write the digits in increasing order, then subtract. Prove that the answer is always divisible by 9. (*Example:* $n = 3479217$. Then $9774321 - 1234779 = 8539542 \equiv 0 \pmod 9$))

53.  Suppose $N = abc \cdots xyz$ is a 26-digit number, and let $R = zyx \cdots cba$ be the number obtained by reversing the order of its digits. Show that $N + R$ is a multiple of 11.

54.  You need to be reimbursed for the 36 light bulbs you bought. Unfortunately, the receipt is smudged and you can only read \$$x82.5y$ where the spacing tells you that $x$ and $y$ are digits. You are sure that the bulbs were less than \$15 each. How much did they cost?

55.  You know that the number $25x54y$ is divisible by 72. Find $x$ and $y$.

## Section 6.4: Linear Congruences

56.  Find all solutions to each of the following:
     (a) $3x \equiv 8 \pmod{11}$    (b) $6x \equiv 7 \pmod 9$
     (c) $4x \equiv 12 \pmod{32}$

57.  Find all solutions to each of the following:
     (a) $5x \equiv 19 \pmod{21}$    (b) $8x \equiv 12 \pmod{19}$
     (c) $91x \equiv 3 \pmod{121}$

58.  Find all solutions to each of the following:
     (a) $4x \equiv 3 \pmod{13}$    (b) $2x \equiv 3 \pmod 5$
     (c) $6x \equiv 18 \pmod{42}$

59.  Find all solutions to each of the following:
     (a) $3x \equiv 11 \pmod{25}$    (b) $4x \equiv 10 \pmod{17}$
     (c) $15x \equiv 10 \pmod{169}$

60.  Find all values of $n$ with $0 \leq n \leq 23$ such that the congruence $10x \equiv n \pmod{24}$ has a solution.

61.  Find all values of $n$ with $0 \leq n \leq 29$ such that the congruence $25x \equiv n \pmod{30}$ has a solution.

62.  Find all values of $n$ with $0 \leq n \leq 17$ such that the congruence $3x \equiv n \pmod{18}$ has a solution.

63.  Find all values of $n$ with $0 \leq n \leq 35$ such that the congruence $24x \equiv n \pmod{36}$ has a solution.

64.  (a) Solve the congruence $3x \equiv 1 \pmod{17}$.
     (b) Use (a) to solve $3x \equiv 5 \pmod{17}$.

65.  (a) Solve $83x \equiv 1 \pmod{100}$.
     (b) Solve $83x \equiv 2 \pmod{100}$.

66.  (a) Solve the congruence $6x \equiv 1 \pmod{19}$.
     (b) Use (a) to solve $6x \equiv 8 \pmod{19}$.

67.  (a) Solve $47x \equiv 1 \pmod{120}$.
     (b) Use (a) to solve $47x \equiv 2 \pmod{120}$.

68.  Show that if $7 \mid a^2 + b^2$, then 7 divides both $a$ and $b$. (*Hint:* If $a \not\equiv 0 \pmod 7$, you can divide by $a$ mod 7.)

69.  Show that $x^2 - 2y^2 = 10$ has no integer solutions. (*Hint:* Work mod 5.)

70. Find all solutions to the pair of congruences $5x - 3y \equiv 4 \pmod{13}$ and $3x - 5y \equiv 5 \pmod{13}$. (*Hint:* Modify the standard way to solve two linear equations in two unknowns.)

71. Find all solutions to the pair of congruences $7x - 11y \equiv 1 \pmod{17}$ and $11x - 5y \equiv 2 \pmod{17}$. (*Hint:* Modify the standard way to solve two linear equations in two unknowns.)

## Section 6.5: The Chinese Remainder Theorem

72. Solve the system of congruences

$$x \equiv 2 \pmod{3}, \quad x \equiv 1 \pmod{7}.$$

73. Solve the system of congruences

$$x \equiv 1 \pmod{3}, \quad x \equiv 2 \pmod{7}, \quad x \equiv 4 \pmod{11}.$$

74. Solve the system of congruences

$$3x \equiv 2 \pmod{5}, \quad 4x \equiv 3 \pmod{7}, \quad x \equiv 2 \pmod{11}.$$

75. (Classical problem: Bhāskara, al-Haitham, Fibonacci) When the eggs in a basket are taken out 2 at a time, there is one egg left in the basket. When they are taken out 3, 4, 5, and 6 at a time there are 2, 3, 4, and 5 eggs, respectively, left in the basket. If 7 at a time are taken, then none are left. What is the least number of eggs in the basket?

76. Find all integers between 100 and 300 that leave a remainder of 2 when divided by 3 and a remainder of 4 when divided by 7.

77. Find all integers between 200 and 400 that leave a remainder of 1 when divided by 3 and a remainder of 4 when divided by 5.

78. Find all integers between 1000 and 2000 that leave a remainder of 2 when divided by 5, a remainder of 4 when divided by 7 and a remainder of 1 when divided by 11.

79. Find the smallest positive integer that leaves a remainder of 9 when divided by 10, 8 when divided by 9, 7 when divided by 8, and 6 when divided by 7. (*Hint:* Use the fact that $x = -1$ is a solution and then find a positive solution without using the Chinese Remainder Theorem.)

80. Find the smallest positive integer that leaves a remainder of 8 when divided by 10, 7 when divided by 9, 6 when divided by 8, and 5 when divided by 7. (*Hint:* Use the fact that $x = -2$ is a solution and then find a positive solution without using the Chinese Remainder Theorem.)

81. A professor is trying to assign grades randomly. If he gives an equal number of A's, B's, and C's, then there is one student left over. If he assigns an equal number of A's, B's, C's, and D's, then there are 3 students left over. If he assigns an equal number of A's, B's, C's, D's, and F's, then there are 3 students left over. What is the smallest possible number of students in the class, and what is the second smallest possible number of students in the class?

82.  The people in a town are lining up for a parade. When they line up 3 to a row, 1 person is left over. When they line up 5 to a row, 2 people are left over. When they line up 13 to a row, 1 is left over. What is the smallest possible population of the town?

83.  (a) Find all $x$ with $2000 < x < 2100$ such that $x \equiv 9 \pmod{13}$ and $x \equiv 2 \pmod{17}$.
(b) A type of cicada appeared in 2011 and appears every 13 years. Another type of cicada appeared in 2008 and appears every 17 years. In what year between 2000 and 2100 will they both appear at the same time?

84.  Suppose that you want to calculate

$$1434661 \cdot 3785648743 - 100020304 \cdot 54300201.$$

You are told that the answer is a positive integer less than 90. Compute the answer mod 10 and mod 9, then combine to get the answer. *Note:* This technique has often been employed to do calculations with large integers because most of the computations use small integers, with only the Chinese Remainder Theorem step requiring large integers. For example, if the answer is known to be less than $10^{36}$, the calculations can be done mod $p$ for each prime less than 100. Since the product of the primes less than 100 is approximately $2.3 \times 10^{36}$, the Chinese Remainder Theorem gives the answer exactly.

85.  Show that there are 100 consecutive integers that are not squarefree. (*Hint:* Take 100 different primes $p$ and arrange for $p^2$ to divide an appropriate member of the sequence.)

86.  If we have the system of congruences

$$x \equiv a \pmod{m} \text{ and } x \equiv b \pmod{n},$$

we know that the Chinese Remainder Theorem guarantees a solution if $\gcd(m, n) = 1$. But what happens otherwise? Show that this system has a solution if and only if $\gcd(m, n) \mid (a - b)$. (*Hint:* If $\gcd(m, n) \mid (a - b)$, express $a - b$ as a linear combination of $m$ and $n$. Use this to explicitly find a solution to both congruences. If there is a solution to both congruences, show how to express $a - b$ as a linear combination of $m$ and $n$.)

87.  Let $m$ and $n$ be positive integers such that $m \mid n$. Let $\gcd(a, m) = 1$. The Chinese Remainder Theorem says that there exists $b$ with $b \equiv a \pmod{m}$ and $b \equiv 1 \pmod{p}$ for every prime $p$ that divides $n$ but does not divide $m$. Show that $\gcd(b, n) = 1$.
This shows that every relatively prime congruence class mod $m$ can be obtained by starting with a relatively prime congruence class mod $n$ and reducing it mod $m$.

88.  Find an integer $n$ that gives a remainder of 1 when divided by 2, 3, 4, 5, and 6, a remainder of 2 when divided by 7, a remainder of 1 when divided by 8, a remainder of 4 when divided by 9, a remainder of 1 when divided by 10 and a remainder of 0 when divided by 11. (*Hint:* Do not use the Chinese Remainder Theorem. Instead, write $n = 11r$ and determine the smallest possible divisor of $r$.)

89.   Find a positive integer $x$ such that $x \equiv 1$ (mod 2), $x \equiv 2$ (mod 3), $x \equiv 3$ (mod 4), and $x \equiv 4$ (mod 5). (*Hint:* First find a negative solution.)

90.   (a) Find a solution to $x \equiv 3$ (mod 13), $x \equiv 0$ (mod 11)
      (b) Find a solution to $y \equiv 0$ (mod 13), $y \equiv 2$ (mod 11).
      (c) Find a positive integer $n$ such that $n/8$ is the 13th power of an integer and $n/9$ is the 11th power of an integer.

91.   Let $m_1, \ldots, m_r$ be pairwise relatively prime integers.
      (a) Let $M_i = m_1 m_2 \cdots m_i$. Show that $\gcd(M_i, m_{i+1}) = 1$ for $1 \leq i \leq r - 1$. (*Hint:* A similar proof is in the second proof of Theorem 6.23.)
      (b) Suppose $n$ is an integer and $m_i \mid n$ for $1 \leq i \leq n$. Show that $M_r \mid n$. (*Hint:* Use Proposition 2.17, induction, and part (a).)

## 6.9.2   Projects

1.   In this chapter, we developed divisibility tests for 2, 3, 4, 5, 8, 9, 10, and 11. All of these tests relied on the base 10 expansion of an integer. In this project, we will prove divisibility tests for bases other than 10.
     (a) Prove that a number is divisible by 2 if and only if its last digit in base 2 is 0.
     (b) Prove that a number is divisible by 2 if and only if the sum of its digits in base 5 is a multiple of 2.
     (c) Prove that a number is divisible by 2 if and only if its last digit in base 8 is 0, 2, 4, or 6.
     (d) Find a test for divisibility by 4 in bases 2, 5, 8, and 12.
     (e) Devise divisibility tests for 2, 3, 4, 5, 7, 13, 14, 15 for base 14.

2.   One of the most useful bases is base 16, which is also called hexadecimal or hex for short. It is used extensively in computer applications. In hex, we have the usual digits $1, 2, 3, \ldots, 9$. Then $A, B, C, D, E$, and $F$ are used to represent 10, 11, 12, 13, 14, and 15. So, $2BD_{16} = 2 \cdot 16^2 + 11 \cdot 16 + 13 = 701_{10}$. Prove the following divisibility tests for base 16:
     (a) $n$ is divisible by 2 if its last digit is 0, 2, 4, 6, 8, A, C, E.
     (b) $n$ is divisible by 3 if the sum of its digits is divisible by 3.
     (c) $n$ is divisible by 4 if its last digit is 0, 4, 8, or C.
     (d) $n$ is divisible by 5 if its digit sum is divisible by 5.
     (e) $n$ is divisible by 8 if its last digit is 0 or 8.
     (f) $n$ is divisible by $A$ if it is divisible by 2 and 5.
     (g) $n$ is divisible by $F$ if the sum of its digits is divisible by $F$.
     (h) In part (e) above, is a hex number always divisible by 5 if its last digit is 0 or 5?

3.   We know that $12^2 = 144$ ends with a repeated last digit. Are there other examples? Can there be more repeated digits? The following answers this.
     (a) Let $x$ be an integer. Show that $x^2 \equiv (x + 50)^2$ (mod 100).
     (b) Let $x$ be an integer. Show that $x^2 \equiv (50 - x)^2$ (mod 100).
     (c) Use (a) and (b) to deduce that $0^2, 1^2, 2^2, 3^2, \ldots, 25^2$ (mod 100) yield all possible squares mod 100. By calculating these squares, show that "44" is the only repeated non-zero digit that a square can end in, and that if $x^2 \equiv 44$ (mod 100) then $x \equiv \pm 12$ (mod 50).

(d) Now we look at the last 3 digits. From (c), we know that the only possible repeated last digits are "444." Write $x = \pm(12 + 50k)$, where $k$ is an integer (possibly negative). Show that if $x^2 \equiv 444 \pmod{1000}$, then $1 + 2k + 5k^2 \equiv 4 \pmod{10}$.

(e) Show that $k \equiv -1 \pmod{10}$, so $x \equiv \pm 38 \pmod{500}$. In fact, $38^2 = 1444$ ends in three 4s.

(f) Let's try for four 4's. Write $x = \pm(38 + 500j)$. Show that if $x^2 \equiv 4444 \pmod{10000}$, then $8j \equiv 3 \pmod{10}$, and then show that this is impossible.

Conclude that it is impossible to have a square end in a nonzero digit that is repeated four times. The largest possible repetition is three 4's, which is achieved by $38^2 = 1444$.

4.  Let $y$, $x$, $n$ be positive integers. Consider the following algorithm. There are three quantities, $a, b, c$. Start with $a = x, b = 1, c = y$.
    1. If $a$ is even, change $a$ to $a/2$, leave $b$ unchanged, and change $c$ to $c^2$ $\pmod n$.
    2. If $a$ is odd, change $a$ to $a - 1$, change $b$ to $bc \pmod n$, and leave $c$ unchanged.
    3. If $a \neq 0$, go to Step 1; if $a = 0$, output $b$.
    (a) Show that $b \cdot c^a \pmod n$ does not change in either Step 1 or Step 2.
    (b) Show that if $a = 0$, then $b \equiv y^x \pmod n$. Therefore, this algorithm computes $y^x \pmod n$.
    (c) Let $y = 3, x = 13, n = 19$. Show that this algorithm computes $3^{13}$ $\pmod{19}$.
    (d) In (c), the binary expansion of 13 is 1101. Show that the values of $b$ through the algorithm are $1, 3^1, 3^{1+4}, 3^{1+4+8}$. Explain how, in general, the values of $b$ are related to the binary expansion of $x$.

# 6.9.3   Computer Explorations

1.  Find all values of $n$ with $2 \leq n \leq 100$ for which $1^3 + 2^3 + 3^3 + \cdots + (n-1)^3 \equiv 0 \pmod n$. If this congruence is true, what congruence conditions does $n$ seem to satisfy? Can you prove this?

2.  For which primes $p$ can you solve the congruence $x^2 \equiv -1 \pmod p$? Compute several examples and see if you can find the general rule.

3.  For which primes $p$ can you solve the congruence $x^2 \equiv 5 \pmod p$? Compute several examples and see if you can find the general rule.

4.  For which primes $p$ can you solve the congruence $x^2 \equiv 2 \pmod p$? Compute several examples and see if you can find the general rule.

5.  Let $a > b$ be fixed integers. We say that a prime $p$ is a primitive divisor of $a^n - b^n$ if $p \mid a^n - b^n$, but $p \nmid a^k - b^k$ for $0 \leq k < n$.
    (a) Compute the primitive divisors of $2^n - 1$ for several values of $n \geq 2$. There is one $n$ for which there are no primitive divisors. Did you find it?
    (b) Compute the primitive divisors of $3^n - 1$ for several values of $n \geq 1$. Are there any values of $n$ for which there are no primitive divisors?
    (c) Compute the primitive divisors of $3^n - 2^n$ for several values of $n \geq 2$.

Are there any values of $n$ for which there are no primitive divisors?
*Remark:* Zsigmondy's Theorem says that $a^n - b^n$ has a primitive divisor
except in the case found in (a) and the case where $n = 1$ and $a - b = 1$.

6. (a) Solve the simultaneous congruences

$$n \equiv 0 \pmod{4}, \quad n \equiv -1 \pmod{9}, \quad n \equiv -2 \pmod{25},$$
$$n \equiv -3 \pmod{49}, \quad n \equiv -5 \pmod{121}$$

to find six consecutive integers that are not squarefree.
(b) Find the smallest positive integer $n$ such that $n, n+1, n+2, n+3, n+4, n+5$ are not squarefree. The answer should be much smaller than the number from part (a).

# 6.9.4  Answers to "CHECK YOUR UNDERSTANDING"

1. (a) True: $17 - 29 = -12$, which is a multiple of 4.
   (b) False: $321 - 123 = 198$, which is not a multiple of 10.
   (c) True: $321 - 123 = 198 = 9 \cdot 22$, which is a multiple of 9.

2. Since $3 \equiv 14 \pmod{11}$, we can rewrite the congruence as $2x \equiv 14 \pmod{11}$. We can divide by 2 because $\gcd(2, 11) = 1$, so $x \equiv 7 \pmod{11}$.

3. We have $\gcd(4, 22) = 2$. Divide the congruence by 2 to get $2x \equiv 3 \pmod{11}$. By the previous problem, $x \equiv 7 \pmod{11}$. Returning to the mod 22 world, we obtain $x \equiv 7, 18 \pmod{22}$.

4. Since 11 is prime, Proposition 6.10 says that either $x - 1 \equiv 0 \pmod{11}$ or $x - 2 \equiv 0 \pmod{11}$. Therefore, $x \equiv 1$ or $2 \pmod{11}$.

5. $987654321 \equiv 9+8+7+6+5+4+3+2+1 \equiv 45 \equiv 4+5 \equiv 0 \pmod{9}$.

6. $343434 \equiv 4-3+4-3+4-3 \equiv 3 \pmod{11}$ and $58 \equiv 8-5 \equiv 3 \pmod{11}$, so $343434 \equiv 58 \pmod{11}$.

7. Change the problem to solving $15x - 73y = 1$. Use the Extended Euclidean Algorithm:

   | 73 | 1 | 0 |
   |----|----|-----|
   | 15 | 0 | 1 |
   | 13 | 1 | -4 |
   | 2 | -1 | 5 |
   | 1 | 7 | -34. |

   We conclude that $1 = 73(7) - 15(34)$, so $15(-34) \equiv 1 \pmod{73}$. Therefore, the multiplicative inverse of 15 is $-34 \equiv 39 \pmod{73}$.

8. From the previous problem, we know that the multiplicative inverse of 15 (mod 73) is 39. Multiply both sides of $15x \equiv 11 \pmod{73}$ by 39 to get $x \equiv 11 \cdot 39 \equiv 429 \equiv 64 \pmod{73}$.

9. The numbers congruent to 3 mod 11 are 3, 14, 25, 36, 47, 58, .... These are 3, 0, 4, 1, 5, 2 mod 7. Therefore, $x \equiv 58 \pmod{77}$ is the answer.

# Chapter 7

# Classical Cryptosystems

## 7.1 Introduction

Bob wants to send a message to Alice so that Alice, *and only Alice*, can read it. For example, Alice may own a business and Bob may need to send her his credit card number so that he may make a purchase. This message is usually called the **plaintext**. To keep the message secret, he encrypts it to a secret, unreadable form called the **ciphertext**, which he sends to Alice. Alice then decrypts to obtain the original plaintext.

For a simple example, suppose Bob's original message is *ZEBRA*. He decides to shift each letter forward by four positions to obtain the ciphertext *DIFVE* (note that the *Z* wraps around to the beginning):

$$ZEBRA \longrightarrow DIFVE.$$

Alice receives the ciphertext *DIFVE* and shifts each letter back by four positions to obtain the plaintext, *ZEBRA*:

$$DIFVE \longrightarrow ZEBRA.$$

Of course, Alice needs to know how many positions Bob shifted each letter in order to decrypt. In this example, the number of positions is called the **key**, and it is shared by Alice and Bob. In general, the key is the secret information that is used to encrypt and decrypt the message. In the present case, there are 26 possible keys (this includes no shift, which is the same as sending an unencrypted message). Cryptosystems that are used in real-life situations are much more complicated and have hundreds of millions of possible keys. In fact, most of the systems used in practice today have more than $10^{30}$ possible keys.

In classical cryptography, Alice and Bob share the same key. This is

155

called a **symmetric cryptosystem**, since the sender and receiver use exactly the same information to encrypt and decrypt. The workhorses of modern cryptography, namely the Data Encryption Standard (DES) and its successor, the Advanced Encryption Standard (AES), are symmetric cryptosystems that are used to transmit massive amounts of data electronically. For these, as with the classical systems, the sender and receiver need to share a key in order to encrypt and decrypt. These systems start with the message written as a sequence of bits or bytes and then use various operations to mix up and modify the bits to obtain the ciphertext. We do not discuss DES and AES in this book, since they have no relation to number theory. If you're interested in seeing how they work, details of these algorithms are described in many books on cryptography.

In this chapter, we discuss some classical symmetric systems (shift and affine ciphers, Vigenère ciphers, Hill ciphers, linear feedback shift registers) that use number theory concepts that we have treated, along with a system (transposition ciphers) that is less number theoretic. (*Note:* You may have noticed that many of the cryptosystems are called "ciphers." This is a word that is often used for methods of encrypting messages.) Later, in Chapter 9, we introduce the RSA cryptosystem, which is used in electronic commerce. It is an asymmetric system, where the information used to encrypt is not the same as the information used to decrypt.

Not every cryptographic protocol involves sending secret messages. In various sections of this chapter and Chapters 9 and 12, we discuss ways to flip coins over the telephone, play mental poker, sign electronic documents, and share secrets. These all use number theory in essential ways.

## 7.2    Shift and Affine Ciphers

Suppose you have a message that you want to send to someone. That message may be dinner plans, what time your last class is, or your opinion of a song you've just heard. In these cases, you don't really care if anyone else reads your message. But if you want to send the password for your bank account to your spouse or battle plans during wartime to one of your officers in the field, then it becomes very important that the message be read only by the intended recipient. When you have information that you want to distribute selectively, you need to use some method to ensure that only the intended reader can understand it. This is where cryptography comes into play.

Let's say that your online bank account's password is your dog's name, Sparky. You want to e-mail this to your spouse, but are (justifiably) concerned that someone else may read it. (By the way, you should never use your pet's name as your password.) As discussed in the first section, *SPARKY* is the *plaintext*, or the unaltered message, you want to send. We need to encrypt, or disguise, this message by altering it. We'll show one way to do this using a very simple (and very insecure) method, replacing each plaintext letter by the letter immediately to its right. In this case, the key is +1. So, *S* becomes *T*, *P* becomes *Q*, *A* becomes *B*, etc., and *SPARKY* becomes *TQBSLZ*.

The cryptosystem discussed above is an example of a **shift cipher**. There's nothing special about shifting by one position to the right. We could have shifted 3 to the right (a key of +3) or 5 to the left (a key of −5), or by any predesignated number. In order to make this simpler to discuss, we're going to assign a number to each letter of the alphabet. From a mathematical perspective, it doesn't really matter how we do this. We're going to keep things simple and begin with $A = 0$.

| A | B | C | D | E | F | G | H | I | J | K | L | M |
|---|---|---|---|---|---|---|---|---|---|---|---|---|
| 0 | 1 | 2 | 3 | 4 | 5 | 6 | 7 | 8 | 9 | 10 | 11 | 12 |
| N | O | P | Q | R | S | T | U | V | W | X | Y | Z |
| 13 | 14 | 15 | 16 | 17 | 18 | 19 | 20 | 21 | 22 | 23 | 24 | 25 |

**Warning:** Starting with *A* as 0 is a standard convention. Since people think of *A* as being the first letter of the alphabet, they often think of *A* as being 1. Be careful not to make this mistake.

Using this, $S = 18$, $P = 15$, etc., so we write *SPARKY* as *18 15 00 17 10 24*. Now, we can think of our shift of 1 to the right, or more explicitly as our key being +1 in the following way:

| PLAINTEXT | S | P | A | R | K | Y |
|---|---|---|---|---|---|---|
| Numerical Translation | 18 | 15 | 0 | 17 | 10 | 24 |
| Right Shift by One | 19 | 16 | 1 | 18 | 11 | 25 |
| CIPHERTEXT | T | Q | B | S | L | Z |

Mathematically, we've shifted each letter by 1 unit under the rule

$$x \mapsto x + k \pmod{26},$$

where $x$ is the numerical equivalent of the letter and the key is $k = 1$.
If we had our key be $k = 7$, we would have

| PLAINTEXT | S | P | A | R | K | Y |
|---|---|---|---|---|---|---|
| Numerical Translation | 18 | 15 | 0 | 17 | 10 | 24 |
| Right Shift by Seven | 25 | 22 | 7 | 24 | 17 | 5 |
| CIPHERTEXT | Z | W | H | Y | R | F |

Once the intended recipient gets the ciphertext, it can be decrypted by subtracting the key.

The Roman historian Suetonius reported that Julius Caesar used a shift by 3 to encrypt, and this shift cipher is often called the **Caesar cipher**. His nephew, Augustus Caesar, used a shift by 1, but apparently didn't know about modular arithmetic, since he used $AA$ instead of $A$ when the shift went past the end of the alphabet.

As we said before, there are 26 possible keys for a shift cipher, making it very easy to break the system. A malicious individual who intercepts the ciphertext could decrypt by trying all 26 keys, one at a time, stopping when a readable message is found. Because of this, we tweak a shift cipher to get a similar cryptosystem, called an **affine cipher**, that has 312 possible keys, making it a little bit more secure. In an affine cipher, instead of simply performing a shift, we first multiply by a nonzero value that is relatively prime to 26 and then shift. Mathematically, an affine cipher works by having

$$x \mapsto ax + b \pmod{26} \tag{7.1}$$

where $\gcd(a, 26) = 1$. (It will soon be clear why this gcd requirement is necessary.) In this situation we will call the ordered pair $(a, b)$ the key for the affine cipher. Since there are twelve numbers (1, 3, 5, 7, 9,

11, 15, 17, 19, 21, 23, 25) relatively prime to 26, we get 12 choices for
$a$ and 26 choices for $b$, and therefore $12 \cdot 26 = 312$ possible keys.

Here's our $SPARKY$ ($= 18\ 15\ 00\ 17\ 10\ 24$) example with key $(a, b) = (3, 5)$:

| PLAINTEXT | S | P | A | R | K | Y |
|---|---|---|---|---|---|---|
| Numerical Translation | 18 | 15 | 0 | 17 | 10 | 24 |
| $3x + 5$ (mod 26) | 7 | 24 | 5 | 4 | 9 | 25 |
| CIPHERTEXT | H | Y | F | E | J | Z |

The ciphertext is $HYFEJZ$.

For example, we encrypt the letter $P$ by first finding its numerical value
of 15. We then take

$$3 \cdot 15 + 5 \equiv 50 \equiv 24 \pmod{26}.$$

Finally, since 24 corresponds to $Y$, we get that

$$P \mapsto Y$$

under the affine cipher whose key is (3, 5).

Decryption is easy. Let's say the ciphertext of $U$ is received and the
key is $(3, 5)$. Since the numerical value of $U$ is 20, we have to solve the
congruence

$$3x + 5 \equiv 20 \pmod{26},$$

so $3x \equiv 15 \pmod{26}$. Notice that because $\gcd(3, 26) = 1$, we can divide
by 3 and we then get $x = 5$. This means that the plaintext corresponding to $U$ is $F$, which is the letter corresponding to the number 5.

If we return to our example, let's say that we received the ciphertext
$HYFEJZ$. In order to decrypt this, we first change it into its numerical
equivalent: $H = 7$,  $Y = 24$,  $F = 5$,  $E = 4$,  $J = 9$, and $Z = 25$. We
could then use the method of Section 6.4 to solve the six congruences

$$
\begin{aligned}
3x + 5 &\equiv 7 \pmod{26} \\
3x + 5 &\equiv 24 \pmod{26} \\
3x + 5 &\equiv 5 \pmod{26} \\
3x + 5 &\equiv 4 \pmod{26} \\
3x + 5 &\equiv 9 \pmod{26} \\
3x + 5 &\equiv 25 \pmod{26}.
\end{aligned}
$$

However, there is an easier way. Write $y \equiv 3x + 5 \pmod{26}$. Multiply both sides by 9 to get

$$9y \equiv 27x + 45 \equiv x + 19 \pmod{26},$$

since $27 \equiv 1$ and $45 \equiv 19$. This can be rearranged to

$$x \equiv 9y - 19 \equiv 9y + 7 \pmod{26}.$$

Therefore, $9y + 7$ is our decryption function. Note that multiplying by 9 is the same as dividing by 3 in the "mod 26 world," since 9 and 3 are multiplicative inverses mod 26. So our calculation is really the same as changing $y = 3x + 5$ into $x = (1/3)y - (5/3)$ and then converting the fractions to integers mod 26.

The decryption is now easy:

$$7 \mapsto 9 \cdot 7 + 7 \pmod{26} \equiv 18 = S$$
$$24 \mapsto 9 \cdot 24 + 7 \pmod{26} \equiv 15 = P$$
$$5 \mapsto 9 \cdot 5 + 7 \pmod{26} \equiv 0 = A$$
$$4 \mapsto 9 \cdot 4 + 7 \pmod{26} \equiv 17 = R$$
$$9 \mapsto 9 \cdot 9 + 7 \pmod{26} \equiv 10 = K$$
$$25 \mapsto 9 \cdot 25 + 7 \pmod{26} \equiv 24 = Y$$

so we recover our plaintext, *SPARKY.*

In general, in order to be able to decrypt a letter that was encrypted using an affine cipher with key $(a, b)$, we have to solve the linear congruence

$$y \equiv ax + b \pmod{26}$$

for $x$ in terms of $y$. This is equivalent to solving

$$ax \equiv y - b \pmod{26},$$

and in order to be able to solve this uniquely we need to have $\gcd(a, 26) = 1$ so we can divide by $a$.

Here's what can go wrong if we choose an $a$ with $\gcd(a, 26) > 1$. Let's say our key is $(13, 5)$ so that we encrypt by having

$$x \mapsto 13x + 5 \pmod{26}.$$

Then the ciphertext corresponding to the plaintext *LOVE* is

$$
\begin{array}{ccccccccc}
L & = & 11 & \mapsto & 13 \cdot 11 + 5 & (\mathrm{mod}\ 26) \equiv & 18 & = & S \\
O & = & 14 & \mapsto & 13 \cdot 14 + 5 & (\mathrm{mod}\ 26) \equiv & 5 & = & F \\
V & = & 21 & \mapsto & 13 \cdot 21 + 5 & (\mathrm{mod}\ 26) \equiv & 18 & = & S \\
E & = & 4 & \mapsto & 13 \cdot 4 + 5 & (\mathrm{mod}\ 26) \equiv & 5 & = & F,
\end{array}
$$

while the ciphertext corresponding to the plaintext *HATE* is

$$
\begin{array}{ccccccc}
H & = & 7 \mapsto 13 \cdot 7 + 5 & (\mathrm{mod}\ 26) \equiv & 18 & = & S \\
A & = & 0 \mapsto 13 \cdot 0 + 5 & (\mathrm{mod}\ 26) \equiv & 5 & = & F \\
T & = & 19 \mapsto 13 \cdot 19 + 5 & (\mathrm{mod}\ 26) \equiv & 18 & = & S \\
E & = & 4 \mapsto 13 \cdot 4 + 5 & (\mathrm{mod}\ 26) \equiv & 5 & = & F.
\end{array}
$$

This could cause some problems on Valentine's Day.

---

## CHECK YOUR UNDERSTANDING

1. The function $9x + 2$ was used to obtain the ciphertext *KMI*. Find the plaintext.

---

# 7.3   Vigenère Ciphers

Before introducing the Vigenère cipher, let's take another look at shift ciphers. We know that we can read a ciphertext by simply looking at all shifts of the ciphertext. But we can also decrypt by counting how often each letter occurs in the ciphertext, a method known as **frequency analysis**. This method is slower, but it allows us to develop an idea that will be useful later.

Here's how frequency analysis works. It is well known that $E$ is the most common letter in standard English texts, and that $Z$ is rather uncommon (unless you have weekly quizzes). Counts of frequencies of letters in English yield the data in Figure 7.1.

For instance, $E$ occurs with frequency around 12.7%, and $A$ occurs around 8.2% of the time.

Let's say we intercept the following ciphertext:

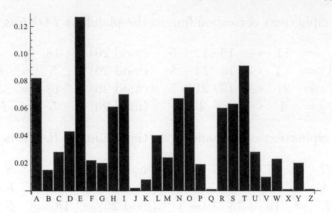

**FIGURE 7.1: Frequencies of letters in English.**

YMJRTXYHTRRTSQDZXJIXNCHMFWFHYJWUFXXBTWINXTSJYBTYMW
JJKTZWKNAJXNCNYNXSTYWJHTRRJSIJIYMFYDTZZXJNYFXDTZWX

The frequency of letters in this ciphertext is given in Figure 7.2.

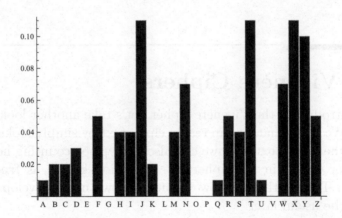

**FIGURE 7.2: Frequencies of letters in ciphertext.**

For example, of the 100 characters in the ciphertext, 3 are $D$'s, so $D$ has
frequency 0.03. Notice that this graph looks similar to the frequency
graph for English, but it's shifted 5 units to the right. (Of course, there
are variations in height, but that is to be expected in small samples.)
So we guess that the ciphertext was obtained by shifting the plaintext

by 5 places. If we shift back by 5 places, we obtain (with spaces and punctuation inserted to make reading easier)

THE MOST COMMONLY USED SIX-CHARACTER PASSWORD IS ONE TWO THREE FOUR FIVE SIX. IT IS NOT RECOM-MENDED THAT YOU USE IT AS YOURS.

Why did this work? The 100-character plaintext is probably a representative sample of English, so it probably has around 8% $A$'s and 12% $E$'s. When the letters are shifted by 5 places, the $A$'s become $F$'s and the $E$'s become $J$'s. This means that the frequency for the plaintext $A$ is the same as the frequency for the ciphertext $F$, and the frequency for the plaintext $E$ is the same as the frequency for the ciphertext $J$. So, the ciphertext frequency distribution is simply a shift 5 places to the right of the plaintext frequency distribution. This is exactly what we saw in Figures 7.1 and 7.2.

Well into the 20th century, sophisticated versions of frequency counts were used to attack many cryptosystems. The **Vigenère cipher** was an attempt to thwart this type of attack. This cryptosystem seems to have been invented around 1553 by Giovan Battista Bellaso, but in the 1800s it was mistakenly named after Blaise de Vigenère (1523–1596), another 16th century cryptographer who devised similar but stronger cryptosystems. It was commonly regarded as unbreakable and was widely used up through World War I, even though the Prussian cryptographer Friedrich Wilhelm Kasiski had published a method for breaking it in 1863. Most likely, Charles Babbage had also developed the method in his work for the British intelligence agency.

In the Vigenère cipher, the key is a sequence of shifts. The key should be of reasonable length and chosen so that someone can remember it (like today's passwords). For example, let's use 2, 5, 4, 7. Apply these shifts cyclically to the message:

$$\begin{array}{ccccccccccc} T & H & I & S & I & S & A & S & E & C & R & E & T \\ 2 & 5 & 4 & 7 & 2 & 5 & 4 & 7 & 2 & 5 & 4 & 7 & 2 \\ V & M & M & Z & K & X & E & M & G & H & V & L & V \end{array}$$

The last line is the encrypted message.

If Bob receives this ciphertext, he shifts in the reverse direction to recover the plaintext.

Notice that in the example the letter $E$ is encrypted once to $G$ and once to $L$. Also, the letter $M$ in the ciphertext comes from both $H$ and $I$. The overall effect is that the counts for the common letters get spread among several letters and therefore the frequencies of the letters in the ciphertext are more uniform than from a simple shift cipher. This is why the Vigenère cipher was regarded as secure.

The best way to describe how to break the system is to consider an example. Suppose we intercept the following ciphertext:

```
VHXOEMJOWWSXFFHTTAGPKGPTTAMKOGCNWTETFIGIOYEOWGMXUSTIEL
KSLKMINEBPTAGEQVRXOETPDTVTAGSTOEMKMXKMIQSLKBEGOYVRTPSE
CTBQNNPLXUSMJEDGYBUKGQWGVHXGALGWBVHPJIVJTAGKXAMTABXEHT
PGXFILCNHVHXTPHKNMKNYCVHTOYVHXCDHRTBQNHHTAKSVQDXDYMJOL
GDXUIKKNZVOMTAGUMBVIFROKVAGVMXUSTIELYIMJONVTAGSEKGAVEL
VDTPGXTOYVHXKRFGSLCGXUBXKNZTETFBRROEKTBEAEQRUWSBPELURB
XAEUEME
```

The first thing we need to know is the key length. One method is due to Kasiski. Observe that the three-letter sequence *VHX* occurs five times. The first one starts with the $V$ as the first letter, the second has $V$ as the 133rd letter of the ciphertext, the third has $V$ as the 172nd letter, the fourth has $V$ as the 190th letter, and the fifth has $V$ as the 280th letter. Each occurrence of *VHX* corresponds to three letters of the plaintext, shifted by amounts corresponding to the shifts required by the key. A reasonable guess is that each *VHX* comes from the same three letters shifted by the same amount. Since the key is applied cyclically, this means that the distances between the occurrences of *VHX* are multiples of the key length. These distances are

$$133 - 1 = 132, \quad 172 - 133 = 39, \quad 190 - 172 = 18, \quad 280 - 190 = 90.$$

If these are multiples of the key length, then

$$\gcd(132, 39, 18, 90) = 3$$

is a multiple of the key length. Therefore, a good guess for the key length is 3. As we'll see, this guess is correct.

Now that we suspect the key length is three, we know that the letters of the ciphertext in positions 1, 4, 7, 10, 14, ... were all shifted by the same amount. It is reasonable to assume that the corresponding letters of the plaintext represent a random sample of English letters. Therefore, we can guess the shift by the same method as we used at the beginning of this section; namely, we'll use frequency counts.

Extract the letters in positions 1, 4, 7, ... from the ciphertext to get

VOJWF ··· .

The frequencies for these letters are given in the following graph:

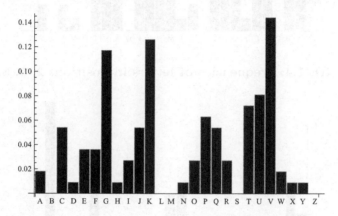

**FIGURE 7.3: Frequencies of letters in positions 1, 4, 7, ···.**

This is similar to the frequency graph for English given in Figure 7.1, except that it is shifted by 2 places. Therefore, we suspect that the first shift in the key is 2.

Now we do the same for the letters in positions 2, 5, 8, 11, ···. Figure 7.4 gives the graph of frequencies. It is essentially the frequency graph for English, but with no shift. So we suspect that the second shift of the key is 0.

Finally we look at positions 3, 6, 9, 12, ··· and obtain the graph in Figure 7.5. This one is a little harder to analyze, but the best guess is that it is a shift by 19.

We now conjecture that the key is 2, 0, 19. When we realize that this corresponds to the easily remembered word *CAT*, we're sure that we're right. But the only way to be certain is to decrypt, which we want to do anyway because that was our objective from the beginning.

Shift the ciphertext back by 2, 0, 19, applied cyclically. The result is (with spaces and punctuation added)

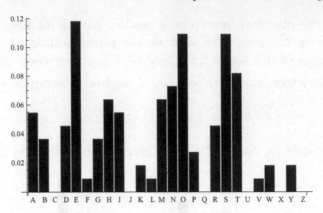

**FIGURE 7.4:** Frequencies of letters in positions 2, 5, 8, ⋯.

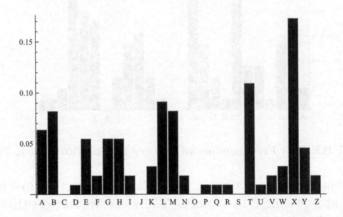

**FIGURE 7.5:** Frequencies of letters in positions 3, 6, 9, ⋯.

THE METHOD USED FOR THE PREPARATION AND READ-
ING OF CODE MESSAGES IS SIMPLE IN THE EXTREME
AND AT THE SAME TIME IMPOSSIBLE OF TRANSLATION
UNLESS THE KEY IS KNOWN. THE EASE WITH WHICH
THE KEY MAY BE CHANGED IS ANOTHER POINT IN FA-
VOR OF THE ADOPTION OF THIS CODE BY THOSE DESIR-
ING TO TRANSMIT IMPORTANT MESSAGES WITHOUT THE
SLIGHTEST DANGER OF THEIR MESSAGES BEING READ
BY POLITICAL OR BUSINESS RIVALS ETC.

This text is from a 1917 *Scientific American* article describing the advantages of using the Vigenère cipher. As we have demonstrated, the author's analysis was not quite correct.

After World War I, the weaknesses inherent in the Vigenère cipher became more well known and its use decreased markedly.

# 7.4    Transposition Ciphers

Can you read the following ciphertext?

IARATICNEDHS.

This was encrypted by a method that has been around for thousands of years and is often called zig-zag writing or a rail fence cipher. Here's how to read it. Split it in half to get `IARATI   CNEDHS`, and put the first half in one column and the second half in another column:

IC

AN

RE

AD

TH

IS.

You get the message by reading the letters row-by-row (the path you follow is definitely a zig-zag):

I CAN READ THIS.

It was encrypted by writing the plaintext in a series of rows and then reading the columns.

More generally, suppose Alice starts with a plaintext such as

THIS IS THE PLAINTEXT THAT WE ARE ENCRYPTING.

Alice decides to use five columns. The message has 37 characters, so she writes it in rows of 5 letters until the last row, which gets only 2 letters:

```
THISI
STHEP
LAINT
EXTTH
ATWEA
REENC
RYPTI
NG
```

To get the ciphertext, she reads this column-by-column:

TSLEARRNHTAXTEYGIHITWEPSENTENTIPTHACI.

Bob receives this ciphertext and counts that there are 37 characters. The *key* that Alice and Bob share beforehand is the number of columns. Since he knows that there were five columns, he computes $37/5 = 7.4$, which means he wants to group the letters into groups of 8 and 7 (because when 37 letters are put into 5 columns, some columns will have 8 letters and some will have 7). In fact, two 8s and three 7s yields 37. Therefore, he writes the message in 5 columns of lengths 8, 8, 7, 7, 7:

```
THISI
STHEP
LAINT
EXTTH
ATWEA
REENC
RYPTI
NG
```

Reading row by row yields the message.

Suppose Eve intercepts the ciphertext but does not know the number of columns. She knows that there will be an $r$-by-$c$ array with $r$ rows and *c* columns, and that $r \times c \geq 37$. Some possibilities are $3 \times 13$, $4 \times 10$, $8 \times 5$, and $7 \times 6$. For example, if Eve guesses that there are 3 rows and 13 columns, then she puts the ciphertext into the array

```
TERTTGIEEEIHI
SANAEITPNNPA
LRHXYHWSTTTC.
```

This does not yield a meaningful plaintext, so she tries other possibilities for the number of columns until she gets to 5 columns, which gives her the correct answer.

To increase the difficulty of Eve's attack, Alice can permute the columns according to some prearranged key word. For example, suppose the *key word* for the day is *water*. First notice that *w* is the first letter in water, *a* is the second letter, *t* the third, *e* the fourth, and *r* is the fifth. Next, put the letters of the key in alphabetical order: *aertw*. Finally, Alice puts the columns in the order 2, 4, 5, 3, 1 corresponding to the alphabetization of *water* together with where each letter occurs in the key word. For example, since *a* is the second letter of water and the first letter in the alphabetical order, she moves the second column to the first position. Similarly, since *e* is the 4th letter in *water* and is the 2nd letter in alphabetical order, she moves the 4th column to the 2nd position. This yields the following array:

```
HSIIT
TEPHS
ANTIL
XTHTE
TEAWA
ENCER
YTIPR
G   N
```

The columns yield the ciphertext:

```
HTAXTEYGSENTENTIPTHACIIHITWEPTSLEARRN.
```

This is much harder for Eve to break. Not only does she need to guess how many columns there are, but she also needs to guess the order of the columns. Otherwise, she doesn't know which are missing letters, and this affects where the breaks are between columns. In our example, if she guesses that the order was 2, 4, 5, 3, 1, then, using the fact that there are two columns of 8 and three of seven, she deduces that the plaintext should be arranged into columns of lengths 8, 7, 7, 7, 8. Once this is done, the columns can be rearranged to yield the plaintext by reversing the moves that put them in order 2, 4, 5, 3, 1.

The above procedure can be applied again to the ciphertext to yield a double encryption of the original plaintext. This was used briefly by the German military in World War I (until the French broke the system) and by the Dutch Resistance in World War II. Breaking this double encryption system is much more difficult than breaking a single encryption.

The above methods are called **columnar transpositions**. More generally, a **transposition cipher** keeps the letters of the plaintext unchanged, but puts them in a different order. Since the frequency counts for the text do not change, a cryptanalyst should notice that the frequency counts resemble those for the language used by the sender, which will lead to a suspicion that such a transposition cipher is being used.

---

## CHECK YOUR UNDERSTANDING

2. The ciphertext `MMSTIEEALSEIIOTNNU` was encrypted by the method of this section. Decrypt it.

---

# 7.5   Stream Ciphers

A stream cipher changes a plaintext message one bit (or one letter) at a time. Let's say that Alice writes her message as a sequence of 0s and 1s. There are many ways of doing this. One of the most basic methods is called *ASCII*, which, for example, represents the letter *A* as 1000001 and *B* as 1000010. We won't worry about what method is used. The key is also a sequence of 0s and 1s and has the same length as the plaintext. To encrypt, Alice adds the key to the plaintext, bit by bit mod 2. This process is called **XOR** and is often denoted ⊕. For example, if the plaintext is 101100111 and the key is 001110101, the ciphertext 100010010 is obtained as follows:

$$101100111$$
$$\underline{001110101}$$
$$100010010.$$

Note that this uses the mod 2 addition laws $0 + 0 = 0$, $0 + 1 = 1$, $1 + 1 = 0$ without any carrying. All the operations are done one bit at a time.

Because adding mod 2 and subtracting mod 2 are the same, decryption is the exact same procedure with the exact same key. In our example,

$$100010010 \oplus 001110101 = 101100111,$$

so *XOR*ing the key with the ciphertext yields the plaintext.

A stream cipher encryption machine XORs the key bits one at a time with the plaintext bits, thus producing the ciphertext bits. Here's a picture of what's happening:

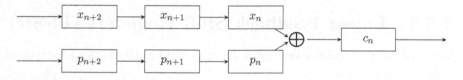

**FIGURE 7.6**
**Stream cipher encryption: $p_n$ = plaintext bit, $x_n$ = key bit, $c_n$ = ciphertext bit.**

This encryption method is called a **stream cipher** since it uses a stream of 0s and 1s to encrypt. Stream ciphers are used in many situations. Some are very sophisticated and have high security. In the following, we describe two methods, one which is very secure and the other a simple one that is not very secure.

## 7.5.1    One-Time Pad

For a **one-time pad**, Alice uses a random process to generate a random sequence of 0s and 1s as a key. This is then used to encrypt by XORing it with the plaintext, as described above. If the key bits are truly random, this encryption method is completely secure. For example, suppose Eve intercepts the ciphertext 100010010 and is trying to decide whether the plaintext is 101100111 or 111110000. The first possibility corresponds to the key 001110101, and the second corresponds to the key 011100010. Since the key was chosen randomly, both possibilities are possible. In this way, we see that the ciphertext gives no information about the plaintext.

During the Cold War, the hotline between Washington, D.C., and Moscow used a one-time pad. The random sequences of 0s and 1s were written on paper tape and carried securely by couriers (probably as in the movies, with an attaché case handcuffed to the courier's wrist). The random sequence was then available for use whenever needed.

Although the one-time pad is very secure, it is very expensive to use. Massive numbers of random 0s and 1s need to be transmitted in advance. Therefore, in practice it is desirable to use methods that quickly generate random-looking sequences but which can be transmitted more efficiently between Alice and Bob. One such method is described next.

## 7.5.2   Linear Feedback Shift Registers (LFSR)

Our goal is to create a key for a stream cipher. Here's an example of one way to do this.

We start off with two bits — we'll call them $x_1$ and $x_2$ — and arbitrarily choose $x_1 = x_2 = 1$. The values of $x_1$ and $x_2$ are called the seed because we'll grow the rest of the bits used in the key from these seed values. From these two bits we need to create a third, $x_3$. We do so by adding $x_1$ and $x_2$ (mod 2):

$$x_3 \equiv x_1 + x_2 \equiv 1 + 1 \equiv 0.$$

We now need to get the fourth key bit, $x_4$. We do this by adding $x_2$ and $x_3$ (mod 2):

$$x_4 \equiv x_2 + x_3 \equiv 1 + 0 \equiv 1 \pmod 2.$$

In general, we have

$$x_{n+2} \equiv x_n + x_{n+1} \pmod 2.$$

So, we see that

$$x_5 \equiv x_3 + x_4 \equiv 0 + 1 \equiv 1 \quad \text{and} \quad x_6 \equiv x_4 + x_5 \equiv 1 + 1 \equiv 0 \pmod{2}.$$

In this way, we generate a sequence

$$1, 1, 0, 1, 1, 0, \ldots.$$

(It's the Fibonacci sequence mod 2.) These 0s and 1s can be used as the key for a stream cipher.

The equation

$$x_{n+2} \equiv x_n + x_{n+1} \pmod{2}$$

is called a **recurrence relation**. In general, we choose a length $\ell$, and then coefficients $c_0, c_1, \ldots, c_{\ell-1} \in \{0, 1\}$, and form the recurrence relation

$$x_{n+\ell} \equiv c_0 x_n + c_1 x_{n+1} + \cdots + c_{\ell-1} x_{n+\ell-1} \pmod{2}.$$

In the example above, we had $\ell = 2$, $c_0 = 1$, $c_1 = 1$.

As another example, suppose we have the recurrence relation

$$x_{n+5} \equiv x_n + x_{n+3} \pmod{2},$$

so $\ell = 5$, $c_0 = 1$, $c_1 = 0$, $c_2 = 0$, $c_3 = 1$, $c_4 = 0$. If we choose initial seed values $x_1 = 0, x_2 = 1, x_3 = 0, x_4 = 1, x_5 = 1$, then we can calculate $x_6, x_7, x_8, \ldots$. For example,

$$x_6 \equiv x_1 + x_4 \equiv 0 + 1 \equiv 1.$$

Similarly, we find that

$$x_7 = 0, \quad x_8 = 1, \ldots.$$

We again get a sequence of 0s and 1s.

If we have initial values $x_1, x_2, \ldots, x_\ell$, then we can use a recurrence relation to produce a sequence of bits to use in a stream cipher. Such a sequence is called a **Linear Feedback Shift Register sequence**, or **LFSR sequence**.

The reason for the name is the way a machine can produce the LFSR bits. Look at Figure 7.7. The boxes, which are called registers, are labeled $x_n$, $x_{n+1}$, etc. They contain bits. For example, suppose $n = 7$, so the bits in the boxes are $x_{10}$, $x_9$, $x_8$, and $x_7$. At the click of the machine's clock, the entries of a register are shifted to the right. The bit $x_7$ moves to the output position, and $x_8$ moves so that it occupies the space previously occupied by $x_7$, and similarly for the other bits. Moreover, $x_7$ and $x_9$ are also fed back into the system; namely, they follow the vertical arrows and are added together mod 2, and the result, which we'll call $x_{11}$, is put into the place previously occupied by $x_{10}$. The rectangles then contain $x_{11}$, $x_{10}$, $x_9$, and $x_8$. At the next click of the clock, $x_8$ is outputted, and an $x_{12}$ is put into the leftmost box. Since the recurrence relation tells us that $x_{11} \equiv x_7 + x_9$ and $x_{12} \equiv x_8 + x_{10}$, we see that the feedback mechanism is designed so that the new bit that enters at the left is exactly the one corresponding to the recurrence relation.

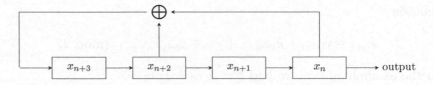

**FIGURE 7.7:** LFSR for $x_{n+4} \equiv x_n + x_{n+2} \pmod{2}$.

An advantage of an LFSR sequence is that only a few bits, namely the coefficients and the initial values, need to be sent to produce a long sequence of bits. The disadvantage is that the method is not very secure. For example, suppose Eve knows the initial several plaintext bits and the corresponding ciphertext bits. She XORs these to get several key bits. Suppose, for example, that these key bits are

$$0, 1, 1, 1, 0, 0, 1, 0, 1, 1, 1, 0, 0, 1, 0.$$

She guesses that the length $\ell$ is 3, which means she guesses that the recurrence relation is

$$x_{n+3} \equiv c_0 x_n + c_1 x_{n+1} + c_2 x_{n+2} \pmod{2}.$$

Letting $n = 1, 2, 3$, and substituting in the values of the $x_i$, she obtains

$$1 \equiv c_0 \cdot 0 + c_1 \cdot 1 + c_2 \cdot 1$$
$$0 \equiv c_0 \cdot 1 + c_1 \cdot 1 + c_2 \cdot 1$$
$$0 \equiv c_0 \cdot 1 + c_1 \cdot 1 + c_2 \cdot 0.$$

For example, the first line is simply the relation $x_4 \equiv c_0 x_1 + c_1 x_2 + c_2 x_3$. Adding the second and third lines yields $c_2 = 0$. Then the first line yields $c_1 = 1$ and the second line yields $c_0 = 1$. So the recurrence relation is

$$x_{n+3} \equiv x_n + x_{n+1} \pmod 2.$$

Once Eve knows this, she can use the $x_i$'s she knows as initial values and then generate as many terms of the sequence as she desires. In our example, she used only the first six members of the sequence to obtain the relation. She can check that the relation generates the remaining key bits that she knows. If it does not, then she guesses a different length and again solves for the coefficients. In this way, Eve eventually can find the key stream and thereby decrypt the remainder of the message. Therefore, this method of LFSR sequences should be used only for very low-level security. In practice, when high security is desired, much more complicated methods are used to produce the key stream.

---

## CHECK YOUR UNDERSTANDING

3. An LFSR machine produces the key bits 110110110110. You receive the ciphertext 011101011101, which was encrypted with a stream cipher using these key bits. What is the plaintext?

---

# 7.6   Block Ciphers

As we discussed, stream ciphers encrypt bit by bit. **Block ciphers**, which form the second type of symmetric key cryptosystems, encrypt several bits at a time. More precisely, the plaintext is divided into blocks of a fixed length. Each block of plaintext is then encrypted, yielding a block of ciphertext. In the simplest version of a block cipher, Alice breaks her plaintext into blocks, all of the same length, and encrypts each one by the same encryption algorithm. The ciphertext is therefore a series of blocks of characters. More complicated implementations use techniques such as feedback, so that the encryption of one block depends on the encryption of preceding blocks. See Exercise 35.

For example, if the plaintext is

ATTACK AT DAWN,

and our block length is 2, we can break up the message into

AT TA CK AT DA WN,

and encrypt each pair of letters separately. If, instead, our block length is five, our plaintext blocks would be

ATTAC KATDA WNXXX,

where we have filled in the empty spaces with $X$'s.

One of the earliest block ciphers, the **Hill cipher**, was invented by Lester Hill in 1929. The key is an $n \times n$ integer matrix $M$ that has an inverse mod 26. (For a brief review of matrices, see the Appendix.) Although the Hill cipher works for any block size $n$, to simplify matters we look only at $n = 2$, so all our matrices are $2 \times 2$. For example, suppose

$$M = \begin{pmatrix} 5 & 7 \\ 2 & 11 \end{pmatrix}.$$

Let

$$N = \begin{pmatrix} -1 & 3 \\ 12 & 9 \end{pmatrix}.$$

Then

$$MN = \begin{pmatrix} 5 & 7 \\ 2 & 11 \end{pmatrix} \begin{pmatrix} -1 & 3 \\ 12 & 9 \end{pmatrix} = \begin{pmatrix} 79 & 78 \\ 130 & 105 \end{pmatrix} \equiv \begin{pmatrix} 1 & 0 \\ 0 & 1 \end{pmatrix} \pmod{26},$$

so $N$ is the inverse of $M$ mod 26. Therefore, $M$ is a suitable key for the Hill cipher.

How did we find $N$? The formula for the inverse of a $2 \times 2$ matrix is

$$\begin{pmatrix} a & b \\ c & d \end{pmatrix}^{-1} = \frac{1}{ad - bc} \begin{pmatrix} d & -b \\ -c & a \end{pmatrix}.$$

For our example, we obtain

$$\begin{pmatrix} 5 & 7 \\ 2 & 11 \end{pmatrix}^{-1} = \frac{1}{41} \begin{pmatrix} 11 & -5 \\ -2 & 5 \end{pmatrix}.$$

Since we are working mod 26, we replace $1/41$ with 7, since 7 is the multiplicative inverse of 41 mod 26. (We discussed multiplicative inverses mod $n$ and how to calculate them in Section 6.4.) This gives us

$$7 \begin{pmatrix} 11 & -5 \\ -2 & 5 \end{pmatrix} \equiv \begin{pmatrix} -1 & 3 \\ 12 & 9 \end{pmatrix},$$

which is $N$. In general, the formula yields an inverse matrix whenever $\gcd(ad - bc, 26) = 1$. In fact, it can be shown that $\gcd(ad - bc, 26) = 1$ if and only if the matrix has an inverse mod 26.

Let's continue with the encryption. Alice breaks her ciphertext into blocks of $n$ letters (where $n$ is the size of the matrix) and converts each letter to a number using $A = 0, B = 1, \ldots, Z = 25$. Each block of $n$ letters is then regarded as an $n$-dimensional vector. The encryption multiplies the matrix $M$ times the vectors mod 26. These are then converted back to letters.

For example, suppose the plaintext is *HIDETHIS*. This becomes the vectors

$$\begin{pmatrix} 7 \\ 8 \end{pmatrix}, \begin{pmatrix} 3 \\ 4 \end{pmatrix}, \begin{pmatrix} 19 \\ 7 \end{pmatrix}, \begin{pmatrix} 8 \\ 18 \end{pmatrix}$$

(the first vector uses $H = 7, I = 8$, for example). The encryption is

$$\begin{pmatrix} 5 & 7 \\ 2 & 11 \end{pmatrix} \begin{pmatrix} 7 \\ 8 \end{pmatrix} \equiv \begin{pmatrix} 13 \\ 24 \end{pmatrix}$$

$$\begin{pmatrix} 5 & 7 \\ 2 & 11 \end{pmatrix} \begin{pmatrix} 3 \\ 4 \end{pmatrix} \equiv \begin{pmatrix} 17 \\ 24 \end{pmatrix}$$

$$\begin{pmatrix} 5 & 7 \\ 2 & 11 \end{pmatrix} \begin{pmatrix} 19 \\ 7 \end{pmatrix} \equiv \begin{pmatrix} 17 \\ 24 \end{pmatrix}$$

$$\begin{pmatrix} 5 & 7 \\ 2 & 11 \end{pmatrix} \begin{pmatrix} 8 \\ 18 \end{pmatrix} \equiv \begin{pmatrix} 10 \\ 6 \end{pmatrix}.$$

Using $13 = N, 24 = Y$, etc., converts this to the letters *NYRYOLKG*, which is the ciphertext.

When Bob receives the ciphertext, he decrypts by multiplying by the inverse of $M$ mod 26. In our example, this is the matrix $N$. So Bob computes

$$\begin{pmatrix} -1 & 3 \\ 12 & 9 \end{pmatrix} \begin{pmatrix} 13 \\ 24 \end{pmatrix} \equiv \begin{pmatrix} 7 \\ 8 \end{pmatrix}$$

$$\begin{pmatrix} -1 & 3 \\ 12 & 9 \end{pmatrix} \begin{pmatrix} 17 \\ 24 \end{pmatrix} \equiv \begin{pmatrix} 3 \\ 4 \end{pmatrix}$$

$$\begin{pmatrix} -1 & 3 \\ 12 & 9 \end{pmatrix} \begin{pmatrix} 17 \\ 24 \end{pmatrix} \equiv \begin{pmatrix} 19 \\ 7 \end{pmatrix}$$

$$\begin{pmatrix} -1 & 3 \\ 12 & 9 \end{pmatrix} \begin{pmatrix} 10 \\ 6 \end{pmatrix} \equiv \begin{pmatrix} 8 \\ 18 \end{pmatrix}.$$

Bob changes this to letters and obtains *HIDETHIS*.

If Eve discovers the plaintext that corresponds to some ciphertext, she can set up linear congruences and solve for the entries in the matrix. Because of this, the Hill cipher should not be used in situations that require high levels of security. But the real reason it was never used widely is that encryption and decryption are rather slow, especially if larger matrices (for example, $6 \times 6$) are used.

The block ciphers that are used for situations where security is important use bigger blocks and are faster than Hill ciphers. One of the most widely used block ciphers was DES (Data Encryption Standard), which was invented in the 1970s and was the standard cryptographic block cipher for more than 25 years. It used plaintext blocks of 64 bits and produced ciphertext blocks also of 64 bits. In 2001, AES (Advanced Encryption Standard) was introduced as a replacement for DES. It uses blocks of 128 bits. Both DES and AES have been widely used in Internet commerce.

---

## CHECK YOUR UNDERSTANDING

4. Use the Hill cipher with matrix $\begin{pmatrix} 3 & 2 \\ 13 & 1 \end{pmatrix}$ to encrypt the message *GOLD*.

# 7.7   Secret Sharing

You are the manager of a bank and you want to skip work tomorrow. Even though you won't be there, the bank's safe will still need to be opened. But, you don't trust any individual employee enough to allow one person to have the lock's combination. So, you need a scheme where any two employees together can deduce the combination, but any single employee obtains no information. Your first idea is to give the first number of the combination to half the employees and the last two numbers to the other half. This works well if a first-number employee gets together with a second/third-number employee, but it doesn't work with two first-number employees. So you need something else. While musing over the requirement

two people determine a secret,

you think back to your high school geometry days and recall that

two points determine a line.

Aha! If you represent the combination as the slope of a line and give each employee a point on the line, then one employee alone cannot determine the slope, but any two employees can find the slope and then determine the combination.

Let's try an example. The lock's combination is 17-34-02, which you write as 173402. We could work with real numbers, but, to avoid slight problems with round-off errors, it is better to work mod a prime. Let's assume that the largest possibility for each number of the combination is 40, so we need a prime larger than 404040. We choose the prime $p = 404051$. Now we choose a line. The slope will be 173402. Choose a random intercept, say 141421, and form the line

$$L: \quad y \equiv 173402x + 141421 \pmod{404051}.$$

Let's say you have five employees: $A$, $B$, $C$, $D$, $E$. Give them the following points on $L$:

$$A : \quad (1, 314823)$$
$$B : \quad (2, 84174)$$
$$C : \quad (3, 257576)$$
$$D : \quad (4, 26927)$$
$$E : \quad (5, 200329).$$

For example, $B$'s point is calculated by putting $x = 2$ into the equation for $L$ and getting $y \equiv 173402 \times 2 + 141421 \equiv 84174 \pmod{404051}$.

Suppose that $C$ and $D$ want to open the safe. The slope of the line through their points is

$$\frac{26927 - 257576}{4 - 3} \equiv -230469 \equiv 173402 \pmod{404051}.$$

They have recovered the combination!

Let's try another pair. Suppose that $B$ and $E$ want to open the safe. The slope of the line through their points is

$$\frac{200329 - 84174}{5 - 2} \equiv \frac{116155}{3} \pmod{404051}.$$

What do we do now? Since we need to get rid of the 3 in the denominator, we use the Extended Euclidean Algorithm to find the multiplicative inverse of 3 mod 404051:

$$\frac{1}{3} \equiv 134684 \pmod{404051}.$$

Then the slope is

$$\frac{116155}{3} \equiv 116155 \times 134684 \equiv 173402 \pmod{404051}.$$

Again, they have recovered the combination.

In general, suppose you have a secret $s$ expressed as an integer. Choose a prime $p$ larger than the largest possibility for the $s$. (Note that $p$ does not need to be a cryptographic-sized prime of 300 digits.) Pick a random integer $r$ and form the line $L : y \equiv sx + r \pmod{p}$. Give each person one of the points $(1, s + r)$, $(2, 2s + r)$, $(3, 3s + r)$, .... Then any two people together can determine the secret $s$. However, one point gives no information about the slope, so one person does not obtain any information.

This method can be generalized. If we replace the line $L$ with a parabola $y \equiv ax^2 + bx + s \pmod{p}$ with randomly chosen coefficients $a$ and $b$ and give each person a point on this parabola, then any three people can determine $s$, but two people obtain no information about $s$ (as long as no one is given a point with $x$-coordinate 0). More generally, we can use polynomials of degree $k$ with one coefficient equal to the secret and the other coefficients randomly chosen. This yields a scheme where any $k + 1$ people can determine a secret but $k$ people do not obtain any information.

---

## CHECK YOUR UNDERSTANDING

5. In a secret sharing scheme mod 13 where the slope is the secret, person $A$ has the point $(2,4)$, person $B$ has $(5,0)$, and person $C$ has $(8,9)$. Find the secret.

---

# 7.8    Generating Random Numbers

Random numbers are useful in many situations. For example, they are used in computer simulations of physical phenomena, and they are needed to generate random keys for cryptography.

Generating truly random numbers is quite difficult. Flipping coins, for example, is slow and can potentially be biased by physical characteristics of a coin. Therefore, deterministic methods that can be run on a computer are often used. Their outputs are called **pseudo-random numbers**. The LFSR method in Section 7.5 is such a method. In this section, we describe two methods based on congruences.

A method that has been used frequently is called a **linear congruential generator**. Let $m$ be a large integer and choose two integers $a$ and $b$. Choose a starting value $x_0$, called a **seed**, and define a sequence by

$$x_i \equiv ax_{i-1} + b \pmod{m}$$

for $i = 1, 2, 3, \ldots$ . Using the least non-negative residue mod $n$ for each $x_i$, we obtain a sequence of integers $x_1, x_2, x_3, \ldots$ .

**Example.** Let $m = 123$, let $a = 5$, and let $b = 2$. Choose $x_0 = 73$. Then

$$x_1 \equiv (5)(73) + 2 \equiv 121 \pmod{123}$$
$$x_2 \equiv (5)(121) + 2 \equiv 115 \pmod{123}$$
$$x_3 \equiv (5)(115) + 2 \equiv 85 \pmod{123}$$
$$x_4 \equiv (5)(85) + 2 \equiv 58 \pmod{123}.$$

In many situations, what is needed is a string of random bits (0s and 1s). We can obtain such a sequence by using a linear congruential generator and reducing the output mod 2. In the example, this yields the sequence $b_1 = 1, b_2 = 1, b_3 = 1, b_4 = 0$.

Linear congruential generators are not regarded as suitable for cryptographic purposes since knowledge of sufficiently many output bits can be used to predict future bits with better than $1/2$ probability. It is a subject of debate as to how dependable they are for simulating physical experiments.

Another method of generating random bits is called the **Blum–Blum–Shub pseudo-random bit generator**. Choose two distinct large (that is, hundreds of digits) primes $p$ and $q$ and let $n = pq$. Choose a seed value $x_0$ with $\gcd(x_0, n) = 1$. Define

$$x_i \equiv x_{i-1}^2 \pmod{n}$$

for $i = 1, 2, 3, \ldots$ . Using the least non-negative residue mod $n$ for each $x_i$, we obtain a sequence of integers $x_1, x_2, x_3, \ldots$ . Now let $b_i = x_i \pmod 2$. We now have a sequence of 0s and 1s.

**Example.** Let $p = 11$ and $q = 19$, so $n = 209$. Let $x_0 = 100$. Then

$$x_1 \equiv 100^2 \equiv 177 \pmod{209}$$
$$x_2 \equiv 177^2 \equiv 188 \pmod{209}$$
$$x_3 \equiv 188^2 \equiv 23 \pmod{209}$$
$$x_4 \equiv 23^2 \equiv 111 \pmod{209}.$$

This yields the sequence $b_1 = 1, b_2 = 0, b_3 = 1, b_4 = 1$.

The security of the Blum–Blum–Shub generator is closely tied to the difficulty of factoring $n$. It is generally regarded as secure, but it is not efficient enough to be used in high-speed situations. In those cases, random number generators based on cryptographic functions called *hash functions* are often used.

# 7.9   Chapter Highlights

1.  Shift and affine ciphers

2.  The Vigenère cipher

3.  The RSA algorithm

4.  Stream and block ciphers

5.  Secret sharing

# 7.10   Problems

## 7.10.1   Exercises

### Section 7.2: Shift and Affine Ciphers

1.  The plaintext *ATDAWN* was encrypted using the affine function $9x + 13$. What is the ciphertext? What is the decryption function? Verify that it works.

2.  The plaintext *THEBEACH* was encrypted using the affine function $7x + 23$. What is the ciphertext? What is the decryption function? Verify that it works.

3.  The ciphertext *QBULF* was encrypted using the affine function $3x + 11$. What is the plaintext?

4.  The ciphertext *QPMZ* was encrypted using the affine function $17x + 9$. What is the plaintext?

5.  The ciphertext *YRFE* was encrypted using the affine function $5x + 17$. What is the plaintext?

6.  The ciphertext *LFICJBOPS* was encrypted using the affine function $7x + 5$. What is the plaintext?

7. You intercept the ciphertext *KRON*, which was encrypted using an affine function. You know that the plaintext starts *DO*. Find the encryption function and find the plaintext.

8. You intercept the ciphertext *YFWD*, which was encrypted using an affine function. You know that the plaintext starts *ST*. Find the encryption function and find the plaintext.

9. Suppose you try to use the affine function $2x + 1$ to encrypt. Find two letters that encrypt to the same ciphertext letter.

10. Suppose you try to use the affine function $13x + 7$ to encrypt. Find two different three-letter words that encrypt to *HHH*.

11. The ciphertext *XVASDW* was encrypted using an affine function $ax + 1$. Determine $a$ and decrypt the message.

12. If you receive a message encrypted by a shift cipher, usually you can decrypt it by trying all possibilities.
    (a) Decrypt *ITSJ*, which is a ciphertext for a shift cipher.
    (b) Show that *KZXNTS* has two different decryptions (it and *BMPIBT* are the longest known examples in English).
    (c) Find some two-letter ciphertexts that have more than one decryption to English words.

## Section 7.3: Vigenère Ciphers

13. Encrypt the message *THISISSECRET* using the Vigenère cipher with key *0,1,2*.

14. (a) Encrypt the message *EVEEATSEGGSEVERYDAY* using the Vigenère cipher with key *7, 4, 13* (= *HEN*).
    (b) There are six *E*'s in the plaintext. What are the seven most common letters in the ciphertext, and how often does each occur?

15. The ciphertext *FOBMHDLMLLHUNH* was encrypted using the Vigenère cipher with key *7, 0* (= *HA*). Decrypt it.

16. The ciphertext *XIZGNXTEBUNHVSXEUKG* was encrypted using the Vigenère cipher with key *2, 0, 19* (= *CAT*). Decrypt it.

17. In the Land of Nod, they have the English alphabet of 26 letters, but they use only the letter *Z* in their words. Suppose they use the Vigenère cipher to encrypt a plaintext consisting of some of their words and then send the ciphertext *RKDDORKDDORKDDORKDDORKDDORKDDORKD-DORKDDO*. Determine the likely key and find the plaintext.

18. Suppose a language has an alphabet with only the letters *A, B*. The letter frequencies are 90% *A* and 10% *B*. A message is encrypted using the Vigenère method (with shifts mod 2) with key length 1 or 2. The ciphertext is *ABAAABABAB*. Determine whether the likely key length is 1 or 2 and find the plaintext.

## Section 7.4: Transposition Ciphers

19. The following were encrypted by a columnar transposition cipher with 3 columns, with no permutation of the columns. Read the messages.

    (a) `MHFAIUTSN`.

    (b) `ERNNSAVYEOTTEOKWH`.

20.   The following were encrypted by a columnar transposition cipher, with no permutation of the columns. Read the messages.

    (a) `IMNOENEEDY`.

    (b) `TEECHIDUISILSMFTOOFNRI`.

21.   Suppose Alice decides to use a columnar transposition cipher with only one column. What can you say about the ciphertext?

22.   The following were encrypted by columnar transportation ciphers, with no permutation of the columns. Read the messages.

    (a) `TROAHSONBTETBOEIQIETTSUOOTHTEN`.

    (b) `TNESEOOTTSBTHHTETAEIOOTQORBIUN`.

23.   The following was encrypted by a columnar transportation cipher, with no permutation of the columns. Read the message.

    `ICASKPAFADTATPYNTHCOHORHARGEUEINYRR`.

24.   The following was encrypted by a columnar transportation cipher, with no permutation of the columns. Read the message.

    `TIEEWHSNREADAGTOIOTFNLHTBDEHO`.

## Section 7.5: Stream Ciphers

25.   An LFSR machine produces the key bits 10101010. You encrypt the plaintext 11111111 using these key bits. What is the ciphertext?

26.   An LFSR machine produces the key bits 10101010. You receive the ciphertext 11111111, which was encrypted with a stream cipher using these key bits. What is the plaintext?

27.   An LFSR machine uses the recurrence relation $x_{n+4} \equiv x_n + x_{n+1}$ (mod 2) and initial values $x_1 = x_2 = x_3 = x_4 = 1$. Find $x_5, x_6, \ldots, x_{10}$.

28.   An LFSR machine uses the recurrence relation $x_{n+3} \equiv x_n + x_{n+1} + x_{n+2}$ (mod 2) and initial values $x_1 = x_2 = 1$, $x_3 = 0$. Find $x_4, x_5, \ldots, x_{10}$.

29.   The sequence $0, 1, 1, 1, 0, 1, 0, 0, 1, 1, 1, 0$ was produced by a recurrence relation $x_{n+3} \equiv c_0 x_n + c_1 x_{n+1} + c_2 x_{n+2}$ (mod 2). Find $c_0, c_1, c_2$.

30.   The sequence $0, 1, 0, 0, 1, 0, 0, 1$ was produced by a recurrence relation $x_{n+3} \equiv c_0 x_n + c_1 x_{n+1} + c_2 x_{n+2}$ (mod 2). Find $c_0, c_1, c_2$.

## Section 7.6: Block Ciphers

31.   Use the Hill cipher with matrix $\begin{pmatrix} 3 & 2 \\ 12 & 9 \end{pmatrix}$ to encrypt the message *HILL*.

32.   Use the Hill cipher with matrix $\begin{pmatrix} 1 & 7 \\ 4 & 11 \end{pmatrix}$ to encrypt the message *BUSY*.

33.   A Hill cipher with matrix $\begin{pmatrix} 11 & 2 \\ 12 & 3 \end{pmatrix}$ was used to encrypt a message. The ciphertext is *WYED*. What is the plaintext?

34.   A Hill cipher with matrix $\begin{pmatrix} 2 & 3 \\ 5 & 12 \end{pmatrix}$ was used to encrypt a message. The ciphertext is *ALNS*. What is the plaintext?

35.  A common method of using block ciphers to encrypt is called *cipher block chaining*, or *CBC*. Suppose Bob has an encryption machine that takes an input $B$ of 128 bits and encrypts it to a block $C$ of 128 bits. We'll write this as $C = E(B)$. Alice knows how to decrypt $E$. That is, Alice knows the function $D$ such that $D(C) = B$. Bob wants to send a message $m$ to Alice. He breaks $m$ into 128-bit blocks $B_1, B_2, B_3, \ldots$. Then he chooses a random 128-bit seed $C_0$. He computes $C_1 = E(B_1 \oplus C_0)$ to get the first block of ciphertext. Then he computes $C_2 = E(B_2 \oplus C_1)$ to get the second block. In general, he computes $C_n = E(C_{n-1} \oplus B_n)$. The initial seed and the ciphertext blocks $C_0, C_1, C_2, \ldots$ are sent to Alice. When Alice receives this ciphertext, how does she decrypt it in order to recover $B_1, B_2, \ldots$?

**Section 7.7: Secret Sharing**

36.  Suppose we have a secret sharing scheme that works mod 53 and where $A, B, C, D$ get the points $(1, 6), (2, 48), (3, 37), (4, 26)$, respectively. The slope of the line through these points is the secret. Find the secret.

37.  Suppose we have a secret sharing scheme that works mod 13 and where $A, B, C, D$ get the points $(1, 12), (2, 4), (3, 9), (4, 1)$, respectively. The slope of the line through these points is the secret. Find the secret.

38.  Suppose we have a secret sharing scheme that works mod 37 and where $A, B, C, D$ get the points $(1, 19), (2, 30), (3, 22), (4, 15)$, respectively. The slope of the line through three of these points is the secret, but the remaining point is a forgery and is not correct. Find which point is the forgery and find the secret.

39.  Suppose we have a secret sharing scheme that works mod 37 and where $A, B, C, D$ get the points $(1, 29), (2, 15), (3, 28), (4, 9)$, respectively. The slope of the line through three of these points is the secret, but the remaining point is a forgery and is not correct. Find which point is the forgery and find the secret.

40.  You have a secret sharing scheme mod 29, as in Section 7.7. Your point is $(2, 18)$ and the secret is $s = 13$. Person $Q$'s point is $(17, *)$. Find the $y$-coordinate of $B$'s point.

# 7.10.2   Answers to "CHECK YOUR UNDERSTANDING"

1.  To obtain the decryption function, solve $y \equiv 9x + 2$ for $x$ in terms of $y$. Since 3 is the inverse of 9 mod 26, we have

$$y \equiv 9x + 2 \implies x \equiv (1/9)(y - 2) \equiv 3(y - 2) \equiv 3y - 6 \pmod{26}.$$

(We could also use $3y+20$, but using $3y-6$ lets us use smaller numbers.)
*KMI* is 10, 12, 8 in numbers. These decrypt as follows:

$$K = 10 \mapsto 3(10) - 6 = 24 \mapsto Y$$
$$M = 12 \mapsto 3(12) - 6 \equiv 4 \mapsto E$$
$$I = 8 \mapsto 3(8) - 6 \equiv 18 \mapsto S.$$

The plaintext is *YES*.

2.  There are 18 letters in the ciphertext, so there could be 2, 3, 4, 5, ... columns. Try them. Eventually you get to 4 columns, which yields

    MEET

    MEIN

    SAIN

    TLOU

    IS

So the plaintext is MEET ME IN SAINT LOUIS.

3.  To get the plaintext, compute ciphertext $\oplus$ key = 011101011101 $\oplus$ 110110110110 = 101010101010. Therefore, 101010101010 is the plaintext.

4.  Break the plaintext into 2-letter blocks: *GO LD*. Change this to numbers: (6, 14), (11, 3). Now compute

$$\begin{pmatrix} 3 & 2 \\ 13 & 1 \end{pmatrix} \begin{pmatrix} 6 \\ 14 \end{pmatrix} = \begin{pmatrix} 46 \\ 92 \end{pmatrix} \equiv \begin{pmatrix} 20 \\ 14 \end{pmatrix} \quad (\mathrm{mod}\ 26)$$

$$\begin{pmatrix} 3 & 2 \\ 13 & 1 \end{pmatrix} \begin{pmatrix} 11 \\ 3 \end{pmatrix} = \begin{pmatrix} 39 \\ 146 \end{pmatrix} \equiv \begin{pmatrix} 13 \\ 16 \end{pmatrix} \quad (\mathrm{mod}\ 26).$$

Change this back to letters to obtain the ciphertext: *UONQ*.

5.  Choose any two points, for example, (5,0) and (8,9). The line through these points has slope $(9 - 0)/8 - 5) = 3$. So the secret is 3. If we had chosen (2,4) and (8,9), we would obtain slope $(9 - 4)/(8 - 2) = 5/6 \equiv 3$ (mod 13).

# Chapter 8

# Fermat, Euler, and Wilson

## 8.1 Fermat's Theorem

Look at the following statements:

$$2^2 - 2 \equiv 0 \pmod{2}$$

$$2^3 - 2 \equiv 0 \pmod{3}$$

$$2^4 - 2 \not\equiv 0 \pmod{4}$$

$$2^5 - 2 \equiv 0 \pmod{5}$$

$$2^6 - 2 \not\equiv 0 \pmod{6}$$

$$2^7 - 2 \equiv 0 \pmod{7}$$

$$2^8 - 2 \not\equiv 0 \pmod{8}$$

$$2^9 - 2 \not\equiv 0 \pmod{9}$$

$$2^{10} - 2 \not\equiv 0 \pmod{10}$$

$$2^{11} - 2 \equiv 0 \pmod{11}.$$

Do you see a pattern? A reasonable guess is that $2^n - 2 \equiv 0 \pmod{n}$ if and only if $n$ is prime. A computer search soon produces the counterexample $2^{341} - 2 \equiv 0 \pmod{341}$, where $341 = 11 \cdot 31$ is not prime. So part of our conjecture falls apart. But the other part remains, namely that if $n$ is prime then $2^n - 2 \equiv 0 \pmod{n}$. In fact, there is a more general statement.

**Theorem 8.1.** *(Fermat's Theorem) Let p be prime.*
*(a) For every integer b,*

$$b^p - b \equiv 0 \pmod{p}.$$

*(b) If $b \not\equiv 0 \pmod{p}$, then*

$$b^{p-1} \equiv 1 \pmod{p}.$$

**Remark.** Sometimes Fermat's Theorem is called "Fermat's Little Theorem" in order to distinguish it from Fermat's Last Theorem.

We'll prove the theorem (twice!) at the end of the section. First, we give some examples.

**Example.** $2^6 = 64 \equiv 1 \pmod 7$, and $2^7 \equiv 2 \pmod 7$. Notice that the second statement can be obtained from the first by multiplying by 2.

**Example.** Fermat's theorem can be used for exponents that are multiples of $p - 1$: $3^{28} \equiv (3^4)^7 \equiv 1^7 \equiv 1 \pmod 5$.

**Example.** Divide 23 into $7^{200}$. What is the remainder? By Fermat's theorem, $7^{22} \equiv 1 \pmod{23}$. Therefore,

$$7^{200} \equiv (7^{22})^9 \cdot 7^2 \equiv 1^9 \cdot 49 \equiv 3 \pmod{23}.$$

This example demonstrates an important point. The original exponent 200 was reduced to 2 using the relation $200 = 22 \cdot 9 + 2$. In other words, $200 \equiv 2 \pmod{22}$. The general statement is the following.

**Corollary 8.2.** *Let p be prime and let $b \not\equiv 0 \pmod p$. If $x \equiv y \pmod{p-1}$, then*

$$b^x \equiv b^y \pmod p.$$

*Proof.* Since $x \equiv y \pmod{p-1}$, we can write $x = y + (p-1)k$ for some $k$. Therefore,

$$b^x = b^y (b^{p-1})^k \equiv b^y (1)^k \equiv b^y \pmod p,$$

where we used Fermat's theorem to change $b^{p-1}$ to 1 mod $p$.    □

**Warning.** Be sure to notice that $x$ and $y$ are congruent mod $p-1$, not mod $p$. In other words, congruences mod $p-1$ in the exponent lead to congruences mod $p$ overall. Do not try to use congruences mod $p$ in the exponent. They lead to wrong answers. For example, $6 \equiv 1 \pmod 5$ but $2^6 = 64$ is not congruent to $2^1 = 2$ mod 5. However, $5 \equiv 1 \pmod 4$ and $2^5 \equiv 2^1 \pmod 5$.

Fermat's theorem says that if $p$ is an odd prime then $2^{p-1} \equiv 1 \pmod p$. The contrapositive of this statement says that if $n$ is odd and $2^{n-1} \not\equiv 1 \pmod n$, then $n$ is not prime. This allows us to prove some integers are composite without factoring. For example, let's show that 77 is composite. We have

$$2^2 \equiv 4 \pmod{77}$$
$$2^4 \equiv (2^2)^2 \equiv 4^2 \equiv 16$$
$$2^8 \equiv (2^4)^2 \equiv 16^2 \equiv 256 \equiv 25$$
$$2^{16} \equiv (2^8)^2 \equiv 25^2 \equiv 625 \equiv 9$$
$$2^{32} \equiv (2^{16})^2 \equiv 9^2 \equiv 4$$
$$2^{64} \equiv (2^{32})^2 \equiv 4^2 \equiv 16$$
$$2^{76} \equiv 2^{64}2^8 2^4 \equiv 16 \cdot 25 \cdot 16 \cdot \equiv 6400 \equiv 9.$$

Since $2^{76} \not\equiv 1 \pmod{77}$, we conclude that 77 is not prime. This means that we have proved that 77 factors without factoring it. This might seem rather silly for a number as small as 77. But the same method can be used to show that much larger numbers are composite, even in cases where their factorization is currently impossible. For example, the 7th Fermat number

$$F_7 = 2^{2^7} + 1$$

was proved to be composite in 1905, but it wasn't until 1970, with the development of modern computer-oriented techniques, that its factorization was found. The Fermat number $F_{24} = 2^{2^{24}} + 1$, a number with more than 5 million digits, is known to be composite, but no prime factors are known.

Now it's time to prove Fermat's theorem. We give two different proofs, each with its own flavor.

**Proof 1.** We'll prove that $b^p \equiv b \pmod p$ by induction on $b$. We start with a lemma.

**Lemma 8.3.** *Let p be a prime. Then the binomial coefficient*

$$\binom{p}{j} = \frac{p!}{j!(p-j)!} \equiv 0 \pmod{p}$$

*for $1 \le j \le p-1$.*

*Proof.* We have

$$p! = \binom{p}{j}(p-j)!j!.$$

Clearly, $p$ divides the left-hand side, so $p$ must divide one of the factors on the right-hand side. Since all the factors of $(p-j)!$ are between 1 and $p-j$, we see that $p$ does not divide $(p-j)!$. Similarly, $p$ does not divide $j!$ since $j < p$. Therefore, $p$ must divide the binomial coefficient. $\qquad\square$

Returning to the proof of Fermat's theorem, we want to prove that $b^p \equiv b \pmod{p}$ for all $b$. We will do this using induction on $b$. Note that $0^p \equiv 0 \pmod{p}$ and $1^p \equiv 1 \pmod{p}$, so the statement is true for $b = 0$ and $b = 1$. Now assume that it's true for $b = k$, so $k^p \equiv k$. Then

$$(k+1)^p \equiv k^p + \binom{p}{1}k^{p-1} + \binom{p}{2}k^{p-2} + \cdots + \binom{p}{p-1}k + 1$$

$$\equiv k^p + 1 \equiv k + 1 \pmod{p},$$

where we have used the lemma to get rid of the binomial coefficients mod $p$ and we have used the induction assumption to change $k^p$ to $k$ mod $p$. Therefore, the statement is true for $b = k+1$. By induction, we conclude that $b^p \equiv b \pmod{p}$ for all positive integers $b$.

Now let $-c < 0$ be a negative integer. Since $c$ is positive, we already know $c^p \equiv c$, so

$$(-c)^p \equiv -c^p \equiv -c \pmod{p}$$

if $p$ is odd. If $p = 2$, we have $(-c)^2 \equiv +c \equiv -c \pmod 2$, so the result also holds for $p = 2$.

We have now proved that $b^p \equiv b \pmod{p}$ for all integers $b$.

If $b \not\equiv 0 \pmod{p}$, we can divide the congruence by $b$ and obtain $b^{p-1} \equiv 1 \pmod{p}$. $\qquad\square$

**Proof 2.** Of course, we start with a lemma.

**Lemma 8.4.** *Let $b \not\equiv 0 \pmod{p}$. Then the set of numbers*

$$b, \ 2b, \ 3b, \ \ldots, \ (p-1)b \pmod{p}$$

*contains each nonzero congruence class mod $p$ exactly once.*

**Example.** Let $p = 7$ and $b = 2$. Then the numbers $2, 4, 6, 8, 10, 12 \pmod 7$ are the same as $2, 4, 6, 1, 3, 5 \pmod 7$, so every nonzero congruence class mod 7 occurs exactly once.

*Proof.* Let $a \not\equiv 0 \pmod p$. Then $bx \equiv a \pmod p$ has a solution $x$, and $x \not\equiv 0$ since $a \not\equiv 0$, so we may assume that $1 \le x \le p - 1$. Therefore, $a$ occurs in the set $\{b, 2b, \ldots, (p-1)b)\}$ mod $p$. Since $a$ was arbitrary, every nonzero congruence class occurs.

If $bi \equiv bj \pmod p$ with $1 \le i < j \le p - 1$, we can divide by $b$ and get $i \equiv j \pmod p$. The bounds on $i$ and $j$ make this impossible, so we conclude that the multiples of $b$ are distinct mod $p$. Therefore, each nonzero congruence class occurs exactly once among the multiples of $b$. □

Now we can prove Fermat's theorem. Lemma 8.4 says that

$$\prod_{i=1}^{p-1} i \equiv \prod_{i=1}^{p-1} bi \pmod{p},$$

because the second product is a rearrangement of the first product. The first product is $(p-1)!$. If we factor $b^{p-1}$ from the second product, we see that the second product is $b^{p-1}(p-1)!$. Therefore,

$$(p-1)! \equiv b^{p-1}(p-1)! \pmod{p}.$$

Since $p$ does not divide $(p-1)!$ (because it doesn't divide any of the factors), Proposition 6.7 says that we can divide by $(p-1)!$ to obtain

$$b^{p-1} \equiv 1 \pmod{p}.$$

Multiplying by $b$ gives the other form: $b^p \equiv b \pmod p$. If $b \equiv 0$, then $b^p \equiv 0^p \equiv 0 \equiv b$, so the congruence holds for all $b$. □

---

## CHECK YOUR UNDERSTANDING

10. Compute $2^{104} \pmod{101}$.

## 8.2   Euler's Theorem

In the 1700s, Euler studied Fermat's work on number theory and provided proofs for many of his statements. In addition, he gave a very useful generalization of Fermat's theorem. This is the subject of the present section.

**Definition 8.5.** *Let $n$ be a positive integer. Define the* **Euler $\phi$-function** $\phi(n)$ *to be the number of integers $j$ with $1 \leq j \leq n$ such that $\gcd(j, n) = 1$.*

**Examples.**

1. $\phi(3) = 2$ since $j = 1$ and $j = 2$ have $\gcd(j, 3) = 1$.

2. $\phi(4) = 2$.

3. $\phi(12) = 4$ (the numbers are 1, 5, 7, 11).

4. $\phi(1) = 1$ (the only value of $j$ is $j = 1$; this is the reason we have $j \leq n$ rather than $j < n$ in the definition).

5. If $p$ is prime, then $\phi(p) = p - 1$.

The following is a very useful property of $\phi$.

**Proposition 8.6.** *Let $m, n$ be positive integers. If $\gcd(m, n) = 1$ then*

$$\phi(mn) = \phi(m) \cdot \phi(n).$$

We give the proof at the end of the section.

For prime powers, we have the following.

**Proposition 8.7.** *If $p$ is a prime and $k \geq 1$, then $\phi(p^k) = p^k - p^{k-1}$.*

*Proof.* There are $p^k$ integers $j$ with $1 \leq j \leq p^k$. Among these, there are $p^{k-1}$ multiples of $p$, namely, $p, 2p, 3p, \ldots, p^k$. These multiples of $p$ are exactly the values of $j$ with $\gcd(j, p^k) \neq 1$. So we have $p^k - p^{k-1}$ values of $j$ with $\gcd(j, p^k) = 1$. $\qquad\square$

Combining the previous two propositions yields the following.

**Theorem 8.8.** *Let $n = p_1^{a_1} p_2^{a_2} \cdots p_r^{a_r}$ with distinct primes $p_i$ and with exponents $a_i \geq 1$. Then*

$$\phi(n) = \prod_{i=1}^{r} \left( p_i^{a_i} - p_i^{a_i-1} \right) = n \prod_{p|n} \left( 1 - \frac{1}{p} \right),$$

*where the second product is over the primes dividing $n$.*

*Proof.* From Propositions 8.6 and 8.7,

$$\phi(n) = \prod_i \phi(p_i^{a_i}) = \prod_i (p_i^{a_i} - p_i^{a_i-1}).$$

This is the first formula. For the second, note that $p^a - p^{a-1} = p^a(1 - 1/p)$. Therefore,

$$\phi(n) = \prod_i p_i^{a_i} \left( 1 - \frac{1}{p_i} \right) = \left( \prod_i p_i^{a_i} \right) \prod_i \left( 1 - \frac{1}{p_i} \right)$$

$$= n \prod_i \left( 1 - \frac{1}{p_i} \right).$$

This is the second formula. $\qquad\qquad\qquad\qquad\qquad\qquad\qquad\qquad\qquad\square$

**Example.** Let's evaluate $\phi(100)$. The first formula says that

$$\phi(100) = \phi(2^2)\phi(5^2) = (2^2 - 2)(5^2 - 5) = 40.$$

We could also use the second formula to obtain

$$\phi(100) = 100 \left( 1 - \frac{1}{2} \right) \left( 1 - \frac{1}{5} \right) = 100 \left( \frac{1}{2} \right) \left( \frac{4}{5} \right) = 40.$$

Finally, we come to the main result.

**Theorem 8.9.** *(Euler's Theorem) Let $n$ be a positive integer and let $b$ be an integer with $\gcd(b, n) = 1$. Then*

$$b^{\phi(n)} \equiv 1 \pmod{n}.$$

**Remark.** Notice that Euler's theorem generalizes Fermat's theorem since $\phi(p) = p - 1$ and if $b$ is not a multiple of $p$, then $\gcd(b, p) = 1$.

*Proof.* We mimic the second proof of Fermat's theorem in order to prove this result. Again, we first need a lemma.

**Lemma 8.10.** *Let $T$ be the set of $j$ with $1 \le j \le n$ and $\gcd(j, n) = 1$. Choose any $b$ with $\gcd(b, n) = 1$ and let $bT \bmod n$ be the set of numbers of the form $bt \bmod n$ with $t \in T$. Then each element of $T$ is congruent to exactly one element of $bT$.*

**Example.** Let $n = 12$ and $b = 5$. Then

$$T = \{1, 5, 7, 11\} \text{ and } bT = \{5, 25, 35, 55\} \equiv \{5, 1, 11, 7\} = T.$$

*Proof.* First, suppose $t \in T$. Then $\gcd(t, n) = 1$ and $\gcd(b, n) = 1$, so $\gcd(bt, n) = 1$ by Proposition 2.15. Let $bt \equiv c \pmod{n}$. Proposition 6.9 implies that $\gcd(c, n) = 1$, so $c \in T$. This proves that every element of $bT \bmod n$ is in $T$.

We now know that $bT \bmod n$ is a proper subset of $T$. We'll show that the two sets have the same number of elements, hence must be equal. Suppose $t_1$ and $t_2$ are two elements in $T$ with $bt_1 \equiv bt_2 \pmod{n}$. Since $\gcd(b, n) = 1$, we can divide by $b$ and get $t_1 \equiv t_2 \pmod{n}$. Since the elements of $T$ are distinct mod $n$, this implies that $t_1 = t_2$. This proves that distinct element of $T$ remain distinct when they are multiplied by $b \bmod n$. Therefore, $bT \bmod n$ has the same number of elements as $T$, therefore equals $T$. This proves the lemma.                               □

Now we can prove Euler's theorem. Let $T$ be as in Lemma 8.10. Then

$$\prod_{i \in T} i \equiv \prod_{i \in T} bi \equiv b^{\phi(n)} \prod_{i \in T} i \pmod{n},$$

because Lemma 8.10 says that the second product is a rearrangement of the first product. There are $\phi(n)$ elements in $T$ so $b^{\phi(n)}$ factors out to give the third expression. Every factor in the first product is relatively prime to $n$, so the product is relatively prime to $n$. Therefore, we can divide by the first product to obtain

$$1 \equiv b^{\phi(n)} \pmod{n}.$$

This is Euler's theorem.                                                   □

**Examples.**

1.  $\phi(15) = \phi(3)\phi(5) = 8$, so $2^8 = 256 \equiv 1 \pmod{15}$.

2.  $\phi(10) = 4$ and $3^4 = 81 \equiv 1 \pmod{10}$.

3.  What are the last two digits of $3^{84}$? The last two digits of a number are given by the number mod 100, so we need to compute $3^{84} \pmod{100}$. Since $\phi(100) = 40$, Euler's theorem tells us that $3^{40} \equiv 1 \pmod{100}$. Therefore,

$$3^{84} \equiv (3^{40})^2 3^4 \equiv (1)^2 3^4 \equiv 81 \pmod{100}.$$

This means that the last two digits are 81.

The last example is a case of a more general phenomenon. Notice that when the modulus $n$ in the following corollary is prime, we get Corollary 8.2.

**Corollary 8.11.** *Let $n$ be a positive integer and suppose $b$ is an integer with $\gcd(b, n) = 1$. If $x \equiv y \pmod{\phi(n)}$, then*

$$b^x \equiv b^y \pmod{n}.$$

*Proof.* Write $x = y + \phi(n)k$ for some integer $k$. Then

$$b^x \equiv b^y (b^{\phi(n)})^k \equiv b^y (1)^k \equiv b^y \pmod{n},$$

where we used Euler's theorem to change $b^{\phi(n)}$ to 1.                □

**Examples.**

1.  Let $n = 15$. Then $\phi(n) = 8$. Since $9 \equiv 1 \pmod{8}$, we have $2^9 \equiv 2^1 \pmod{15}$. (You can check this. It says that $512 \equiv 2 \pmod{15}$.)

2.  Let $n = 10$. Then $\phi(n) = 4$. Since $5 \equiv 1 \pmod{4}$, it follows that $b^5 \equiv b \pmod{10}$ when $\gcd(b, 10) = 1$. This says that $b^5$ and $b$ have the last digit $b$ when $b$ is 1, 3, 7, or 9. A quick check shows that this is also true when $\gcd(b, 10) > 1$ (for example, $2^5 = 32$ has 2 as its last digit).

3.  Suppose $m$ is an integer with $\gcd(m, 77) = 1$. Let $c \equiv m^7$ (mod 77). What is $c^{43}$ (mod 77)? Since $\phi(77) = \phi(7)\phi(11) = 6 \cdot 10 = 60$, and $301 \equiv 1$ (mod 60), we have

$$c^{43} \equiv (m^7)^{43} \equiv m^{301} \equiv m^1 \pmod{77},$$

so $c^{43}$ (mod 77) gives back the original number $m$. This is an example of RSA encryption and decryption (see Chapter 9).

4.  What is the last digit of $3^{7^5}$? (*Note:* $a^{b^c}$ means $a^{(b^c)}$; the other possible interpretation, namely $(a^b)^c$, is better written in the easier form $a^{bc}$.) Let's start at the top. Since $\phi(4) = 2$ and $5 \equiv 1$ (mod 2), Corollary 8.11 says that $7^5 \equiv 7^1 \equiv 3$ (mod 4). Since $\phi(10) = 4$, Corollary 8.11 now says that

$$3^{7^5} \equiv 3^3 \equiv 27 \equiv 7 \pmod{10}.$$

This means that the last digit is 7.

Finally, let's return to Proposition 8.6 and give its proof. We want to show that if $\gcd(m, n) = 1$ then $\phi(mn) = \phi(m)\phi(n)$.

We begin by looking at two sets:

$S_1(m, n) =$
$\qquad \{(a, b) \mid 1 \le a \le m,\ 1 \le b \le n,\ \gcd(a, m) = 1,\ \text{and}\ \gcd(b, n) = 1\}$

and

$$S_2(m, n) = \{x \mid 1 \le x \le mn,\ \gcd(x, mn) = 1\}.$$

For example, let $m = 4$ and $n = 5$. Then

$$S_1(4, 5) = \{(1, 1), (3, 1), (1, 2), (3, 2), (1, 3), (3, 3), (1, 4), (3, 4)\}$$

(namely, all pairs with first entry equal to 1 or 3 and second entry equal to 1, 2, 3, or 4), and

$$S_2(4, 5) = \{1, 3, 7, 9, 11, 13, 17, 19\}.$$

We will continue with this example later in order to illustrate the main idea of the proof.

Next, we match up the elements in $S_1(m, n)$ with elements in $S_2(m, n)$. This will show that the two sets have the same number of elements.

Let $(a, b) \in S_1(m, n)$. By the Chinese Remainder Theorem, there is a unique element $x$ (mod $mn$) with the property that $x \equiv a$ (mod $m$) and $x \equiv b$ (mod $n$). We'll always choose $1 \le x \le mn$ and we'll refer to this $x$ as $x_{a,b}$. Since $x_{a,b}$ is important in this proof, we restate its definition:

$$x_{a,b} \text{ satisfies } x_{a,b} \equiv a \text{ (mod } m) \text{ and } x_{a,b} \equiv b \text{ (mod } n).$$

For example (with $m = 4$ and $n = 5$), $x_{3,1} = 11$ because $11 \equiv 3$ (mod 4) and $11 \equiv 1$ (mod 5). Let's list all of the numbers $x_{a,b}$ in this example:

$$x_{1,1} = 1, \quad x_{3,1} = 11, \quad x_{1,2} = 17, \quad x_{3,2} = 7$$
$$x_{1,3} = 13, \quad x_{3,3} = 3, \quad x_{1,4} = 9, \quad x_{3,4} = 19.$$

Notice that the numbers $x_{a,b}$ that we obtain are exactly the numbers in $S_2(4, 5)$, in a different order. There are $\phi(4)\phi(5)$ pairs in $S_1(4, 5)$ and $\phi(20)$ numbers in $S_2(4, 5)$. If we match a pair $(a, b)$ with $x_{a,b}$, then the sets $S_1(4, 5)$ and $S_2(4, 5)$ are matched up exactly, so $S_1(4, 5)$ and $S_2(4, 5)$ must have the same numbers of elements. This says that $\phi(4)\phi(5) = \phi(20)$.

We now return to the general case. If we start with $(a, b) \in S_1(m, n)$, we obtain $x_{a,b}$. We need to show that $x_{a,b} \in S_2(m, n)$. Because $x \equiv a$ (mod $m$), we know from Proposition 6.9 that $\gcd(x_{a,b}, m) = \gcd(a, m)$. Since $\gcd(a, m) = 1$, we see that $\gcd(x_{a,b}, m) = 1$. Similarly, $\gcd(x_{a,b}, n) = 1$. Proposition 2.14 then implies that $\gcd(x_{a,b}, mn) = 1$. This shows that $x_{a,b} \in S_2$.

Up to now, we have matched each pair $(a, b) \in S_1(m, n)$ with a number $x_{a,b} \in S_2(m, n)$. There are two questions that need to be answered:
(1) Can two different pairs $(a, b)$ be matched with the same $x_{a,b}$?
(2) Is every $x \in S_2(m, n)$ matched with some pair $(a, b) \in S_1(m, n)$?

First, let's consider (1). If $(a_1, b_1)$ and $(a_2, b_2)$ are both matched with the same number in $S_2(m, m)$, this means $x_{a_1, b_1} = x_{a_2, b_2}$. But

$$a_1 \equiv x_{a_1, b_1} = x_{a_2, b_2} \equiv a_2 \text{ (mod } m),$$

so $a_1 \equiv a_2$ (mod $m$). Since $1 \le a_i \le m$ for $i = 1, 2$, we must have

$a_1 = a_2$. Similarly, by looking at congruences mod $n$, we find that $b_1 = b_2$. Therefore, if two pairs are matched with the same $x \in S_2(m, n)$, then the pairs are equal.

For (2), let $x \in S_2$. Define $(a, b)$ with $1 \le a \le m$ and $1 \le b \le n$ by $a \equiv x$ (mod $m$) and $b \equiv x$ (mod $n$). Since $m \mid mn$, Exercise 61 in Chapter 2 tells us that $\gcd(x, m) = 1$. Since $x \equiv a$ (mod $m$), Proposition 6.9 implies that $\gcd(a, m) = 1$. Similarly, $\gcd(b, n) = 1$. So $(a, b) \in S_1(m, n)$.

Note that $x$ satisfies the conditions that define $x_{a,b}$. Since the $x_{a,b}$ obtained from the Chinese Remainder Theorem is unique mod $mn$, we must have $x_{a,b} \equiv x$ (mod $mn$). But $1 \le x \le mn$ and similarly for $x_{a,b}$. Therefore, $x = x_{a,b}$.

We have now proved that the pairs in $S_1(m, n)$ are matched up exactly with the numbers in $S_2(m, n)$. This means that the sets $S_1(m, n)$ and $S_2(m, n)$ have the same numbers of elements. Since $S_1(m, n)$ has $\phi(m)\phi(n)$ pairs, and $S_2$ has $\phi(mn)$ numbers, we have $\phi(m)\phi(n) = \phi(mn)$. This completes the proof. $\square$

---

**CHECK YOUR UNDERSTANDING**

11. Compute $2^{67}$ (mod 77).

---

# 8.3   Wilson's Theorem

In 1770, Edward Waring wrote the book *Meditationes Arithmeticae*, in which he stated a property of prime numbers that his student John Wilson had discovered but neither Waring nor Wilson could prove. Henceforth, it has been known as Wilson's Theorem. However, the result was known to Bhāskara about 1000 years before Wilson, and also to Ibn al-Haytham (around 1000) and to Leibniz (around 1700). The first published proof was by Lagrange in 1771.

**Theorem 8.12.** *(Wilson's Theorem) Let $p$ be prime. Then*

$$(p - 1)! \equiv -1 \pmod{p}.$$

**Example.** Let $p = 7$. Then

$$(p-1)! = 6! = 720 \equiv -1 \pmod 7.$$

*Proof.* Let's see why the example works. Rearrange the factorial as

$$6! \equiv (6)(5 \cdot 3)(4 \cdot 2)(1) \equiv (-1)(1)(1)(1) \pmod 7.$$

Note that we have paired numbers with their multiplicative inverses. For example, 5 is paired with 3 and $5 \cdot 3 \equiv 1 \pmod 7$. Since 6 and 1 are their own inverses ($6 \cdot 6 \equiv 1 \pmod 7$, and $1 \cdot 1 \equiv 1$), they did not get paired with anything else.

The same idea works in general. Start with the integers $1 \le b \le p-1$. Recall that the equation $bx \equiv 1 \pmod p$ always has a unique solution $x$ with $1 \le x \le p-1$, so we can pair each integer with its multiplicative inverse mod $p$. If $y$ is the multiplicative inverse of $x$, then $x$ is the multiplicative inverse of $y$ (this says that $xy \equiv 1 \pmod p$ implies $yx \equiv 1 \pmod p$). Therefore, if $x$ is paired with $y$, then $y$ is paired with $x$. A number $b$ is paired with itself exactly when $b$ is its own inverse, which says that $b \equiv b^{-1}$, hence $b^2 \equiv 1 \pmod p$. By Corollary 6.11, the only solutions of this are $b \equiv \pm 1 \pmod p$, which means that $b = 1$ and $b = p - 1$ are the only numbers that get paired with themselves. Now rearrange $(p-1)!$ so that each number except 1 and $p-1$ is next to its multiplicative inverse, just as in the example for $p = 7$. All the pairs in the product combine to yield factors of 1, leaving only the unpaired factors 1 and $p - 1$. Therefore,

$$(p-1)! \equiv 1 \cdot (p-1) \equiv p - 1 \equiv -1 \pmod p.$$

This completes the proof.                                            $\square$

**Example.** Here is another example of what happened in the proof of Wilson's theorem. Let $p = 11$. Then

$$(p-1)! = 10! \equiv (10)(9 \cdot 5)(8 \cdot 7)(6 \cdot 2)(4 \cdot 3)(1)$$
$$\equiv (10)(1)(1)(1)(1)(1) \equiv 10 \equiv -1 \pmod{11}.$$

Wilson's theorem yields a primality test. Unfortunately, this test is not practical since there is no known method for computing factorials quickly.

**Corollary 8.13.** *Let $n \geq 2$ be an integer. Then $n$ is prime if and only if $(n-1)! \equiv -1 \pmod{n}$.*

*Proof.* If $n$ is prime then $(n-1)! \equiv -1 \pmod{n}$ is just Wilson's theorem. Conversely, suppose $(n-1)! \equiv -1 \pmod{n}$. If $n$ is composite then $n = ab$ with $1 < a < n$. This implies that $a$ is a factor of $(n-1)!$, so $(n-1)! \equiv 0 \pmod{a}$. But $a$ is also a factor of $n$, so

$$(n-1)! \equiv -1 \pmod{n} \implies (n-1)! \equiv -1 \pmod{a}.$$

The last two sentences tell us that $0 \equiv -1 \pmod{a}$, which is impossible. Therefore, $n$ cannot factor, so $n$ is prime. $\qquad\qquad\square$

**Example.** Let $n = 6$. Then

$$(n-1)! = 5! = 120 \equiv 0 \pmod{6},$$

which implies that 6 is not prime.

In general, when $n$ is composite, $(n-1)! \equiv 0 \pmod{n}$ if $n \neq 4$. See Exercise 62. Examples can be deduced from what we saw in Chapter 5. Exercise 39 in that chapter says that 1000! ends in 249 zeros. Therefore, 999! ends in 246 zeros. In particular, 999! is a multiple of 1000.

---

# 8.4   Chapter Highlights

1. Fermat's Theorem: If $p$ is prime and $a \not\equiv 0 \pmod{p}$, then $a^{p-1} \equiv 1 \pmod{p}$.

2. Euler's Theorem: If $\gcd(a, n) = 1$, then $a^{\phi(n)} \equiv 1 \pmod{n}$.

3. Wilson's Theorem: If $p$ is prime, then $(p-1)! \equiv -1 \pmod{p}$.

# 8.5    Problems

## 8.5.1    Exercises

### Section 8.1: Fermat's Theorem

1.    Compute $5^{72}$ (mod 73).
2.    Compute $2^{100}$ (mod 97).
3.    Compute $2^{1234}$ (mod 11).
4.    Compute $3^{75}$ (mod 73).
5.    Compute $7^{50}$ (mod 47).
6.    Evaluate $111^{222} + 222^{111}$ (mod 13).
7.    Show that $2222^{5555} + 5555^{2222}$ is a multiple of 7.
8.    Find an integer $n$ with $0 \le n \le 22$ such that $7^{90} \equiv n$ (mod 23).
9.    Find an integer $n$ with $0 \le n \le 16$ such that $5^{90} \equiv n$ (mod 17).
10.   Show that $5^{12} + 12^5 \equiv 0$ (mod 13).
11.   Show that $11^{16} + 16^{11} \equiv 0$ (mod 17).
12.   Show that $3333^{4444} + 4444^{3333}$ is divisible by 7.
13.   Show that for every integer $n$, the number $n^{23} + 6n^{13} + 4n^3$ is a multiple of 11.
14.   Show that if $p$ is a prime and $a^p + b^p = c^p$, then $a + b - c \equiv 0$ (mod $p$). (*Remark:* Fermat's Last Theorem (see Chapter 21) says that if $p \ge 3$ then at least one of $a, b, c$ must be 0. However, do not assume this result for this problem.)
15.   Divide $111^{222} + 222^{111}$ by 11. What is the remainder?
16.   Using the fact that $2^{11} \equiv 1$ (mod 89), evaluate $2^{123456789}$ (mod 89).
17.   Divide $221^{654}$ by 109. What is the remainder? (*Note:* 109 is prime.)
18.   Let $n$ be an integer. Show that $n^{13} - n$ is a multiple of 2730.
19.   If $p$ is a prime number and $a$ is a positive integer, show that $a^{(p-1)!+1} \equiv a$ (mod $p$).
20.   Show that $n^{17} - n \equiv 0$ (mod 510) for all integers $n$. (*Hint:* Factor 510 and use Fermat's theorem for each factor.)
21.   Use Fermat's theorem to show that $1^n + 2^n + 3^n + 4^n$ is divisible by 5 if and only if $n$ is *not* divisible by 4.
22.   Let $p$ be an odd prime and let $i \ge 0$. Show that $2^i \not\equiv 2^{p+i}$ (mod $p$). (*Remark:* This shows that having the exponents congruent mod $p$ does not yield an overall congruence mod $p$.)
23.   Let $p$ be prime and let $a \not\equiv 0$ (mod $p$). Let $b \equiv a^{p-2}$ (mod $p$). Show that $b$ is a multiplicative inverse for $a$ mod $p$.
24.   (a) Show that if $p \equiv 3$ (mod 4) is prime then $x^2 \equiv -1$ (mod $p$) has no solution. (*Hint:* Raise both sides of the congruence to the $(p-1)/2$

power.)

(b) Suppose $p$ is an odd prime and $n$ is an integer. Show that if $p \mid n^2 + 1$ then $p \equiv 1 \pmod{4}$.

(c) Show that there are infinitely many primes $p \equiv 1 \pmod{4}$. (*Hint:* Look at $(2p_1 p_2 \cdots p_r)^2 + 1$.)

## Section 8.2: Euler's Theorem

25.  Calculate (a) $\phi(77)$,   (b) $\phi(200)$,   (c) $\phi(61)$,   (d) $\phi(165)$.

26.  Calculate (a) $\phi(65)$,   (b) $\phi(99)$,   (c) $\phi(31)$,   (d) $\phi(385)$.

27.  Compute $5^{60} \pmod{21}$.

28.  Compute $3^{240} \pmod{385}$.

29.  Compute $2^{35} \pmod{35}$.

30.  Compute $7^{30} \pmod{15}$.

31.  Compute $11^{100} \pmod{35}$.

32.  What are the possible remainders when the 42nd power of an integer is divided by 49?

33.  What are the possible remainders when the 110th power of an integer is divided by 121?

34.  Find the last two digits of $123^{403}$.

35.  Find the last three digits of $101^{40404}$.

36.  Find the last two digits of $3^{444}$.

37.  Find the last 3 digits of $401^{801} + 801^{401}$.

38.  (a) Show that $13^{13} \equiv 1 \pmod{4}$.
     (b) Find the last digit of $13^{13^{13}}$.

39.  Find the last four digits of $207^{8004002}$.

40.  Let $n$ be a positive integer,
     (a) What are the possible values of the last digit of $3^n$?
     (b) What are the possible values of the last digit of $4^n$?
     (c) What are the possible values of the last digit of $6^n$?
     (d) What are the possible values of the last digit of $7^n$?

41.  Find all solutions to the following. If the equation has no solutions, give an explanation.
     (a) $\phi(x) = 2$
     (b) $\phi(x) = 3$
     (c) $\phi(x) = 4$
     (d) $\phi(x) = 10$
     (e) $\phi(x) = 14$

42.  (a) Show that if $p$ is a prime dividing $n$ then $p - 1$ is a factor of $\phi(n)$.
     (b) Show that there is no integer solution to $\phi(n) = 26$.
     (c) Show that there is no integer solution to $\phi(n) = 98$.

43.  Find all values of $n$ for which $\phi(n) = n - 2$.

44.  Let $n \geq 2$.
     (a) Show that if $n$ is even then $\phi(n) \leq n/2$.
     (b) Show that $\phi(n) = n/2$ if and only if $n$ is a power of 2.

45.  Let $n$ be a positive integer. Show that $n\phi(n) = \phi(n^2)$.

46.  Let $n$, $a$, and $b$ be positive integers. Show that $n^{a+b}\phi(n^a)\phi(n^b) = \left(\phi(n^{a+b})\right)^2$.

47.  Let $a$ and $b$ be relatively prime positive integers. Show that $a^{\phi(b)} + b^{\phi(a)} \equiv 1 \pmod{ab}$.

48.  Decide whether each of the following is true or false. If true, give a proof; if false, give a counterexample.
     (a) If $d \mid n$, then $\phi(d) \mid \phi(n)$.
     (b) If $\phi(d) \mid \phi(n)$, then $d \mid n$.
     (c) If $d \mid n$, then $\phi(dn) = d\phi(n)$.

49.  Decide whether each of the following is true or false. If true, give a proof; if false, give a counterexample.
     (a) If $n$ has $k$ distinct prime factors, then $2^k \mid \phi(n)$.
     (b) If $n$ has $k$ distinct odd prime factors, then $2^k \mid \phi(n)$.

50.  (a) Let $n$ be an integer with $\gcd(n, 10) = 1$. Show that there is a number $N = 1111 \cdots 111$, whose base 10 expression is a string of 1's, such that $n \mid N$.
     (b) Part (a) can be done using the fact that there are only finitely many possibilities mod $n$ for a string of 1's and then subtracting. It can also be done via Euler's theorem. Use a method that you didn't use in (a) to give a second proof of (a).

51.  Here is another proof that there are infinitely many primes. Suppose $2, 3, 5, \ldots, p_n$ are all the primes. Let $N = 2 \cdot 3 \cdot 5 \cdots p_n$.
     (a) Show that $\phi(N) \geq 8$.
     (b) Show that if $1 < a \leq N$, then $\gcd(a, N) \neq 1$. (*Hint:* Lemma 2.6.)
     (c) Use (a) and (b) to obtain a contradiction and therefore conclude that there are infinitely many primes.
     Even though this proof looks easy, it uses the multiplicativity of the $\phi$-function, which is fairly complicated.

52.  (a) Find $x$ such that $x \equiv 0 \pmod 9$ and $x \equiv 41 \pmod{100}$.
     (b) Divide $987654321^{60402}$ by 900. What is the remainder? (*Hint:* Use part (a). You cannot use Euler's theorem here for mod 900 calculations because the gcd condition is not satisfied.)

53.  (a) Find $x$ such that $x \equiv 0 \pmod{11}$ and $x \equiv 69 \pmod{100}$.
     (b) Divide $3113^{20202}$ by 1100. What is the remainder? (*Hint:* Use part (a). You cannot use Euler's theorem here for mod 900 calculations because the gcd condition is not satisfied.)

## Section 8.3: Wilson's Theorem

54.  Show how the numbers match up in the proof of Wilson's theorem for $p = 13$.

55.  Let $p$ be prime. Show that $p \mid (p - 2)! - 1$.

56.  Let $p$ be prime and let $0 < s < p$. Show that

$$(s - 1)! \, (p - s)! \equiv (-1)^s \pmod p.$$

57.  Show that for every prime number $p$ and every integer $m$, the number $m^p + (p-1)!\, m$ is divisible by $p$.

58.  Show that if $p \geq 7$ is a prime, then $(p-1)^2$ divides $(p-1)!$. (*Hint: $p-1$ is divisible by 2 and $p-1$ is divisible by $(p-1)/2$.*)

59.  Let $p$ be prime. Let $N$ be the binomial coefficient $\binom{2p}{p}$. Show that $N \equiv 2$ (mod $p$).

60.  Show that $1000000! \equiv 500001 \pmod{1000003}$. You may use the fact that 1000003 is prime.

61.  Find the least positive residue of $2009! \pmod{2011}$. You may use the fact that 2011 is prime.

62.  Let $n \neq 4$ be composite. Show that $(n-1)! \equiv 0 \pmod{n}$ if $n \neq 4$. (*Hint: Let $1 < b < n$ be a factor of $n$. Consider the cases $b \neq n/b$ and $b = n/b$ separately. In the latter case, if $b > 2$ then both $b$ and $2b$ occur in $(n-1)!$.*)

63.  Let $p$ be an odd prime. Let $x = ((p-1)/2)!$.
     (a) Using the fact that $(p-1)! =$
     $$(1(p-1))\,(2(p-2))\cdots(((p-1)/2)((p+1)/2)),$$
     show that $-1 \equiv (-1)^{(p-1)/2} x^2 \pmod{p}$.
     (b) Suppose that $p \equiv 1 \pmod 4$. Show that $x^2 \equiv -1 \pmod{p}$.

64.  Show that each of the numbers
     $$100! + 1, 100! + 2, 100! + 3, \ldots, 100! + 100$$
     is composite. (*Note:* The proof that the first number on the list is composite is harder than the proof for the other 99 numbers.)

65.  Let $p$ be prime. Show that if $a$ and $b$ are positive integers such that $a + b \geq 2p - 1$, then $a!\, b! \equiv 0 \pmod{p}$.

66.  What is the smallest prime $p$ that divides $n = 46! + 1$? You must say why $p$ is a divisor of $n$ and why no smaller prime divides $n$.

## 8.5.2  Projects

1.  This problem will show how to deduce Euler's theorem from Fermat's theorem.
    (a) Let $p$ be prime and $k \geq 1$. Let $a$ and $b$ be integers with $a \equiv b$ (mod $p^k$). Show that $a^p \equiv b^p \pmod{p^{k+1}}$. (*Hint:* Write $a = b + cp^k$ and use the Binomial Theorem.)
    (b) Let $p$ be prime and let $m \geq 1$. Assume $\gcd(b, p) = 1$. Use Fermat's theorem, part (a), and induction to show that
    $$b^{(p-1)p^{m-1}} \equiv 1 \pmod{p^m}.$$
    This is Euler's theorem for prime powers.
    (c) Let $n = p_1^{m_1} p_2^{m_2} \cdots p_r^{m_r}$ be the prime factorization of $n$. Assume that $\gcd(b, n) = 1$. Use part (b) and the multiplicativity of $\phi$ to show that
    $$b^{\phi(n)} \equiv 1 \pmod{p_i^{m_i}}$$

for $1 \le i \le r$.

(d) Let $n$ and $b$ be as in part (c). Use part (c) to show that $b^{\phi(n)} \equiv 1$ (mod $n$).

## 8.5.3    Computer Explorations

1. (a) A prime $p$ is called a *Wieferich prime* if $2^{p-1} \equiv 1$ (mod $p^2$). There are only two such primes known. Both are less than 4000. Can you find them?

   (b) A prime $p$ is called a *Mirimanoff prime* if $3^{p-1} \equiv 1$ (mod $p^2$). There are two such primes known. One is small and the other is slightly larger than 1 million. Can you find them?

   (c) Can you find other examples of primes $p$ where $r^{p-1} \equiv 1$ (mod $p^2$) for a small prime $r$?

   Wieferich primes and Mirimanoff primes arose in connection with Fermat's Last Theorem.

## 8.5.4    Answers to "CHECK YOUR UNDERSTANDING"

1. Fermat's theorem tells us that $2^{100} \equiv 1$ (mod 101). Therefore, $2^{104} \equiv 2^{100}2^4 \equiv 2^4 \equiv 16$ (mod 101).

2. Since $\phi(77) = \phi(7)\phi(11) = 6 \cdot 10 = 60$, Euler's theorem tells us that $2^{60} \equiv 1$ (mod 77). Therefore, $2^{67} \equiv 2^{60}2^7 \equiv 2^7 \equiv 128 \equiv 51$ (mod 77).

# Chapter 9

# RSA

Symmetric systems work well if Alice and Bob have met previously and have agreed on a key. But what happens if they haven't? For example, suppose Alice is in central Australia (at Alice Springs) and wants to communicate with the Bank of Baltimore (Bob). Alice and Bob cannot agree on a secret key by e-mail or by talking on the telephone, since eavesdroppers could easily intercept their communications. (If they could securely agree on a secret key by mail, e-mail, or telephone, they would have no need of a cryptosystem!) Alice could send a courier to Bob with the secret key locked in a box. But that might take a few days. Fortunately, there is a solution, called **public key cryptography**. Alice generates two keys, one called the **public key** and one called the **private key**. Alice sends the public key to Bob and this allows Bob to encrypt messages. But decryption requires the private key, which only Alice knows. So Bob can send a secret message to Alice that only Alice can read, even though every communication between Alice and Bob is over public channels. Because Alice and Bob have different information about the keys, public key cryptography is an **asymmetric cryptosystem**.

The concept of public key cryptography was invented by James Ellis in 1970, and an implementation was found by Clifford Cocks in 1973. However, they worked for the British cryptographic agency and the discovery was kept secret until 1997. Meanwhile, in the unclassified world, Whitfield Diffie and Martin Hellman published an influential paper in 1976 describing public key cryptography. Then, in 1977, Ron Rivest, Adi Shamir, and Leonard Adleman discovered the implementation (essentially the same as the one by Cocks) that is known as RSA. This algorithm is the focus of this chapter.

The influence of RSA on number theory has been enormous. It showed that number theory, long assumed to be a purely theoretical branch of mathematics, has important practical applications. For example, since the security of the RSA system is closely tied to the difficulty of factoring, the development of factorization algorithms has received a lot of attention, and computational problems have become much more popular. Of course, you are still permitted to study number theory because it is fun, but now people also study it because it is useful.

One drawback of public key systems is that they are relatively slow, making it impractical to send massive amounts of data. For this reason, the process of sending an encrypted message is usually broken up into two steps. First, a key is transmitted using RSA or another public key method. Once both parties have this key, the much faster symmetric systems such as DES and AES are then used to send the data. For example, public key systems are used to initiate communications whenever you buy things over the Internet. This is natural, since you probably have not met the online company, but you still need to transmit your credit card number or other confidential data.

In Chapter 12, we discuss another method (Diffie-Hellman) for establishing a secret key for Alice and Bob.

# 9.1   RSA Encryption

Let's say that Alice owns a bookstore and, in order to generate revenue, needs to accept online orders. Bob sees a book he wants, but in order to purchase it, he needs to send Alice his credit card number. He's concerned that if he does so, someone may intercept his communication with Alice's store and steal his number. Being well read, he knows that in symmetric key cryptosystems, the key used to encrypt a message is essentially the same as the key used to decrypt the message. This means that the person who wants to send a secret message has to protect the encryption key as well as the decryption key. Since Alice wants lots of people to have her encryption key, this can cause logistical problems no matter how strong a symmetric key cryptosystem she possesses. This dilemma was addressed in the groundbreaking paper by Diffie and Hellman, which set the stage for the rise of practical and commercially feasible public key cryptosystems.

All public key cryptosystems have a common theme: A public encryption key is produced using secret or private numbers. This key is distributed without any concerns about concealment. It could be printed in every local and national newspaper or distributed to every e-mail address in the world. With this key, anyone can encrypt a message and send it to you. The security of this system rests on the difficulty of finding the secret numbers that were used to create it, because they are the numbers used to decrypt.

The most widely used public key cryptographic system is RSA, named after Ron Rivest, Adi Shamir, and Len Adleman. It was developed in 1977, a year after the Diffie-Hellman paper appeared. The public encryption key uses the product of two large primes, and its security rests on the difficulty of being able to factor that product and retrieve the primes. The following is a step-by-step description of how Alice uses RSA in order to allow Bob and others to communicate securely with her.

---

### RSA Setup

1. Alice chooses two primes, $p$ and $q$, and calculates $n = pq$ and $\phi(n) = (p-1)(q-1)$.

2. Alice chooses an $e$ (the encryption key) so that $\gcd(e, \phi(n)) = 1$.

3. Alice calculates $d$ (the decryption key) with the property that $ed \equiv 1 \pmod{\phi(n)}$.

4. Alice makes $n$ and $e$ public and keeps $d, p, q$ secret.

---

**Public Information:** $n$ and $e$

**Private Information:** $p, q,$ and $d$

---

### RSA Encryption

1. Bob looks up Alice's $n$ and $e$.
2. Bob writes his message as $m$ (mod $n$).
3. Bob computes $c \equiv m^e$ (mod $n$).
4. Bob sends $c$ to Alice.

---

### RSA Decryption

1. Alice receives $c$ from Bob.
2. Alice computes $m \equiv c^d$ (mod $n$).

---

**Example.** Alice, realizing that she can't make money selling books, decides to become a spy. She needs to be able to communicate securely with the other spies in her network. To do so, she chooses two primes $p$ and $q$ and then multiplies them together to get her modulus $n$:

$$p = 3598279, \quad q = 7815629, \text{ and } n = p \cdot q = 28122813702491.$$

She next chooses an encryption exponent $e$, which has to be relatively prime to $\phi(n) = (p-1)(q-1)$, and calculates her decryption exponent $d$. The number $d$ is a solution to the congruence $ed \equiv 1$ (mod $(p-1)(q-1)$) and is kept secret. Alice's choice of

$$e = 233$$

forces

$$d = 27519308677241,$$

which she calculates using the Extended Euclidean Algorithm. Alice makes $n$ and $e$ public while keeping $p$, $q$, and $d$ private.

Bob needs to set up a meeting with Alice. Out of their prearranged spots, he decides that they'll meet at his car. (They're not very good spies.) So, he needs to send the secret message $CAR$. In order to do so, he converts each letter to a number. Previously, we started with $a = 0$. We don't want to do that here because any message starting with the letter $A$ would then start with the number 0 and the 0 would get dropped during the RSA process. So, he starts with $A = 01$, $B = 02$,..., $Z = 26$. This makes his message ($CAR$) have the numerical value

$$m = 030118 = 30118$$

since $C = 03$, $A = 01$, and $R = 18$. Bob calculates his ciphertext with the following calculation:

$$c \equiv m^e \equiv 30118^{233} \equiv 21666077416496 \pmod{28122813702491}.$$

After Alice receives the ciphertext $c = 21666077416496$, she decrypts it with her secret exponent $d$:

$$m \equiv c^d \equiv 21666077416496^{27519308677241}$$
$$\equiv 30118 \pmod{28122813702491}$$

(a calculation such as this is done with a modular exponentiation algorithm; see Section 6.2). Alice now has Bob's original message and knows to meet him at the car.

**Remarks:**

1.  In Step 1, both of the two primes chosen should be large. For secure cryptographic applications, they are more than 100 digits long. Section 14.2 will describe methods for finding them. In Exercise 21, you'll see that computing $\phi(n)$ is equivalent to factoring $n$.

2.  In Step 2, somewhat surprisingly, $e$ does not have to be large. Sometimes $e = 3$ has been used, although it is usually recommended that larger values be used. In practice, the most common value is $e = 65537$. The advantages are that 65537 is a reasonably large prime, so it's easier to arrange that $\gcd(e, \phi(n)) = 1$, and that it is one more than a power of 2, so raising a number to the 65537th power consists mostly of squaring, plus only one multiplication, when the algorithm of Section 6.2 is used.

3. In Step 3, $d$ is calculated using the Extended Euclidean Algorithm and is kept secret. To do this, Alice uses her knowledge of $p$ and $q$ to compute $\phi(n) = (p-1)(q-1)$, which allows her to solve $de \equiv 1 \pmod{\phi(n)}$ by the Extended Euclidean Algorithm.

4. We will see in Section 14.3 that if someone knows $e$, $d$, and $n$, then it is easy to factor $n$. Therefore, finding $d$ is equivalent to factoring $n$. If we believe that factoring $n$ is hard, then we conclude that there is no easy way for an eavesdropper to find $d$. Therefore, attacks on an RSA system often try to decrypt a plaintext without finding $d$. One such attack is given later in this section.

5. If Bob's message is larger than $n$, then it has to be broken up into pieces (or blocks), with each piece smaller than $n$. Since we're primarily concerned with the mathematics and not with cryptographic implementation issues, we'll ignore this and assume that $0 \le m < n$.

Your first question might be, "Why does $c^d \pmod{n}$ give back $m$?" This "magic" is a consequence of Euler's theorem, as the following proposition shows.

**Proposition 9.1.** *Let $n = pq$ be the product of two distinct primes, and let $d, e$ be positive integers satisfying $ed \equiv 1 \pmod{(p-1)(q-1)}$. Then, for all integers $m$,*

$$m^{ed} \equiv m \pmod{n}.$$

*Therefore, if $c \equiv m^e \pmod{n}$, we have $m \equiv c^d \pmod{n}$.*

*Proof.* First, we'll treat the case where $\gcd(m, n) = 1$. Since $ed \equiv 1 \pmod{\phi(n)}$, we can write

$$ed = 1 + k\phi(n) = 1 + k(p-1)(q-1)$$

for some $k$. By Euler's theorem,

$$m^{\phi(n)} \equiv 1 \pmod{n},$$

which implies that

$$m^{ed} \equiv m^{1+k\phi(n)} \equiv m \left( m^{\phi(n)} \right)^k \equiv m(1)^k \equiv m \pmod{n}.$$

This is the heart of the proof, but we still need to treat the cases where $\gcd(m, n) > 1$. In practice, $n$ is the product of two extremely large primes, so the chances of this actually occurring are infinitesimally small. Since $\gcd(m, n)$ is a divisor of $n = pq$, we then must have $\gcd(m, n) = p$, $q$, or $pq$.

The case $\gcd(m, n) = pq$ is easy. In this case $n \mid m$, so $m \equiv 0 \pmod{n}$. Therefore, $m^{ed} \equiv 0^{ed} \equiv 0 \equiv m \pmod{n}$, as desired.

Now assume that $\gcd(m, n) = p$. Then $p \mid m$, so $m \equiv 0 \pmod{p}$. Therefore, $m^{ed} \equiv 0^{ed} \equiv 0 \equiv m \pmod{p}$. But $q \nmid m$, so $m^{q-1} \equiv 1 \pmod{q}$ by Fermat's theorem. Therefore,

$$m^{ed} \equiv m^{1+k(p-1)(q-1)} \equiv m \left(m^{q-1}\right)^{k(p-1)}$$
$$\equiv m(1)^{k(p-1)} \equiv m \pmod{q}.$$

We have now proved that

$$p \mid m^{ed} - m, \text{ and } q \mid m^{ed} - m.$$

Therefore, $pq \mid m^{ed} - m$, which is the same as $m^{ed} \equiv m \pmod{pq}$.

The case where $\gcd(m, n) = q$ uses the same argument, with $p$ and $q$ switched. Since we have treated all cases, the proof is complete.    □

Why are people so confident in the security of this system? If we can factor $n$, then we can use the knowledge of $p$ and $q$ to compute $d$. In fact, we'll show (Section 14.3) that knowing $d$ is computationally equivalent to factoring $n$. Factoring integers has been a topic of research for hundreds of years. In our situation, where the number that needs to be factored is the product of two very large primes, this is generally believed to be computationally infeasible.[1] Therefore, the public information does not allow an attacker to find $d$.

One of the uses of RSA is to transmit keys for use in symmetric cryptosystems. In this case, $n$ could be a few hundred digits, while the message could have only a few digits. For example, suppose the message $m$ has 16 or fewer digits. In this situation, with a relatively short message, the plaintext can be retrieved without factoring $n$ or finding $d$. Here's how to accomplish this. An eavesdropper intercepts the ciphertext $c$, which probably has approximately as many digits as $n$. Next, the eavesdropper makes two lists:

---

[1]This term is rather time dependent. It means that the computation would take thousands of years on the fastest computers currently available.

1.    $x^e \pmod{n}$ for all $x$ with $1 \leq x \leq 10^9$.

2.    $cy^{-e} \pmod{n}$ for all $y$ with $1 \leq y \leq 10^9$.

Suppose there is a match between the two lists: $x^e \equiv cy^{-e} \pmod{n}$. Then

$$(xy)^e \equiv c \pmod{n}.$$

Since $1 \leq xy \leq 10^{18} < n$, the original message must have been $m = xy$. Why can we hope to get a match? Since $m$ is at most a 16-digit number, there is a reasonable chance that it factors as the product of two numbers of at most 9 digits (this is certainly not guaranteed, but it happens some of the time). If $m = xy$ with $1 \leq x, y < 10^9$, then $c \equiv m^e \equiv (xy)^e$, so $x^e \equiv cy^{-e}$. This yields a match. The way to avoid this "small message attack" is to append some random digits at the beginning and end of the message to make it into a much larger number.

**Example.** Let's take another look at the previous example in this section. We know that the modulus is $n = 28122813702491$, the encryption exponent is $e = 233$, and the ciphertext is $c = 21666077416496$. We suspect that the message satisfies $m < 10^5$, and we hope that it factors as the product of two numbers less than 1000. We make two lists,

1.    $x^{233} \pmod{28122813702491}$ for $1 \leq x < 1000$

2.    $21666077416496 \cdot y^{-233} \pmod{28122813702491}$
      for $1 \leq y < 1000$,

and discover that there is a match:

$$74^{233} \equiv 21666077416496 \cdot 407^{-233} \pmod{28122813702491}.$$

Therefore,

$$21666077416496 \equiv (74 \cdot 407)^{233} \equiv 30118^{233},$$

which means (because we can decrypt uniquely) that

$$m = 30118 \quad (= \ car).$$

However, suppose Bob changes his message to $m = 111030118111$. Then $m$ cannot factor as $m = xy$ unless we allow $x$ or $y$ to have at least 6 digits, so this attack cannot succeed without longer lists. Bob can even tell everyone that he has added the extra digits.

In real-life situations, if someone knows that the message satisfies $m <$ $10^5$, then an eavesdropper simply encrypts all $m < 10^5$ and sees which one gives the ciphertext. This takes a very short amount of time to do. That is why we talked about a 16-digit message when describing the method. However, a 16-digit message could also be discovered by encrypting all $m < 10^{16}$. But it would have to be a very important message before this would be worthwhile because an $m$ of this size would take a relatively long time to discover with this technique. Making and comparing two lists of length $10^9$ is much more feasible.

---

## CHECK YOUR UNDERSTANDING

1. Suppose an RSA public key is $n = 55$ and $e = 27$. If the ciphertext is $c = 4$, find the plaintext (you may use the factorization $55 = 5 \cdot 11$).

---

# 9.2    Digital Signatures

Alice applies for insurance. She calls up her broker, Bob, they negotiate, and he sends her a contract. Normally this contract is printed, put in an envelope, and mailed to Alice. She then reads it, signs it, and mails it back to Bob. This process may take more than a week and adds cost (paper, ink, and postage) to both Bob and Alice. Bob would like to send this contract via e-mail and have Alice approve it, saving them both time and money. He doesn't have to worry about encrypting the document itself, since it's a standard policy and there's nothing secretive about it. But what guarantee does he have that the person who replies, "Yes, this looks great" is really Alice? Maybe it's someone else in her household with access to her e-mail account. Or maybe it's malicious Eve who knows how to masquerade electronically as Alice. What's needed is a way for Alice to say, "Yes, this looks good. Signed, Alice" and have the recipient be certain that Alice's response really came from Alice. There are several methods that can be used to guarantee that Alice's response is as legally binding as a written signature. We'll discuss one that allows a digital signature to be attached to a message using RSA-type techniques.

Let's call the document $m$. In our example, $m$ is the insurance policy that Bob has written for Alice. Just as with RSA, Alice chooses two large primes $p$ and $q$, multiplies them together to get $n = pq$, and then chooses an $e$ with $\gcd(e, \phi(n)) = 1$ and $1 < e < \phi(n)$. Next, she uses the Extended Euclidean Algorithm to calculate a $d$ with the property that $ed \equiv 1 \pmod{\phi(n)}$. She publishes $n$ and $e$, and keeps $p$, $q$, and $d$ private.

After reading the policy, Alice creates her signature by raising $m$ to the $d$th power mod $n$.

$$\text{Alice's Signature: } s \equiv m^d \pmod{n}.$$

Alice can now send $m$ and $s$ back to Bob. How does Bob know that it's really Alice who has agreed to the contract? He can go to the secure site where Alice's public information is kept and get her $e$ and $n$. He can then raise the $s$ he received to the $e$th power:

$$s^e \equiv (m^d)^e \equiv m^{ed} \equiv m \pmod{n}.$$

Why is $m^{ed} \equiv m \pmod{n}$? The scheme is simply RSA encryption and decryption of $m$, applied in reverse order.

---

### RSA Signatures

**Alice's Public Information:** $n$ and $e$

**Alice's Private Information:** $p, q,$ and $d$

**Alice's Signature:** $s = m^d \pmod{n}$

**Signed Document:** $(m, s)$

**Verification:** $s^e \overset{?}{=} m \pmod{n}$

---

Suppose Eve wants to subvert this arrangement. She tries to draw up a different contract, say $m_1$, and forge Alice's signature. In order to fool Bob, she needs to find a signature $s_1$ with the property that $s_1^e \equiv m_1 \pmod{n}$, and this is believed to be computationally infeasible. This is the same as starting with the RSA "ciphertext" $m_1$ and decrypting it to find the "plaintext" $s_1$. Since RSA is generally expected to be difficult to decrypt, Eve should not be able to forge Alice's signature.

As a result, if a document is signed with Alice's signature, Alice cannot successfully deny that she signed it.

Of course, Eve could start with $s_1$ and compute $m_1 \equiv s_1^e \pmod{n}$. Then $s_1$ is a legitimate signature for $m_1$. But it is very unlikely that $m_1$ is a meaningful message.

For another method of digitally signing a document, see Project 1 in Chapter 12.

---

**CHECK YOUR UNDERSTANDING:**
2. Alice has public RSA modulus $n = 91 = 7 \cdot 13$ and public exponent $e = 5$. She wants to sign the document $m = 7$.
(a) What is the signed document?
(b) Show that Bob's calculation verifies that the document you produced in part (a) is valid.

---

# 9.3   Chapter Highlights

1. The RSA algorithm for encrypting messages.

2. The security of RSA encryption is closely related to the difficulty of factorization.

3. Digital signatures using the RSA algorithm.

---

# 9.4   Problems

## 9.3.1   Exercises

**Section 9.1: RSA**

1. In each of the following, you are given two primes $p$ and $q$ to use for RSA. Decide whether the given encryption exponent can be used. If it can be used, find the decryption exponent $d$. If it cannot be used, explain why.

    (a) $p = 7, q = 11, e = 6$.
    (b) $p = 11, q = 13, e = 7$.
    (c) $p = 13, q = 23, e = 22$.
    (d) $p = 13, q = 29, e = 11$.

2.    For the following, encrypt the message $m$ using RSA with modulus $n = pq$ and exponent $e$ to obtain the ciphertext $c$. Then find the decryption exponent $d$, and verify that $c^d \equiv m \pmod{n}$:
    (a) $p = 7, q = 13, e = 5, m = 5$.
    (b) $p = 7, q = 13, e = 7, m = 7$.
    (c) $p = 3, q = 19, e = 7, m = 5$.
    (d) $p = 11, q = 17, e = 7, m = 9$.
    (e) $p = 11, q = 17, e = 13, m = 3$.

3.    A general at the Pentagon tells the lieutenant colonels (labeled (a), (b), (c), (d)) in the field to move $m$ tanks into position for an attack by using RSA with modulus $pq$. The lieutenant colonels know their values of $p$ and $q$ and $e$ the encryption exponent. The ciphertext is $c$. Use this information to find the decryption exponent $d$ and the number of tanks for each lieutenant colonel.
    (a) $p = 3, q = 11, e = 3, c = 7$.
    (b) $p = 5, q = 11, e = 3, c = 9$.
    (c) $p = 3, q = 17, e = 3, c = 5$.
    (d) $p = 13, q = 17, e = 11, c = 7$.

4.    In the following, a message $m$ was encrypted using RSA with modulus $pq$ and encryption exponent $e$, yielding the ciphertext $c$. Find $m$.
    (a) $p = 3, q = 23, e = 3, c = 9$.
    (b) $p = 3, q = 29, e = 3, c = 5$.
    (c) $p = 5, q = 17, e = 5, c = 3$.

5.    Romeo sends Juliet the hour at which they will meet. He uses RSA with $p = 3, q = 17, e = 5$. The ciphertext is $c = 3$. When will they meet?

6.    (a) Show that if $\gcd(a, 24) = 1$ then $a^2 \equiv 1 \pmod{24}$. (*Hint:* It's fastest to work mod 3 and mod 8.)
    (b) Suppose Alice and Bob use RSA with modulus $n = 35$. Show that the encryption exponent $e$ equals the decryption exponent $d$.

7.    Alice chooses primes $p = 149$ and $q = 317$ and encryption exponent $e = 71$. What public modulus does she publish? What is her decryption exponent?

8.    Alice chooses primes $p = 3491$ and $q = 8219$ and encryption exponent $e = 97$. What public modulus does she publish? What is her decryption exponent?

The following 7 problems are easiest to do with computer software that can handle up to 25-digit numbers and calculate mods.

9.    Alice chooses primes $p = 8923$ and $q = 6581$. She publishes $n = p \cdot q = 58722263$ and her encryption exponent $e = 307$. Using this, Bob sends her the message 21511484. What did Bob tell her?

10.    (a) Suppose you use $n = 29 \times 41$ as your RSA modulus and $e = 9$ as your

encryption exponent. What is your decryption exponent $d$?

(b) Suppose you try using $e = 7$ as your encryption exponent in part (a). What goes wrong?

11. Alice is sending a message to the systems administrator, Bob, using RSA with modulus $n = pq$ and encryption exponent $e$. However, Bob is lazy and uses $p = q$. What congruence should he use to determine the decryption exponent $d$?

12. Let's say that Bob wants to send Alice the message *HELP*. Alice chooses her secret primes $p = 43753$ and $q = 87149$ and encryption exponent $e = 6043$. She calculates $n = p \cdot q = 3813030197$ and sends Bob $n$ and $e$. We'll help Bob get started by writing out the number *HELP* corresponds to. Since $H = 08$, $E = 05$, $L = 12$, and $P = 16$ (remember that $A = 01$ when we use RSA), his message $m = 8051216$.

(a) What ciphertext $c$ does Alice receive?

(b) What is Alice's decryption exponent $d$?

(c) Verify that Alice can decrypt Bob's message.

13. Let's say that Bob wants to send Alice the message *HELP*. Alice chooses her secret primes $p = 658943$ and $q = 357389$ and encryption exponent $e = 2347$. She calculates $n = p \cdot q = 235498979827$ and sends Bob $n$ and $e$.

(a) What ciphertext $c$ does Alice receive?

(b) What is Alice's decryption exponent $d$?

(c) Verify that Alice can decrypt Bob's message.

14. Bob wants to send Alice the message *GOOD LUCK* using her encryption key $(e, n) = (5, 21631)$. He writes his message as a single number: $m = 715150412210311$ and encrypts it using Alice's public key.

(a) What message does Bob send to Alice?

(b) What are Alice's "secret" $p$ and $q$? What is her decryption exponent?

(c) What message does Alice receive after she decrypts? What went wrong?

(d) Can you come up with two ways to fix the problem?

15. When RSA is used with encryption key $(e, n) = (17, 31439)$, the ciphertext received is

$$3807 \quad 17063 \quad 25522 \quad 12107 \quad 26663$$

(each block was encrypted separately). What is the plaintext?

16. When RSA is used with encryption key $(e, n) = (5, 5352499)$, the ciphertext received is

$$4784648 \quad 1933497 \quad 4437506$$

(each block was encrypted separately). If $n = 1237 \cdot 4327$, find the decryption key and then the plaintext.

17. Sometimes, when decrypting a message, a cryptographer can speed up the process by decrypting a piece of the ciphertext and then making educated guesses about what follows. For example, if a letter is decrypted as $Q$, we can be pretty certain that the next letter is $U$. Suppose RSA is being used with encryption key $(e, n) = (23, 58951)$. The plaintext is encrypted in

blocks of two letters at a time and the received ciphertext is the following sequence of blocks:

$$40658||35980||31131||02677||12202||37240||52650.$$

You are told that the plaintext begins with

$$2015||0205||1518.$$

(a) Can you guess the rest of the message? (If you can't, you must decide whether 'tis nobler to factor $n$ or keep trying.)

(b) Show that your guess is correct by encrypting it and comparing with the ciphertext.

18. Explain why it would be silly to use an RSA encryption exponent of 1. Why would an encryption exponent of 2 never be used?

19. Alice and Alicia each set up an RSA public key cryptosystem with the same modulus $n$ but different, relatively prime, encryption exponents so that Alice's key is $(e_1, n)$ and Alicia's is $(e_2, n)$. Bob encrypts the same message, sending $m^{e_1}$ (mod $n$) to Alice and $m^{e_2}$ (mod $n$) to Alicia. Show that if Eve intercepts both ciphertexts she can find the plaintext $m$.

20. Choose two primes $p$ and $q$ for your RSA cryptosystem and an encryption exponent $e$ and let $\lambda(n) = \phi(n)/\gcd(p-1, q-1)$. Find $d'$ by solving

$$d' \cdot e \equiv 1 \pmod{\lambda(n)}.$$

Show that $d'$ can also be used as the private decryption exponent.

21. Suppose that you know $n$ can be factored as $pq$, where $p$ and $q$ are prime numbers, but you are not given $p$ and $q$. In this exercise, we show that knowledge of both $n$ and $\phi(n)$ allows you to factor $n$.

(a) Let $r = n - \phi(n) + 1$. Show that $r = p + q$.

(b) Show that $p$ and $q$ are the two solutions to the quadratic equation $x^2 - rx + n = 0$.

(c) If $n = pq = 218957$ and $\phi(n) = 217980$, find $p$ and $q$ using the quadratic formula.

22. Suppose Alice chooses three distinct primes $p, q, r$ and computes $n = pqr$. She chooses encryption and decryption exponents $e$ and $d$. What requirements should $d$ and $e$ satisfy in order for Alice to encrypt and decrypt using $c \equiv m^e$ (mod $n$) and $m \equiv c^d$ (mod $n$) ?

(There seems to be no advantage to using three primes in RSA. The difficulty of factoring $n$ often depends on the smallest prime factor of $n$, so using three primes might make factoring $n$ easier than factoring a standard RSA modulus of the same size.)

23. Suppose that Eve wants Alice to decrypt $c$ but Alice refuses. Eve computes $c_1 \equiv 123^e c$ (mod $n$) and gives $c_1$ to Alice. Alice decrypts $c_1$ to get $m_1$ and, since the message $m_1$ does not appear to be anything meaningful, she gives $m_1$ to Eve. How does Eve recover $m$ from $m_1$?

24. Suppose that Bob knows about the short message attack. He has a message $m$ with $m < 10^{12}$, so he appends 75 zeros at the end of $m$ to get $m_1 = 10^{75}m$. He then encrypts $m_1$ to get $c_1$. Show how Eve can still use the short message attack to obtain $m$.

## Section 9.2: Digital Signatures

In the next four problems, Alice uses the RSA signature with public modulus $n$ and public encryption exponent $e$. She sends three documents to Bob with her signature attached. Bob creates a fourth document and unsuccessfully tries to forge Alice's signature. Which of the four documents is the forgery? (Each ordered pair is of the form $(m, s)$, where $m$ is the document and $s$ is the signature.) Do not factor $n$ and do not find the private exponent used to sign. The first two problems can be done on a calculator. The third and fourth problems are better done with computer software that does mods and handles larger integers.

25. Public modulus $n = 55$ and public encryption modulus $e = 7$.
    Signed documents:
    $(6, 51)$,   $(7, 13)$,   $(8, 19)$,   $(9, 14)$.

26. Public modulus $n = 51$ and public encryption modulus $e = 5$.
    Signed documents:
    $(2, 32)$,   $(5, 20)$,   $(7, 24)$,   $(9, 42)$.

27. Public modulus $n = 10379$ and public encryption exponent $e = 17$.
    Signed documents:

    $(921, 636)$,   $(209, 8690)$,   $(347, 5120)$,   $(1059, 5909)$.

28. Public modulus $n = 443617$ and public encryption exponent $e = 23$.
    Signed documents:

    $(983, 72702)$, $(76, 347115)$, $(2731, 191056)$, $(3711, 65782)$.

29. Alice uses the RSA signature scheme with primes $p = 3, q = 11$, and $e = 7$.
    (a) What is $n$, the public modulus she will use?
    (b) What is her private exponent $d$?
    (c) What is the signature $s$ for the message $m = 17$?

30. Alice is using the RSA signature scheme with primes $p = 5, q = 11$, and $e = 9$.
    (a) What is $n$, the public modulus she will use?
    (b) What is her private exponent $d$?
    (c) What is the signature $s$ for the message $m = 7$?

31. Alice is using the RSA signature scheme with primes $p = 3, q = 17$, and $e = 5$.
    (a) What is $n$, the public modulus she will use?
    (b) What is her private exponent $d$?
    (c) What is the signature $s$ for the message $m = 14$?

32. Alice is using the RSA signature scheme with primes $p = 3, q = 17$, and $e = 7$.
    (a) What is $n$, the public modulus she will use?
    (b) What is her private exponent $d$?
    (c) What is the signature $s$ for the message $m = 10$?

In the next four problems, Alice wants to sign a message $m$ that she sends to Bob and plans to use the RSA signature scheme. Her RSA modulus is $n$ and her public encryption exponent is $e$. Bob receives the pair $(m, s)$ where $s$ is Alice's signature. In each problem, determine whether Bob should believe that $m$ was really sent by Alice.

33.  $n = 77$, $m = 31$, $e = 13$. Bob receives $(31, 59)$.

34.  $n = 95$, $m = 47$, $e = 7$. Bob receives $(47, 63)$.

35.  $n = 2911$, $m = 91$, $e = 11$. Bob receives $(91, 794)$.

36.  $n = 5141$, $m = 2136$, $e = 19$. Bob receives $(2136, 1424)$.

37.  Alice is using the RSA signature scheme with primes $p = 7, q = 13$, and $e = 5$. She decides that $s = 11$ is a nice signature, but she needs a message to sign. What message $m$ has $s = 11$ as its valid signature?

38.  Alice is using the RSA signature scheme with primes $p = 7, q = 13$, and $e = 7$. Since 9 rhymes with "sign," she decides that $s = 9$ is a good signature, but she needs a message to sign. What message $m$ has $s = 9$ as its valid signature?

39.  Alice is using the RSA signature scheme with primes $p = 3, q = 19$, and $e = 11$. She wants to sign the message $m = 987654321$ (because she wants to certify that this is the correct launch sequence for a rocket). Since $m$ is much larger than the modulus $n = 57$, she computes $m_0 = m \pmod{11}$, with $0 \le m_0 \le 10$ and signs $m_0$. Show that her signature is $s = 47$. (*Note:* Usually, if a document $m$ is larger than the modulus $n$, Alice does not sign $m$. Instead, she compresses $m$ into a number less than $n$ using what is known as a *hash function*. In this exercise, she uses a much simpler function, namely, reduction mod 11.)

## 9.4.2    Projects

1.  Suppose that you are a cryptologist and you discover, to your amazement, that your RSA encryption of the message $m$ is $m$ itself. In other words, the ciphertext and the plaintext are identical. Let's say that your encryption exponent is $e = 3$ and your modulus $n = pq$.

    (a) If $n = 15$ (a *very* poor choice since factorization of $n$ is then easy), find all messages that remain unchanged.

    (b) If $n = 55$ (also a *very* poor choice), find all messages that remain unchanged.

    (c) Show that for every $n = pq$ (with $p, q$ distinct odd primes), there are always 9 messages that remain unchanged.

2.  In our discussion of RSA, we translated words to numbers by assigning 01 to the letter $A$, 02 to the letter $B$,..., and 26 to the letter $Z$. In our example, the word *CAR* became 30118. Another way to translate words to numbers is to convert each word to a base 26 representation. Here's how to do this with the word *NOW*.

    First find the numerical equivalent of the letters: $N = 14$, $O = 15$, and $W = 23$. Then $NOW = 14 \cdot 26^2 + 15 \cdot 26 + 23 = 9877$. You can translate back from a number to a word by using the Division Algorithm to divide by 26, but you need to allow 26 but not 0 as remainders because of $z = 26$. So, if you received 217 you would get

$$217 = 8 \cdot 26 + 9,$$

which corresponds to the word *HI*, since *H* is the eighth letter and *I* is the ninth.

(a) Change 11971077 to a word (this is tricky; observe that when you use 26 as a remainder instead of 0, the quotient decreases by 1).

(b) Show that *JAZZ* becomes 177138.

(c) Alice chooses an RSA modulus: $n = 153583 = 383 \cdot 401$. She uses $e = 7$. Compute her value of $d$.

(d) Let's say that Bob wants to send Alice the message *HELP*. He writes the plaintext as $m = 8 \cdot 26^3 + 5 \cdot 26^2 + 12 \cdot 26 + 16 = 144316$. Compute the ciphertext $c$ that Bob sends, and verify that Alice's decryption of $c$ yields $m$.

(e) Suppose Bob decides to write his message *HELP* using the standard base 10 method: $HELP = 08051216$. Bob encrypts to get $c \equiv 8051216^7$ (mod $n$). When Alice decrypts $c$, why doesn't she recover $m$?

(One way to fix this is to break the message into the blocks $HE = 805$ and $LP = 1216$ and encrypt each block separately. Another way is to use the base 26 representation of the message.)

3. Bob decides that RSA is not secure enough, so he decides to devise a public key cryptosystem where the encryption modulus is a product of three primes instead of two primes.

(a) Choose three primes, each with at least three digits. What restriction must apply to the encryption exponent $e$? What will your decryption exponent $d$ be?

(b) Using your numbers from (a), encrypt the plaintext *TO* and show that you are able to convert the ciphertext to plaintext.

(c) If you have three arbitrary primes $p_1, p_2, p_3$ so that $n = p_1 \cdot p_2 \cdot p_3$, what must be true of the encryption exponent $e$? How would you find the decryption exponent $d$? Explain how you would encrypt and decrypt messages and show why decryption works.

(d) In many factorization methods, the difficulty of factoring an integer $n$ depends on the size of the smallest prime factor of $n$. Assuming this, explain why using RSA with three primes should give no advantage over using two primes.

## 9.4.3    Computer Explorations

1. (a) For RSA, let $p = 167$ and $q = 251$, so $n = 41917$. Let $e = 3$. Encrypt $ban = 20114$, $bat = 20120$, and $bay = 20125$. Can you tell from the ciphertexts that only one letter has been changed in the plaintexts? (In a good cryptosystem, a small change in the plaintext makes a large change in the ciphertext.)

(b) Find the decryption exponent $d$.

(c) Bob tries to send you the ciphertext $c = 27120$, but there is a transmission error and you receive 27121. Do you obtain anything close to the intended message?

2. Suppose Bob sends the same message $m$ to three different people. The ciphertexts are

$$c_1 = 257261 \pmod{303799}, \quad c_2 = 117466 \pmod{289279},$$
$$c_3 = 260584 \pmod{410503},$$

and their RSA moduli are

$$n_1 = 303799, \quad n_2 = 289279, \quad n_3 = 410503.$$

Each person has encryption exponent $e = 3$ (so $m^3 \equiv c_1 \pmod{n_1}$, and similarly for the other two).
(a) Find an $x$ with $0 \leq x < n_1 n_2 n_3$ and

$$x \equiv c_1 \pmod{n_1}, \quad x \equiv c_2 \pmod{n_2}, \quad x \equiv c_1 \pmod{n_2}.$$

(b) Show that $0 \leq m^3 < n_1 n_2 n_3$. (*Hint: $m < n_i$.*)
(c) Show that $x = m^3$ (not just congruent, but actually equal).
(d) Find the message $m$.
This shows a disadvantage of using a small encryption exponent. (An advantage is that encryption is fast.)

## 9.4.4    Answers to "CHECK YOUR UNDERSTANDING"

1. First, we need the decryption key. To do this, solve $de \equiv 1 \pmod{(p-1)(q-1)}$, which means solve $27d \equiv 1 \pmod{40}$. The solution (find it by trial and error or by the Extended Euclidean Algorithm) is $d = 3$. Now we can decrypt:

$$m \equiv c^d \equiv 4^3 \equiv 64 \equiv 9 \pmod{55}.$$

The plaintext is 9 (if you want to change it to letters, the plaintext could be *I*).

2. (a) Alice's secret RSA decryption exponent satisfies $5d \equiv 1 \pmod{6 \cdot 12}$. Therefore, $d = 29$ (use the Extended Euclidean Algorithm). Alice's signature is $s \equiv m^d \equiv 7^{29} \equiv 63 \pmod{91}$. The signed document is $(m, s) = (7, 63)$.
(b) Bob verifies by computing $s^e \equiv 63^5 \equiv 7 \pmod{91}$. Since $m = 7$, the signature is valid.

# Chapter 10

# Polynomial Congruences

Solving linear congruences is fundamental in many parts of number theory. The generalization, solving polynomial congruences, is perhaps not as basic but is still an important topic. This chapter covers some concepts that will be needed in later chapters, along with some ideas that are interesting in their own right.

## 10.1   Polynomials Mod Primes

We start by studying polynomial congruences mod primes. For example, maybe we want to solve $x^2 - 3x + 17 \equiv 0 \pmod{37}$, or $3x^4 - 2x^3 + x - 18 \equiv 0 \pmod{23}$, or $x^2 - 1 \equiv 0 \pmod{31}$.

An example that we saw in Corollary 6.11 is solving $x^2 - 1 \equiv 0 \pmod{p}$, where $p$ is prime. Since $x^2 - 1 = (x+1)(x-1)$, this congruence becomes $(x+1)(x-1) \equiv 0 \pmod{p}$. Recall that if a product of two numbers is 0 mod a prime, then one of the factors is 0 mod the prime. So we have either $x + 1 \equiv 0 \pmod{p}$ or $x - 1 \equiv 0 \pmod{p}$. This means that $x \equiv \pm 1 \pmod{p}$, so we have two solutions if $p$ is odd, and one solution if $p = 2$ (since $+1 \equiv -1 \pmod{2}$).

Perhaps the most basic question is "How many solutions are there?" For the polynomials you have worked with in the past, namely polynomials with rational, real, or complex coefficients, the number of solutions in complex numbers is at most the degree of the polynomial. This is exactly what happened when we solved $x^2 - 1 \equiv 0 \pmod{p}$: there are at most two solutions mod $p$ of this degree two polynomial congruence.

The following result, due to Lagrange, says that a similar result holds for any polynomial mod a *prime*.

**Proposition 10.1.** *Let $p$ be a prime and let $f(x)$ be a non-zero polynomial mod $p$ of degree $n \geq 1$. Let $r_1, r_2, \ldots, r_s$ be distinct integers mod $p$ such that $f(r_i) \equiv 0 \pmod{p}$. Then $s \leq n$. In other words, a polynomial of degree $n$ has at most $n$ roots.*

**Remark.** The lemma is not true for composite moduli. For example, the degree 2 polynomial congruence $x^2 - 1 \equiv 0 \pmod 8$ has four solutions, namely $x \equiv 1, 3, 5, 7$.

*Proof.* If $s = 0$ (for example, if the polynomial has no roots mod $p$), then automatically $s \leq n$, so henceforth we assume that $s > 0$.

We prove the lemma by induction on $n$, the degree of the polynomial. For $n = 1$, the polynomial $f(x)$ is linear: $f(x) = a_1 x + a_0$ with $a_1 \not\equiv 0 \pmod p$. The congruence $a_1 x + a_0 \equiv 0 \pmod p$ has a unique solution mod $p$, so the lemma is true for $n = 1$.

We now assume that the lemma is true for $n = k$ and we'll use this to show that it is true for $n = k + 1$. So, suppose $f(x)$ has degree $k + 1$. Since

$$f(r_s) \equiv 0 \pmod{p},$$

we want to factor off $(x - r_s)$ and obtain a degree $k$ polynomial, just like we do when working with real numbers or integers. But, since we're working mod $p$, we need to justify this step. Write

$$f(x) = a_{k+1} x^{k+1} + a_k x^k + \cdots + a_1 x + a_0.$$

Then, working mod $p$, we have

$$
\begin{aligned}
f(x) &\equiv f(x) - 0 \\
&\equiv f(x) - f(r_s) \\
&= a_{k+1} x^{k+1} + \cdots + a_1 x + a_0 - (a_{k+1} r_s^{k+1} + \cdots + a_1 r_s + a_0) \\
&= a_{k+1}(x^{k+1} - r_s^{k+1}) + a_k(x^k - r_s^k) + \cdots + a_1(x - r_s).
\end{aligned}
$$

We can factor $(x - r_s)$ from each term because

$$x^j - r_s^j = (x - r_s)(x^{j-1} + x^{j-2} r_s + \cdots + x r_s^{j-2} + r_s^{j-1}).$$

This yields

$$f(x) \equiv (x - r_s) g(x) \pmod{p}$$

for some polynomial $g(x)$ of degree $k$. (*Technical point:* A congruence between polynomials means that the coefficients of one polynomial are congruent to the corresponding coefficients of the other polynomial.)

If $s = 1$, then, since $k \geq 1$, we automatically have $s \leq k$. Therefore, assume from now on that $s \geq 2$. Let $1 \leq i \leq s - 1$. Since we have assumed that $r_i \not\equiv r_s \pmod{p}$, and

$$0 \equiv f(r_i) \equiv (r_i - r_s)g(r_i) \pmod{p},$$

we must have

$$g(r_i) \equiv 0 \pmod{p}$$

for $1 \leq i \leq s - 1$. (This is where we need $p$ to be a prime. We're using Proposition 6.10, which says that if $p$ is prime and $ab \equiv 0 \pmod{p}$, then $a \equiv 0 \pmod{p}$ or $b \equiv 0 \pmod{p}$.) Since $g(x)$ has degree $k$, the induction assumption implies that $s - 1 \leq k$. Adding 1 yields $s \leq k+1$, as desired. Therefore, the lemma is true for $n = k + 1$. By induction, the lemma is true for all $n$. $\qquad\qquad\qquad\qquad\qquad\qquad\qquad\square$

It often happens that a polynomial congruence has no solutions. For example, $x(x-1)(x-2)+1 \equiv 0 \pmod{3}$ has no solutions since plugging $x = 0,\ 1,\ 2$ into the polynomial yields 1 for each value of $x$. Therefore, Proposition 10.1 gives only an upper bound on the number of solutions. It says nothing about the existence of solutions. A similar situation arises when you work with polynomials whose coefficients are rational or real numbers. A quadratic polynomial has at most two rational or real roots, but may have none. For example $x^2 - 2 = 0$ has no rational roots while $x^2 + 1 = 0$ has no real roots.

---

## CHECK YOUR UNDERSTANDING

1. Using Proposition 6.10, but not quoting Proposition 10.1, explain why the only solutions to $(x - 1)(x - 3) \equiv 0 \pmod{19}$ are $x \equiv 1, 3 \pmod{19}$.

## 10.2    Solutions Modulo Prime Powers

In the Newton–Raphson Method from calculus, you have a function $f(x)$ and are trying to solve $f(x) = 0$. You start with an approximate solution $x_0$ and try to make it closer to an actual solution by calculating

$$x_1 = x_0 - \frac{f(x_0)}{f'(x_0)}.$$

This process can be continued with the hope of getting even better approximations to solutions. The same idea works when solving polynomial congruences mod prime powers. Namely, we start with $x_0$ that solves $f(x_0) \equiv 0 \pmod{p}$. Then we improve on $x_0$ to get $x_1$ that is a solution mod $p^2$. This process can be continued to get solutions mod higher powers of $p$.

The first thing we need to think about is what is a derivative? When working mod $p$, we can't take limits. Instead, we simply define the derivative of a polynomial to be given by the same formula as we get in calculus.

**Definition 10.2.** *If* $f(x) = a_n x^n + a_{n-1} x^{n-1} + \cdots + a_1 x + a_0$, *then the* **derivative** *of* $f(x)$ *is given by*

$$f'(x) = n a_n x^{n-1} + (n-1) a_{n-1} x^{n-2} + \cdots + 2 a_2 x + a_1.$$

Suppose we want to solve $f(x) = x^2 + 5x + 7 \equiv 0 \pmod{13^3}$. Using trial and error, we see that $x_0 \equiv 7 \pmod{13}$ is a solution mod 13. Let's modify it to get a solution mod $13^2$. Write $x_1 = 7 + 13k$ and substitute into the quadratic:

$$
\begin{aligned}
0 \equiv f(x_1) &\equiv (7 + 13k)^2 + 5(7 + 13k) + 7 \\
&\equiv 7^2 + 2 \cdot 7 \cdot 13k + 13^2 k^2 + 5 \cdot 7 + 5 \cdot 13k + 7 \\
&\equiv f(7) + 2 \cdot 7 \cdot 13k + 5 \cdot 13k \pmod{13^2}
\end{aligned}
$$

(the remaining term has $13^2$ in it, so it disappears mod $13^2$). This rearranges to

$$13(14 + 5)k \equiv -f(7) \equiv -91 \pmod{13^2}.$$

Dividing by 13 yields

$$(14 + 5)k \equiv -7 \pmod{13},$$

which gives $k \equiv 1 \pmod{13}$. Notice that we left the coefficient of $k$ as $14 + 5$ and did not add to get 19. This is because $f(x) = x^2 + 5x + 7$, so $f'(x) = 2x + 5$, and we want to point out that $f'(7) = 2 \cdot 7 + 5 = 14 + 5$. As we'll see, $f'(x)$ plays an essential role in solving these congruences just as it did when looking for real number solutions to polynomial equations in calculus.

Because $x_1 = 7 + 13k$ and $k \equiv 1 \pmod{13}$, we can choose $k = 1$ and get $x_1 = 20$. You can check that $f(20) \equiv 0 \pmod{13^2}$.

Now let's try to use the same techniques to modify $x_1 = 20$ to get a solution $x_2$ mod $13^3$. Write $x_2 = 20 + 13^2 j$ and substitute into the quadratic:

$$0 \equiv (20 + 13^2 j)^2 + 5(20 + 13^2 j) + 7$$
$$\equiv f(20) + 2 \cdot 20 \cdot 13^2 j + 5 \cdot 13^2 j \pmod{13^3}.$$

This rearranges to

$$13^2 (2 \cdot 20 + 5)j \equiv -f(20) \equiv -507 \pmod{13^3}.$$

Notice that we have a factor of $2 \cdot 20 + 5$, which is $f'(20)$.

Dividing by $13^2$ and solving yields $j \equiv 6 \pmod{13}$, so we have $x_2 \equiv 20 + 13^2 j \equiv 1034 \pmod{13^3}$. Therefore, $x_2 = 1034$ is a solution to our problem: $f(1034) \equiv 0 \pmod{13^3}$.

The following describes the general procedure, which is often known as *Hensel's Lemma*. As we pointed out, it is the analogue of the *Newton–Raphson Method* from calculus, and the formula is the same.

**Proposition 10.3.** *Let $f(x)$ be a polynomial with integer coefficients, let $p$ be a prime, and let $x_0$ be an integer with $f(x_0) \equiv 0 \pmod{p}$. Assume that $f'(x_0) \not\equiv 0 \pmod{p}$. For $n = 0, 1, 2, \ldots,$ define integers $x_{n+1} \pmod{p^{2^{n+1}}}$ by*

$$x_{n+1} \equiv x_n - \frac{f(x_n)}{f'(x_n)} \pmod{p^{2^{n+1}}}.$$

*Then $x_{n+1} \equiv x_0 \pmod{p}$ and*

$$f(x_{n+1}) \equiv 0 \pmod{p^{2^{n+1}}}.$$

*Proof.* We start with a lemma that is essentially the beginning of the Taylor expansion for polynomials.

**Lemma 10.4.** *Let $f(x)$ be a polynomial with integer coefficients and let $b$ be an integer. Then there is a polynomial $g(x)$ with integer coefficients such that*

$$f(x) = f(b) + (x - b)f'(b) + (x - b)^2 g(x).$$

*Proof.* Let $h(x) = f(x + b)$. Then $h(x)$ is a polynomial with integer coefficients. Write

$$h(x) = h_0 + h_1 x + h_2 x^2 + \cdots + h_n x^n.$$

Then $h_0 = h(0)$ and $h_1 = h'(0)$. Let $g_1(x) = h_2 + h_3 x + \cdots + h_n x^{n-2}$, which is a polynomial with integer coefficients. Then

$$h(x) = h(0) + x h'(0) + x^2 g_1(x),$$

which is the lemma for $h(x)$ and $b = 0$. But $h(0) = f(b)$ and $h'(0) = f'(b)$. Therefore,

$$
\begin{aligned}
f(x) = h(x - b) &= h(0) + (x - b)h'(0) + (x - b)^2 g_1(x - b) \\
&= f(0) + (x - b)f'(b) + (x - b)^2 g(x),
\end{aligned}
$$

with $g(x) = g_1(x - b)$. This proves the lemma.                              □

We now can prove the proposition. We'll prove by induction that

$$f(x_n) \equiv 0 \pmod{p^{2^n}} \text{ and } x_n \equiv x_0 \pmod{p} \tag{10.1}$$

for all $n \geq 0$. By assumption, (10.1) is true for $n = 0$. Assume that it is true for $n = k$, so we are assuming that

$$f(x_k) \equiv 0 \pmod{p^{2^k}} \text{ and } x_k \equiv x_0 \pmod{p}.$$

First, note that $x_k \equiv x_0 \pmod{p}$ implies that $f'(x_k) \equiv f'(x_0) \not\equiv 0 \pmod{p}$. Therefore, $f'(x_k)$ has a multiplicative inverse mod powers of $p$, so the definition of $x_{k+1}$ makes sense.

By the lemma, we can write

$$f(x) = f(x_k) + (x - x_k)f'(x_k) + (x - x_k)^2 g(x),$$

where $g(x)$ has integer coefficients. Substitute $x = x_{k+1}$ to get

$$f(x_{k+1}) = f(x_k) + (x_{k+1} - x_k)f'(x_k) + (x_{k+1} - x_k)^2 g(x_k).$$

The definition of $x_{k+1}$ says that $f(x_k) + (x_{k+1} - x_k)f'(x_k) \equiv 0$ (mod $p^{2^{n+1}}$), so we have

$$f(x_{k+1}) \equiv (x_{k+1} - x_k)^2 g(x_k) \pmod{p^{2^{k+1}}}.$$

But $f(x_k) \equiv 0 \pmod{p^{2^k}}$ implies that

$$x_{k+1} - x_k \equiv 0 \pmod{p^{2^k}},$$

hence

$$(x_{k+1} - x_k)^2 g(x_k) \equiv 0 \pmod{p^{2^{k+1}}}.$$

Therefore, $f(x_{k+1}) \equiv 0 \pmod{p^{2^{k+1}}}$, so (10.1) is true for $n = k+1$. By induction, it is true for all $n$. ◻

**Example.** Let's use the formulas of Proposition 10.3 and revisit the congruence $f(x) = x^2 + 5x + 7 \equiv 0 \pmod{13^2}$, which we considered at the beginning of this section. We start with $x_0 = 7$, which is a solution of $f(x) \equiv 0 \pmod{13}$ and compute $f(7) = 91$. We also have $f'(x) = 2x + 5$, so $f'(7) = 19$. Therefore,

$$x_1 \equiv x_0 - \frac{f(x_0)}{f'(x_0)} \equiv 7 - \frac{91}{19} \equiv 7 + 13 \equiv 20 \pmod{13^2}.$$

(Note that we changed $-91/19$ to 13 mod $13^2$. This can be done by using the Extended Euclidean Algorithm to compute $1/19 \pmod{13^2}$ and then multiplying by 91. Actually, we can compute $1/19 \equiv 11 \pmod{13}$. Since $13 \mid 91$, multiplying by 91 gives the desired result mod $13^2$.) Since $f(20) = 507$ and $f'(20) = 45$, we have

$$x_2 \equiv 20 - \frac{507}{45} \equiv 7625 \pmod{13^4}.$$

A calculation shows that $f(7625) \equiv 0 \pmod{13^4}$ and $7625 \equiv x_0 = 7$ (mod 13). Moreover, $7625 \equiv 1034 \pmod{13^3}$, so we recover the solution we found previously.

We could find another solution of the congruence $f(x) \equiv 0 \pmod{13^3}$ by the same method, starting with $x_0 = 1$. But it's more fun to use a trick. First, note that the sum of the two solutions of $x^2 + bx + c = 0$ is $-b$. Why? The quadratic formula says that the roots are $x = (-b \pm \sqrt{b^2 - 4c})/2$, and the sum of these two roots is $-b$. Therefore we guess that the same fact works mod powers of 13, which means that $1034 + x \equiv -5 \pmod{13^3}$, where $x$ is the other solution. This gives $x \equiv -1039 \equiv 1158 \pmod{13^3}$. If you want, you can check that this works.

What happens when $f'(x_0) \equiv 0 \pmod{p}$? Things get much more technical. For example, let $f(x) = x^2 - 5$ and let $p = 2$. Then $x_0 = 1$ gives a solution mod 2. The formula gives $x_1 = x_0 + 2 = 3$, which gives a solution mod 4. But the process must stop here because there is no solution mod 8 since the square of an odd integer is always 1 mod 8.

Of course, sometimes there is a solution mod $p^n$ for all $n$. For a version of Proposition 10.3 that works in some cases where $f'(x_0) \equiv 0 \pmod{p}$, see Project 2.

---

## CHECK YOUR UNDERSTANDING

2. Starting from the fact that $2^2 + 1 \equiv 0 \pmod{5}$, find $x_2 \equiv 2 \pmod{5}$ such that $x_2^2 + 1 \equiv 0 \pmod{25}$.

---

# 10.3   Composite Moduli

Now that we know how to solve polynomial congruences mod primes and mod prime powers, the Chinese Remainder Theorem allows us to solve polynomial congruences for composite moduli.

Suppose we want to solve $f(x) = x^2 + 2x + 7 \equiv 0 \pmod{175}$. We use a "factor and conquer" strategy. First, factor 175 as $5^2 \cdot 7$. Next, solve $f(x) \equiv 0 \pmod{25}$, getting $x \equiv 11$ or $12 \pmod{25}$, and solve $f(x) \equiv 0 \pmod{7}$, getting $x \equiv 0$ or $5 \pmod{7}$. The Chinese Remainder Theorem allows us to combine these to get four solutions: $x \equiv 12, 61, 112, 161 \pmod{175}$.

As another example, suppose we want to solve $x^2 + 3x + 4 \equiv 0 \pmod{77}$. There is only one solution mod 7, namely $x \equiv 2 \pmod{7}$. There are two solutions mod 11, namely $x \equiv 3$ and $5 \pmod{11}$. The solutions mod 7 and mod 11 combine to give $x \equiv 16$ and $58 \pmod{77}$.

As a final example, consider $x^2 + 1 \equiv 0 \pmod{75}$. There are two solutions mod 25, namely $x \equiv 7$ and $18 \pmod{25}$. However, there are no solutions mod 3. Therefore, there are no solutions mod 75.

---

## CHECK YOUR UNDERSTANDING

3. Find all solutions to $x^2 - 1 \equiv 0 \pmod{35}$.

---

# 10.4   Chapter Highlights

1. A polynomial of degree $n \geq 1$ has at most $n$ roots mod a prime $p$.

2. The Newton–Raphson Method for solving polynomial congruences mod prime powers.

3. Polynomial congruences mod composite numbers can be solved using the Chinese Remainder Theorem.

---

# 10.5   Problems

## 10.4.1   Exercises

1. (a) Solve $x^2 - 5x + 6 \equiv 0 \pmod{19}$. (*Hint:* Try factoring.)
   (b) Solve $x^2 - 5x + 25 \equiv 0 \pmod{19}$. (*Hint:* Relate this to part (a).)
   (c) Factor $x^2 - 5x + 25$ mod 19. (Note that it doesn't factor if you don't use mods).

2. Find two solutions of $x^2 + 2x + 2 \equiv 0 \pmod{5^3}$.

3. Find two solutions of $x^2 - 2 \equiv 0 \pmod{7^2}$.

4. Find a solution of $x^3 + 6x^2 + x - 1 \equiv 0 \pmod{7^3}$.

5. Find all solutions to $x^4 + x^3 + x^2 + 2x + 4 \equiv 0 \pmod{9}$.

6. Find all solutions to $x^5 + 23x \equiv 10 \pmod{25}$.

7. Can you solve $x^3 + x^2 \equiv 25 \pmod{27}$? If so, how many solutions are there?

8. Can you solve $x^3 + 13x^2 + 3x \equiv 11 \pmod{25}$? If so, how many solutions are there?

9. (a) Find all solutions of $x^2 + 3x - 4 \equiv 0 \pmod{21}$.
   (b) Find all solutions of $x^2 + 3x - 4 \equiv 0 \pmod{35}$.

10. Let $p$ be prime.
   (a) Show that $f(x) = x^{p-1} - 1$ has $p - 1$ distinct roots mod $p$.
   (b) Show that $f(x) - \prod_{j=1}^{p-1}(x - j) \pmod{p}$ has degree less than $p - 1$ but has $p - 1$ roots.
   (c) Show that $f(x) \equiv \prod_{j=1}^{p-1}(x - j) \pmod{p}$.
   (d) Let $x = 0$ in part (c) and deduce Wilson's theorem.

## 10.5.2    Projects

1. Let $p$ be a prime. If $a$ is a nonzero integer, let $p^r$ the largest power of $p$ dividing $a$. Define the **$p$-adic absolute value** to be

$$|a|_p = p^{-r}.$$

In addition, define $|0|_p$ to be 0.

(a) Show that $|12|_2 = 1/4$ and $|12|_3 = 1/3$.

(b) Show that $x \equiv y \pmod{p^r}$ if and only if $|x - y|_p \leq p^{-r}$.

(c) Show that if $a$ and $b$ are integers, then $|ab|_p = |a|_p \cdot |b|_p$ and $|-a|_p = |a|_p$.

(d) Let $|a|$ be the usual (real numbers) absolute value. Show that if $a$ is a non-zero integer, then (the product is over all primes)

$$|a| \prod_p |a|_p = 1.$$

(e) Show that if $a$ and $b$ are integers, then $|a+b|_p \leq \mathrm{Max}(|a|_p, |b|_p)$. (*Hint:* Figure out why this is just Corollary 2.4 applied to a power of $p$.)

(f) We can define a "disk" in the integers using the $p$-adic absolute value: Let $a$ be an integer and let $R > 0$ be a real number. Then

$$D_R(a) = \{x \mid x \text{ is an integer with } |x - a|_p \leq R\}.$$

Suppose $R_1 \leq R_2$ and that $D_{R_1}(a_1) \cap D_{R_2}(a_2)$ is non-empty. Show that $D_{R_1}(a_1) \subseteq D_{R_2}(a_2)$. This says that two disks are either disjoint or one is contained in the other. (*Hint:* Let $x \in D_{R_1}(a_1)$ and let $z$ be in the intersection. Then $x - a_2 = (x - a_1) + (a_1 - z) + (z - a2)$.)

(g) Suppose $|a|_p > |b|_p$. Write $|a|_p = |(a+b) - b|_p \leq \mathrm{Max}(|a+b|_p, |-b|_p)$. Show that the maximum on the right cannot be $|b|_p$, and conclude that $|a|_p \leq |a + b|_p$ when $|a|_p > |b|_p$.

(h) Use parts (e) and (g) to show that if $|a|_p > |b|_p$ then $|a + b|_p = |a|_p$.

(i) Show that every triangle is isosceles with respect to the $p$-adic absolute value. (A "triangle" is given by three integers, $a$, $b$, $c$, and the sides have lengths $|a - b|_p$, $|b - c|_p$, and $|c - a|_p$.)

2.  The following gives a version of Proposition 10.3 that can be used when $f'(x_0) \equiv 0 \pmod{p}$. Let $p$ be prime and let $f(x)$ be a polynomial with integer coefficients. Let $x_0$ be an integer such that $f(x_0) \equiv 0 \pmod{p^r}$ for some $r \geq 1$ and such that $f'(x_0) \equiv 0 \pmod{p^s}$ and $f'(x_0) \not\equiv 0 \pmod{p^{s+1}}$ for some $s \geq 0$. Let $c = r - 2s$ and suppose that $c > 0$. Define $x_1, x_2, \cdots$ by

$$x_{n+1} \equiv x_n - \frac{f(x_n)}{f'(x_n)} \pmod{p^{2^{n+1}c+s}}.$$

We'll prove by induction that

$$x_n \equiv x_0 \pmod{p^{s+1}} \text{ and } f(x_n) \equiv 0 \pmod{p^{2^n c + 2s}}. \tag{10.2}$$

(a) Show that (10.2) is true for $n = 0$.
(b) Assume that (10.2) is true for $n = k$. Show that $f'(x_k) \equiv 0 \pmod{p^s}$ and $f'(x_k) \not\equiv 0 \pmod{p^{s+1}}$.
(c) Show that the fraction $f(x_k)/f'(x_k)$, when written in the form $u/v$ with $\gcd(u, v) = 1$, is such that $p \nmid v$ and $p^{2^k c + s} \mid u$. Therefore, $f(x_k)/f'(x_k) \bmod p^{2^k c+s}$ makes sense, and $f(x_k)/f'(x_k) \equiv 0 \pmod{p^{2^k c+s}}$.
(d) Show that $f(x_{k+1}) \equiv 0 \pmod{p^{2^{k+1}c+2s}}$. (*Hint:* Lemma 10.4.) Therefore, (10.2) is true for $n = k+1$. By induction, (10.2) is true for all $n \geq 0$.
(e) Show that if $r > 2s$ then $f(x) \equiv 0 \pmod{p^m}$ has a solution for all $m \geq 1$. (*Note:* A solution for a large $m$ is automatically a solution for smaller $m$.)
(f) Let $p = 2$ and let $f(x) = x^2 - 17$. Let $x_1 = 1$. Find a solution to $f(x) \equiv 0 \pmod{2^{10}}$.

## 10.5.3    Computer Explorations

1.  (a) Compute two solutions to $x^2 + 1 \equiv 0 \pmod{5^{100}}$.
    (b) Let $r$ and $s$ be the two solutions from part (a). Find the base 5 expansions

    $$r = a_0 + a_1 5 + a_2 5^2 + \cdots + a_{99} 5^{99}$$
    $$s = b_0 + b_1 5 + b_2 5^2 + \cdots + b_{99} 5^{99}$$

    with $0 \leq a_i \leq 4$ and $0 \leq b_i \leq 4$ for all $i$. Do you see any relation between $a_i$ and $b_i$?
    (c) Count how many times $a_i = 0$, how many times $a_i = 1$, etc. Are the values of $a_i$ distributed reasonably uniformly among the five possible values 0, 1, 2, 3, 4?
    (d) You could answer the question of part (c) for the distribution of $b_i$, but you should already know the answer from (b) and (c). Why?

## 10.5.4    Answers to "CHECK YOUR UNDERSTANDING"

1.  By Proposition 6.10, since $(x - 1)(x - 3) \equiv 0 \pmod{19}$, we must have $x - 1 \equiv 0 \pmod{19}$ or $x - 3 \equiv 0 \pmod{19}$. This means that $x \equiv 1, 3 \pmod{19}$.

2.  We have $f(x) = x^2 + 1$ and $x_1 = 2$. Therefore, since $f(2) = 5$ and $f'(2) = 4$,

$$x_2 \equiv 2 - \frac{5}{4} \equiv 2 + 5 \equiv 7 \pmod{25}.$$

Let's check this: $f(7) = 7^2 + 1 = 50 \equiv 0 \pmod{25}$.

3.  Break into $x^2 - 1 \equiv 0 \pmod 5$ and $x^2 - 1 \equiv 0 \pmod 7$. The solutions to the first are $x \equiv \pm 1 \pmod 5$ and the solutions to the second are $x \equiv \pm 1 \pmod 7$. These can be put together as follows:

$$x \equiv +1 \pmod 5, \ x \equiv +1 \pmod 7 \Rightarrow x \equiv +1 \pmod{35}$$
$$x \equiv -1 \pmod 5, \ x \equiv -1 \pmod 7 \Rightarrow x \equiv -1 \pmod{35}$$
$$x \equiv +1 \pmod 5, \ x \equiv -1 \pmod 7 \Rightarrow x \equiv +6 \pmod{35}$$
$$x \equiv -1 \pmod 5, \ x \equiv +1 \pmod 7 \Rightarrow x \equiv -6 \pmod{35}.$$

Therefore, the solutions are $x \equiv \pm 1, \pm 6 \pmod{35}$. This may also be written as $x \equiv 1, 6, 29, 34 \pmod{35}$.

# Chapter 11

# Order and Primitive Roots

## 11.1 Orders of Elements

Let $n$ be a positive integer and let $\gcd(a, n) = 1$. The **order** of $a$ mod $n$, denoted $\text{ord}_n(a)$, is the smallest positive integer $m$ such that

$$a^m \equiv 1 \pmod{n}.$$

**Example.**

$$2^6 \equiv 1 \pmod 9, \quad 2^i \not\equiv 1 \pmod 9 \text{ for } 1 \leq i \leq 5,$$

so $\text{ord}_9(2) = 6$.

Here's another way to look at the order of a number. The powers of 2 mod 9 are

$$2^0 \equiv 1, \ 2^1 \equiv 2, \ 2^2 \equiv 4, \ 2^3 \equiv 8, \ 2^4 \equiv 7, \ 2^5 \equiv 5,$$
$$2^6 \equiv 1, \ 2^7 \equiv 2, \ 2^8 \equiv 4, \ldots.$$

The powers form a periodic sequence of period 6 (that is, they repeat after 6 steps). In general, the powers of $a$ mod $n$ are periodic with period equal to $\text{ord}_n(a)$.

Before we go any farther, there's a fundamental question. How do we know the order of a number exists? In other words, is there *always* some exponent $m \geq 1$ such that $a^m \equiv 1 \pmod{n}$? The answer is "Yes," because Euler's theorem says that $a^{\phi(n)} \equiv 1 \pmod{n}$. Therefore, the order exists and is less than or equal to $\phi(n)$.

One of the most useful results about the order of a number is the following.

**Theorem 11.1.** *Let $n$ be a positive integer, let $a$ be an integer with* $\gcd(a, n) = 1$, *and let $m$ be an integer. Then*

$$a^m \equiv 1 \pmod{n} \Longleftrightarrow \mathrm{ord}_n(a) \mid m.$$

*Proof.* For ease of notation, let $m_0 = \mathrm{ord}_n(a)$. If $m_0 \mid m$, then $m = m_0 k$ for some $k$, so

$$a^m \equiv (a^{m_0})^k \equiv (1)^k \equiv 1 \pmod{n},$$

where we have used the fact that $a^{m_0} \equiv a^{\mathrm{ord}_n(a)} \equiv 1$, by the definition of $\mathrm{ord}_n(a)$.

Conversely, suppose $a^m \equiv 1 \pmod{n}$. Write $m = m_0 q + r$ with $0 \leq r < m_0$. Then

$$a^r \equiv a^r (1)^q \equiv a^r (a^{m_0})^q \equiv a^{m_0 q + r} \equiv a^m \equiv 1 \pmod{n}.$$

Since $m_0$ is the smallest positive exponent that yields 1 and $r$ is smaller than $m_0$, we must have $r = 0$. Therefore, $m = m_0 q$, so $m_0 \mid m$.    □

**Corollary 11.2.** *(a) Let $p$ be prime and let $a$ be an integer such that $a \not\equiv 0 \pmod{p}$. Then $\mathrm{ord}_p(a) \mid p - 1$.*
*(b) Let $n$ be a positive integer and let $a$ be an integer with $\gcd(a, n) = 1$. Then $\mathrm{ord}_n(a) \mid \phi(n)$.*

*Proof.* Part (a) is a special case of (b), but we single it out since it is so useful. However, it suffices to prove (b). By Euler's theorem, $a^{\phi(n)} \equiv 1 \pmod{n}$. The theorem with $m = \phi(n)$ implies that $\mathrm{ord}_n(a) \mid \phi(n)$.    □

**Example.** Let's compute $\mathrm{ord}_{23}(3)$. We could compute $3^i \pmod{23}$ for $i = 1, 2, 3, \ldots$ until we get $3^i \equiv 1$, but it's faster to use the fact that $\mathrm{ord}_{23}(3) \mid 22$ by Corollary 11.2. Therefore, $\mathrm{ord}_{23}(3) = 1, 2, 11, 22$. Since $3^1 \not\equiv 1$ and $3^2 \not\equiv 1 \pmod{23}$, the next possibility is $3^{11}$. By successive squaring, as in Section 6.2, we compute

$$3^2 \equiv 9, \quad 3^4 \equiv 9^2 \equiv 81 \equiv 12, \quad 3^8 \equiv 12^2 \equiv 6$$
$$3^{11} \equiv 3^8 3^2 3^1 \equiv 6 \cdot 9 \cdot 3 \equiv 1 \pmod{23}.$$

So $\mathrm{ord}_{23}(3) = 11$.

**CHECK YOUR UNDERSTANDING:**
1. Compute $\mathrm{ord}_{13}(5)$.
2. We know that $2^{16} \equiv 1 \pmod{17}$. Does this imply that $\mathrm{ord}_{17}(2) = 16$?

## 11.1.1    Fermat Numbers

We claimed that Theorem 11.1 is useful. Let's back up our claim. Back in the 1600s, Fermat looked at the *Fermat numbers*

$$F_n = 2^{2^n} + 1.$$

The first few are $F_0 = 3$, $F_1 = 5$, $F_2 = 17$, $F_3 = 257$, $F_4 = 65537$. Fermat observed that all of these are prime and conjectured that $F_n$ is prime for all $n \geq 0$. But in 1732, Euler showed that

$$F_5 = 4294967297 = 641 \times 6700417.$$

Here's how he did this.

**Proposition 11.3.** *Let $n \geq 2$ and let $p$ be a prime dividing $F_n$. Then $p \equiv 1 \pmod{2^{n+2}}$.*

*Proof.* If $p \mid 2^{2^n} + 1$, then $2^{2^n} \equiv -1 \pmod{p}$. Squaring yields

$$2^{2^{n+1}} \equiv (-1)^2 \equiv 1 \pmod{p}.$$

By Theorem 11.1, $\mathrm{ord}_p(2) \mid 2^{n+1}$, which means that $\mathrm{ord}_p(2) = 2^j$ for some $j \leq n + 1$. If $j \leq n$, then

$$2^{2^n} \equiv \left(2^{2^j}\right)^{2^{n-j}} \equiv 1^{2^{n-j}} \equiv 1 \pmod{p}.$$

Since $2^{2^n} \equiv -1$, this is a contradiction. Therefore, $\mathrm{ord}_p(2) = 2^{n+1}$. By Corollary 11.2, $2^{n+1} \mid p - 1$.

We now have $p \equiv 1 \pmod{2^{n+1}}$. But we want $p \equiv 1 \pmod{2^{n+2}}$. First, $p \equiv 1 \pmod 8$ since $n \geq 2$. In Exercise 31 and in Chapter 13, we'll show that there exists $b$ such that $b^2 \equiv 2 \pmod{p}$ when $p \equiv 1$ (mod 8). We now repeat the above steps with $b$ in place of 2:

$$b^{2^{n+1}} \equiv \left(b^2\right)^{2^n} \equiv 2^{2^n} \equiv -1 \pmod{p},$$
$$b^{2^{n+2}} \equiv \left(b^2\right)^{2^{n+1}} \equiv 2^{2^{n+1}} \equiv 1 \pmod{p}.$$

This means that $\mathrm{ord}_p(b)$ divides $2^{n+2}$ and does not divide $2^{n+1}$, so $\mathrm{ord}_p(b) = 2^{n+2}$. Corollary 11.2 tells us that $2^{n+2} \mid p-1$.                    □

Here's the method Euler used to factor $F_5$. Let's say we're looking for prime factors of $F_5$. By Proposition 11.3, any such prime must be congruent to 1 mod 128. The first few primes satisfying $p \equiv 1$ (mod 128) are

$$257, \quad 641, \quad 769, \quad 1153, \quad 1409.$$

Division by hand (Euler didn't have an electronic calculator, only his brain) immediately shows that

$$F_5 = 641 \cdot 6700417.$$

Additionally, 6700417 is not a multiple of any of the primes listed and these are the only primes $p \equiv 1$ (mod 128) that are less than $\sqrt{6700417}$. But a prime factor of 6700417 is a prime factor of $F_5$ and therefore must be 1 mod 128, so 6700417 has no prime factors less than $\sqrt{6700417}$. Therefore, 6700417 must be prime, and we have the complete factorization.

**Remark.** Euler proved only the congruence $p \equiv 1 \pmod{2^{n+1}}$, so he had to consider a few more primes than we did. The improvement to $p \equiv 1 \pmod{2^{n+2}}$ is due to Lucas in the 1800s.

What happens if we try Euler's idea on the Fermat numbers that actually are prime? Let's look at $F_4 = 65537$. If $p$ is a prime factor of $F_4$, then $p \equiv 1 \pmod{64}$. The first two such primes are 193 and 257. But $193 \nmid 65537$ and $257 > \sqrt{65537}$, so $F_4$ is prime.

For $F_3 = 257$, any prime factor must be 1 mod 32. The first such prime is 97, which is greater than $\sqrt{257}$. So we find that $F_3$ is prime because there are no possible prime factors up to its square root. The same happens for $F_2$.

We now have some insight into why the first few Fermat numbers are prime: $F_2, F_3$ are not allowed any prime factors up to their square roots by Proposition 11.3. Moreover, $F_4$ is allowed only one possibility and it doesn't work. Finally, $F_5$ is allowed several and one of them is a factor. As $n$ gets larger, there are many more potential divisors, so there is a higher chance that there is a factor. In fact, $F_n$ is known to be composite for $5 \le n \le 32$ (but complete factorizations are known only up to $n = 11$).

## 11.1.2    Mersenne Numbers

If $p$ is prime, the $p$th **Mersenne number** is

$$M_p = 2^p - 1.$$

The first few are

$$M_2 = 3, \quad M_3 = 7, \quad M_5 = 31, \quad M_7 = 127,$$
$$M_{11} = 2047 = 23 \times 89, \quad M_{13} = 8191.$$

As in the case of Fermat numbers, there is a restriction on the possible prime factors of Mersenne numbers.

**Proposition 11.4.** *Let $p$ and $q$ be primes and suppose that $q \mid 2^p - 1$. Then $q \equiv 1 \pmod{p}$.*

*Proof.* If $2^p \equiv 1 \pmod{q}$, then Theorem 11.1 implies that $\mathrm{ord}_q(2) \mid p$. But $p$ is prime, so $\mathrm{ord}_q(2) = 1$ or $p$. If $\mathrm{ord}_q(2) = 1$, then $2^1 \equiv 1 \pmod{q}$, which is impossible. Therefore, $\mathrm{ord}_q(2) = p$. Corollary 11.2 says that $p \mid q - 1$, so $q \equiv 1 \pmod{p}$. $\qquad\square$

As of March 2017, there were 49 known Mersenne primes. The largest has $p = 74207281$ and is an integer with more than 22 million decimal digits. In Chapter 14, we'll discuss how to test Mersenne numbers for primality.

Proposition 11.4 gives another proof that there are infinitely many primes: If there are only finitely many primes, let $p$ be the largest. Let $q$ be a prime divisor of $2^p - 1$. Since $q \equiv 1 \pmod{p}$, we must have $q > p$, contradicting the fact that $p$ is largest. Therefore, there cannot be a largest prime, so there are infinitely many primes.

## 11.2    Primitive Roots

We know that the order of an integer mod $p$ divides $p - 1$. If the order of $g$ equals $p - 1$, we say that $g$ is a **primitive root** for $p$.

**Example.** $\text{ord}_5(2) = 4$, so 2 is a primitive root for 5.

**Example.** $\text{ord}_7(2) = 3$, so 2 is not a primitive root for 7.

There is another way to characterize primitive roots. The powers of 2 mod 5 are

$$2^0 \equiv 1, \ 2^1 \equiv 2, \ 2^2 \equiv 4, \ 2^3 \equiv 3, \ 2^4 \equiv 1, \ 2^5 \equiv 2, \ 2^6 \equiv 4, \ \ldots.$$

So $1, 2, 3, 4$ are congruent to powers of 2 mod 5. However, the powers of 2 mod 7 are

$$1, \ 2, \ 4, \ 1, \ 2, \ 4, \ \ldots,$$

so 3, 5, 6 are not congruent to powers of 2 mod 7.

**Proposition 11.5.** *Let $p$ be a prime and assume $\gcd(g, p) = 1$. The following are equivalent:*
*(a) $g$ is a primitive root for $p$ (so $\text{ord}_p(g) = p - 1$)*
*(b) every integer that is non-zero mod $p$ is congruent to a power of $g$ mod $p$.*

*Proof.* Let $g$ be a primitive root for $p$. Look at the numbers $1, g, g^2, g^3, \ldots g^{p-2} \pmod{p}$. We claim they are distinct. Suppose, instead, that $g^i \equiv g^j \pmod{p}$, with $0 \leq i < j \leq p - 2$. Then $g^{j-i} \equiv 1 \pmod{p}$. By Theorem 11.1, $p - 1 = \text{ord}_p(g)$ divides $j - i$. Since $0 < j - i \leq p - 2 < p - 1$, this is impossible. Therefore, the powers of $g$ mod $p$ give $p - 1$ distinct nonzero congruence classes. Since there are only $p - 1$ nonzero congruence classes mod $p$, each is represented by a power of $g$. This proves that (a) implies (b).

Conversely, suppose that (b) is true for $g$. Let $m = \text{ord}_p(g)$. The powers of $g$ are

$$1, g, g^2, g^3, \ldots, g^{m-1} \pmod{p},$$

since $g^m \equiv 1 \pmod{p}$, and the powers of $g$ start repeating at this point. Therefore, there are exactly $m$ different powers of $g$ mod $p$. Since (b) says that there are $p-1$ numbers mod $p$ that are congruent to powers of $p$, we must have $m = p-1$, so $g$ is a primitive root for $p$.                     □

**Remark.** Part (b) of Proposition 11.5 gives the main reason why primitive roots are useful. The proposition gives the equivalence between part (a), which is a property that is usually easy to verify, and part (b), which is the important property.

**Warning:** If we are given a prime $p$ and a number $g \not\equiv 0 \pmod{p}$, Fermat's Theorem says that $g^{p-1} \equiv 1 \pmod{p}$. But this does not automatically say that $g$ is a primitive root mod $p$. We need to show that $g^k \not\equiv 1 \pmod{p}$ when $1 \le k < p-1$. Fortunately, Corollary 11.2 says that we have to check only those $k$ that are divisors of $p-1$, which is usually faster than checking the criterion of part (b) of the proposition.

Here is a property of primitive roots that will be useful in later chapters.

**Proposition 11.6.** *Let $g$ be a primitive root for the odd prime $p$. Then*

$$g^{(p-1)/2} \equiv -1 \pmod{p}.$$

*Proof.* Let $x \equiv g^{(p-1)/2} \pmod{p}$. Then

$$x^2 \equiv g^{p-1} \equiv 1 \pmod{p}.$$

Therefore, $x \equiv \pm 1 \pmod{p}$ by Corollary 6.11. If $x \equiv 1 \pmod{p}$, then we have $g^{(p-1)/2} \equiv x \equiv 1$, which contradicts the fact that the order of $g$ is $p-1$. Therefore, $x \equiv -1$, which is what we wanted to prove.     □

What we just proved is the case $n = p$, $m = p-1$, $i = (p-1)/2$ of the next proposition.

**Proposition 11.7.** *Let $n$ be a positive integer and $\gcd(x, n) = 1$. Let $m = \operatorname{ord}_n(x)$ and let $i$ be an integer. Then*

$$\operatorname{ord}_n(x^i) = \frac{m}{\gcd(i, m)}.$$

*Proof.* Let

$$k = \operatorname{ord}_n(x^i).$$

Then $x^{ik} \equiv 1 \pmod{n}$. By Theorem 11.1, $ik \equiv 0 \pmod{m}$. Let

$$d = \gcd(i, m).$$

Then

$$\frac{i}{d}k \equiv 0 \pmod{\frac{m}{d}}.$$

Since $\gcd(i/d, m/d) = 1$, we can divide the congruence by $i/d$ and get $k \equiv 0 \pmod{m/d}$. In particular, $k \geq m/d$.

Since $i/d$ is an integer,

$$\left(x^i\right)^{m/d} \equiv (x^m)^{i/d} \equiv (1)^{i/d} \equiv 1 \pmod{p}.$$

(Why do we care that $i/d$ is an integer? If it's not, then $1^{i/d}$ might have more than one possible value, and the whole calculation becomes ambiguous.) By Theorem 11.1, $k \mid m/d$, so $k \leq m/d$.

The two inequalities combine to yield $k = m/d$. This is the equation of the proposition.                                                                    □

**Corollary 11.8.** *Let $p$ be prime and let $g$ be a primitive root mod $p$. Let $i$ be an integer. Then*

$$\operatorname{ord}_p(g^i) = \frac{p-1}{\gcd(i, p-1)}. \tag{11.1}$$

*Proof.* In the proposition, let $x = g$ and $m = p - 1$.                □

**Example.** An easy calculation shows that 2 is a primitive root for 13. We have $2^8 \equiv 9 \pmod{13}$. The proposition says that

$$\operatorname{ord}_{13}(9) = \frac{12}{\gcd(8, 12)} = 3.$$

The proposition allows us to find all primitive roots for $p$ as soon as we know one primitive root.

**Corollary 11.9.** *Let $g$ be a primitive root for the prime $p$. The primitive roots for $p$ are the numbers congruent to $g^i \pmod{p}$ for $\gcd(i, p - 1) = 1$.*

*Proof.* Since $g$ is a primitive root, every number that is nonzero mod $p$ is congruent to $g^i$ for some $i$. By Corollary 11.8, $\operatorname{ord}_p(g^i) = p - 1$ if and only if $\gcd(i, p - 1) = 1$.                                    □

**Example.** The numbers relatively prime to 12 are $1, 5, 7, 11$, so the primitive roots for 13 are

$$2, \quad 2^5 \equiv 6, \quad 2^7 \equiv 11, \quad 2^{11} \equiv 7.$$

Why did we stop at $i = 11$? Isn't $\gcd(17, 12) = 1$, for example? Fermat's theorem tells us that everything starts over at $2^{12} \equiv 1$, so

$$2^{17} \equiv 2^{12} \cdot 2^5 \equiv 2^5 \equiv 6 \pmod{13}.$$

Therefore, the exponents past $p - 1 = 12$ give us nothing new to consider.

**Theorem 11.10.** *Let $p$ be prime. There are $\phi(p-1)$ primitive roots $g$ for $p$ with $1 \leq g < p$.*

*Proof.* Let $g$ be a primitive root. The other primitive roots are exactly $g^i \pmod{p}$ with $1 \leq i \leq p-1$ with $\gcd(i, p-1) = 1$. There are $\phi(p-1)$ such values of $i$, so we're done.

Wait! Are we done? We missed an important point. How do we know there are any primitive roots? We needed at least one to get the proof started. In Section 11.6, we'll fill in this gap and show that there is always at least one primitive root for each prime.                    □

**Example.** The first prime after 100000 is 100003. (What happened to 100001? Look back at the test for divisibility by 11.) The number of primitive roots for 100003 is

$$\phi(100002) = 28560,$$

so more than 1/4 of the numbers mod 100003 are primitive roots. The first ten primitive roots are

$$2, 3, 5, 7, 29, 31, 32, 34, 46, 47.$$

(Once we know that 2 is a primitive root, why do we know that 32 is also a primitive root? See Corollary 11.9.)

How do we find a primitive root? Since they are fairly abundant, we can look at 2, 3, 5, 6, 7, 8, ... until we find one. (Why did we omit 4? It's a square, and squares cannot be primitive roots for primes greater than 2. See Exercise 25.) But when we look at our list of *potential* primitive roots, how do we know that a number is *actually* a primitive root without checking successive powers?

**Example.** Suppose we want to show that 6 is a primitive root mod 41. Let $m = \text{ord}_{41}(6)$. Since $m \mid 40$ by Corollary 11.2, we see that $m$ is one of the following:

$$1, 2, 4, 5, 8, 10, 20, 40.$$

A calculation shows that $6^{20} \equiv -1 \pmod{41}$, so $m \neq 20$. Without any more calculation, we can see that $m \neq 5$: If $6^5 \equiv 1 \pmod{41}$ then $6^{20} \equiv \left(6^5\right)^4 \equiv 1^4 \equiv 1$, which is not the case. Similarly, we see that $m$ cannot be a divisor of 20. The only choices left are $m = 8$ and $m = 40$. A calculation shows that $6^8 \equiv 10 \pmod{41}$, so $m \neq 8$. We conclude that $m = 40$, so 6 is a primitive root for 41.

Let's analyze what we did. We needed only two computations: $6^{20} \not\equiv 1$ and $6^8 \not\equiv 1 \pmod{41}$. The first removed all divisors of 20, namely, 1, 2, 4, 5, 10, 20, from consideration. The second removed all divisors of 8, namely, 1, 2, 4, 8, from consideration. The only possibility that remained was 40. What was so special about 20 and 8? They are $40/2$ and $40/5$. In other words, they are $(p-1)/\ell$, where $\ell$ runs through the prime divisors of $p-1$.

**Proposition 11.11.** *Let $p$ be a prime and let $h \not\equiv 0 \pmod{p}$. The following are equivalent:*
*(a) $h$ is a primitive root for $p$.*
*(b) For each prime $\ell$ dividing $p-1$,*

$$h^{(p-1)/\ell} \not\equiv 1 \pmod{p}.$$

*Proof.* If $h$ is a primitive root, then

$$\text{ord}_p(h) = p - 1 > (p-1)/\ell > 0$$

for each $\ell$, so

$$h^{(p-1)/\ell} \not\equiv 1 \pmod{p}.$$

Conversely, suppose (b) is true. Let $m = \text{ord}_p(h)$. Corollary 11.2 says that $m \mid p - 1$. If $m \neq p - 1$, let $\ell$ be a prime dividing $(p-1)/m$, so $\ell k = (p-1)/m$ for some $k$. Then

$$mk = (p-1)/\ell,$$

and

$$h^{(p-1)/\ell} \equiv (h^m)^k \equiv 1^k \equiv 1 \pmod{p},$$

which contradicts (b). Since the assumption that $m \neq p - 1$ produced a contradiction, we must have $m = p - 1$, which means that $h$ is a primitive root for $p$. □

In practice, the proposition allows us to find primitive roots for a prime $p$ fairly quickly, as long as we know the factorization of $p - 1$. Unfortunately, if we cannot factor $p - 1$, there does not seem to be a quick way of verifying that a number is a primitive root.

Up to now, we have been starting with a prime and looking for a primitive root. Let's reverse the procedure, starting with an integer and asking for primes for which it is a primitive root. The number $h = 2$ is a primitive root for the primes

$$3, 5, 11, 13, 19, \ldots,$$

and, in fact, it is a primitive root for 67 of the 168 primes less than 1000. Is 2 a primitive root for infinitely many primes? This type of question was perhaps first raised by Gauss, who asked whether there are infinitely many primes $p$ where the decimal expansion of $1/p$ has period $p - 1$. As we'll see in the next section, this is the same as asking whether 10 is a primitive root for infinitely many primes. In the 1920s, Emil Artin conjectured that if $h$ is not a square or the negative of a square, then it is a primitive root for infinitely many primes. In fact, he conjectured that $h$ is a primitive root for a positive proportion of primes. That is, he predicted that there is a constant $C_h$ such that

$$\lim_{x \to \infty} \frac{\#\{p \mid p \leq x \text{ is prime and } h \text{ is a primitive root for } p\}}{\pi(x)} = C_h.$$

For $h = 2$, the conjectured constant is $C_2 = .3739558 \cdots$, which is close to the ratio $3603/9592 = 0.3756255 \cdots$ for primes up to 100000.

In 1967, Christopher Hooley proved Artin's conjecture on primitive roots, under the assumption of the Generalized Riemann Hypothesis (a generalization of the Riemann Hypothesis [see Section 5.7]; it is still unproved). In 1985, Roger Heath-Brown, inspired by work of Ram Murty and Rajiv Gupta, showed, without any unproved assumptions, that there are at most two primes that are not primitive roots for infinitely many primes. So, for example, at least one of 2, 3, 5 is a primitive root for infinitely many primes, but we don't know whether it's 2, 3, or 5 (but probably it's all three).

---

**CHECK YOUR UNDERSTANDING:**

3. Show that 2 is a primitive root for 19.
4. Find all primitive roots $g$ for 19 with $1 < g < 19$.

---

# 11.3   Decimals

Look at the powers of 10 mod 7:

$$10^1 \equiv 3, \quad 10^2 \equiv 2, \quad 10^3 \equiv 6,$$
$$10^4 \equiv 4, \quad 10^5 \equiv 5, \quad 10^6 \equiv 1.$$

Now, look at the long division calculation of 1/7:

```
            0.142857
     7 ) 1.000000
            7
           ‾‾
           30
           28
           ‾‾
           20
           14
           ‾‾
           60
           56
           ‾‾
           40
           35
           ‾‾
           50
           49
           ‾‾
            1
```

Do you see anything in common? The remainders in the long division are 3, 2, 6, 4, 5, 1, 3,..., and these are the powers of 10 reduced mod 7. For example, the 6 that is obtained as the remainder in the third step

of the long division is the remainder when 1000 is divided by 7. This is
the same as $10^3 \equiv 6 \pmod 7$. Moreover, when we get the remainder 1
after 6 steps, the decimal starts repeating. This corresponds to $10^6 \equiv 1$
$\pmod 7$.

These observations lead to the following. We define the **period length**
of a repeating decimal to be the length of the block that is repeated.
For example, the decimal expansion of $1/7$ has period length 6 and
the decimal expansion of $1/3 = .3333\cdots$ has period length 1. If the
repetition of the decimal begins immediately after the decimal point,
the decimal is called **purely periodic**. So, $1/3 = .3333\cdots$ and $1/7 =$
$.142857142857\cdots$ are purely periodic.

**Proposition 11.12.** *Let $m \geq 3$ satisfy $\gcd(m, 10) = 1$ and let $0 < b <$
$m$ with $\gcd(b, m) = 1$. The period length for the decimal expansion of
$b/m$ is $\mathrm{ord}_m(10)$. Therefore, the period length for the decimal expansion
of $b/m$ divides $\phi(m)$.*

*Proof.* A periodic decimal expansion that repeats every $n$ terms has
the form

$$\frac{A}{10^n} + \frac{A}{10^{2n}} + \frac{A}{10^{3n}} + \cdots = \frac{A/10^n}{1 - 10^{-n}} = \frac{A}{10^n - 1}$$

with $1 \leq A \leq 10^n - 1$. (We have used the formula for the sum of a
geometric series: $a + ar + ar^2 + \cdots = a/(1 - r)$. See Appendix A.) For
example,

$$\frac{3}{11} = .272727\cdots = \frac{27}{10^2} + \frac{27}{10^4} + \frac{27}{10^6} + \cdots = \frac{27}{10^2 - 1}.$$

If the decimal expansion of $b/m$ repeats every $n$ terms, then $b/m =$
$A/(10^n - 1)$ for some $A$. This means that $(10^n - 1)b = Am$, so $m \mid$
$(10^n - 1)b$. Since $\gcd(m, b) = 1$, we must have $m \mid 10^n - 1$ by Proposition
2.16, so $10^n \equiv 1 \pmod m$.

Conversely, suppose that $10^n \equiv 1 \pmod m$. Then we can write $10^n -$
$1 = am$ for some integer $a$. Since $0 < b < m$, we have $0 < ab < am =$
$10^n - 1$. Moreover,

$$\frac{b}{m} = \frac{ab}{am} = \frac{ab}{10^n - 1} = \frac{ab}{10^n} + \frac{ab}{10^{2n}} + \frac{ab}{10^{3n}} + \cdots,$$

so the decimal expansion of $b/m$ repeats every $n$ terms.

We conclude that the decimal expansion of $b/m$ repeats every $n$ terms if and only if $10^n \equiv 1 \pmod{m}$. The smallest such positive $n$ is the period of $b/m$ and is $\text{ord}_m(10)$. The fact that $\text{ord}_m(10)$ divides $\phi(m)$ is Corollary 11.2. $\qquad\square$

**Examples.**

$1/7 = 0.\overline{142857}$ has period length 6, and $\text{ord}_7(10) = 6$. (The notation $0.\overline{142857}$ means that 142857 is infinitely repeated, so $0.\overline{142857} = 0.142857142857\cdots$.)

$1/11 = 0.\overline{09}$ has period length 2, and $\text{ord}_{11}(10) = 2$.

$1/13 = 0.\overline{076923}$ has period length 6, and $\text{ord}_{13}(10) = 6$.

$5/21 = 0.\overline{238095}$ has period length 6, and $\text{ord}_{21}(10) = 6$.

The following well-known result is closely related to the preceding discussion. An **eventually periodic** decimal is one that starts repeating from some point onward. The repetition does not need to start at the beginning of the expansion. For example, $0.27272727\cdots$, $0.123757575\cdots$, and $0.123000000\cdots$ are all eventually periodic (the first is also purely periodic).

**Theorem 11.13.** *A decimal expansion of a real number $x$ is eventually periodic if and only if $x$ is rational.*

*Proof.* If the decimal expansion of $x$ is periodic from the $k$th term after the decimal point onward, then $10^{k-1}x$ is an integer plus a purely periodic decimal. The purely periodic decimal can be evaluated using the formula for the sum of a geometric series and this gives a rational number. Since $10^{k-1}x$ minus an integer is rational, so is $x$. For example,

$$10^2 \left(153.74123123123123\cdots\right)$$
$$= 15374 + .123123123 + \cdots$$
$$= 15374 + \frac{123}{999}.$$

Conversely, if $x$ is rational, then for some $k$ sufficiently large, $10^k x$ can be written in the form $t + r/s$ with $t$ and integer and $0 < r < s$ and such that $\gcd(s, 10) = 1$. (Why? Simply multiply by a high enough power of 10 to remove 2's and 5's from the denominator.) Then $r/s$ has a purely periodic decimal expansion (with period equal to $\text{ord}_s(10)$).

Adding the integer $t$ and dividing by $10^k$ yields a decimal expansion that is eventually periodic for $x$. This completes the proof.    □

---

**CHECK YOUR UNDERSTANDING:**
5. (a) Find the period length of the decimal expansion of $1/37$.
(b) Calculate $\text{ord}_{37}(10)$.

---

## 11.3.1   Midy's Theorem

Look at the following two decimal expansions:

$$\frac{3}{17} = 0.1764705882352941\overline{1764705882352941}\cdots$$

$$\frac{8}{13} = 0.615384\overline{615384615384}\cdots.$$

In the first expansion, take the period 1764705882352941, which has length 16, split it in half, and add the two halves:

$$17647058$$
$$\underline{+82352941}$$
$$99999999.$$

Awesome! Now try it with the other decimal:

$$615$$
$$\underline{+384}$$
$$999.$$

These are examples of Midy's Theorem, which was discovered by the French mathematician E. Midy in 1836.

**Theorem 11.14.** *(Midy) Let $p \neq 2, 5$ be prime and let $0 < a < p$. Let*

$$\frac{a}{p} = 0.\overline{c_1 c_2 c_3 \ldots c_m}$$

*be the decimal expansion of $a/p$, where $m$ is the period (that is, the period is exactly $m$, not smaller). Suppose $m$ is even: $m = 2n$. Then*

$$c_1 c_2 \cdots c_n \ + \ c_{n+1} c_{n+2} \cdots c_{2n} = 99 \cdots 9$$

*where there are $n$ nines (this is addition of integers in base 10, so $c_1 c_2 \cdots c_n$ represents $c_1 10^{n-1} + c_2 10^{n-2} + \cdots + c_n$).*

**Proof.** Let $A = c_1 c_2 \cdots c_n$ and $B = c_{n+1} c_{n+2} \cdots c_{2n}$. Then

$$\frac{a}{p} = \frac{10^n A + B}{10^{2n}} + \frac{10^n A + B}{10^{4n}} + \frac{10^n A + B}{10^{6n}} + \cdots$$

(this is just writing a repeating decimal as a geometric series). The formula for the sum of a geometric series (see Appendix A) implies that

$$\frac{1}{10^{2n}} + \frac{1}{10^{4n}} + \frac{1}{10^{6n}} + \cdots = \frac{10^{-2n}}{1 - 10^{-2n}} = \frac{1}{10^{2n} - 1}.$$

Therefore,

$$\frac{a}{p} = \frac{10^n A + B}{10^{2n} - 1} = \frac{10^n A + B}{(10^n - 1)(10^n + 1)}. \tag{11.2}$$

Multiply by $(10^n + 1)(10^n - 1)p$ to obtain

$$a(10^n + 1)(10^n - 1) = (10^n A + B)p.$$

Since $p \nmid a$, this implies that $p$ divides $10^{2n} - 1 = (10^n - 1)(10^n + 1)$. Since $p$ is prime (this is where we use this hypothesis), $p$ divides $10^n - 1$ or $10^n + 1$.

If $p \mid 10^n - 1$, we have $\operatorname{ord}_p(10) \leq n$, hence Proposition 11.12 implies that

$$2n = \operatorname{ord}_p(10) \leq n,$$

which is impossible. Therefore, $p \nmid 10^n - 1$. We conclude that $p$ divides $10^n + 1$. Therefore, $a(10^n + 1)/p$ is an integer. Since this integer equals $(10^n A + B)/(10^n - 1)$ by Equation 11.2, we have

$$10^n A + B \equiv 0 \pmod{10^n - 1}.$$

But $10^n \equiv 1 \pmod{10^n - 1}$, so $A + B \equiv 0 \pmod{10^n - 1}$. Because $A$ and $B$ are $n$-digit numbers, $0 \leq A, B \leq 10^n - 1$. Moreover, at least one of $A, B$ is nonzero (since $a \neq 0$), so $A + B > 0$. We cannot have

$A = B = 10^n - 1 = 999 \cdots 9$ since that gives $a/p = .999 \cdots = 1$, contradicting $a < p$. Therefore, at least one of $A, B$ is less than $10^n - 1$ and

$$0 < A + B < 2(10^n - 1).$$

The only multiple of $10^n - 1$ in this range is $10^n - 1$, so

$$A + B = 10^n - 1 = 999 \cdots 9.$$

$\square$

## 11.4    Card Shuffling

Many phenomena with cards have mathematical explanations. Here we discuss what happens during a perfect riffle shuffle.

Start with a deck of $2n$ cards. For simplicity, we'll draw our diagrams with $2n = 6$ cards and call them $A, B, C, D, E, F$:

$$
\begin{array}{l}
A \; \text{------} \\
B \; \text{------} \\
C \; \text{------} \\
D \; \text{------} \\
E \; \text{------} \\
F \; \text{------}
\end{array}
$$

To shuffle the cards, split the pile into two piles of height $n$:

$$
\begin{array}{ll}
A \; \text{------} & D \; \text{------} \\
B \; \text{------} & E \; \text{------} \\
C \; \text{------} & F \; \text{------}
\end{array}
$$

and then interlace them so that they alternate from the two piles, with the top card from the second pile, the next card from the first pile, etc. This yields

$$
\begin{array}{l}
D \ \underline{\hspace{2cm}} \\
A \ \underline{\hspace{2cm}} \\
E \ \underline{\hspace{2cm}} \\
B \ \underline{\hspace{2cm}} \\
F \ \underline{\hspace{2cm}} \\
C \ \underline{\hspace{2cm}}
\end{array}
$$

The process can be described mathematically. Number the positions of the cards from 1 to $2n$, with 1 being the top card and $2n$ being the bottom card. If a card starts in position $j$ with $1 \le j \le n$, then it is part of the first pile. After the shuffle, it is in position $2j$. If a card starts in position $j$ with $n + 1 \le j \le 2n$, then it is part of the second pile. After the shuffle, the card that started in position $n+1$ ends up in position 1, the card that started in position $n + 2$ ends up in position 3, and, in general, the card that started in position $n + i$ ends up in position $2i - 1$. If we write $j = n + i$, then

$$2i - 1 = 2(j - n) - 1 = 2j - (2n + 1).$$

The general rule for the change of position is therefore

$$j \mapsto 2j \text{ if } 1 \le j \le n, \text{ and } j \mapsto 2j - (2n + 1) \text{ if } n + 1 \le j \le 2n.$$

This can be stated as

$$j \mapsto 2j \pmod{2n + 1}.$$

How many shuffles of this type does it take to get back to the original position? Since one shuffle is multiplying the positions by 2 mod $2n+1$, the effect of $k$ shuffles is to multiply by $2^k$ mod $2n + 1$. Therefore, after $\mathrm{ord}_{2n+1}(2)$ shuffles, the cards return to their original positions.

For example, with 6 cards, we have $n = 3$ and $2n + 1 = 7$. Since $\mathrm{ord}_7(2) = 3$, it takes three shuffles to return the deck to its original ordering. To check this, observe that a second shuffle orders the cards as $B, D, F, A, C, E$. One more shuffle yields $A, B, C, D, E, F$, as claimed.

With a deck of 52 cards, we have $2n + 1 = 53$. Since $\mathrm{ord}_{53}(2) = 52$, it takes 52 shuffles of this type to return the deck to its original order.

The situation is completely different if we modify the shuffling so that the top card is from the first pile, the next card is from the second pile, etc. After one shuffle of six cards, the order of the cards is $A, D, B, E, C, F$. The top card and the bottom cards always maintain their positions at the top and bottom. In fact, we are really doing a perfect riffle shuffle of the remaining four cards. In general, we are shuffling the interior $2n - 2$ cards, so it takes $\text{ord}_{2n-1}(2)$ shuffles to return the deck to its original order. For six cards, we have $\text{ord}_5(2) = 4$, so it takes four shuffles. For 52 cards, we have $\text{ord}_{51}(2) = 8$, so after eight shuffles we have the original order.

# 11.5    The Discrete Log Problem

We can solve the exponential equation $2^x = 32$ by trial and error (or by recalling the powers of 2) and get that $x = 5$ is a solution. If we have a more difficult problem, say $2^x = 140$ we can still get an idea as to where to look for a solution by noticing that $2^7 = 128$ and $2^8 = 256$ so that $7 < x < 8$. We can do this because $f(x) = 2^x$ is a *strictly increasing* function: as $x$ gets larger, $f(x)$ gets larger as well. More precisely and more generally, if $b > 1$ and $x_1 > x_2$, then $b^{x_1} > b^{x_2}$.

On the other hand, if we fix $b$ and $n$ and look at the least positive residues of $b^x \pmod{n}$, their magnitudes are chaotic and apparently random for increasing values of $x$. As an example, the following table gives the least positive residue of $2^x \pmod{19}$ for $1 \le x \le 17$ (since $2^{18} \equiv 1 \pmod{19}$, we stop at $x = 17$).

| $x$ | 1 | 2 | 3 | 4 | 5 | 6 | 7 | 8 | 9 | 10 | 11 | 12 | 13 | 14 | 15 | 16 | 17 |
|-----|---|---|---|----|----|---|----|---|----|----|----|----|----|----|----|----|----|
| $2^x$ | 2 | 4 | 8 | 16 | 13 | 7 | 14 | 9 | 18 | 17 | 15 | 11 | 3 | 6 | 12 | 5 | 10 |

Notice that once $x > 4$, so $2^x > 19$, there seems to be no pattern in the way that $2^x \pmod{19}$ changes: it increases and decreases in a haphazard way. This leads to the following:

**Discrete Log Problem (DLP).** Given a prime $p$, a primitive root $g$, and $h \not\equiv 0 \pmod{p}$, find $x$ so that $g^x \equiv h \pmod{p}$.

The answer $x$ is called the **discrete log** of $h$.

**Example.** If we want to solve $3^x = 1594323$ without mods, it's easy. We can check that $3^{10} = 59049$ and $3^{15} = 14348907$, so $x$ is between 10 and 15. Quickly we find that $x = 13$ works. But now suppose we want to solve $3^x \equiv 8 \pmod{43}$. Since $3^1 = 3$ and $3^3 = 27$, we might want to guess that $x$ is between 1 and 3 and then see if $x = 2$ works. It doesn't, since the same problem we encountered when looking at $2^x \bmod 19$ is still present: the higher powers of 3 are more than 43, so reducing them mod 43 gives remainders that appear to be random. Soon we'll see slightly faster methods, but, for the present problem, a brute force search through possible exponents yields $x = 39$.

If you had somehow missed this solution, you'd find that $x = 81$ also works:
$$3^{81} \equiv 3^{42} \cdot 3^{39} \equiv 1 \cdot 3^{39} \equiv 8 \pmod{43}.$$
We have used Fermat's theorem to change $3^{42}$ to 1. In fact, $x = 39 + 42k$ works, for any integer $k$.

In general, Fermat's theorem tells us that if $g^x \equiv h \pmod{p}$, then $x + (p-1)k$ is also a solution for any integer $k$. Therefore, the discrete log can be regarded as a number mod $p-1$. We usually take the discrete log to have values from 0 to $p - 2$.

The importance of the Discrete Log Problem for cryptography is the following:

**The Discrete Log Problem is (believed to be) hard.**

There is no proof of this statement. But people have tried to develop efficient algorithms to solve the discrete log problem and so far no one has succeeded. Current technology allows computation of discrete logs for primes up to around 230 decimal digits.

We now describe two methods for computing discrete logs that are faster than brute force. The first works for primes of moderate size, up to around 20 digits. The second contains the basic idea of how discrete logs are currently computed for large primes.

## 11.5.1 Baby Step–Giant Step Method

Let $g$ be a primitive root for the prime $p$ and let $h \not\equiv 0 \pmod{p}$. We want to solve $g^x \equiv h \pmod{p}$.

1.  Let $N = \lceil \sqrt{p-1} \rceil$ (when $x$ is a real number, $\lceil x \rceil$ denotes the smallest integer $n$ with $x \le n$; in other words, it is obtained by taking $x$ and rounding up to get an integer).

2.  Make two lists:

    (a)  $g^i \pmod{p}$ for $0 \le i \le N-1$

    (b)  $h \cdot g^{-Nj} \pmod{p}$ for $0 \le j \le N-1$

3.  Find a match between the two lists: $g^i \equiv h \cdot g^{-Nj} \pmod{p}$.

4.  $x = i + Nj$ solves the Discrete Log Problem.

**Example.** Let's solve $2^x \equiv 9 \pmod{19}$. We take

$$N = \lceil \sqrt{19-1} \rceil = 5$$

and $h = 9$. The lists are

(a) $2^0 \equiv 1, \quad 2^1 \equiv 2, \quad 2^2 \equiv 4, \quad 2^3 \equiv 8, \quad 2^4 \equiv 16$

(b) $9 \cdot 2^{-0} \equiv 9, \quad 9 \cdot 2^{-5} \equiv 8, \quad 9 \cdot 2^{-10} \equiv 5, \quad 9 \cdot 2^{-15} \equiv 15,$
$9 \cdot 2^{-20} \equiv 7.$

Both lists have 8 in common, so a match is $2^3 \equiv 8 \equiv 9 \cdot 2^{-5}$. Therefore, $2^8 \equiv 9$.

If we find a match, then we get a solution to the discrete log problem. But why should we get a match? Since $g$ is a primitive root, there is a solution $x$ with $0 \le x \le p-2$. Write $x$ in base $N$. That is, write

$$x = a_0 + a_1 N + a_2 N^2 + \cdots,$$

with $0 \le a_i < N$ for each $i$. Since $N = \lceil \sqrt{p-1} \rceil \Longrightarrow N^2 \ge p-1 > x$, we must have $a_2 = a_3 = \cdots = 0$. Therefore,

$$x = a_0 + a_1 N$$

for some $a_0$ and $a_1$. Then

$$g^{a_0} \equiv g^{x - a_1 N} \equiv h \cdot g^{-a_1 N} \pmod{p},$$

which means $i = a_0$ and $j = a_1$ yield a match.

The Baby Step–Giant Step Method was developed by Dan Shanks, who named it after a children's game.[1] The baby steps are the first list, where multiplication by $g$ gets from one element to the next. The giant steps are the second list, where multiplication by $g^{-N}$ gets from one element to the next by taking the "Giant Step" of $N$ factors of $g^{-1}$.

Suppose $p$ has 12 decimal digits. Then the two lists each have around $10^6$ entries, so we need only a few million computations, compared to the almost one trillion computations needed for a brute force search.

## 11.5.2    The Index Calculus

When $p$ is large, say around 20 or more decimal digits, the Baby Step–Giant Step Method is too slow. In this case, there is a faster way to solve discrete log problems. We start with an example.

**Example.** Let's solve $2^x \equiv 55 \pmod{101}$. We start by ignoring 55 and compute some other discrete logs instead. First, we choose a set of small primes: $\{3, 5, 7\}$. This set is called a *factor base*. (We'll see this concept again in Subsection 14.3.4.) Our first goal is to compute their discrete logs. To do so, we compute $2^r \pmod{101}$ for randomly chosen values of $r$ and try to factor the results using only the primes 3, 5, 7. We find the following:

$$2^7 \equiv 27 \equiv 3^3 \cdot 5^0 \cdot 7^0 \pmod{101}$$
$$2^9 \equiv \phantom{0}7 \equiv 3^0 \cdot 5^0 \cdot 7^1 \pmod{101}$$
$$2^{17} \equiv 75 \equiv 3^1 \cdot 5^2 \cdot 7^0 \pmod{101}$$
$$2^{24} \equiv \phantom{0}5 \equiv 3^0 \cdot 5^1 \cdot 7^0 \pmod{101}$$
$$2^{47} \equiv 63 \equiv 3^2 \cdot 5^0 \cdot 7^1 \pmod{101}.$$

Relations such as $2^{22} \equiv 77 \pmod{101}$ are excluded since 77 is not a product of numbers in the factor base.

It's convenient to use the notation $\log(h)$ for the discrete log of $h$ when the choices of $p$ and $g$ are understood. In the present example, $\log(h) = x$ means that $2^x \equiv h \pmod{101}$. We would like to find $\log(n)$ for the three values of $n$ (that is, 3, 5, 7) in our factor base. Since $2^9 \equiv 7$ and

---

[1]This game was often played by children in the 1960s and earlier and was also called "Mother, may I?" The players in turn were instructed to take a certain number of baby steps or giant steps (and sometimes backward steps and sideward steps). Whoever reached the front first won.

$2^{24} \equiv 5$, we have

$$\log(7) = 9, \quad \log(5) = 24.$$

To get $\log(3)$, we observe from the prime factorizations that we already have listed that

$$\begin{aligned}
3 &\equiv \left(3^3 \cdot 5^0 \cdot 7^0\right) \left(3^0 \cdot 5^0 \cdot 7^1\right) \left(3^2 \cdot 5^0 \cdot 7^1\right)^{-1} \\
&\equiv 2^7 \cdot 2^9 \cdot 2^{-47} \\
&\equiv 2^{-31} \equiv 2^{69} \pmod{101},
\end{aligned}$$

where we have used Fermat's theorem to multiply by $2^{100} \equiv 1$. Therefore, $\log(3) = 69$.

After this preparatory work, we're finally ready to find $\log(55)$. We compute $55 \cdot 2^r \pmod{101}$ for some random values of $r$ until we obtain a number that can be factored using only the primes in the factor base. Eventually, we obtain

$$55 \cdot 2^{25} \equiv 45 \equiv 3^2 \cdot 5 \pmod{101},$$

which implies that

$$55 \equiv 2^{-25} \cdot 3^2 \cdot 5 \pmod{101}.$$

Since $2^{69} \equiv 3$ and $2^{24} \equiv 5$, we get

$$55 \equiv 2^{-25} \cdot 2^{2 \cdot 69} \cdot 2^{24} \equiv 2^{137} \equiv 2^{37} \pmod{101},$$

where the last congruence follows, again, from Fermat's theorem. We conclude that $x = 37$.

The method presented in this example is called the Index Calculus, because "index" is the classical term (used by Euler and Gauss and many later mathematicians) for discrete logs. In general, here's how we proceed.

Let $g$ be a primitive root for the prime $p$ and let $h \not\equiv 0 \pmod{p}$. We want to solve $g^x \equiv h \pmod{p}$.

1.  Choose a factor base $B$ of small primes.

2.  Compute $g^r \pmod{p}$ for many random values of $r$ and try to factor the results using only primes from $B$.

3.  Use combinations of the successes from Step 2 to evaluate $\log(q)$ for all primes $q$ in $B$.

4. Compute $h \cdot g^r \pmod{p}$ for random values of $r$ and try to factor these using only primes from $B$, If this happens, evaluate $\log(h)$ using the values of $\log(q)$ for $q \in B$.

Let's look at step 3 more closely. In the example, we had five successes. Put the exponents of these into a matrix:

$$\begin{pmatrix} 3 & 0 & 0 \\ 0 & 0 & 1 \\ 1 & 2 & 0 \\ 0 & 1 & 0 \\ 2 & 0 & 1 \end{pmatrix}.$$

In terms of discrete logs, the matrix says that

$$\begin{pmatrix} 3 & 0 & 0 \\ 0 & 0 & 1 \\ 1 & 2 & 0 \\ 0 & 1 & 0 \\ 2 & 0 & 1 \end{pmatrix} \begin{pmatrix} \log(3) \\ \log(5) \\ \log(7) \end{pmatrix} \equiv \begin{pmatrix} 7 \\ 9 \\ 17 \\ 24 \\ 47 \end{pmatrix} \pmod{100}.$$

Why mod 100? As mentioned earlier, discrete logs are really numbers that exist mod $p-1$, which is mod 100 in the present case.

The matrix has more rows than columns, so results from linear algebra tell us that there is a good chance that there are 3 independent rows and we can solve for $\log(3), \log(5), \log(7)$. For example,

$$\text{1st row } + \text{ 2nd row } - \text{ 5th row } = (1, 0, 0),$$

which corresponds to

$$\log(3) \equiv 7 + 9 - 47 \equiv -31 \equiv 69 \pmod{100}.$$

This is the log form of the relation we used earlier to evaluate $\log(3)$.

Since we are working mod 100 in the example and mod $p-1$ in general, we have to be careful. For example, 5th row $-$ 2nd row $= (2, 0, 0)$, which yields $2\log(3) \equiv 38 \pmod{100}$. This yields $\log(3) \equiv 19 \pmod{50}$, so $\log(3) = 19$ or $69$. It takes more relations to determine the actual value. Of course, in the present situation, we could also compute $2^{19}$ and $2^{69}$ $\pmod{101}$ and see which is congruent to 3, rather than using more relations.

When the index calculus is used in practice, there are several modifications made to the procedure outlined above. For example, the random values of $r$ in Step 2 are not really random. There are ways to choose them to increase the chances of obtaining factorizations using the factor base $B$. In addition, the choice of $B$ is very delicate. If $B$ is small, it is very hard to find factorizations using only elements of $B$, but the smaller matrix makes solving for the values $\log(q)$ easier. On the other hand, if $B$ is large, the relations are much easier to find, but the matrix becomes unwieldy. In 2007, when Thorsten Kleinjung solved a discrete log problem for a prime of 160 decimal digits, he used a matrix with 423671492 rows. Removing unnecessary rows and columns led to a matrix with 2177226 rows and 2177026 columns. At this point, fairly sophisticated linear algebra methods were used to find $\log(q)$ for the primes $q$ in the factor base.

## 11.6   Existence of Primitive Roots

In this section, we prove Theorem 11.10, which asserts that if $p$ is a prime then there exist primitive roots for $p$. If you know about fields, then you might notice that the proof we give also shows that a finite multiplicatively closed subset of a field forms a cyclic group. For another proof of the existence of primitive roots, see Project 2 in Chapter 16.

We start with a technical lemma, which we need to handle the case where $a$ and $b$ are not relatively prime in Lemma 11.16 below.

**Lemma 11.15.** *Let $a$ and $b$ be positive integers. Then there exist integers $a'$ and $b'$ such that*

- $a' \mid a$ *and* $b' \mid b$

- $\gcd(a', b') = 1$

- $a'b' = \text{lcm}(a, b)$.

*Proof.* Write

$$a = 2^{a_2} 3^{a_3} 5^{a_5} \cdots, \qquad b = 2^{b_2} 3^{b_3} 5^{b_5} \cdots.$$

Let

$$a'_p = \begin{cases} a_p & \text{if } a_p \geq b_p \\ 0 & \text{if } a_p < b_p \end{cases}$$

and

$$b'_p = \begin{cases} b_p & \text{if } b_p > a_p \\ 0 & \text{if } b_p \geq a_p. \end{cases}$$

Let

$$a' = 2^{a'_2} 3^{a'_3} 5^{a'_5} \cdots, \qquad b' = 2^{b'_2} 3^{b'_3} 5^{b'_5} \cdots.$$

The definition of $a'_p$ and $b'_p$ shows that

$$a'_p + b'_p = \max(a'_p, b'_p) = \max(a_p, b_p),$$

which is the exponent of $p$ in the factorization of $\operatorname{lcm}(a, b)$, by Proposition 4.5. Since $a'_p + b'_p$ is the exponent of $p$ in $a'b'$, we obtain

$$a'b' = \operatorname{lcm}(a', b') = \operatorname{lcm}(a, b).$$

For each $p$, at least one of $a'_p$ and $b'_p$ is 0, so $\gcd(a', b') = 1$. Since $a'_p \leq a_p$ and $b'_p \leq b_p$ for each $p$, Proposition 4.4 says that $a' \mid a$ and $b' \mid b$. This proves the lemma. $\qquad\square$

For example, let $a = 2^3 3^4 5^3$ and $b = 2^5 3^4 5^2 7^3$. Since $2^3$ occurs in $a$ but the larger power $2^5$ occurs in $b$, we give $2^5$ to $b'$. We similarly assign the larger power of each prime to the corresponding $a'$ or $b'$ while discarding the lower power. When there are equal powers, we arbitrarily assign them to $a'$. This is what we do, for example, with the $3^4$ in both $a$ and $b$ We obtain $a' = 3^4 5^3$ and $b' = 2^5 7^3$.

The next lemma is a key step. Recall that we are trying to find a number of the largest possible order, namely $p - 1$. If we have a number whose order is not $p - 1$, this lemma helps us find one of a larger order. We do not yet need to be working mod a prime, so we state the lemma for arbitrary $n$. It's not until Lemma 11.17 that we need to be working modulo a prime.

**Lemma 11.16.** *Let $n$ be a positive integer, let $x$ and $y$ be integers, and assume that $\gcd(x, n) = \gcd(y, n) = 1$. Let $a = \operatorname{ord}_n(x)$ and $b = \operatorname{ord}_n(y)$. Then there exists $z$ with $\gcd(z, n) = 1$ and $\operatorname{ord}_n(z) = \operatorname{lcm}(a, b)$.*

*Proof.* Let $a'$ and $b'$ be as in Lemma 11.15. Let $x' = x^{a/a'}$ and $y' = y^{b/b'}$. By Exercise 32(a), $\operatorname{ord}_n(x') = a'$ and $\operatorname{ord}_n(y') = b'$.

Since $\gcd(a', b') = 1$, there are integers $u$ and $v$ such that $a'u + b'v = 1$. Let

$$z = (x')^v (y')^u.$$

We have $((y')^u)^{b'} = ((y')^{b'})^u \equiv 1^u \equiv 1 \pmod{n}$. Since $b'v = 1 - a'u$, we have

$$(x'^v)^{b'} = (x')^1 (x'^{a'})^{-u} \equiv x'(1)^{-u} \equiv x' \pmod{n}.$$

Therefore,

$$z^{b'} \equiv ((x')^v)^{b'} ((y')^u)^{b'} \equiv (x')(1) \equiv x' \pmod{n}.$$

By Exercise 32(b), $a' = \operatorname{ord}_n(x') = \operatorname{ord}_n(z^{b'})$ divides $\operatorname{ord}_n(z)$. Similarly, $b' \mid \operatorname{ord}_n(z)$. By Exercise 30 in Chapter 4,

$$a'b' = \operatorname{lcm}(a', b') \mid \operatorname{ord}_n(z). \tag{11.3}$$

Since $z^{a'b'} = x'^{a'b'v} y'^{a'b'u} \equiv 1^{b'v} 1^{a'u} \equiv 1 \pmod{n}$, Theorem 11.1 says that $\operatorname{ord}_n(z) \mid a'b'$. Combining this with (11.3), we find that $a'b' = \operatorname{ord}_n(z)$. Since $a'b' = \operatorname{lcm}(a, b)$, we have proved the lemma. $\qquad\square$

The next result is the second key step. It is where we need to be working mod a prime, since a polynomial congruence of degree $m$ has at most $m$ solutions mod a prime, but can have more than $m$ solutions mod a composite number.

**Lemma 11.17.** *Let $p$ be prime and let $x \not\equiv 0 \pmod{p}$. Suppose $a = \operatorname{ord}_p(x)$. Let $a_1$ be a divisor of $a$. If $w \not\equiv 0 \pmod{p}$ has order $a_1$, then $w \equiv x^k \pmod{p}$ for some $k$.*

*Proof.* Let $a_2 = a/a_1$ and let $y = x^{a_2}$. Exercise 32(a) says that $\operatorname{ord}_p(y) = a_1$, and Exercise 18 says that the numbers $y^j$ for $0 \le j \le a_1 - 1$ are distinct mod $p$. Since

$$(y^j)^{a_1} \equiv (y^{a_1})^j \equiv 1^j \equiv 1 \pmod{p},$$

we now have $a_1$ distinct solutions of the congruence $X^{a_1} - 1 \equiv 0 \pmod{p}$. By Proposition 10.1, these must be all of the solutions of this congruence. But $w$ has order $a_1$, which implies that $w^{a_1} \equiv 1 \pmod{p}$, so $w$ is a solution. Since we showed that all solutions have the form $y^j$, it follows that

$$w \equiv y^j \equiv x^{ja_2} \pmod{p}$$

for some $j$.                                                                      □

We can now prove the theorem. Let $g \not\equiv 0 \pmod{p}$ be an integer mod $p$ with largest possible order. Let's call this order $d$. If $d = p - 1$, we're done. If not, there is a number $y$ mod $p$ that is not a power of $g$ mod $p$. By Lemma 11.17, $m = \mathrm{ord}_p(y)$ cannot divide $d$. By Exercise 34 in Chapter 4, $\mathrm{lcm}(d, m) \neq d$, so $\mathrm{lcm}(d, m) > d$. Lemma 11.5 says that there exists a number $z \not\equiv 0 \pmod{p}$ with order equal to $\mathrm{lcm}(d, m) > d$. This contradicts the maximality of $d$. Therefore, every number mod $p$ is a power of $g$. By Proposition 11.5, $g$ is a primitive root mod $p$. This proves the theorem. □

# 11.7   Chapter Highlights

1. $a^m \equiv 1 \pmod{n} \Longleftrightarrow \mathrm{ord}_n(a) \mid m$ (Theorem 11.1).

2. Primitive roots for primes exist.

# 11.8   Problems

## 11.7.1   Exercises

### Section 11.1: Orders of Elements

1.  Find the order of 5 mod 13.
2.  Find the order of 4 mod 11.
3.  Find the order of 14 mod 15.
4.  Find the order of 3 mod 10.
5.  Find the order of 3 mod 121.
6.  (a) Show that the possible orders of numbers mod 11 are 1, 2, 5, and 10.
    (b) Find the orders of 2 and of 5 mod 11.
7.  (a) Show that the possible orders of numbers mod 17 are 1, 2, 4, 8, 16.
    (b) Find the order of 2 mod 17.
8.  Let $p$ be an odd prime and suppose $b$ is an integer whose order mod $p$ is 7. Show that $-b$ has order 14.

9. Let $n$ be a positive integer and let $\gcd(a,n) = 1$. Show that $\text{ord}_n(a) = 1$ if and only if $a \equiv 1 \pmod{n}$.

10. Let $p$ be a prime and let $q$ be an odd prime dividing $(3^p - 1)/2$. Show that $q \equiv 1 \pmod{p}$.

11. Let $p$ be an odd prime and suppose that $p \mid 3^{32} + 1$.
    (a) Show that the order of 3 mod $p$ is 64.
    (b) Show that $p \equiv 1 \pmod{64}$.

12. Suppose $a^x \equiv 1 \pmod{m}$ and $a^y \equiv 1 \pmod{m}$. Let $d = \gcd(x,y)$. Show that $a^d \equiv 1 \pmod{m}$.

13. Let $p$ be prime and let $R_p = (10^p - 1)/9 = 111 \cdots 11$ (the number consisting of a string of $p$ ones, often called a *repunit*).
    (a) Let $q \geq 7$ be a prime dividing $R_p$. Show that $q \equiv 1 \pmod{p}$.
    (b) Use part (a) to factor 11111.

14. Let $n$ be a positive integer and let $\gcd(b,n) = 1$. Show that $\text{ord}_n(b^{-1}) = \text{ord}_n(b)$.

15. If $a > 1$ and $n \geq 1$, show that $n \mid \phi(a^n - 1)$. (*Hint:* What is the order of $a \bmod a^n - 1$?)

16. Let $n$ be a positive integer and suppose that $\text{ord}_n(a) = m$. Let $k \mid m$. Show that there exists $b$ such that $\text{ord}_n(b) = k$.

17. Let $m$ and $n$ be positive integers with $\gcd(m,n) = 1$. Let $\gcd(a,mn) = 1$. Suppose $\text{ord}_m(a) = b$ and $\text{ord}_n(a) = c$. Show that $\text{ord}_{mn}(a) = \text{lcm}(b,c)$. (*Hint:* $a^j \equiv 1 \pmod{mn}$ if and only if $a^j \equiv 1 \pmod{m}$ and $a^j \equiv 1 \pmod{n}$.)

18. Let $n$ be a positive integer and let $\gcd(b,n) = 1$. Let $d = \text{ord}_n(b)$. Show that the numbers $1, b, b^2, \cdots, b^{d-1}$ are distinct mod $n$.

19. Let $n$ be a positive integer and let $\gcd(b,n) = 1$. Show that $b^x \equiv b^y \pmod{n}$ if and only if $x \equiv y \pmod{\text{ord}_n(b)}$. (*Hint:* Use Theorem 11.1 applied to $x - y$.)

## Section 11.2: Primitive Roots

20. (a) Show that 3 is a primitive root mod 7.
    (b) Show that 3 is not a primitive root mod 11.

21. (a) Show that 2 is a primitive root mod 13.
    (b) Find all primitive roots mod 13.

22. (a) Find the order of 2 mod 31.
    (b) Is 2 a primitive root mod 23? Why or why not?

23. Find all primitive roots mod 17.

24. Given that 2 is a primitive root mod 101, find a number $x$ mod 101 that has order 10.

25. Show that a square mod $p$ cannot be a primitive root mod $p$ when $p \geq 3$ is prime. (*Hint:* Show that a square has order dividing $(p-1)/2$.)

26. Let $p$ be a prime and let $g$ and $k$ be integers. Show that if $g^k$ is a primitive root for $p$, then $g$ is a primitive root for $p$.

27. Let $p > 3$ be prime. Show that the product of the primitive roots $g$ with $1 < g < p-1$ is 1 mod $p$. (*Hint:* See Exercise 14.)

28.  Let $p$ be an odd prime and let $g$ be a primitive root mod $p$.
     (a) Suppose that $g^j \equiv \pm 1 \pmod p$. Show that $j \equiv 0 \pmod{(p-1)/2}$.
     (b) Show that $\text{ord}_p(-g) = (p-1)/2$ or $p-1$.
     (c) If $p \equiv 1 \pmod 4$, show that $-g$ is a primitive root mod $p$.
     (d) If $p \equiv 3 \pmod 4$, show that $-g$ is not a primitive root mod $p$.

29.  Let $p$ be an odd prime.
     (a) Show that
     $$0 + 1 + 2 + \cdots + (p-2) \equiv -(p-1)/2 \equiv (p-1)/2 \pmod{p-1}.$$
     (b) Let $g$ be a primitive root mod $p$. Show that
     $$g^{0+1+2+\cdots+(p-2)} \equiv -1 \pmod p.$$
     (c) Use (b) to deduce Wilson's theorem.

30.  Let $p$ be prime and let $j$ be an integer with $j \not\equiv 0 \pmod{p-1}$. Let $S \equiv \sum_{a=1}^{p-1} a^j \pmod p$.
     (a) Let $b \not\equiv 0 \pmod p$. Show that $S \equiv \sum_{a=1}^{p-1}(ba)^j \pmod p$. (*Hint:* Lemma 8.4.)
     (b) If $b$ is a primitive root mod $p$, then $b^j - 1 \not\equiv 0 \pmod p$. Use this to prove that $S \equiv 0 \pmod p$.
     (c) Show that if $j \equiv 0 \pmod{p-1}$, then $\sum_{a=1}^{p-1} a^j \equiv -1 \pmod p$.

31.  Let $p \equiv 1 \pmod 8$ be prime and let $g$ be a primitive root mod $p$. Let $y \equiv g^{(p-1)/8} \pmod p$.
     (a) Show that $y^4 \equiv -1 \pmod p$.
     (b) Let $x \equiv y + y^{-1}$. Show that $x^2 \equiv 2 \pmod p$.

32.  Let $n$ be a positive integer and let $\gcd(b, n) = 1$. Let $m = \text{ord}_n(b)$. Let $i$ be an integer.
     (a) Let $m_1 \mid m$. Show that $\text{ord}_n(b^{m_1}) = m/m_1$.
     (b) Let $i$ be an integer. Show that $\text{ord}_n(b^i) \mid m$.

33.  Let $p$ be an odd prime and let $g$ be a primitive root for $p$. Let $a \not\equiv 0 \pmod p$, so $a \equiv g^i \pmod p$ for some $i$ with $0 \le i < p-1$.
     (a) Show that $x^2 \equiv a \pmod p$ has a solution if and only if $i$ is even. (*Hint:* If a solution $x$ exists, it is congruent to a power of $g$ mod $p$.)
     (b) Show that there are exactly $(p-1)/2$ nonzero squares mod $p$. (This was also proved as Exercise 34 in Chapter 6.)

## Section 11.3: Decimals

34.  (a) Compute the decimal expansion of $1/39$.
     (b) Compute $\text{ord}_{39}(10)$.

35.  (a) Compute the decimal expansion of $1/37$.
     (b) Compute $\text{ord}_{37}(10)$.

36.  (a) Compute the decimal expansion of $11/21$.
     (b) Compute $\text{ord}_{21}(10)$.

37.  Verify Midy's Theorem for each of the following fractions.
     (a) 3/7   (b) 5/7   (c) 7/17   (d) 5/17   (e) 9/13

38.  Verify Midy's Theorem for each of the following fractions.
     (a) 4/7   (b) 3/19   (c) 3/23   (d) 9/17   (e) 10/17

39. Find three examples for Midy's theorem with primes that are different from the ones in the text.

40. Show by example that Midy's theorem can fail for $1/n$ when $n$ satisfies $\gcd(10, n) = 1$ but $n$ is composite.

41. Let $m$ and $n$ be integers $\geq 2$ with $\gcd(mn, 10) = 1$ and $\gcd(m, n) = 1$. Let $a$ be the period length of the decimal expansion of $1/m$ and let $b$ be the period length for $1/n$. Show that the period length for $1/mn$ is $\operatorname{lcm}(a, b)$.

## Section 11.4: Card Shuffling

42. How many shuffles will it take to get back to the original position if a deck has
    (a) 10 cards?    (b) 58 cards?

43. How many shuffles will it take to get back to the original position if a deck has
    (a) 12 cards?    (b) 62 cards?

## Section 11.5: The Discrete Log Problem

44. Solve $2^x \equiv 3 \pmod{11}$.

45. Solve $3^x \equiv 5 \pmod{17}$.

46. (a) Suppose you know that $7^{57} \equiv 11 \pmod{101}$ and that $2^9 \equiv 7 \pmod{101}$. Find $x$ such that $2^x \equiv 11 \pmod{101}$.
    (b) Solve $7^y \equiv 2 \pmod{101}$.

47. Given that $3^{12} \equiv 4 \pmod{17}$, find $x$ such that $3^x \equiv 2 \pmod{17}$. Check your answer. If it doesn't work, look at Proposition 11.6.

48. (a) Solve $125x \equiv 1 \pmod{256}$.
    (b) Starting from the fact that $3^{125} \equiv 19 \pmod{257}$, find $y$ such that $19^y \equiv 3 \pmod{257}$. You may use the fact that 257 is prime. If you use part (a), explain how and why you are using it.

49. Let $g$ be a primitive root for the odd prime $p$. Suppose $g^x \equiv h \pmod{p}$. Show that $x$ is even if $h^{(p-1)/2} \equiv 1 \pmod{p}$ and $x$ is odd if $h^{(p-1)/2} \equiv -1 \pmod{p}$. (This shows that it is possible to determine the parity of a discrete log without evaluating the log.)

50. Solve $3^x \equiv 5 \pmod{17}$ using the Baby Step–Giant Step Method.

51. Solve $5^x \equiv 3 \pmod{23}$ using the Baby Step–Giant Step Method.

52. Try to solve $2^x \equiv 3 \pmod{17}$ using the Baby Step–Giant Step Method. Why doesn't the method work?

# 11.8.2    Projects

Let $n$ be a positive integer. A number $g$ is called a *primitive root* mod $n$ if each $b$ with $\gcd(b, n) = 1$ is congruent to a power of $g$ mod $n$. The first three problems determine which $n$ have primitive roots.

1. Let $p$ be an odd prime and let $g$ be a primitive root mod $p$.
   (a) Show that $p - 1$ divides the order of $g$ mod $p^2$. (*Hint:* If $g^j \equiv 1$ (mod $p^2$), then $g^j \equiv 1$ (mod $p$).)
   (b) Show that $\mathrm{ord}_{p^2}(g) = p - 1$ or $p(p - 1)$.
   (c) Show that $\mathrm{ord}_{p^2}(g) = p - 1$ if and only if $g^p \equiv g$ (mod $p^2$).
   (d) Show that if $g^p \equiv g$ (mod $p^2$), then $(g+p)^p \not\equiv g+p$ (mod $p^2$).(*Hint:* Use the binomial theorem.)
   (e) Show that either $\mathrm{ord}_{p^2}(g)$ or $\mathrm{ord}_{p^2}(g + p)$ equals $p(p - 1)$.
   (f) Show that there exists $b$ such that every $c$ with $\gcd(c, p) = 1$ is congruent to a power of $b$ mod $p^2$.

2. Let $p$ be an odd prime. The previous problem showed that there is a primitive root mod $p^2$. Now we'll show that a primitive root mod $p^2$ is automatically a primitive root mod $p^n$ for all $n \geq 1$. Let $g$ be a primitive root mod $p^2$. Suppose that $g$ is a primitive root mod $p^n$. We'll show that $g$ is a primitive root mod $p^{n+1}$.
   (a) Let $b$ and $c$ be such that $bc \not\equiv 0$ (mod $p$) and $b^p \equiv c^p$ (mod $p^{n+1}$). Show that $b \equiv c$ (mod $p^n$). (*Hint:* By Fermat's theorem, $b \equiv c$ (mod $p$). Write $b = c + ap^k$ with $p \nmid a$ and with $k \geq 1$. Expand $(c + ap^k)^p$ by the binomial theorem.)
   (b) Show that $\mathrm{ord}_{p^{n+1}}(g) = \phi(p^n)$ or $\phi(p^{n+1})$.
   (c) Suppose $g^{\phi(p^n)} \equiv 1$ (mod $p^{n+1}$). Show that $g^{\phi(p^{n-1})} \equiv 1$ (mod $p^n$). (*Hint:* Use part (a) with $c = 1$.)
   (d) Show that $\mathrm{ord}_{p^{n+1}}(g) = \phi(p^{n+1})$, which means that $g$ is a primitive root mod $p^{n+1}$. Explain why $g$ is a primitive root mod $p^n$ for all $n$.

3. (a) Let $m$ and $n$ be positive integers with $m \mid n$. Suppose $g$ is a primitive root for $n$. Show that $g$ is a primitive root for $m$. (*Hint:* Use Exercise 87 in Chapter 6.)
   (b) Show that there is no primitive root mod 8.
   (c) Let $p$ and $q$ be distinct odd primes and let $\gcd(b, pq) = 1$. Show that $b^{(p-1)(q-1)/2} \equiv 1$ (mod $pq$). (*Hint:* It suffices to work mod $p$ and mod $q$ separately. Note that $(q - 1)/2$ and $(p - 1)/2$ are integers, which is important since you should not use fractional exponents.)
   (d) Let $p$ and $q$ be distinct odd primes. Show that there is no primitive root mod $pq$.
   (e) Let $p$ be an odd prime. Let $\gcd(b, 4p) = 1$. Show that $b^{p-1} \equiv 1$ (mod $4p$).
   (f) Let $p$ be an odd prime. Show that there is no primitive root mod $4p$.
   (g) Let $p$ be an odd prime, let $n \geq 1$, and let $g$ be a primitive root mod $p^n$. Let $h \equiv g$ (mod $p^n$) and $h \equiv 1$ (mod 2). Show that $\mathrm{ord}_{2p^n}(h) = \phi(p^n) = \phi(2p^n)$. Conclude that $h$ is a primitive root mod $2p^n$.
   (h) Show that there is a primitive root mod $n$ exactly when $n$ is one of the following: one, two, four, a power of an odd prime, twice a power of an odd prime.

4. Generalize and prove Midy's theorem for an arbitrary base $b$.

## 11.8.3    Computer Explorations

1.  How many primes less than 1000 have 2 as a primitive root? How well does this compare with the estimate given at the end of Section 11.2? What if you count up to 10000? up to 100000?

2.  (a) Factor $2^{100} - 1$.
    (b) There are exactly three primes $p$ with $\text{ord}_p(2) = 100$. Find them and show that your list is complete.

3.  Find five different composite Mersenne numbers $M_p = 2^p - 1$ with $p$ prime and factor them. Show that Proposition 11.4 holds for these examples.

4.  The prime $p = 1093$ has 5 as a primitive root. Solve $5^x \equiv 489 \pmod{1093}$
    (a) by brute force (that is, trying all values of $x$),
    (b) by the Baby Step–Giant Step Method.

## 11.8.4    Answers to "CHECK YOUR UNDERSTANDING"

1.  Since $\text{ord}_{13}(5) \mid 12$, the possibilities are $1, 2, 3, 4, 6, 12$. Calculate $5^j \not\equiv 1$ (mod 13) for $j = 1, 2, 3$. However, $5^2 = 25 \equiv -1 \pmod{13}$, so squaring yields $5^4 \equiv 1 \pmod{13}$. Therefore, $\text{ord}_{13}(5) = 4$.

2.  No. The order $\text{ord}_{17}(2)$ is the smallest positive $k$ with $2^k \equiv 1 \pmod{17}$. The congruence $2^{16} \equiv 1 \pmod{17}$ says only that 16 is a candidate to be the smallest $k$. However, since $2^8 \equiv 1 \pmod{17}$, 16 is not the smallest. A quick calculation shows that 8 is the smallest $k$, so $\text{ord}_{17}(2) = 8$.

3.  The possible orders of 2 mod 19 are divisors of $19 - 1 = 18$, namely 1, 2, 3, 6, 9, 18. Since $2^j \not\equiv 1 \pmod{19}$ for $j = 1, 2, 3, 6, 9$, we must have $\text{ord}_{13}(2) = 18$, so 2 is a primitive root.

4.  By Corollary 11.9, the primitive roots for 19 are $2^i \pmod{19}$ for $i = 1, 5, 7, 11, 13, 17$ (these are the numbers less than 18 that are relatively prime to 18). Since $2^5 \equiv 13$, $2^7 \equiv 14$, $2^{11} \equiv 15$, $2^{13} \equiv 3$, and $2^{17} \equiv 10$ (mod 19), the primitive roots for 19 are 2, 3, 10, 13, 14, 15.

5.  (a) $1/37 = .027027027 \cdots$, so the period length is 3.
    (b) Since $10 \not\equiv 1 \pmod{37}$ and $10^2 \not\equiv 1 \pmod{37}$, but $10^3 \equiv 1 \pmod{37}$, we have $\text{ord}_{37}(10) = 3$, which agrees with the answer from part (a).

# Chapter 12

# More Cryptographic Applications

## 12.1 Diffie–Hellman Key Exchange

Imagine for a moment that you have the perfect cryptography system, completely impenetrable and totally unreadable by anyone who intercepts an encrypted message that you send. If you need to send classified information to agents thousands of miles away, they need the cryptographic key in order to read the message. This is a serious problem. You can't just call them up or email them or text them — if you trusted the phone or Internet you could just tell them the message and you would have no need of the cryptography. This is called the *Key Exchange Problem* and, in the past, couriers often used a diplomatic pouch system to solve it. Alternatively, there could be a meeting with all relevant people present. Keys would then be distributed and the agents would go out into the field. The problem is that these methods are expensive and time consuming, and they require a lot of planning. In 1976, Whitfield Diffie and Martin Hellman proposed a key exchange method that can be done publicly but nevertheless maintains the secrecy of the key. (Their idea is more accurately a key establishment system. Nevertheless, since it is universally referred to as a key exchange protocol, we will continue to use this terminology.) We now describe their idea, which is based on simple properties of exponents and congruences. As with RSA, which is another solution of the key exchange problem, all of the communications are over public channels, but we'll see that in the present case they are protected by the difficulty of computing discrete logarithms.

Here is the protocol. As usual, the two parties are Alice and Bob.

---

### Diffie–Hellman Key Exchange

1. Alice and Bob agree on a large prime $p$ and a primitive root $g$ mod $p$.

2. Alice chooses a secret integer $a$ and then calculates $h_1 \equiv g^a$ mod $p$.

3. Bob chooses a secret integer $b$ and calculates $h_2 \equiv g^b$ mod $p$.

4. Alice sends $h_1$ to Bob and Bob sends $h_2$ to Alice.

5. Alice computes $k \equiv h_2^a \pmod{p}$.

6. Bob computes $k \equiv h_1^b \pmod{p}$.

---

Alice has computed $k \equiv h_2^a \equiv g^{ba}$ and Bob has computed $k \equiv h_1^b \equiv g^{ab}$. Since $ba = ab$, both Alice and Bob have the same number for $k$, which is now their shared key. You can see why this is really a key establishment protocol. There is no key that Alice or Bob initially have and can then exchange. Rather, working together, they create a key with the implicit assumption that this will be the key that they will use.

**Remark:** Eavesdroppers can intercept $g$, $g^a \pmod{p}$, and $g^b \pmod{p}$. If they can solve the Discrete Log Problem, they can use $g$ and $g^a$ to find $a$, and then raise $g^b$ to the $a$ power to compute $g^{ba} \pmod{p}$. It is not known whether $g^{ba} \pmod{p}$ can be computed without computing a discrete logarithm. The **Computational Diffie–Hellman Problem** asks for $g^{ba} \pmod{p}$, given $p$, $g$, $g^a \pmod{p}$, and $g^b \pmod{p}$.

**Example.** Alice and Bob need to create a key that they can use to send each other financial documents. They agree on the prime $p = 31981$ and a primitive root $g = 6$. Alice chooses her secret number to be 817 and Bob chooses his to be 23971.

$$\text{Alice calculates } 6^{817} \equiv 16978 \pmod{31981}$$
$$\text{and sends 16978 to Bob.}$$

$$\text{Bob calculates } 6^{23971} \equiv 28349 \pmod{31981}$$
$$\text{and sends 28349 to Alice.}$$

Alice now has 28349, which Bob has sent her, and Bob has 16978, which Alice has sent him.

$$\text{Alice calculates} \quad 28349^{817} \quad \equiv \quad 7857 \pmod{31981}.$$
$$\text{Bob calculates} \quad 16978^{23971} \quad \equiv \quad 7857 \pmod{31981}.$$

So, 7857 is the key that they will both use.

---

**CHECK YOUR UNDERSTANDING:**

1. In the Diffie–Hellman Key Exchange, Alice and Bob agree on $p = 17$ and $g = 3$. Alice chooses $a = 5$ and Bob chooses $b = 4$. What number does Alice send to Bob, what number does Bob send to Alice, and what is the secret that they establish?

---

# 12.2   Coin Flipping over the Telephone

Bob and Alice decide to have dinner together. Bob wants to have Italian food, but Alice wants to go to a Thai restaurant. They decide to flip a coin to choose between the two. Unfortunately, they're talking about this over the phone and each one is concerned that the other will not be honest about the result of the coin flip. The idea behind the Diffie–Hellman Key Exchange can be used to solve this problem, enabling one person to do the coin flip and have the other be certain that it's being done fairly. Here's how the process works.

1.  Alice and Bob agree on a prime $p$ and share the factorization of $p - 1$. Of course, if they can't factor $p - 1$, they just choose a different prime.

2.  Alice chooses two distinct primitive roots mod $p$, call them $g_1$ and $g_2$, and sends them to Bob.

3.  Bob chooses a number $n$ that is relatively prime to $p - 1$ and keeps it secret.

4.  Bob now randomly picks either $g_1$ or $g_2$. (He can flip a real coin to help him choose.) Let's say he chooses $g_i$.

5. Bob sends $a \equiv g_i{}^n \pmod{p}$ to Alice.

6. Because Alice doesn't know Bob's choice of $n$, she can't know which primitive root he chose. For either choice of $g_i$, there is some $n$ such that $g_i^n \equiv a \pmod{p}$, and Bob is keeping $n$ secret. So, Alice makes a guess, randomly choosing either $g_1$ or $g_2$. She tells Bob her guess, and if she's correct she wins the coin toss. If she's wrong, Bob wins. Of course, if Alice doesn't trust Bob, how does she know he's not lying about her guess? (If that's the case, you may be wondering why she wants to have dinner with him.) To ensure that does not occur, Bob sends $n$ to Alice. Let's say that Alice's guess was $g_2$. She can calculate $g_1^n \pmod{p}$. If this is $a$ then Alice knows Bob was being honest.

**Remarks.** 1. If Bob claims that Alice loses, then it is a good idea for Alice to check that $\gcd(n, p-1) = 1$. See Exercise 7.

2. If Bob can find an $x$ with $g_1^x \equiv g_2 \pmod{p}$, then he can always successfully claim to win: He chooses $n$ and lets $g_2^n \equiv a \pmod{p}$. Then he knows that $g_1^{xn} \equiv a \pmod{n}$. After Alice makes her choice, he tells her that she guessed wrong. When she asks for the exponent, he sends $n$ if she guessed $g_1$ or $xn$ if she guessed $g_2$. Therefore, the prime $p$ must be chosen large enough that Bob cannot solve the discrete log problem.

3. In the spirit of the previous remark, if Bob knows $n_1$ with $g_1^{n_1} \equiv a \pmod{p}$ and $n_2$ with $g_2^{n_2} \equiv a \pmod{p}$, then he finds all solutions to $xn_1 \equiv n_2 \pmod{p-1}$. One of these values of $x$ (we know there is at least one value, so there are $\gcd(n_1, p-1)$ of them by Theorem 6.18) solves the discrete log problem $g_1^x \equiv g_2 \pmod{p}$. Therefore, if Bob cannot compute discrete logs, he should not be able to have two values of $n$, one for each of the two possible guesses by Alice.

4. Here is why Alice, not Bob, chooses the primitive roots. If Bob chooses them, he can let $g_1$ be any primitive root, choose $x$ with $\gcd(x, p-1) = 1$, then let $g_2 \equiv g_1^x \pmod{p}$. The choice of $x$ makes $g_2$ a primitive root, by Corollary 11.9. Bob can now use the method of the previous remark to cheat.

5. Suppose Bob decides that he wants to lose (there are many possible situations where this could be a prudent thing to do). Alice suspects this and asks for $n$. He could simply send a random exponent $r \neq n$, but then Alice could calculate $g_1^r$ and $g_2^r$ and see that neither is congruent

to $a \bmod p$. Bob would then be in trouble. But if Bob sends the true value of $n$, Alice can check which primitive root is correct. So, Bob is stuck. If he wins, he has to admit it.

**Example.** Alice and Bob agree on the prime $p = 675216457$. Alice chooses her two primitive roots,

$$g_1 = 30 \text{ and } g_2 = 39,$$

and sends them to Bob. Bob chooses 39 as his primitive root and $n = 35189$ as his number relatively prime to $p - 1 = 675216456$. He then sends Alice

$$a \equiv 39^{35819} \equiv 75914980 \pmod{675216457}.$$

Alice now has 75914980, but has no idea as to which primitive root it came from. She guesses 39 and Bob tells her she's correct, so she wins the coin flip. What would have happened if Alice had guessed 39 and Bob had told her she was wrong, that his primitive root had actually been 30? Then, as we said, Alice could ask for the $n$ Bob had used and then she could check his honesty. If Bob sends her 35819, Alice would then calculate
$$30^{35819} \pmod{675216457}$$
and get 319436617 which is not the number that Bob had sent. Alice would then know that Bob was being dishonest.

Or, Bob could try sending her an $n$ that was not the one he had actually used, say $n = 62137$. Then Alice would compute

$$30^{62137} \pmod{675216457}$$

and get 450455581 which is also not the number Bob has sent.

For a different method of flipping coins over the telephone, see Section 14.4.

# 12.3    Mental Poker

After flipping coins over the telephone for a few years, Alice and Bob decide to try something different. Bob suggests that they play poker and Alice agrees. After Bob deals the first hand and tells Alice that she got a full house, Alice figures she's won. But rotten luck for her when Bob says that he got a royal flush. At this point, Alice realizes that they need a different way to play.

Alice goes to a local office supply store and buys 52 safes. She puts one card from a new deck of cards into each safe, locks them, puts them in a big box and mails the box to Bob. When Bob gets the safes, he has no idea as to what cards are in which safe. He randomly chooses five safes, puts his locks on each of them, and mails those five safes back to Alice. When Alice receives the five safes, she takes her locks off so the only remaining locks are Bob's. So, when Alice mails these safes back to him, he can unlock each one and discover what cards he's been dealt. Bob, who still has 47 safes with Alice's locks on them now sends five of them to Alice who can open them to discover her cards.

This is the end of the first round of dealing: Bob and Alice each have five cards, Alice has no locked safes at all while Bob has 42 locked safes, each locked by Alice.

When Bob looks at his five cards, he decides to discard two. He puts the two cards in an empty safe, puts his lock on it and mails it to Alice. Bob now chooses two of the remaining 42 boxes he has (all with Alice's lock on them), puts his lock on them, and sends these two to Alice. Alice takes her lock off these two and sends them to Bob, who can remove his locks to get his cards.

If Alice wants to replace three cards, she puts them in a safe, locks it, and mails it to Bob, who then takes three of his remaining 40 safes (with Alice's lock on them) and sends them to Alice. She can now unlock them to see her three new cards. They can then compare their cards and see who wins.

After the first game is over, Alice and Bob have spent thousands of dollars on safes and postage and realize that there has to be a cheaper way to play. They contact their local university and speak to the resident cryptographer, who suggests the following method.

Alice and Bob agree on a large prime $p$. Alice chooses a secret number $\alpha$ that is relatively prime to $p-1$ and then calculates $\alpha'$ with the property that

$$\alpha\alpha' \equiv 1 \bmod (p-1).$$

Bob does the same thing, choosing a secret number $\beta$ relatively prime to $p - 1$ and then $\beta'$ with

$$\beta\beta' \equiv 1 \bmod (p - 1).$$

> **Alice's Secret:** $\alpha$ and $\alpha'$ with $\alpha\alpha' \equiv 1 \bmod (p - 1)$.
> **Bob's Secret:** $\beta$ and $\beta'$ with $\beta\beta' \equiv 1 \bmod (p - 1)$.

If we think of $\alpha$ as Alice's secret key, then $\alpha'$ is plays the same role as the decryption exponent $d$ did in the RSA cryptosystem. (Of course the same goes for Bob's $\beta'$.) To see this, we first note that $\alpha\alpha' \equiv 1$ (mod $p - 1$) means that $\alpha\alpha' = 1 + k(p - 1)$ for some integer $k$. So, if $M$ is any integer that is not a multiple of $p$,

$$M^{\alpha\alpha'} = M^{1+k(p-1)} = M^1(M^{p-1})^k \equiv M \cdot 1^k \equiv M \pmod{p}. \quad (12.1)$$

We've just used Fermat's theorem (again!), which says that if $\gcd(a, p) = 1$, then $a^{p-1} \equiv 1 \pmod{p}$. In our calculation, we assumed that $M \not\equiv 0 \pmod{p}$. If $M \equiv 0 \pmod{p}$, then we still have $M^{\alpha\alpha'} \equiv M \pmod{p}$ since both sides are $0 \pmod{p}$.

Alice and Bob now assign a distinct number to each of the 52 cards in a deck and share this assignation. So, they may decide that the ace of hearts is $n_1$, the ace of spades is $n_2$, ..., and the king of diamonds is $n_{52}$. Alice raises each of the 52 numbers to the $\alpha$th power mod $p$ so that she now has 52 numbers $c_1, c_2, ..., c_{52}$, where

$$c_i \equiv n_i^{\alpha} \pmod{p}.$$

This is the mathematical version of putting her locks on the 52 safes. Then, after randomly permuting these numbers, she sends them to Bob.

Bob now has 52 numbers and does not know what cards they correspond to. This is analogous to him holding, face down, a deck of 52 cards. He chooses five of the numbers (cards), $c_{i_1}, c_{i_2}, c_{i_3}, c_{i_4}, c_{i_5}$, raises each to the $\beta$th power mod $p$ (the mathematical version of putting his locks on the safes) and sends these five numbers back to Alice:

$$\text{Alice has } c_{i_1}^{\alpha\beta}, \quad c_{i_2}^{\alpha\beta}, \quad c_{i_3}^{\alpha\beta}, \quad c_{i_4}^{\alpha\beta}, \quad c_{i_5}^{\alpha\beta}. \quad (12.2)$$

At this point it looks like things are a total mess. But, in fact, we're almost done because Alice can now raise each of her five numbers to the $\alpha'$ power. This will, in effect, strip away the $\alpha$ in the exponent, since, with $c_{i_1}$ as an example, (12.1) tells us that

$$\left(c_{i_1}^{\alpha\beta}\right)^{\alpha'} \equiv c_{i_1}^{\alpha\beta\alpha'} \equiv c_{i_1}^{\alpha\alpha'\beta} \equiv \left(c_{i_1}^{\alpha\alpha'}\right)^{\beta} \equiv c_{i_1}^{\beta} \pmod{p}$$

(so Alice has removed her locks from the mathematical safes). After Alice does this, she has

$$c_{i_1}^{\beta}, \quad c_{i_2}^{\beta}, \quad c_{i_3}^{\beta}, \quad c_{i_4}^{\beta}, \quad c_{i_5}^{\beta}.$$

She now sends these five numbers to Bob, who can decrypt by raising each one to the $\beta'$ power, and by (12.1) with $\beta$ in place of $\alpha$,

$$c_{i_1}^{\beta\beta'} \equiv c_{i_1} \pmod{p}$$

(so Bob has removed his locks). Finally, Bob has been dealt his five cards. Using a similar process, Alice gets her five cards and they can begin.

**Example.** Let's see how this works if Alice and Bob play a simpler game. There are only four cards (Jack, Queen, King, and Ace), each player is dealt only one, and the winner is the person with the higher card. We'll first change the cards to numbers using $a = 01, b = 02, c = 03, ..., y = 25$, and $z = 26$. This gives us:

| jack | queen | king | ace |
|------|-------|------|-----|
| 10010311 | 1721050514 | 11091407 | 10305. |

Alice and Bob agree on the prime $p = 12047353607$. Alice chooses her secret $\alpha = 11041$ and Bob chooses his secret $\beta = 70399$. (They both check that their respective secret numbers are relatively prime to $12047353607 - 1 = 12047353606$.) Using the Extended Euclidean Algorithm, Alice solves

$$11041\alpha' \equiv 1 \pmod{12047353606}$$

and Bob solves

$$70399\beta' \equiv 1 \pmod{12047353606}.$$

They get

$$\alpha' = 6467228043 \quad \text{and} \quad \beta' = 11651873071.$$

Alice now encrypts each of the four cards using their numerical values and her secret $\alpha$. In this encryption, all congruences are modulo $p = 12047353607$:

$$
\begin{aligned}
10010311^{\alpha} &\equiv 5674830805 \\
1721050514^{\alpha} &\equiv 4004945206 \\
11091407^{\alpha} &\equiv 11022499208 \\
10305^{\alpha} &\equiv 6975049274.
\end{aligned}
$$

These cards get "shuffled" and sent to Bob, so he receives

$$11022499208, \quad 4004945206, \quad 6975049274, \quad \text{and} \quad 5674830805.$$

Bob now chooses a card by choosing one of these four numbers (say the first), raises it to the $\beta$ power, and then sends the number to Alice.

Alice receives:
$$11022499208^{70399} \equiv 10159065536 \pmod{12047353607}.$$

In terms of our locked safe analogy, Alice now has a safe with two locks on it. She takes her lock off by raising 10159065536 to the $\alpha'$ power and sends the result to Bob:

Bob receives:
$$10159065536^{6467228043} \equiv 3733881457 \pmod{12047353607}.$$

Bob removes his lock by raising 3733881457 to the $\beta'$ power:

Bob's card:
$$3733881457^{11651873071} \equiv 11091407 \pmod{12047353607}.$$

If we check, we see that Bob got the king.

Bob now chooses Alice's card by taking one of the three remaining numbers, say 6975049274, and sending it to Alice. Alice calculates

$$
\begin{aligned}
6975049274^{\alpha'} &\equiv 6975049274^{6467228043} \\
&\equiv 10305 \pmod{12047353607},
\end{aligned}
$$

so her card is the ace and she wins this game.

If anyone suspects that someone lied about his or her cards, the outcomes can be checked by having Alice and Bob reveal $\alpha$ and $\beta$. The encrypted values of the cards can then be calculated and compared with the cards used during the game. However, there is another way to cheat; see Computer Exploration 2.

# 12.4   The ElGamal Public Key Cryptosystem

As we saw in Section 9.1, the security of RSA rests on the computational infeasibility of being able to factor numbers that are the product of two primes. In 1984, Tahar ElGamal designed a public key cryptosystem whose security is closely related to the difficulty of solving the discrete log problem.

Let's say that once again Alice has a business and would like her customers to be able to communicate securely with her. Here's how the process works, with her customer again being Bob.

---

**ElGamal Setup**

1. Alice chooses a prime $p$ and a primitive root $g$ (mod $p$).

2. Alice chooses an $x$ with $2 \leq x \leq p - 2$ and calculates $h \equiv g^x$ (mod $p$).

---

**Alice's Public Information:** $p, g,$ and $h$

**Alice's Private Information:** $x$

---

### ElGamal Encryption

1. Bob chooses a random number $y$ with $2 \leq y \leq p-2$ and calculates $r \equiv g^y \pmod{p}$.

2. Bob takes his message $m$ and calculates $c \equiv h^y m \pmod{p}$ and sends $r$ and $c$ to Alice.

---

### ElGamal Decryption

1. Alice computes $m \equiv c \cdot r^{-x} \pmod{p}$.

---

The decryption works because

$$cr^{-x} \equiv (h^y m)(g^y)^{-x} \equiv (g^{yx} m) g^{-xy} \equiv mg^0 \equiv m \pmod{p}.$$

Anyone who knows $x$ can also decrypt the message, but finding $x$ requires solving $g^x \equiv h \pmod{p}$, which is a discrete log problem.

The number $h^y \pmod{p}$ is essentially a random number mod $p$. When Bob forms $c$ by multiplying $m$ by $h^y$, he's multiplying $m$ by a random number, which has the effect of hiding $m$. By sending Alice the extra number $r$, he's giving her $y$ in a hidden form, but in a form that she can use to find $m$.

Bob has to be careful to choose a different $y$ every time he wants to communicate securely with Alice. Let's say Bob sends the message $m_1$ to Alice and, a few days later, sends her message $m_2$ using the same $y$ both times. So Alice receives

$$(r, c_1) \text{ and } (r, c_2),$$

where $r$ is the same in both messages because $r \equiv g^y \pmod{p}$ and there was no change in $y$ between messages. If an evil eavesdropper Eve is able to determine $m_1$, she can then figure out $m_2$ (and any other message Bob sends using the same $y$) by noticing the following:

$$\frac{c_1}{m_1} \equiv h^y \pmod{p} \text{ and } \frac{c_2}{m_2} \equiv h^y \pmod{p}.$$

So,

$$\frac{c_1}{m_1} \equiv \frac{c_2}{m_2} \pmod{p}.$$

Now,

$$m_2 \equiv \frac{c_2 m_1}{c_1} \pmod{p}$$

and Eve, knowing $m_1$, $c_1$, and $c_2$, can calculate $m_2$.

**Example.** Let's say that Bob wants to send Alice the message $m = 371$. Alice has already chosen her prime $p = 6581$ and a primitive root for $p$; let's say she chose $g = 14$. She uses her random number generator to pick a secret exponent $x = 2141$, using the requirement that $2 \leq x \leq 6579 = p - 2$. She then calculates $h \equiv 14^{2141} \equiv 1543 \pmod{6581}$ and publishes

$$p = 6581, \ g = 14, \ h = 1543$$

in the online ElGamal Public Key directory. Note that she keeps $x = 2141$ private.

Bob retrieves the information that Alice has published, chooses his random number $y = 6391$, and uses this to compute

$$r \equiv 14^{6391} \equiv 1968 \pmod{6581}$$

and

$$c \equiv 371 \cdot 1543^{6391} \equiv 371 \cdot 6172 \equiv 6205 \pmod{6581}.$$

Bob now sends Alice his ciphertext $(r, c)$.

Recall that Alice needs to calculate $cr^{-x} \pmod{p}$. She does so and gets

$$6205 \cdot 1968^{-2141} \equiv 371 \pmod{6581},$$

which is Bob's message.

---

**CHECK YOUR UNDERSTANDING:**
2. Suppose Alice is using the ElGamal Public Key Cryptosystem. She chooses prime $p = 19$, the primitive root $g = 2$, and the secret exponent $x = 5$, so $h \equiv 2^5 \equiv 13 \pmod{19}$.
(a) Bob wants to send her the message $m = 7$. He chooses $y = 2$. What does Bob send to Alice?
(b) Show that when Alice decrypts what she receives from Bob, she obtains the plaintext.

# 12.5    Chapter Highlights

1. Diffie–Hellman Key Exchange
2. The ElGamal Public Key Cryptosystem

# 12.6    Problems

## 12.6.1    Exercises

### Section 12.1: Diffie–Hellman Key Exchange

The next four exercises refer to the Diffie–Hellman Key Exchange. Alice and Bob agree on the given prime $p$ and the given primitive root $g$. Alice chooses her secret $a$ and Bob his secret $b$. In each exercise, find their agreed-upon key.

1. $p = 7$, $g = 3$, $a = 4$, $b = 5$
2. $p = 11$, $g = 2$, $a = 8$, $b = 4$
3. $p = 19$, $g = 2$, $a = 9$, $b = 11$
4. $p = 23$, $g = 5$, $a = 7$, $b = 12$

5. Why would it be foolish for Alice or Bob to chooses $p - 1$ or 1 as the secret exponent in the Diffie–Hellman Key Exchange algorithm? (That is, explain how an eavesdropper could find the key.)
6. What happens in the Diffie–Hellman Key Exchange if $g$ is chosen so that $\mathrm{ord}_p(g)$ is small, rather than having $g$ be a primitive rood mod $p$? Explain why an eavesdropper can compute $g^{ab}$ easily in this case.

### Section 12.2: Coin Flipping over the Telephone

7. In the protocol for flipping coins, Bob is required to choose $n$ so that $\gcd(n, p - 1) = 1$. This exercise shows why.
   (a) To start with a simple case, suppose Bob chooses $n$ to be a multiple of $p - 1$, so $\gcd(n, p - 1) = p - 1$. Show what $a$ must be and why Bob can follow the protocol to "prove" he was using the primitive root that Alice didn't guess. For definiteness, assume that Bob chooses $g_1^n \equiv a \pmod{p}$ and Alice guesses that Bob used $g_1$.
   (b) More generally, let $\gcd(n, p - 1) = d$. Show that

   $$a^{(p-1)/d} \equiv 1 \pmod{p}. \tag{12.3}$$

   (c) Show that the solutions to (12.3) are given by $g_2^{di} \pmod{p}$, for $0 \leq$

$i < (p-1)/d$. Therefore, $a$ is one of these numbers.

(d) If $(p-1)/d$ is small, how can Bob produce $n'$ with $g_2^{n'} \equiv a \pmod{p}$ and thereby claim that Alice loses?

## Section 12.3: Mental Poker

8. The mental poker protocol uses a technique similar to the following method for encrypting messages. Alice and Bob agree on a large prime $p$. Alice chooses secret $\alpha$ and $\alpha'$ such that $\alpha\alpha' \equiv 1 \pmod{p-1}$ and Bob chooses secret $\beta$ and $\beta'$ such that $\beta\beta' \equiv 1 \pmod{p-1}$. Bob wants to send $m$ to Alice. He computes $c_1 \equiv m^\beta \pmod{p}$ and sends $c_1$ to Alice. Alice computes $c_2 \equiv c_1^\alpha \pmod{p}$ and sends $c_2$ to Bob. Bob now computes a number $c_3$ and sends $c_3$ to Alice, who uses $c_3$ in a computation to recover $m$.

   (a) What computation does Bob perform to produce $c_3$?

   (b) What computation does Alice do with $c_3$ to recover $m$?

   (c) Show that this method produces $m$. Explain how the congruences $\alpha\alpha' \equiv 1$ and $\beta\beta' \equiv 1 \pmod{p-1}$ are used.

## Section 12.4: The ElGamal Public Key Cryptosystem

In the next four exercises, Bob uses the ElGamal Public Key Cryptosystem to send a ciphertext to Alice using Alice's public information. He uses her prime $p$, her primitive root $g$, and her published key $h \equiv g^x \pmod{p}$, where $x$ is Alice's private encryption key. What two numbers $r$ and $c$ (recall that $r \equiv g^y \pmod{p}$ and $c \equiv h^y m \pmod{p}$) does Alice receive?

9. $p = 7$, $g = 3$, $h = 6$. Bob's exponent $y$ is 4 and his message $m$ is 6.

10. $p = 13$, $g = 2$, $h = 5$. Bob's exponent $y$ is 8 and his message $m$ is 10.

11. $p = 46454609$, $g = 3$, $h = 7902328$. Bob's exponent $y$ is 1142987 and his message $m$ is 7601846.

12. $p = 612985319$, $g = 7$, $h = 511830974$. Bob's exponent $y$ is 96407412 and his message $m$ is 5863091.

In the next four problems, Alice receives the pair $(r, c)$ from Bob, where $c$ is the encryption of his message $m$. Assume that the ElGamal system is their agreed-upon encryption method. Using the prime $p$ and Alice's secret exponent $x$, find the plaintext.

13. $p = 23$, $x = 7$, $r = 21$, $c = 2$.

14. $p = 59$, $x = 13$, $r = 9$, $c = 34$.

15. $p = 6581$, $x = 6475$, $r = 1976$, $c = 4214$.

16. $p = 45893$, $x = 6491$, $r = 26642$, $c = 25581$.

17. Alice sets up her ElGamal cryptosystem using $p = 13$, $g = 2$, $h = 6$. Bob sends Alice the ciphertexts $(r, c) = (11, 8)$ and $(11, 9)$. Eve finds out that the plaintext for the first ciphertext is $m = 3$. Show how Eve can find the second plaintext without calculating any discrete logs.

18. Suppose Bob sends Alice a message using Alice's ElGamal cryptosystem. Eve finds out the value $y$ that Bob used. Explain how Eve can use this information to find $m$.

## 12.6.2  Projects

1. (*The ElGamal Signature Scheme*) This method of signing uses ideas from the ElGamal cryptosystem, just as the RSA Signature Scheme uses ideas from the RSA cryptosystem.

    Alice wants to sign her message $m$. She chooses a large prime $p$, a primitive root $g$, and a number $x$ with $1 \le x \le p - 2$. She calculates $h \equiv g^x$ (mod $p$). The values of $p, g, h$ are made public, and Alice keeps $x$ secret. To sign $m$, Alice does the following:

    (i) Chooses a secret random $k$ with $\gcd(k, p - 1) = 1$.

    (ii) Computes $r \equiv g^k$ (mod $p$) and $s \equiv k^{-1}(m - xr)$ (mod $p-1$), where $k^{-1}$ denotes the multiplicative inverse of $k \bmod p - 1$.

    The signed message is $(m, r, s)$.

    Bob verifies the signature by computing $v_1 \equiv h^r r^s$ (mod $p$) and $v_2 \equiv g^m$ (mod $p$). He decides that the signature is valid if $v_1 \equiv v_2$ (mod $p$).

    (a) Show that if Alice signs the message correctly then the signature is valid (that is, $v_1 \equiv v_2$).

    (b) Show that someone who finds out $x$ can produce a valid signature on any desired document.

    (c) Show that someone who finds out the value of $k$ can figure out the value of $x$.

    (*Hint:* There should be $\gcd(r, p - 1)$ choices for $x$. Eve computes $g^x$ (mod $p$) for each of them and stops when she obtains $h$ (mod $p$).)

    (d) The random number $k$ should be different for each signature. Suppose Alice uses the same $k$ for two different messages. How does an eavesdropper recognize this and how can the knowledge of $p, g, h, m_1, m_2, r, s_1, s_2$ be used to find $k$?

## 12.6.3  Computer Explorations

1. The prime 12347 has 2 as a primitive root. Suppose I tell you that $2^x \equiv 8938$ (mod 12347) and $2^y \equiv 9620$ (mod 12347), but I don't tell you $x$ and $y$. Is $2^{xy} \equiv 7538$ (mod 12347)? Is $2^{xy} \equiv 7557$ (mod 12347)?

    The **Decisional Diffie–Hellman Problem** asks the following question: Given $g, g^a, g^b$, and $h$, is $g^{ab} \equiv h$ (mod $p$)? It is not known whether this can be solved without first solving the Computational Diffie–Hellman Problem. (Recall that this asks you, given $g, g^a$, and $g^b$, to compute $g^{ab}$.)

2. Here's how to cheat at poker when playing with the protocol given in Section 12.3. We'll use the simplified version with only four cards (see the end of Section 12.3) in order to keep the numbers small.

    (a) Let $p = 12047353607$. Compute $x^{(p-1)/2}$ (mod $p$) for each card $x$ (so

$x = 10010311$ for the jack, etc.). Notice that the jack and king yield $-1$ (mod $p$) while the queen and ace yield $+1$ (mod $p$).

(b) Compute $y^{(p-1)/2}$ (mod $p$) for each encrypted card (so $y = 5674830805$ for the jack, etc.). What do you observe? Can you explain it? (*Hint:* $\alpha$ is odd.)

(c) Find a new prime $p$ such that the ace can be separated from the other cards by this procedure.

## 12.6.4    Answers to "CHECK YOUR UNDERSTANDING"

1.    Alice computes $3^5 \equiv 5$ (mod 17) and Bob computes $3^4 \equiv 13$ (mod 17), so Alice sends 5 to Bob and Bob sends 13 to Alice. Then Alice computes $13^5 \equiv 13$ (mod 17) and Bob computes $5^4 \equiv 13$ (mod 17). Their common secret is 13.

2.    (a) Bob computes $r \equiv g^d \equiv 2^2 \equiv 4$ (mod 19) and $c \equiv h^d m \equiv 13^2 \cdot 7 \equiv 5$ (mod 19). He sends $r = 4$ and $c = 5$ to Alice.

(b) Alice computes $m \equiv 5 \cdot 4^{-5} \equiv 7$ (mod 19).

# Chapter 13

# Quadratic Reciprocity

## 13.1   Squares and Square Roots Mod Primes

We say that a number $a$ is a square mod $n$ if $x^2 \equiv a \pmod{n}$ has a solution. For example, 2 is a square mod 7 since $3^2 \equiv 2 \pmod 7$, and $-1$ is a square mod 5 since $2^2 \equiv -1 \pmod 5$. However, 2 is not a square mod 3 because $x^2 \not\equiv 2 \pmod 3$ for $x = 0, 1, 2$.

If $a$ is a square mod $n$, we say that $a$ is a **quadratic residue** mod $n$. If not, $a$ is a **quadratic nonresidue**.

**Proposition 13.1.** *Let $p$ be an odd prime and let $a \not\equiv 0 \pmod p$. Then $a^{(p-1)/2} \equiv \pm 1 \pmod p$. Moreover,*

$$a \text{ is a square mod } p \iff a^{(p-1)/2} \equiv 1 \pmod p.$$

*Proof.* Let $b \equiv a^{(p-1)/2} \pmod p$. Then $b^2 \equiv a^{p-1} \equiv 1 \pmod p$ by Fermat's theorem. By Corollary 6.11, $b \equiv \pm 1 \pmod p$, so $a^{(p-1)/2} \equiv \pm 1 \pmod p$.

If $a$ is a square mod $p$, we have $x^2 \equiv a$ for some $x$, so

$$a^{(p-1)/2} \equiv (x^2)^{(p-1)/2} \equiv x^{p-1} \equiv 1 \pmod p,$$

by Fermat's theorem. Conversely, suppose $a^{(p-1)/2} \equiv 1 \pmod p$. Let $g$ be a primitive root mod $p$. Then $g^i \equiv a$ for some $i$, so

$$1 \equiv a^{(p-1)/2} \equiv g^{i(p-1)/2} \pmod p.$$

Since a primitive root has order $p - 1$, Theorem 11.1 says that $p - 1 \mid i(p-1)/2$. This implies that $(p-1)k = i(p-1)/2$ for some $k$, so $i = 2k$. Therefore, $a \equiv g^i \equiv (g^k)^2$, so $a$ is a square mod $p$. $\qquad\square$

**Definition 13.2.** *Let p be an odd prime and let a be an integer with*
$a \not\equiv 0 \pmod{p}$. *Define the* **Legendre symbol**

$$\left(\frac{a}{p}\right) = \begin{cases} +1 & \text{if } x^2 \equiv a \pmod{p} \text{ has a solution} \\ -1 & \text{if } x^2 \equiv a \pmod{p} \text{ has no solution.} \end{cases}$$

**Examples.** The examples at the beginning of this section can be restated as

$$\left(\frac{2}{7}\right) = +1, \quad \left(\frac{-1}{5}\right) = +1, \quad \left(\frac{2}{3}\right) = -1.$$

**Proposition 13.3.** *Let p be an odd prime and let* $a, b \not\equiv 0 \pmod{p}$.
*Then*
*(a) (Euler's Criterion)*

$$\left(\frac{a}{p}\right) \equiv a^{(p-1)/2} \pmod{p}.$$

*(b)*

$$\left(\frac{a}{p}\right)\left(\frac{b}{p}\right) = \left(\frac{ab}{p}\right).$$

*(c) If* $a \equiv b \pmod{p}$ *then* $\left(\frac{a}{p}\right) = \left(\frac{b}{p}\right)$.

*(d)*

$$\left(\frac{-1}{p}\right) = \begin{cases} +1 & \text{if } p \equiv 1 \pmod{4} \\ -1 & \text{if } p \equiv 3 \pmod{4}. \end{cases}$$

*Proof.* If $a$ is a square mod $p$ then Proposition 13.1 says that $a^{(p-1)/2} \equiv +1 = \left(\frac{a}{p}\right) \pmod{p}$. If $a$ is not a square mod $p$, then Proposition 13.1 says that $a^{(p-1)/2} \equiv -1 = \left(\frac{a}{p}\right) \pmod{p}$. This proves (a).

The congruence of (a) also holds for $b$ and for $ab$. Therefore,

$$\left(\frac{a}{p}\right)\left(\frac{b}{p}\right) \equiv a^{(p-1)/2}b^{(p-1)/2} = (ab)^{(p-1)/2} \equiv \left(\frac{ab}{p}\right) \pmod{p}.$$

Both ends of the above string of congruences are $\pm 1$. Since $-1 \not\equiv +1$ (mod $p$) when $p \geq 3$, the two ends must be equal. This proves (b).

If $a \equiv b$ (mod $p$), then $x^2 \equiv a$ (mod $p$) has a solution if and only if $x^2 \equiv b$ (mod $p$) has a solution. This is exactly what (c) says.

For (d), note that $(p-1)/2$ is even if $p \equiv 1$ (mod 4) and is odd if $p \equiv 3$ (mod 4). Therefore,

$$\left(\frac{-1}{p}\right) \equiv (-1)^{(p-1)/2} = \begin{cases} +1 & \text{if } p \equiv 1 \pmod{4} \\ -1 & \text{if } p \equiv 3 \pmod{4}. \end{cases}$$

$\square$

A useful fact is that $\left(\frac{x^2}{p}\right) = 1$ if $p \nmid x$. This is because $x^2$ is automatically a square mod $p$. It also follows immediately from part (b) of the proposition since $\left(\frac{x^2}{p}\right) = \left(\frac{x}{p}\right)^2 = (\pm 1)^2 = 1$.

The following is the major result about the Legendre symbol.

**Theorem 13.4.** *(Quadratic Reciprocity) (a) Let $p$ and $q$ be distinct odd primes. Then*

$$\left(\frac{q}{p}\right) = (-1)^{(p-1)(q-1)/4} \left(\frac{p}{q}\right).$$

*In other words,*

$$\left(\frac{q}{p}\right) = \begin{cases} \left(\dfrac{p}{q}\right) & \text{if at least one of } p, q \text{ is } 1 \pmod{4} \\[4mm] -\left(\dfrac{p}{q}\right) & \text{if } p \equiv q \equiv 3 \pmod{4}. \end{cases}$$

*(b) Let $p$ be an odd prime. Then*

$$\left(\frac{-1}{p}\right) = (-1)^{(p-1)/2} = \begin{cases} +1 & \text{if } p \equiv 1 \pmod{4} \\ -1 & \text{if } p \equiv 3 \pmod{4}. \end{cases}$$

*(c) Let $p$ be an odd prime. Then*

$$\left(\frac{2}{p}\right) = (-1)^{(p^2-1)/8} = \begin{cases} +1 & \text{if } p \equiv 1, 7 \pmod{8} \\ -1 & \text{if } p \equiv 3, 5 \pmod{8}. \end{cases}$$

**Remarks.** Parts (b) and (c) are called the *Supplementary Laws.* We will prove Quadratic Reciprocity and also the Supplementary Law for 2 in Section 13.5. We proved the Supplementary Law for $-1$ in Proposition 13.3. The Law of Quadratic Reciprocity is one of the high points of number theory. It was conjectured by Euler (1744) and Legendre (1785). In 1796, when he was 18 years old, Gauss gave the first proof. Gauss published six proofs during his lifetime, and two more were found in his unpublished manuscripts. Attempts to generalize Quadratic Reciprocity inspired the development of algebraic number theory in the 1800s and 1900s, and greatly influenced modern areas of research such as elliptic curves and modular forms from the 1900s up to the present.

Quadratic Reciprocity is the deepest theorem that we will meet in this book. Let's consider congruences and linear Diophantine equations as being at the easy level. The results are not surprising, although they are quite interesting and useful. Fermat's theorem and Euler's theorem are at the next level and are perhaps the first non-trivial results in number theory. In modern times, they are regarded as special cases of theorems in group theory. Finally, Quadratic Reciprocity is a very non-obvious theorem. There does not seem to be any heuristic reason that $q$ being a square mod $p$ should be related to $p$ being a square mod $q$.

**Example.** Is 23 a square mod 419? Let's check:

$$\left(\frac{23}{419}\right) = -\left(\frac{419}{23}\right) \quad \text{(since } 23 \equiv 419 \equiv 3 \pmod 4\text{)}$$

$$= -\left(\frac{5}{23}\right) \quad \text{(since } 419 \equiv 5 \pmod{23}\text{)}$$

$$= -\left(\frac{23}{5}\right) \quad \text{(since } 5 \equiv 1 \pmod 4\text{)}$$

$$= -\left(\frac{3}{5}\right) \quad \text{(since } 23 \equiv 3 \pmod 5\text{)}$$

$$= -\left(\frac{5}{3}\right) \quad \text{(since } 5 \equiv 1 \pmod 4\text{)}$$

$$= -\left(\frac{2}{3}\right) \quad \text{(since } 5 \equiv 2 \pmod 3\text{)}$$

$$= -(-1) = +1 \text{ (by the Supplementary Law for 2).}$$

Therefore, 23 has a square root mod 419. In the next section, we'll see how to discover that $182^2 \equiv 23 \pmod{419}$, but for the moment, let's note that Quadratic Reciprocity is non-constructive. We find out that a square root exists, without any information about how to find it.

**Example.** Let's see whether 246 is a square mod the prime 257. By Proposition 13.3,

$$\left(\frac{246}{257}\right) = \left(\frac{2}{257}\right)\left(\frac{3}{257}\right)\left(\frac{41}{257}\right).$$

The Supplementary Law tells us that

$$\left(\frac{2}{257}\right) = +1.$$

By Quadratic Reciprocity,

$$\left(\frac{3}{257}\right) = \left(\frac{257}{3}\right) = \left(\frac{2}{3}\right) = -1$$

and

$$\left(\frac{41}{257}\right) = \left(\frac{257}{41}\right) = \left(\frac{11}{41}\right) = \left(\frac{41}{11}\right)$$
$$= \left(\frac{8}{11}\right) = \left(\frac{2}{11}\right)^3 = (-1)^3 = -1.$$

Therefore,

$$\left(\frac{246}{257}\right) = (+1)(-1)(-1) = +1,$$

which means that 246 is a square mod 257.

Since $246 \equiv -11 \pmod{257}$, we could have done the calculation another way:

$$\left(\frac{246}{257}\right) = \left(\frac{-11}{257}\right) = \left(\frac{-1}{257}\right)\left(\frac{11}{257}\right).$$

Since $257 \equiv 1 \pmod 4$, the Supplementary Law for $-1$ says that

$$\left(\frac{-1}{257}\right) = +1.$$

By Quadratic Reciprocity,

$$\left(\frac{11}{257}\right) = \left(\frac{257}{11}\right) = \left(\frac{4}{11}\right) = +1,$$

since 4 is a square mod anything, in particular mod 11. Therefore,

$$\left(\frac{246}{257}\right) = (+1)(+1) = +1.$$

**Example.** Is 295 a square mod 401?

$$\left(\frac{295}{401}\right) = \left(\frac{5}{401}\right)\left(\frac{59}{401}\right).$$

By Quadratic Reciprocity,

$$\left(\frac{5}{401}\right) = \left(\frac{401}{5}\right) = \left(\frac{1}{5}\right) = +1$$

and

$$\left(\frac{59}{401}\right) = \left(\frac{401}{59}\right) = \left(\frac{47}{59}\right) = -\left(\frac{59}{47}\right) = -\left(\frac{12}{47}\right)$$

$$= -\left(\frac{4}{47}\right)\left(\frac{3}{47}\right) = -\left(\frac{3}{47}\right) = +\left(\frac{47}{3}\right) = \left(\frac{2}{3}\right) = -1.$$

Therefore,

$$\left(\frac{295}{401}\right) = (+1)(-1) = -1,$$

so 295 is not a square mod 401.

In the preceding examples, we factored numbers in order to apply quadratic reciprocity. In Section 13.4, we'll show that factorization is not necessary. In essence, we can pretend that every odd integer is prime and use this to evaluate Legendre symbols. For details, see Section 13.4.

Quadratic Reciprocity is a useful computational tool, as we have seen, but it has a deeper significance. Let's consider an example. For which primes $p$ is 5 a square mod $p$? We can look at 5 mod $p$ for each $p$ and get a list of such primes. Will there be any pattern for such $p$? Is there an easy way to list them? Quadratic Reciprocity gives the answer:

$$\left(\frac{5}{p}\right) = \left(\frac{p}{5}\right) = \begin{cases} +1 \text{ if } p \equiv \pm 1 \pmod 5 \\ -1 \text{ if } p \equiv \pm 2 \pmod 5. \end{cases}$$

The primes for which 5 is a quadratic residue form two congruence classes mod 5, namely, 1 mod 5 and 4 mod 5. The miracle of Quadratic

Reciprocity is that it takes a problem that depends on 5 mod $p$ and changes it to a problem that depends on congruence conditions for $p$ mod 5.

Let's look at another example. For which primes $p$ is 3 a square mod $p$? It might be surprising, but the answer depends on $p$ mod 12, not just $p$ mod 3:

If $p \equiv 1 \pmod{12}$, we have

$$\left(\frac{3}{p}\right) = \left(\frac{p}{3}\right) = \left(\frac{1}{3}\right) = 1.$$

The first equality is because $p \equiv 1 \pmod{12}$ implies that $p \equiv 1 \pmod 4$. The second is because also $p \equiv 1 \pmod 3$.

If $p \equiv 5 \pmod{12}$, we have

$$\left(\frac{3}{p}\right) = \left(\frac{p}{3}\right) = \left(\frac{2}{3}\right) = -1.$$

If $p \equiv 7 \pmod{12}$, we have

$$\left(\frac{3}{p}\right) = -\left(\frac{p}{3}\right) = -\left(\frac{1}{3}\right) = -1.$$

In this case, there is a "−" in the first equality because both 3 and $p$ are 3 mod 4.

If $p \equiv 11 \pmod{12}$, we have

$$\left(\frac{3}{p}\right) = -\left(\frac{p}{3}\right) = -\left(\frac{2}{3}\right) = 1.$$

As we see, it is necessary to consider the congruence class of $p$ both mod 3 and mod 4, which means that we are looking at $p$ mod 12 (by the Chinese Remainder Theorem). This was not necessary in the previous case where we asked for the primes $p$ for which 5 is a square mod $p$. This is because 5 is 1 mod 4 while 3 is 3 mod 4. Therefore, when we apply Quadratic Reciprocity, the "−" never occurs for 5 (because one of the primes, namely 5, is 1 mod 4) but it can occur for 3, depending on the congruence class of $p$ mod 4.

More generally, if $a$ is a nonzero integer and we ask for which primes $p$ is $a$ a square mod $p$, the answer depends only on the congruence class of $p$ mod $4a$. This is one of the key steps (Lemma 13.11) in the proof of Quadratic Reciprocity. When $a \equiv 1 \pmod 4$, this can be simplified to a congruence on $p$ mod $a$, as in the above case with $a = 5$.

If $f(x) = ax^2 + bx + c$ is a polynomial with integer coefficients, Quadratic Reciprocity implies that the set of primes $p$ for which $f(x) \equiv 0 \pmod p$ has a solution can be described by congruence conditions on $p$. This is because $f(x) \equiv 0 \pmod p$ has a solution when $b^2 - 4ac$ is a square mod $p$ (see Section 13.3), and Quadratic Reciprocity implies that the set of primes $p$ for which this is true can be given by congruence conditions.

---

**CHECK YOUR UNDERSTANDING:**

1. Show that $\left(\frac{3}{13}\right) = 1$ by (a) finding a solution of $x^2 \equiv 3 \pmod{13}$ and (b) using Quadratic Reciprocity.

---

# 13.2    Computing Square Roots Mod $p$

Suppose we have used Quadratic Reciprocity and proved that a certain number is a square mod $p$. How do we find a square root? In this section, we treat two cases where there are simple formulas.

**Proposition 13.5.** *Let $p \equiv 3 \pmod 4$ be prime. Let $x \not\equiv 0 \pmod p$ be an integer. Then exactly one of $x$ and $-x$ is a square mod $p$. Let*

$$y \equiv x^{(p+1)/4} \pmod p.$$

*Then $y^2 \equiv \pm x \pmod p$.*

*Proof.* Since $p \equiv 3 \pmod 4$, Proposition 13.3 says that $\left(\frac{-1}{p}\right) = -1$, so

$$\left(\frac{-x}{p}\right) = \left(\frac{-1}{p}\right)\left(\frac{x}{p}\right) = -\left(\frac{x}{p}\right).$$

Therefore, one of $\left(\frac{x}{p}\right)$ and $\left(\frac{-x}{p}\right)$ is $+1$ and the other is $-1$. This means that exactly one of $x$ and $-x$ is a square mod $p$.

Let $y \equiv x^{(p+1)/4} \pmod{p}$. Then

$$y^2 \equiv \left(x^{(p+1)/4}\right)^2 \equiv x^{(p+1)/2} \equiv x^{(p-1)/2}x \equiv (\pm 1)x \pmod{p},$$

since $x^{(p-1)/2} \equiv \pm 1$ by Proposition 13.1.                           □

**Proposition 13.6.** *Let $p \equiv 5 \pmod 8$ be prime and let $x \not\equiv 0 \pmod p$ be an integer. If $x \equiv y^2 \pmod p$ then*

$$y \equiv \begin{cases} \pm x^{(p+3)/8} & \text{if } x^{(p-1)/4} \equiv 1 \pmod{p} \\ \pm 2^{(p-1)/4}x^{(p+3)/8} & \text{if } x^{(p-1)/4} \equiv -1 \pmod{p}. \end{cases}$$

*Proof.* Note that $x^{(p-1)/4} \equiv y^{(p-1)/2} \equiv \pm 1 \pmod{p}$, so the two cases of the theorem are the only possibilities.

First, let's assume that $x^{(p-1)/4} \equiv 1$. Then

$$\left(x^{(p+3)/8}\right)^2 \equiv x^{(p+3)/4} \equiv x^{(p-1)/4}x \equiv 1 \cdot x \equiv y^2 \pmod{p},$$

which implies that $\pm x^{(p+3)/8} \equiv y \pmod{p}$.

Now, suppose that $x^{(p-1)/4} \equiv -1$. Then

$$\left(2^{(p-1)/4}x^{(p+3)/8}\right)^2 \equiv 2^{(p-1)/2}x^{(p-1)/4}x$$

$$\equiv \left(\frac{2}{p}\right)(-1)y^2 \equiv y^2 \pmod{p},$$

since the Supplementary Law for 2 says that $(2/p) = -1$ when $p \equiv 5 \pmod 8$. Again, this gives the formula in the proposition.                           □

**Example.** We can check via Quadratic Reciprocity that 10 is a square mod 13:

$$\left(\frac{10}{13}\right) = \left(\frac{2}{13}\right)\left(\frac{5}{13}\right) = (-1)\left(\frac{13}{5}\right) = -\left(\frac{3}{5}\right) = +1.$$

Let's find a square root: $10^{(p-1)/4} \equiv 1000 \equiv -1 \pmod{13}$, so a square root is given by

$$2^{(p-1)/4}10^{(p+3)/8} \equiv 8 \cdot 100 \equiv 7 \pmod{13}.$$

If $p \equiv 1 \pmod 8$, there are algorithms for computing square roots mod $p$, but they are more complicated. See Project 2.

---

**CHECK YOUR UNDERSTANDING:**

2. Solve $y^2 \equiv 5 \pmod{19}$.

---

# 13.3   Quadratic Equations

Recall the quadratic formula: If $ax^2 + bx + c = 0$ with $a \neq 0$, then

$$x = \frac{-b \pm \sqrt{b^2 - 4ac}}{2a}.$$

This formula is derived by completing the square, and the same method works mod $p$.

Let $p$ be an odd prime. We want to solve

$$ax^2 + bx + c \equiv 0 \pmod p,$$

where $a, b, c$ are integers with $a \not\equiv 0 \pmod p$. Since $p \nmid a$, we can divide by $a$ and get

$$x^2 + \frac{b}{a}x + \frac{c}{a} \equiv 0.$$

Since $p$ is odd, we can divide by 2 and write

$$\left(x + \frac{b}{2a}\right)^2 + \left(\frac{c}{a} - \frac{b^2}{4a^2}\right) \equiv 0.$$

This rearranges to

$$\left(x + \frac{b}{2a}\right)^2 \equiv \frac{b^2 - 4ac}{4a^2}.$$

There is a solution exactly when $b^2 - 4ac$ is a square mod $p$. If it is a square mod $p$, we can take the square root. This yields the following.

**Proposition 13.7.** *Let $p$ be an odd prime and let $a, b, c$ be integers with $p \nmid a$. Then $ax^2 + bx + c \equiv 0 \pmod{p}$ has a solution mod $p$ if and only if $b^2 - 4ac$ has a square root mod $p$. If this is the case, the solutions are*

$$x \equiv \frac{-b \pm \sqrt{b^2 - 4ac}}{2a} \pmod{p}.$$

**Example.** Consider

$$x^2 + 5x + 7 \equiv 0 \pmod{13}.$$

Then $b^2 - 4ac = -3 \equiv 6^2 \pmod{13}$. The solutions are

$$x \equiv \frac{-5 \pm 6}{2} \equiv \frac{1}{2}, \frac{-11}{2} \equiv 7,\ 1 \pmod{13}.$$

As explained in Section 6.6, we change $1/2 \pmod{13}$ to $7 \pmod{13}$ because $2 \cdot 7 \equiv 1 \pmod{13}$.

**Example.** Consider $x^2 + 11x + 7 \equiv 0 \pmod{1009}$ (note: 1009 is prime). We have $b^2 - 4ac = 93$. We could try to find a square root of 93 mod 1009, but we should first use Quadratic Reciprocity to see if this is even possible:

$$\left(\frac{93}{1009}\right) = \left(\frac{3}{1009}\right)\left(\frac{31}{1009}\right) = \left(\frac{1009}{3}\right)\left(\frac{1009}{31}\right) = \left(\frac{1}{3}\right)\left(\frac{17}{31}\right)$$

$$= 1 \cdot \left(\frac{31}{17}\right) = \left(\frac{-3}{17}\right) = \left(\frac{-1}{17}\right)\left(\frac{3}{17}\right) = 1 \cdot \left(\frac{17}{3}\right) = \left(\frac{2}{3}\right) = -1.$$

Fortunately, we didn't try to look for a square root since our calculation shows that there isn't one. So this quadratic equation doesn't have any solutions mod 1009.

---

**CHECK YOUR UNDERSTANDING:**

3. Solve $x^2 + 3x - 2 \equiv 0 \pmod{13}$.

# 13.4    The Jacobi Symbol

Let $m$ be a positive odd integer. Write the prime factorization of $m$ as

$$m = p_1^{a_1} p_2^{a_2} \cdots p_r^{a_r}.$$

Let $\gcd(b, m) = 1$. Define the **Jacobi symbol**

$$\left(\frac{b}{m}\right) = \left(\frac{b}{p_1}\right)^{a_1} \left(\frac{b}{p_2}\right)^{a_2} \cdots \left(\frac{b}{p_r}\right)^{a_r},$$

where $\left(\frac{b}{p_i}\right)$ is the Legendre symbol. For example,

$$\left(\frac{2}{15}\right) = \left(\frac{2}{3}\right)\left(\frac{2}{5}\right) = (-1)(-1) = +1.$$

Since 2 is not a square mod 3, it cannot be a square mod 15. Therefore, we issue the following:

**Warning:** $(b/m) = +1$ does not imply that $b$ is a square mod $m$.

What, then, is it good for? For our purposes, it allows us to compute Legendre symbols without factoring, as we'll see shortly as a consequence of Theorem 13.8 below. But first, we note that the definition implies that

$$\left(\frac{a}{m}\right) = \left(\frac{b}{m}\right) \quad \text{if } a \equiv b \pmod{m}.$$

and

$$\left(\frac{ab}{m}\right) = \left(\frac{a}{m}\right)\left(\frac{b}{m}\right).$$

Also, note that if $m$ is prime, then the Jacobi symbol is the same as the Legendre symbol.

**Theorem 13.8.** *(Reciprocity for Jacobi symbols) Let $m, n$ be positive odd integers with $\gcd(m, n) = 1$.*
*(a)*

$$\left(\frac{m}{n}\right) = (-1)^{(m-1)(n-1)/4} \left(\frac{n}{m}\right).$$

*In other words,*

$$\left(\frac{m}{n}\right) = \begin{cases} \left(\dfrac{n}{m}\right) & \textit{if at least one of } m, n \textit{ is } 1 \pmod 4 \\[3mm] -\left(\dfrac{n}{m}\right) & \textit{if } m \equiv n \equiv 3 \pmod 4. \end{cases}$$

*(b) (Supplementary Law for $-1$)*

$$\left(\frac{-1}{m}\right) = (-1)^{(m-1)/2} = \begin{cases} +1 \textit{ if } m \equiv 1 \pmod 4 \\ -1 \textit{ if } m \equiv 3 \pmod 4. \end{cases}$$

*(c) (Supplementary Law for 2)*

$$\left(\frac{2}{m}\right) = (-1)^{(m^2-1)/8} = \begin{cases} +1 \textit{ if } m \equiv 1, 7 \pmod 8 \\ -1 \textit{ if } m \equiv 3, 5 \pmod 8. \end{cases}$$

*Proof.* Let's start with (a). Write $m = m_1 m_2^2$ and $n = n_1 n_2^2$ with $m_1$ and $n_1$ squarefree. Then

$$\left(\frac{m}{n}\right) = \left(\frac{m_1}{n}\right)\left(\frac{m_2}{n}\right)^2 = \left(\frac{m_1}{n}\right)$$

$$= \left(\frac{m_1}{n_1 n_2^2}\right) = \left(\frac{m_1}{n_1}\right)\left(\frac{m_1}{n_2}\right)^2 = \left(\frac{m_1}{n_1}\right)$$

and similarly

$$\left(\frac{n}{m}\right) = \left(\frac{n_1}{m_1}\right).$$

We'll prove (a) for $m_1$ and $n_1$. Since the square of an odd number is 1 mod 8, we have $m \equiv m_1 \pmod 8$ and $n \equiv n_1 \pmod 8$. This means that

$$(m-1)(n-1)/4 \equiv (m_1-1)(n_1-1)/4 \pmod 2$$

and

$$(-1)^{(m-1)(n-1)/4} = (-1)^{(m_1-1)(n_1-1)/4},$$

so the pluses and minuses work the same for $m_1$ and $n_1$ as for $m$ and

$n$. Therefore, if we prove (a) for $m_1$ and $n_1$ then

$$\left(\frac{m}{n}\right) = \left(\frac{m_1}{n_1}\right)$$

$$= (-1)^{(m_1-1)(n_1-1)/4}\left(\frac{n_1}{m_1}\right)$$

$$= (-1)^{(m-1)(n-1)/4}\left(\frac{n_1}{m_1}\right)$$

$$= (-1)^{(m-1)(n-1)/4}\left(\frac{n}{m}\right),$$

so we obtain (a) for $m, n$.

Let the prime factorizations of $m_1$ and $n_1$ be

$$m_1 = p_1 p_2 \cdots p_r, \qquad n_1 = q_1 q_2 \cdots q_s.$$

Then

$$\left(\frac{m_1}{n_1}\right) = \prod_{j=1}^{s}\left(\frac{m_1}{q_j}\right) = \prod_{i=1}^{r}\prod_{j=1}^{s}\left(\frac{p_i}{q_j}\right)$$

and

$$\left(\frac{n_1}{m_1}\right) = \prod_{i=1}^{r}\left(\frac{n_1}{p_i}\right) = \prod_{i=1}^{r}\prod_{j=1}^{s}\left(\frac{q_j}{p_i}\right).$$

To change from $(m_1/n_1)$ to $(n_1/m_1)$, we use Quadratic Reciprocity for primes to change each $(p_i/q_j)$ to $(q_j/p_i)$. We pick up a factor of $-1$ exactly when $p_i \equiv q_j \equiv 3 \pmod{4}$. Therefore,

$$\left(\frac{m_1}{n_1}\right) = (-1)^t \left(\frac{n_1}{m_1}\right),$$

where $t$ is the number of pairs $(p_i, q_j)$ where $p_i \equiv q_j \equiv 3 \pmod{4}$. The number of such pairs is $t_1 t_2$, where $t_1$ is the number of primes $p_i \equiv 3 \pmod{4}$ dividing $m_1$ and $t_2$ is the number of primes $q_j \equiv 3 \pmod{4}$ dividing $n_1$.

Suppose $t_1$ and $t_2$ are odd. The product of an odd number of primes that are 3 mod 4 is 3 mod 4, so in this case, $m_1 \equiv n_1 \equiv 3 \pmod{4}$. Also, $t = t_1 t_2$ is odd. Therefore,

$$\left(\frac{m_1}{n_1}\right) = -\left(\frac{n_1}{m_1}\right).$$

Now suppose that at least one of $t_1$ and $t_2$ is even. The product of an even number of primes that are 3 mod 4 is 1 mod 4, so at least one of $m_1$ and $n_1$ is 1 mod 4. Also, $t = t_1 t_2$ is even. Therefore,

$$\left(\frac{m_1}{n_1}\right) = \left(\frac{n_1}{m_1}\right).$$

We conclude that $(m_1/n_1) = (n_1/m_1)$ if at least one of $m_1$ and $n_1$ is 1 mod 4, and $(m_1/n_1) = -(n_1/m_1)$ if both are 3 mod 4. This is exactly statement (a) for $m_1$ and $n_1$. As we showed above, this yields the result for $m$ and $n$.

The reasoning for (b) is similar. Write $m = m_1 m_2^2$ as in the proof of (a) and let

$$m_1 = p_1 p_2 \cdots p_r$$

be the prime factorization of $m_1$. Then $m \equiv m_1 \pmod 4$ and

$$\left(\frac{-1}{m}\right) = \left(\frac{-1}{m_1}\right) = \prod_{i=1}^{r}\left(\frac{-1}{p_i}\right) = (-1)^{t_1},$$

where $t_1$ is the number of prime factors $p_i \equiv 3 \pmod 4$. As in the proof of (a), we see that $t_1$ is even if and only if $m_1 \equiv 1 \pmod 4$, so

$$\left(\frac{-1}{m_1}\right) = (-1)^{(m_1-1)/2} = (-1)^{(m-1)/2},$$

which yields (b).

The proof of (c) is very similar to that of (b), but we need the following easily verified facts. A product of primes $p \equiv \pm 1 \pmod 8$ is $\pm 1 \pmod 8$. A product of an even number of primes $p \equiv \pm 3 \pmod 8$ is $\pm 1 \pmod 8$. A product of an odd number of primes $p \equiv \pm 3 \pmod 8$ is $\pm 3 \pmod 8$. By the reasoning in (b), we have

$$\left(\frac{2}{m}\right) = \left(\frac{2}{m_1}\right) = \prod_{i=1}^{r}\left(\frac{2}{p_i}\right) = (-1)^{u_1},$$

where $u_1$ is the number of prime factors $p_i \equiv \pm 3 \pmod 8$. Since $u_1$ is odd exactly when $m_1$ (and therefore $m$) is $\pm 3 \pmod 8$, we obtain the equality of the first and third parts of statement (c). The facts that $(-1)^{(m^2-1)/8} = +1$ if $m \equiv \pm 1 \pmod 8$ and $(-1)^{(m^2-1)/8} = -1$ if $m \equiv \pm 3 \pmod 8$ are easily verified (for example, $m \equiv \pm 3 \pmod 8 \Rightarrow m^2 \equiv 9 \pmod{16} \Rightarrow (m^2 - 1)/8 \equiv 1 \pmod 2$).                 $\square$

**Example.** Let's calculate $\left(\frac{246}{257}\right)$. We did this in Section 13.1, but we needed to factor 123 as $3\times41$. With the Jacobi symbol, this factorization is unnecessary. The only factorization involves 2:

$$\left(\frac{246}{257}\right) = \left(\frac{2}{257}\right)\left(\frac{123}{257}\right)$$

$$= (+1)\left(\frac{123}{257}\right) \quad \text{(since } 257 \equiv 1 \pmod 8)$$

$$= \left(\frac{257}{123}\right) \quad \text{(by Reciprocity for the Jacobi symbol)}$$

$$= \left(\frac{11}{123}\right) \quad \text{(since } 257 \equiv 11 \pmod{123})$$

$$= -\left(\frac{123}{11}\right) \quad \text{(since } 123 \equiv 11 \equiv 3 \pmod 4)$$

$$= -\left(\frac{2}{11}\right) = -(-1) = +1.$$

Notice how much easier this was than when we did the calculation in Section 13.1 using only the Legendre symbol. With larger numbers, the advantage becomes even greater since we do not need to check whether odd numbers are primes and we do not need to factor odd integers, which can be extremely difficult for large numbers. Just remember not to assign too much meaning to the results of intermediate steps. For example, our calculation shows that $(11/123) = +1$. This does not mean that 11 is a square mod 123. In fact, it is not since 11 is not a square mod 3.

---

**CHECK YOUR UNDERSTANDING:**

4. Compute $\left(\frac{501}{1013}\right)$.

5. Show that $\left(\frac{3}{35}\right) = +1$ but $x^2 \equiv 3 \pmod{35}$ has no solutions.

## 13.5   Proof of Quadratic Reciprocity

The proof of Quadratic Reciprocity is rather technical and can easily be skipped without impairing the understanding of the rest of the book.

Recall the proof of Fermat's theorem. We had a prime $p$ and an $a \not\equiv 0 \pmod{p}$. We computed $a^{p-1} \cdot (p-1)!$ and showed that this was a rearrangement of $(p-1)! \bmod p$. Dividing by $(p-1)!$ yielded $a^{p-1} \equiv 1 \pmod{p}$. In order to work with Legendre symbols, Proposition 13.3 says that we need to be able to work with $a^{(p-1)/2}$. The natural thing to do is compute $a^{(p-1)/2} \cdot ((p-1)/2)!$ and see if this is a rearrangement of $((p-1)/2)! \bmod p$. This is almost the case. Let's analyze what occurs when $p = 11$ and $a = 7$.

$$7^5 \cdot 1 \cdot 2 \cdot 3 \cdot 4 \cdot 5 \equiv 7 \cdot 14 \cdot 21 \cdot 28 \cdot 35 \pmod{11}$$
$$\equiv 7 \cdot 3 \cdot 10 \cdot 6 \cdot 2$$
$$\equiv (-4) \cdot 3 \cdot (-1) \cdot (-5) \cdot 2$$
$$\equiv (-1)^3 \cdot 1 \cdot 2 \cdot 3 \cdot 4 \cdot 5 \pmod{11}.$$

Divide by $1 \cdot 2 \cdot 3 \cdot 4 \cdot 5$ to get

$$7^5 \equiv (-1)^3 \pmod{11}.$$

By Proposition 13.3, this says that

$$\left(\frac{7}{11}\right) \equiv 7^5 \equiv (-1)^3 \equiv -1 \pmod{11}.$$

Observe that we changed the signs of the numbers in the second line that were between $11/2$ and $11$, and we changed the product to $(-1)^3 \cdot 5!$, which allowed us to compute $\left(\frac{7}{11}\right)$. The following result shows that this computation is part of a general phenomenon.

**Lemma 13.9.** *(Gauss's Lemma) Let $p$ be an odd prime and let $a \not\equiv 0 \pmod{p}$. Let $n$ be the number of integers in the set*

$$\{a, 2a, 3a, \ldots, (p-1)a/2\}$$

*that are congruent to an integer between $p/2$ and $p$. Then*

$$\left(\frac{a}{p}\right) = (-1)^n.$$

For example, when $p = 11$ and $a = 7$, we have

$$\{7,\, 2 \cdot 7,\, 3 \cdot 7,\, 4 \cdot 7,\, 5 \cdot 7\}.$$

Reducing the numbers in this set modulo 11, we get

$$\{7, 3, 10, 6, 2\}.$$

Since $11/2 < k < 11$ for $k = 7, 10, 6$, we have $n = 3$.

*Proof.* Let $0 < i < p/2$. We claim that there exists a unique $j$ with $0 < j < p/2$ such that

$$aj \equiv \pm i \pmod{p}.$$

Since $\gcd(a, p) = 1$, there exists $j'$ with $0 < j' < p$ such that $aj' \equiv i$ $\pmod{p}$. If $0 < j' < p/2$, we're done: let $j = j'$. If $p/2 < j' < p$, let $j = p - j'$. Then $0 < j < p/2$ and

$$aj \equiv a(p - j') \equiv -aj' \equiv -i \pmod{p}.$$

Therefore, $j$ exists.

Now suppose that $aj_1 \equiv \pm aj_2 \pmod{p}$, with $0 < j_1, j_2 < p/2$. Divide by $a$ to get

$$j_1 \equiv \pm j_2 \pmod{p}.$$

Since $0 < j_1 + j_2 < p$, we cannot have $j_1 + j_2 \equiv 0 \pmod{p}$, so $j_1 \not\equiv -j_2$ $\pmod{p}$. Therefore, $j_1 \equiv j_2 \pmod{p}$. Since $0 < j_1, j_2 < p$, we must have $j_1 = j_2$. Therefore, the solution of $aj \equiv \pm i \pmod{p}$ is unique. This proves the claim.

Consider the numbers $a, 2a, 3a, \ldots, (p-1)a/2$. The claim and the definition of $n$ imply that if we change the signs of $n$ of them, we obtain numbers congruent to each of the numbers from 1 to $(p-1)/2$. Therefore,

$$a \cdot 2a \cdot 3a \cdots (p-1)a/2 \equiv (-1)^n((p-1)/2)! \pmod{p}.$$

Separating out the factors of $a$ yields

$$a^{(p-1)/2} \cdot ((p-1)/2)! \equiv (-1)^n((p-1)/2)! \pmod{p}.$$

Dividing by $((p-1)/2)!$ yields $a^{(p-1)/2} \equiv (-1)^n \pmod{p}$. Proposition 13.3 implies

$$\left(\frac{a}{p}\right) \equiv a^{(p-1)/2} \equiv (-1)^n \pmod{p}.$$

This completes the proof.                                    $\square$

Gauss's Lemma immediately allows us to compute $(2/p)$. The numbers in the lemma are

$$2, 4, 6, \ldots, (p-1).$$

Approximately half are less than $p/2$ and the others lie between $p/2$ and $p$. We need to determine the middle point precisely.

If $p = 1+4k$, then $2, 4, 6, \ldots, 2k < p/2$ and $p/2 < 2k+2, 2k+4, \ldots, 4k < p$. Therefore, $n = k = (p-1)/4$. If $p \equiv 1 \pmod 8$, then $n = (p-1)/4$ is even, and if $p \equiv 5 \pmod 8$, then $n = (p-1)/4$ is odd, so

$$\left(\frac{2}{p}\right) = \begin{cases} +1 \text{ if } p \equiv 1 \pmod 8 \\ -1 \text{ if } p \equiv 5 \pmod 8. \end{cases}$$

If $p = 3+4k$, then $2, 4, 6, \ldots, 2k < p/2$ and $p/2 < 2k+2, 2k+4, \ldots, 4k+2 < p$. Therefore, $n = k+1 = (p+1)/4$. If $p \equiv 3 \pmod 8$, then $n = (p+1)/4$ is odd, and if $p \equiv 7 \pmod 8$, then $n = (p+1)/4$ is even, so

$$\left(\frac{2}{p}\right) = \begin{cases} -1 \text{ if } p \equiv 3 \pmod 8 \\ +1 \text{ if } p \equiv 7 \pmod 8. \end{cases}$$

This completes the proof of the Supplementary Law for 2.

We're now ready to tackle the main Quadratic Reciprocity Law. We need another expression for $n$.

**Lemma 13.10.** *Let $a$ be a positive integer and let $j_0$ be the largest even integer less than or equal to $a$. Let $n$ be as in Lemma 13.9. Then $n$ is the number of integers in the union of the intervals*

$$(j-1)p/(2a) < x < jp/(2a)$$

*as $j$ runs through the even integers $2 \le j \le j_0$.*

**Example.** Let $p = 11$ and $a = 7$. Then $j_0 = 6$, so $j$ takes the values $2, 4, 6$. We have the intervals

$$11/14 < x < 22/14, \quad 33/14 < x < 44/14, \quad 55/14 < x < 66/14.$$

The integers 1, 3, 4 are in these intervals, so $n = 3$.

*Proof.* To evaluate $n$, we need to find when $ai$ is congruent to some $k$ with $p/2 < k < p$. If this is the case, then

$$\frac{p}{2} < ai - \ell p < p$$

for some $\ell$. These inequalities are equivalent to

$$\frac{2\ell + 1}{2} < ai < \frac{2\ell + 2}{2}p.$$

Letting $j = 2\ell + 2$, we find that $ai$ is congruent to some $k$ with $p/2 < k < p$ if and only if

$$(j - 1)p/2 < ai < jp/2$$

for an even integer $j$.

Note that

$$(j - 1)p/2 < ai < jp/2 \iff (j - 1)p/(2a) < i < jp/(2a).$$

Therefore, recalling that $j$ must be even, we must count integers $i$ that lie in intervals $(j - 1)p/(2a) < x < jp/(2a)$ for even integers $j$. But which integers $j$ do we use?

We have $ai \le a(p - 1)/2$. Recall that $j_0$ is the largest even integer less than or equal to $a$, so

$$j_0 \le a \le j_0 + 1$$

($a = j_0$ if $a$ is even, and $a = j_0 + 1$ if $a$ is odd). Then

$$a(p - 1)/2 < (j_0 + 1)p/2,$$

so the last $ai$, namely $(p - 1)a/2$, lies before the interval $(j_0 + 1)p/2 < x < (j_0 + 2)p/2$. Therefore, we can stop at $j = j_0$ or before. Since

$$a(p + 1)/2 > j_0 p/2,$$

the next multiple of $a$ past those we are counting, namely $a(p + 1)/2$, lies past $pj_0/2$. The conclusion is that if we stop at the interval $(j_0 - 1)p/2 < x < j_0 p/x$, then we count all the $ai$ up to $a(p - 1)/2$ that lie in the desired intervals, but we do not count $aj$ for $j \ge (p + 1)/2$. This completes the proof of the lemma.                                                    □

The following lemma is a key step in the proof of Quadratic Reciprocity.

**Lemma 13.11.** *Let $p$ and $q$ be odd primes and let $a$ be a positive integer with $\gcd(a, pq) = 1$. If $p \equiv \pm q \pmod{4a}$, then*

$$\left(\frac{a}{p}\right) = \left(\frac{a}{q}\right).$$

This result says something rather surprising. We know that $(a/p)$ depends only on $a \pmod p$, but the lemma says that it also depends only on $p \pmod{4a}$. This is a congruence condition for the lower argument, not the usual upper argument. For example, we know that $(3/7) = -1$. Since $127 \equiv 7 \pmod{4 \cdot 3}$, the lemma immediately tells us that $(3/127) = -1$, too. Why should the fact that 3 is not a square mod 7 imply that 3 is also not a square mod 127? This is the mystery of Quadratic Reciprocity.

*Proof.* First, suppose that $p \equiv q \pmod{4a}$. Write $q = p + 4am$ for some integer $m$. Let the notation be as in the previous lemma and let $n_p$ be the $n$ for $a$ and $p$, as in Lemma 13.9, and let $n_q$ be the $n$ for $a$ and $q$. Consider an interval

$$(j - 1)p/(2a) < i < jp/(2a).$$

The corresponding interval for $q$ is

$$(j - 1)q/(2a) < i < jq/(2a).$$

Substituting $q = p + 4am$, we obtain

$$(j - 1)p/(2a) + 2(j - 1)m < i < jp/(2a) + 2jm.$$

This is just the interval for $p$ shifted to the right by $2(j - 1)m$ and then lengthened by $2m$. If there are $n_{p,j}$ integers $i$ in the interval for $p$, then there are $n_{q,j} = n_{p,j} + 2m$ integers $i$ in the interval for $q$. In particular,

$$n_{p,j} \equiv n_{q,j} \pmod 2$$

for all $j$. Therefore,

$$n_p = n_{p,2} + n_{p,4} + \cdots + n_{p,j_0} \equiv n_{q,2} + n_{q,4} + \cdots + n_{q,j_0} = n_q \pmod 2.$$

It follows that

$$\left(\frac{a}{p}\right) = (-1)^{n_p} = (-1)^{n_q} = \left(\frac{a}{q}\right).$$

Now assume that $p \equiv -q \pmod{4a}$. Write $p + q = 4at$ for some integer $t$. Let $n_{q,j}$ denote the number of integers $i$ satisfying

$$(j-1)q/(2a) < i < jq/(2a).$$

This interval can be rewritten as

$$2t(j-1) - (j-1)p/(2a) < i < 2tj - jp/(2a).$$

We can shift this by $2tj$ and multiply by $-1$ to find that $n_{q,j}$ is the number of integers $i$ satisfying

$$jp/(2a) < i < 2t + (j-1)p/(2a).$$

Since $n_{p,j}$ is the number of integers $i$ satisfying

$$(j-1)p/(2a) < i < jp/(2a),$$

we can put these two together and find that $n_{p,j} + n_{q,j}$ is the number of integers satisfying

$$(j-1)p/(2a) < i < (j-1)p/(2a) + 2t.$$

This is an interval of length $2t$, and neither endpoint is an integer since $(j-1)p$ is odd and $2a$ is even. Therefore, there are $2t$ integers in the interval, so $n_{p,j} + n_{q,j} = 2t$. This implies that

$$n_{p,j} \equiv n_{q,j} \pmod 2.$$

As in the case $p \equiv q \pmod{4a}$, this implies that $n_p \equiv n_q \pmod 2$ and therefore

$$\left(\frac{a}{p}\right) = (-1)^{n_p} = (-1)^{n_q} = \left(\frac{a}{q}\right).$$

$\square$

The proof of Quadratic Reciprocity is almost done. Suppose first that

$p \equiv q \pmod 4$. Write $q = p + 4\ell$. Then

$$\left(\frac{q}{p}\right) = \left(\frac{4\ell}{p}\right) \quad (\text{since } q \equiv 4\ell \pmod p)$$

$$= \left(\frac{4}{p}\right)\left(\frac{\ell}{p}\right) = \left(\frac{\ell}{p}\right) \quad (\text{since 4 is a square})$$

$$= \left(\frac{\ell}{q}\right) \quad (\text{since } p \equiv q \pmod{4\ell}, \text{ so Lemma 13.11 applies})$$

$$= \left(\frac{4\ell}{q}\right)$$

$$= \left(\frac{-p}{q}\right) \quad (\text{since } 4\ell \equiv -p \pmod q).$$

If $p \equiv q \equiv 1 \pmod 4$, then $(-1/q) = 1$, so

$$\left(\frac{q}{p}\right) = \left(\frac{-p}{q}\right) = \left(\frac{-1}{q}\right)\left(\frac{p}{q}\right) = \left(\frac{p}{q}\right).$$

If $p \equiv q \equiv 3 \pmod 4$, then $(-1/q) = -1$, and the same calculation yields

$$\left(\frac{q}{p}\right) = -\left(\frac{p}{q}\right).$$

Now suppose $p \equiv -q \pmod 4$. Write $p + q = 4w$. Then

$$\left(\frac{q}{p}\right) = \left(\frac{4w}{p}\right) \quad (\text{since } q \equiv 4w \pmod p)$$

$$= \left(\frac{4}{p}\right)\left(\frac{w}{p}\right) = \left(\frac{w}{p}\right) \quad (\text{since 4 is a square})$$

$$= \left(\frac{w}{q}\right) \quad (\text{by Lemma 13.11})$$

$$= \left(\frac{4w}{q}\right) \quad (\text{since 4 is a square})$$

$$= \left(\frac{p}{q}\right) \quad (\text{since } p \equiv 4w \pmod q).$$

This completes the proof of Quadratic Reciprocity. $\square$

## 13.6    Chapter Highlights

1. The Legendre symbol: $\left(\frac{a}{p}\right) = +1$ if $a$ is a square mod $p$ and $= -1$ if $a$ is not a square mod $p$.

2. Quadratic Reciprocity (Theorem 13.4).

3. The Jacobi symbol.

## 13.7    Problems

### 13.7.1    Exercises

**Section 13.1: Squares and Square Roots Mod Primes**

1. Show that $\left(\frac{3}{11}\right) = 1$ by finding a solution of $x^2 \equiv 3 \pmod{11}$.

2. Evaluate $\left(\frac{38}{79}\right)$.

3. Evaluate $\left(\frac{3}{17}\right)$.

4. Evaluate $\left(\frac{31}{103}\right)$.

5. Does $x^2 \equiv 19 \pmod{101}$ have any solutions? (Justify your answer; do not find the solutions if they exist.)

6. Does $x^2 \equiv 23 \pmod{79}$ have any solutions? (Justify your answer; do not find the solutions if they exist.)

7. (a) Does $x^2 \equiv 2 \pmod{209}$ have solutions? Note that $209 = 11 \cdot 19$ is not prime. Justify your answer.
   (b) Does $x^2 \equiv 47 \pmod{209}$ have solutions? Justify your answer.

8. (a) Show that $\left(\frac{7}{107}\right) = -1$.

   (b) Show that $7^{53} \equiv -1 \pmod{107}$. (*Hint:* $(107 - 1)/2 = 53$.)
   (c) Find the order of 7 mod 107.

9. Suppose $p = 2^{(2^m)} + 1$ is prime, with $m \geq 1$.
   (a) Show that $p \equiv 2 \pmod 3$.
   (b) Show that $\left(\frac{3}{p}\right) = -1$.

(c) Show that the order of 3 mod $p$ is a power of 2.

(d) Show that the order of 3 mod $p$ is $p-1$, so 3 is a primitive root for every Fermat prime.

(e) Use the method of (c) and (d) to show that if $p$ is a Fermat prime and $a$ satisfies $\left(\dfrac{a}{p}\right) = -1$, then $a$ is a primitive root mod $p$.

10. Let $p$ and $q$ be primes with $p \equiv q \pmod{28}$. Use Quadratic Reciprocity (not Lemma 13.11) to show that $\left(\dfrac{7}{p}\right) = \left(\dfrac{7}{q}\right)$.

11. Let $p$ and $q$ be odd primes with $p \equiv q \pmod 5$. Show that $\left(\dfrac{5}{p}\right) = \left(\dfrac{5}{q}\right)$.

12. (a) Let $p$ be a prime with $p \equiv 9 \pmod{28}$. Show that $\left(\dfrac{7}{p}\right) = 1$.

    (b) More generally, let $p$ and $q$ be primes with $p \equiv 9 \pmod{4q}$. Show that $\left(\dfrac{q}{p}\right) = 1$.

13. Let $p \equiv 1 \pmod 3$ be prime and $b \not\equiv 0 \pmod p$. Show that $b$ is a cube mod $p$ if and only if $b^{(p-1)/3} \equiv 1 \pmod p$.

14. Let $p \equiv 2 \pmod 3$ be prime and let $b \not\equiv 0 \pmod p$. Let $x \equiv b^{(2p-1)/3} \pmod p$. Show that $x^3 \equiv b \pmod p$. This shows that every number mod $p$ is a cube when $p \equiv 2 \pmod 3$.

15. Give congruence conditions mod 20 for primes $p$ for which $-5$ is a square mod $p$.

16. (a) Show that if $n \geq 5$ then $1! + 2! + 3! + \cdots + n! \equiv 3 \pmod 5$.

    (b) Show that $1! + 2! + 3! + \cdots + n!$ is a square for $n = 1$ and $n = 3$, but is not a square for all other values of $n \geq 1$.

17. (a) Let $p$ be an odd prime and let $n \geq p$. Suppose that $q = n! + 1$ is prime. Show that $\left(\dfrac{p}{q}\right) = 1$, except when $p = 3$ and $q = 7$.

    (b) There are 11 values of $n$ with $1 \leq n \leq 200$ such that $n! + 1$ is prime. Find them. For each such $n$, except $n = 1, 2, 3$, produce $p$ and $q$ that give an example for part (a). (*Note:* This exercise requires a computer program that can test for primality.)

18. Let $p$ be a prime with $p \equiv 3 \pmod 4$, and suppose that $q = 2p + 1$ is also prime.

    (a) Show that 2 is a square mod $q$.

    (b) Show that $2^p \equiv 1 \pmod q$, so $q \mid 2^p - 1$. (This explains the fact that $23 \mid 2^{11} - 1$.)

    It is not yet proved that there are infinitely many primes $p$ such that $2p + 1$ is prime. It is also not yet proved that there are infinitely many primes $p$ for which $M_p$ is composite.

19. Let $p$ be an odd prime and let $q$ be a prime divisor of $2^p - 1$.

    (a) Show that 2 is a square mod $q$. (*Hint:* What is $\left(\dfrac{2}{q}\right)^p$?)

    (b) Show that $q \equiv \pm 1 \pmod 8$.

20. Let $p$ be an odd prime and let $a$ and $b$ be integers with $ab \not\equiv 0 \pmod p$. Show that $ax^2 \equiv b \pmod p$ has a solution if and only if $\left(\dfrac{a}{p}\right)\left(\dfrac{b}{p}\right) = 1$.

21.  Let $p$ be an odd prime. Suppose $ab \equiv 1 \pmod{p}$. Show that $\left(\frac{a}{p}\right) = \left(\frac{b}{p}\right)$.

22.  Let $p$ be an odd prime such that $2^p - 1$ is also prime. Show that $\left(\frac{3}{2^p - 1}\right) = -1$.

23.  Let $p$ be an odd prime. Show that $\left(\frac{(p-1)!}{p}\right) = (-1)^{(p-1)/2}$.

24.  Prove that there are infinitely many primes $p \equiv 3 \pmod 8$. (*Hint:* Let $N = (p_1 p_2 \cdots p_n)^2 + 2$. Show that $N \equiv 3 \pmod 8$ and that $N$ must have a prime factor $p \not\equiv 1 \pmod 8$.)

25.  Prove that there are infinitely many primes $p \equiv 7 \pmod 8$. (*Hint:* Let $N = (p_1 p_2 \cdots p_n)^2 - 2$. Show that $N \equiv 7 \pmod 8$ and that $N$ must have a prime factor $p \not\equiv 1 \pmod 8$.)

26.  (a) Let $p \equiv 2 \pmod 3$. Show that $\left(\frac{-3}{p}\right) = -1$. (*Hint:* Consider the cases $p \equiv 1 \pmod 4$ and $p \equiv 3 \pmod 4$ separately.)
     (b) Prove that there are infinitely many primes $p \equiv 1 \pmod 3$. (*Hint:* Let $N = (p_1 p_2 \cdots p_n)^2 + 3$.)

27.  Let $n$ be an integer and let $p \neq 3$ be a prime dividing $n^2 + n + 1$.
     (a) Show that $p$ is odd.
     (b) Show that $p \mid (2n + 1)^2 + 3$.
     (c) Use Part (a) of Exercise 26 to conclude that $p \equiv 1 \pmod 3$.

28.  Let $p$ be an odd prime and let $g$ be a primitive root mod $p$. Let

$$S = \sum_{j=1}^{p-1} \left(\frac{j}{p}\right).$$

     (a) Show that

$$S = \sum_{j=1}^{p-1} \left(\frac{gj}{p}\right).$$

     (b) Show that $S = \left(\frac{g}{p}\right) S$.
     (c) Show that $\left(\frac{g}{p}\right) = -1$.
     (d) Show that $S = 0$.
     (e) Show that the number of nonzero squares mod $p$ equals the number of nonzero nonsquares mod $p$. (This was also proved in Exercise 34 of Chapter 6 and Exercise 33 of Chapter 11.)

## Section 13.2: Computing Square Roots Mod $p$

29.  Use the method of Section 13.2 to solve $y^2 \equiv 2 \pmod{23}$.

30.  Use the method of Section 13.2 to solve $y^2 \equiv 11 \pmod{19}$.

31.  Use the method of Section 13.2 to solve $y^2 \equiv 7 \pmod{29}$.

32.  Use the method of Section 13.2 to solve $y^2 \equiv 5 \pmod{29}$.

## Section 13.3: Quadratic Equations

33.  Does $x^2 + 7x - 10 \equiv 0 \pmod{37}$ have any solutions?

34.  Does $x^2 + 5x + 7 \equiv 0 \pmod{23}$ have any solutions?

35.  Find the solutions to $x^2 + 16x + 9 \pmod{23}$.

36.  Find the solutions to $x^2 - x + 5 \equiv 0 \pmod{11}$.

## Section 13.4: The Jacobi Symbol

37.  Compute $\left(\dfrac{35}{223}\right)$.

38.  Compute $\left(\dfrac{35}{73}\right)$.

39.  Compute $\left(\dfrac{203}{3511}\right)$.

40.  Compute $\left(\dfrac{55}{401}\right)$.

41.  (a) Compute $\left(\dfrac{35}{143}\right)$.

(b) Use the result of (a) to show that 35 is not a square mod 143.

(c) Show that 35 is not a square mod 11, and conclude (without using (b)) that 35 is not a square mod 143 ($= 11 \cdot 13$).

42.  (a) Compute $\left(\dfrac{5}{77}\right)$.

(b) Use the result of (a) to show that 5 is not a square mod 77.

(c) Show that 5 is not a square mod 7, and conclude (without using (b)) that 5 is not a square mod 77.

43.  (a) Compute $\left(\dfrac{3}{35}\right)$.

(b) Does the result of (a) show that 3 is a square mod 35?

(c) Show that 3 is not a square mod 5, and conclude that 3 is not a square mod 35.

44.  (a) Compute $\left(\dfrac{13}{55}\right)$.

(b) Does the result of (a) show that 13 is not a square mod 55?

(c) Show that 13 is not a square mod 5, and conclude that 13 is not a square mod 55.

## Section 13.5: Proof of Quadratic Reciprocity

45.  Use Lemma 13.9 (Gauss's Lemma) to calculate $\left(\dfrac{3}{7}\right)$ and $\left(\dfrac{3}{13}\right)$. Check that you get the same answers as you get using Quadratic Reciprocity.

46.  Use Gauss's Lemma to show that if $p$ is an odd prime then $\left(\frac{1}{p}\right) = 1$.

47.  Use Gauss's Lemma to evaluate $\left(\frac{-1}{p}\right)$ for odd primes $p$.

48.  For each of the following primes $p$, make a list consisting of $2, 4, 6, \ldots, (p-1)$ and count how many have their least non-negative residue mod $p$ between $p/2$ and $p$. Then, use Gauss's Lemma to calculate $(2/p)$. If $(2/p) = 1$, find both solutions to $x^2 \equiv 2 \pmod{p}$.
(a) $p = 5$,   (b) $p = 7$,   (c) $p = 29$,   (d) $p = 41$.

49.  For each of the following primes $p$, make a list consisting of $2, 4, 6, \ldots, (p-1)$ and count how many have their least non-negative residue mod $p$ between $p/2$ and $p$. Then, use Gauss's Lemma to calculate $(2/p)$. If $(2/p) = 1$, find both solutions to $x^2 \equiv 2 \pmod{p}$.
     (a) $p = 17$,    (b) $p = 13$,    (c) $p = 19$,    (d) $p = 31$.

50.  Use Gauss's Lemma to calculate

$$\left(\frac{3}{p}\right)$$

for $p = 11, 13, 17, 19$. Do your results agree with the computation of $\left(\dfrac{3}{p}\right)$

at the end of Section 13.1?

51.  Use Gauss's Lemma to calculate

$$\left(\frac{5}{p}\right)$$

for $p = 11, 13, 17, 19$. Do your results agree with the computation of $\left(\dfrac{5}{p}\right)$

at the end of Section 13.1?

## 13.7.2   Projects

1.  Let $f(x) = x^4 - 2x^2 + 9$.
    (a) Show that $\pm(\sqrt{2} + i)$ and $\pm(\sqrt{2} - i)$ are roots of $f(x)$.
    (b) Suppose that $f(x)$ factors as $g(x)h(x)$, where $g(x)$ and $h(x)$ are polynomials with rational coefficients. Show that neither $g(x)$ nor $h(x)$ can have degree 1, so $g(x)$ and $h(x)$ must have degree 2.
    (c) Show that no choice $r_1, r_2$ of roots of $f(x)$ yields a polynomial $(x - r_1)(x - r_2)$ with rational coefficients.
    (d) Show that $f(x)$ does not factor into polynomials of smaller degree with rational coefficients.
    (e) Let $p \equiv 1 \pmod{4}$ be a prime and let $s^2 \equiv -1 \pmod{p}$. Show that

$$f(x) \equiv (x^2 + 2sx - 3)(x^2 - 2sx - 3) \pmod{p}.$$

    (f) Let $p \equiv 3 \pmod{8}$ be a prime. Show that there exists $t$ with $t^2 \equiv -2 \pmod{p}$.
    (g) Let $p$ and $t$ be as in (f). Show that

$$f(x) \equiv (x^2 - 2t - 1)(x^2 + 2t - 1) \pmod{p}.$$

    (h) Let $p \equiv 7 \pmod{8}$ be a prime. Show that there exists $u$ with $u^2 \equiv 2 \pmod{p}$.
    (i) Let $p$ and $u$ be as in (h). Show that

$$f(x) \equiv (x^2 + 2ux + 3)(x^2 - 2ux + 3) \pmod{p}.$$

    (j) Show that $f(x)$ factors mod 2.
    Therefore, $f(x)$ does not factor using rational numbers, but it factors mod $p$ for each prime $p$.

2.  In this chapter we pointed out that Theorem 13.4 (The Law of Quadratic Reciprocity) is non-constructive: it tells us *when* you can solve a quadratic congruence mod a prime, but not *how* to find a solution. In this project we present a method known as the *Tonelli–Shanks Algorithm* which fills in this gap and shows how to construct such a solution. It was first discovered in 1891 by Alberto Tonelli and was put in the present form by Dan Shanks in 1973.

Let $p$ be an odd prime and let $a$ be an integer satisfying $\left(\frac{a}{p}\right) = 1$. We want to solve the congruence

$$x^2 \equiv a \pmod{p}$$

We begin by writing $p - 1 = 2^k m$ with $m$ odd and choosing an $n$ with $\left(\frac{n}{p}\right) = -1$. Now, set

$$c_0 \equiv n^m \pmod{p}, \quad k_0 = k,$$
$$r_0 \equiv a^{(m+1)/2} \pmod{p}, \quad t_0 \equiv a^m \pmod{p}.$$

(a) Show that $c_0^{2^{k_0-1}} \equiv -1 \pmod{p}$.

(b) Show that $r_0^2 \equiv at_0 \pmod{p}$.

(c) Show that $t_0^{2^{k_0-1}} \equiv 1 \pmod{p}$.

Now assume that we have a 4-tuple $(c_i, k_i, r_i, t_i)$ with $k_i \geq 2$ and $t_i \neq 1$ such that (from now on, all congruences are congruences mod $p$)

$$c_i^{2^{k_i-1}} \equiv -1, \quad r_i^2 \equiv at_i, \quad t_i^{2^{k_i-1}} \equiv 1. \tag{13.1}$$

We will construct a 4-tuple $(c_{i+1}, k_{i+1}, r_{i+1}, t_{i+1})$ with $1 \leq k_{i+1} < k_i$ and such that Equation (13.1) is satisfied with $i + 1$ in place of $i$.

(d) Show that $\mathrm{ord}_p(t_i) = 2^{k_{i+1}}$ for some $k_{i+1}$ satisfying $1 \leq k_{i+1} < k_i$.

(e) Show that $t_i^{2^{k_{i+1}-1}} \equiv -1$.

Let

$$b_i \equiv c_i^{2^{k_i - k_{i+1} - 1}}$$

and define

$$c_{i+1} \equiv b_i^2, \quad r_{i+1} \equiv b_i r_i, \quad t_{i+1} \equiv b_i^2 t_i.$$

(f) Show that $c_{i+1}^{2^{k_{i+1}-1}} \equiv -1$.

(g) Show that $r_{i+1}^2 \equiv at_{i+1}$.

(h) Show that $t_{i+1}^{2^{k_{i+1}-1}} \equiv 1$.

Therefore, our new 4-tuple satisfies (13.1) with $i + 1$ in place of $i$.

This process can continue as long as $k_{i+1} \geq 2$ and $t_{i+1} \neq 1$. Since we have $k_0 > k_1 > \cdots \geq 1$, we eventually must stop, which means that, for some $j$, we have either $k_j = 1$ or $t_j = 1$. However, if $k_j = 1$ then

$$t_j \equiv t_j^{2^{k_j-1}} \equiv 1.$$

Therefore, in both cases we have $t_j = 1$, so $r_j^2 \equiv a \pmod{p}$. Therefore, $r_j$ is a square root of $a$ mod $p$.

3. Let $n$ be a positive odd integer. We know from Proposition 13.3 that if $n$ is prime and $\gcd(a, n) = 1$ then

$$a^{(n-1)/2} \equiv \left(\frac{a}{n}\right) \pmod{n}. \tag{13.2}$$

If $n$ is composite and (13.2) holds for $a$ (where the right-hand side is the Jacobi symbol), then $n$ is called an *Euler–Jacobi pseudoprime* for the base $a$.

(a) Show that if $n$ is an Euler–Jacobi pseudoprime for the base $a$ then $n$ is an $a$-pseudoprime.

(b) Show that 561 is an Euler–Jacobi pseudoprime for the base 2 but not for the base 3.

(c) Show that 1729 is an Euler–Jacobi pseudoprime for the bases 2 and 3.

(d) Find a base for which 1729 is not an Euler–Jacobi pseudoprime.

The *Solovay–Strassen Primality Test* consists of choosing several random bases $a$ and checking whether $n$ satisfies (13.2).

## 13.7.3   Answers to "CHECK YOUR UNDERSTANDING"

1. (a) $4^2 \equiv 3 \pmod{13}$, so $\left(\frac{3}{13}\right) = 1$.
   (b) $\left(\frac{3}{13}\right) = \left(\frac{13}{3}\right) = \left(\frac{1}{3}\right) = 1$.

2. Since $19 \equiv 3 \pmod 4$, we compute $5^{(p+1)/2} \equiv 5^5 \equiv 3125 \equiv 9 \pmod{19}$. The solutions are $y \equiv \pm 9 \pmod{19}$.

3. The quadratic formula says that

$$x \equiv \frac{-3 \pm \sqrt{3^2 + 8}}{2} \equiv \frac{-3 \pm \sqrt{4}}{2}$$
$$\equiv -1/2, -5/2 \equiv 6, 4 \pmod{13}.$$

4.

$$\left(\frac{501}{1013}\right) = \left(\frac{1013}{501}\right) = \left(\frac{11}{501}\right) = \left(\frac{501}{11}\right)$$
$$= \left(\frac{6}{11}\right) = \left(\frac{2}{11}\right)\left(\frac{3}{11}\right) = (-1)(-1)\left(\frac{11}{3}\right)$$
$$= \left(\frac{2}{3}\right) = -1.$$

5. $\left(\frac{3}{35}\right) = -\left(\frac{35}{3}\right) = -\left(\frac{2}{3}\right) = 1$. However, $\left(\frac{3}{5}\right) = -1$, which means that $x^2 \equiv 3 \pmod 5$ has no solutions, and therefore $x^2 \equiv 3 \pmod{35}$ has no solutions.

# Chapter 14

# Primality and Factorization

Two fundamental computational problems in number theory are finding large primes and factoring composite numbers. Although these problems are related, they have intrinsic differences. It is relatively easy to prove that a number is composite but, once this has been done, it is much harder to find its factors. With the popularity of cryptographic algorithms that use primes, primality testing and factoring have become very important. In this chapter, we show how to decide whether a number is prime or composite. Then we discuss some factorization methods that are used in practice.

## 14.1 Trial Division and Fermat Factorization

The easiest method for primality testing and factorization is trial division. If we want to factor an integer $n > 1$, we divide $n$ by all the primes $p \leq \sqrt{n}$ until we find a prime factor. If we don't find a factor, $n$ is prime. If we find a factor $p_1$, then we continue to try to factor $n_1 = n/p_1$.

We could start again at the beginning, trying to divide $2, 3, 5, \cdots$ into $n_1$. But we have already taken care of the primes less than $p_1$, so we can start at $p_1$ and continue until either we get to $\sqrt{n_1}$ (by Proposition 2.7) or we find another prime factor $p_2$. Then we try to factor $n_2 = n_1/p_2$. This process eventually terminates and yields the factorization of $n$.

**Example.** Let $n = 43771$. It is not divisible by 2, 3, and 5, but it is

divisible by 7:
$$43771 = 7 \times 6253.$$

Now we divide 7, 11, 13 into 6253 and find that

$$6253 = 13 \times 481.$$

Next, we try to factor 481, starting with 13. In fact,

$$481 = 13 \times 37.$$

Finally, we try to factor 37. Again, we start at 13, but now we are past $\sqrt{37} \approx 6.08$, so we are done. The factorization is

$$6253 = 7 \times 13^2 \times 37.$$

**Example.** Let $n = 401$. Dividing 401 by the primes 2, 3, 5, 7, 11, 13, 17, 19, we find that 401 has no prime factors less than or equal to $\sqrt{401} \approx 20.02$, so 401 is prime.

Doing these divisions is rather time consuming and this method is impractical for large values of $n$. However, there are a few tricks that make the calculations faster. For example, suppose we want to check that $19 \nmid 401$. If you notice that $401 = 420 - 19$, you can use Corollary 2.4 to see that

$$19 \nmid 401 \iff 19 \nmid 420 \iff 19 \nmid 42 \times 10.$$

But 19 is prime and $19 \nmid 10$, so we have to check only that $19 \nmid 42$, which is easy to do. Therefore, $19 \nmid 401$.

A slight extension of this technique handles other situations. For example, to check whether 7 divides 401, we observe that $401 - 21 = 380$, so

$$7 \nmid 401 \iff 7 \nmid 380 \iff 7 \nmid 38 \cdot 10 \iff 7 \nmid 38,$$

and this last fact is easily checked. With a little practice, this type of technique lets you check for factors of 3-digit and 4-digit numbers in your head. If you get bored while driving, this lets you check the license plate number ahead of you for primality.

Trial division is good for searching for small prime factors, but cryptographic applications often use products of large primes. Suppose $n = pq$

and both $p$ and $q$ are primes of 100 digits. The Prime Number Theorem (see Theorem 20.7) says that there are approximately $4 \times 10^{97}$
primes of 100 digits, so trial division might take a long time. For example, if you have a computer that can do $10^{10}$ divisions per second, $10^{97}$
trial divisions take $10^{87}$ seconds, which is approximately $3 \times 10^{79}$ years.
Therefore, better factorization methods are needed. Some of these will
be discussed in later sections.

A factorization method that does not use trial division is due to Fermat
and uses the identity

$$x^2 - y^2 = (x + y)(x - y).$$

As an example, suppose you want to factor $n = 247$. Observe that
$247 = 256 - 9 = 16^2 - 3^2$, so

$$247 = 16^2 - 3^2 = (16 + 3)(16 - 3) = 19 \times 13.$$

There are two ways to use this method in general to factor an integer
$n$, and it is instructive to compare them. One way is to look at $n + 1$,
$n + 4$, $n + 9$, etc., until you find a square:

$$n + k^2 = s^2.$$

Then $n = (s + k)(s - k)$. The method works well when there is a factor
of $n$ that is very close to $\sqrt{n}$. But is it less effective in other cases. For
example, if we want to factor 43771, it takes 45 steps to get to

$$43771 + 45^2 = 45796 = 214^2,$$

which yields the factorization

$$43771 = 259 \times 169$$

(this is not the complete factorization since $43771 = 7 \times 37 \times 13^2$). This
is slower than the trial division used above; however, trial division is
slower on $n = 247$, so the speed depends on the nature of the prime
factors.

There is a better way, and this is what we'll call the Fermat Factorization Method. When writing $n = x^2 - y^2$, instead of running through
values of $y$, use $x$. For example, if we want to write $247 = x^2 - y^2$, then
we need $x \geq \sqrt{247} \approx 15.7$, so the first possible $x$ is 16. We check that

$$16^2 - 247 = 9 = 3^2,$$

so we have $247 = 16^2 - 3^2 = 19 \times 13$. In general, we can start with $x = x_0 = \lceil \sqrt{n} \rceil$, namely the largest integer greater than or equal to $\sqrt{n}$. We then check $x_1 = x_0 + 1, x_2 = x_0 + 2, \ldots$ until we find an $x_j$ such that $x_j^2 - n$ is a square.

Let's try this speed-up on $n = 43771$. Since $\sqrt{43771} \approx 209.2$, we start with $x_0 = 210$. Since $210^2 - 43771 = 329$, which is not a square, we move to $x_1 = 211$ and compute $211^2 - 43771 = 750$, which is not a square. Continuing, we soon get to $214^2 - 43771 = 2025 = 45^2$. This yields

$$43771 = 214^2 - 45^2 = 259 \times 169.$$

This improved method took 5 steps, compared to the 45 steps in the previous version. The reason is that the squares larger than 43771 are much farther apart than the small squares. Therefore, going from one value of $x$ to the next bypasses several values of $y$. However, even this better method is too slow to be useful for large numbers.

When the RSA Algorithm was invented, it was suggested that the two primes, $p$ and $q$, should be chosen of slightly different sizes, say 100 digits and 104 digits. The idea was to keep Fermat Factorization from factoring $n = pq$ and breaking the system. However, even if $p$ and $q$ are of the same length, it is extremely unlikely that Fermat Factorization factors $n$. For example, suppose $p$ and $q$ are each 100 digits long and that they have the same initial 20 digits (very unlikely, unless your random number generator is malfunctioning), but the last 80 digits differ. Then the Fermat Factorization would take around $10^{60}$ steps (Exercise 8). In other words, Fermat Factorization has limited applicability with large numbers.

---

## CHECK YOUR UNDERSTANDING:

1. Factor 851 by the Fermat Factorization method by considering (a) $851 + y^2$ for $y = 1, 2, 3, \ldots$ and (b) $x^2 - 851$ for $x > \sqrt{851}$.

# 14.2  Primality Testing

Suppose that you need a 100-digit prime. The Prime Number Theorem (Theorem 20.7) says that the density of primes near a large number $x$ is approximately $1/\ln x$, so the chance that a random 100-digit number is prime is around

$$\frac{1}{\ln(10^{100})} \approx \frac{1}{230}.$$

If we look only at odd numbers, then the chances are 1 in 115 that a number is prime. A good strategy is to look at the odd numbers, starting at a random starting point, and test each one. Since we expect to encounter many more composites that primes, we need a fast way to recognize composites. Fermat's theorem (Theorem 8.1) comes to the rescue.

## 14.2.1  Pseudoprimes

**Proposition 14.1.** *(Fermat Primality Test) Let $n > 1$ be an odd integer. Choose an integer $b$ with $1 < b < n$. (Often, we choose $b = 2$.) If $b^{n-1} \not\equiv 1 \pmod{n}$ then $n$ is composite.*

*Proof.* If $n$ is prime, then Fermat's theorem says that $b^{n-1} \equiv 1 \pmod{n}$, which we have assumed does not happen. Therefore, $n$ cannot be prime.                                                                                      □

**Example.** Let's show that 209 is not prime. We start by noting that $2^4 \equiv 16 \pmod{209}$. By successive squaring, we have

$$
\begin{aligned}
2^8 &\equiv 16^2 \equiv 47 &&\pmod{209} \\
2^{16} &\equiv 47^2 \equiv 119 &&\pmod{209} \\
2^{32} &\equiv 119^2 \equiv 158 &&\pmod{209} \\
2^{64} &\equiv 158^2 \equiv 93 &&\pmod{209} \\
2^{128} &\equiv 93^2 \equiv 80 &&\pmod{209}.
\end{aligned}
$$

Therefore,

$$2^{208} \equiv 2^{128} \cdot 2^{64} \cdot 2^{16} \equiv 80 \cdot 93 \cdot 119 \equiv 36 \not\equiv 1 \pmod{209}.$$

By Proposition 14.1, we see that 209 is composite.

The fact that we showed 209 is composite without factoring it or finding any prime divisors is the key to why this method is fast. Generally, for large numbers, factorization is much slower than modular exponentiation. Let's use the Fermat Test to find the first prime after $10^{22}$. We'll do so by using 14.1 with $b = 2$ to test $n = 10^{22} + k$ for primality for $k = 1, 3, 7,$ and 9. (If $k = 2, 4$ or 8, then $n$ is even and if $k = 5$, then $n$ is divisible by 5.) We calculate (with a computer)

$$2^{10^{22}} \equiv 3852143514621151784591 \pmod{10^{22} + 1}$$
$$2^{10^{22}+2} \equiv 2467293414617665173233 \pmod{10^{22} + 3}$$
$$2^{10^{22}+6} \equiv 7533933929676167258456 \pmod{10^{22} + 7}$$
$$2^{10^{22}+8} \equiv 1 \pmod{10^{22} + 9}.$$

We now know that $10^{22} + k$ is composite for $k = 1, 2, 3, 4, 5, 6, 7, 8$ and we suspect that $10^{22} + 9$ is prime. In fact, it is prime; however, actually proving this takes more work. See the examples following the Pocklington–Lehmer Test (Proposition 14.6) below.

In practice, the Fermat Test is quite accurate. If it says a number $n$ is composite, then $n$ is guaranteed to be composite. However, if a number $n$ satisfies $b^{n-1} \equiv 1 \pmod{n}$ for some $b$, it is possible that $n$ is composite. The first example with $b = 2$ is $n = 341$, which was discovered in 1819 by Sarrus:

$$2^{340} \equiv 1 \pmod{341}, \quad \text{but} \quad 341 = 11 \times 31.$$

**Definition 14.2.** *A composite integer* $n > 1$ *is called a* **b-pseudoprime** *if* $b^{n-1} \equiv 1 \pmod{n}$.

For example, 341 is a 2-pseudoprime. The number 91 is a 3-pseudoprime because $3^{90} \equiv 1 \pmod{91}$ but $91 = 7 \times 13$. However, we can use the Fermat Test to check that 341 and 91 are composite, without factoring, by computing

$$3^{340} \equiv 56 \pmod{341} \quad \text{and} \quad 2^{90} \equiv 64 \pmod{91}.$$

The question now arises, if $n$ is composite, is there some $b$ such that $b^{n-1} \not\equiv 1 \pmod{n}$? The easy answer is "yes" since we can take $b$ to be a prime divisor of $n$ and then $b^{n-1}$ is 0 mod $b$, so it cannot be 1 mod $n$. We therefore ask the question with the extra condition that $\gcd(b, n) = 1$. Of course, if we happen to have a value of $b$ such that $\gcd(b, n) \neq 1$, then we have a factor of $n$, so $n$ is not prime. But we are mostly worrying about whether we can pick a few values of $b$ with the hope of finding one that will fail the Fermat Test.

**Definition 14.3.** *A composite $n > 1$ is called a* **Carmichael number** *if $b^{n-1} \equiv 1 \pmod{n}$ for all integers $b$ with $\gcd(b, n) = 1$.*

**Example.** 561 and 1729 are Carmichael numbers. Let's prove this for $561 = 3 \times 11 \times 17$. Fermat's theorem says that if $b \not\equiv 0 \pmod{17}$ then $b^{16} \equiv 1 \pmod{17}$. Therefore,

$$b^{560} \equiv \left(b^{16}\right)^{35} \equiv (1)^{35} \equiv 1 \pmod{17}.$$

Similarly, when $b \not\equiv 0 \pmod{11}$, Fermat's theorem tells us that $b^{10} \equiv 1 \pmod{11}$, so

$$b^{560} \equiv \left(b^{10}\right)^{56} \equiv (1)^{56} \equiv 1 \pmod{11}.$$

Finally, when $b \not\equiv 0 \pmod{3}$, Fermat tells us that $b^2 \equiv 1 \pmod{3}$, which implies that $b^{560} \equiv 1 \pmod{3}$. We now have, whenever $\gcd(b, 561) = 1$, that $b^{560} - 1$ is a multiple of 3, 11, and 17. Therefore, it is a multiple of $561 = 3 \times 11 \times 17$. This is exactly the statement that

$$b^{560} \equiv 1 \pmod{561}$$

for all integers $b$ with $\gcd(b, 561) = 1$, so 561 is a Carmichael number. The proof for $1729 = 7 \times 13 \times 19$ is similar.

In 1994, Alford, Granville, and Pomerance proved that there are infinitely many Carmichael numbers. However, these numbers are fairly rare. Pinch calculated that there are 20138200 Carmichael numbers less than $10^{21}$. That seems like a lot, but it means that the chances are only 1 in $5 \times 10^{13}$ that a randomly chosen integer of this size is a Carmichael number.

There is an improvement of the Fermat Primality Test, and it comes with no extra effort. If a number $n$ is a Carmichael number, the test often can be used to verify that $n$ is not a prime.

**Proposition 14.4.** *(Strong Fermat Test) Let $n > 1$ be an odd integer. Write $n - 1 = 2^k s$ with $s$ odd and $k \geq 1$. Choose an integer $b$ with $1 < b < n$. (Again, we often use $b = 2$.) Define*

$$b_0 \equiv b^s \pmod{n}$$
$$b_1 \equiv b_0^2 \pmod{n}$$
$$b_2 \equiv b_1^2 \pmod{n}$$
$$\vdots \quad \vdots$$
$$b_k \equiv b_{k-1}^2 \pmod{n}.$$

*Suppose $n$ is prime. Then $b_k \equiv 1 \pmod{n}$, and if $j$ is the smallest index for which $b_j \equiv 1 \pmod{n}$, then either $j = 0$ or $b_{j-1} \equiv -1 \pmod{n}$.*

*Therefore, if $b_k \not\equiv 1 \pmod{n}$, or if $b_{j-1} \not\equiv -1 \pmod{n}$, then $n$ is not prime.*

*Proof.* Since $n$ is prime, Fermat's theorem says that $b^{n-1} \equiv 1 \pmod{n}$. Since

$$b_k \equiv (b^s)^{2^k} \equiv b^{n-1} \pmod{n},$$

we have $b_k \equiv 1 \pmod{n}$. If $j > 0$, then, by the choice of $j$, $b_{j-1}^2 \equiv 1 \pmod{n}$. Since $n$ is prime, the only solutions to $x^2 \equiv 1 \pmod{n}$ are $x \equiv \pm 1$. Since $b_j$ is the first that is 1 mod $n$, we must have $b_{j-1} \equiv -1 \pmod{n}$. $\qquad\square$

**Remark:** As in the Fermat Test, this test does not *prove* primality, but it can be used to prove that a number is composite. If $b_k \equiv 1 \pmod{n}$ and $b_{j-1} \equiv -1 \pmod{n}$, then we say that $n$ passes the Strong Fermat Test. In practice, an integer that passes the test is very likely prime.

**Example.** Let $n = 41$ and $b = 2$. Then $n - 1 = 2^3 \cdot 5$, so $s = 5$ and $k = 3$. We compute

$$b_0 = 32, \quad b_1 \equiv 32^2 \equiv 40, \quad b_2 \equiv 40^2 \equiv 1, \quad b_3 \equiv 1^2 \equiv 1 \pmod{41}.$$

Note that $b_k = b_3 = 1$. The smallest index with $b_j = 1$ is $j = 2$. Then $b_{j-1} = b_1 = 40 \equiv -1 \pmod{41}$. We conclude that 41 is probably prime.

**Example.** Let $n = 91$ and $b = 3$. Then $90 = 2^1 \cdot 45$, so $s = 45$ and $k = 1$. We have

$$b_0 \equiv 3^{45} \equiv 27, \quad b_1 \equiv 27^2 \equiv 1 \pmod{91}.$$

The fact that $b_1 = 1$ should not be surprising: we saw earlier that 91 is a 3-pseudoprime. However, note that $b_{i-1} = b_0 = 27 \not\equiv -1 \pmod{91}$. Therefore, the Strong Fermat Test says that 91 cannot be prime, since a prime would require $b_0 \equiv -1$ in this situation.

As we see in the second example, by looking at the intermediate computations, the Strong Fermat Test sometimes finds composites when the Fermat Test does not. The calculation of $b_0, b_1, b_2, \ldots$ can be done as part of the successive squaring used to calculate $b^{n-1} \pmod{n}$. Moreover, as soon as we find some $b_i = 1$, we can stop and look at the previous step to see if $b_{i-1} \equiv -1$, so we might not need to do the last few squarings. However, in practice, when $n$ is composite we find that $b_k \neq 1$, so the simple Fermat Test suffices. The advantage of the improved algorithm is that it lets very few composites slip through as possible primes. Moreover, as we'll see in the next subsection, a $b$-pseudoprime that fails the Strong Fermat Test can easily be factored.

**Definition 14.5.** *A composite integer* $n > 1$ *is a* **strong** $b$-**pseudoprime** *if* $n$ *passes the Strong Fermat Test for base* $b$. *That is, in the notation of Proposition 14.4,* $b_k \equiv 1 \pmod{n}$, *and if* $j$ *is the smallest index for which* $b_j \equiv 1 \pmod{n}$, *then either* $j = 0$ *or* $b_{j-1} \equiv -1 \pmod{n}$.

**Example.** Let's show that $4033 = 37 \times 109$ is a strong 2-pseudoprime. Write $4032 = 2^6 \cdot 63$, so $k = 6$ and $s = 63$. Then

$$b_0 \equiv 2^{63} \equiv 3521 \pmod{4033}$$
$$b_1 \equiv 3521^2 \equiv -1 \pmod{4033}$$
$$b_2 \equiv (-1)^2 \equiv 1 \pmod{4033}.$$

Clearly, $b_3, b_4, b_5, b_6$ are also 1 mod 4033. Since $b_2 \equiv 1$ and the previous $b_1 \equiv -1$, we see that 4033 is a strong 2-pseudoprime.

**Example.** Let's show that $121 = 11^2$ is a strong 3-pseudoprime. Write $120 = 2^3 \cdot 15$, so $k = 3$ and $s = 15$. Then

$$b_0 \equiv 3^{15} \equiv 1 \pmod{121}.$$

We stop here, since $b_j \equiv 1 \pmod{121}$ for all $j$, so $b_3 \equiv 1$. In the notation of the definition, we have $i = 0$, which is one of the allowed possibilities for a strong pseudoprime.

Strong pseudoprimes are rare. If you look at integers less than $10^8$, 488 of them are strong 2-pseudoprimes, 582 are strong 3-pseudoprimes, and 475 are strong 5-pseudoprimes. Of the 488 strong 2-pseudoprimes, 21 are *also* strong 3-pseudoprimes, and of these there is only one strong 5-pseudoprime: 25326001. We can quickly check that 25326001 is *not* a strong 7-pseudoprime. In fact, computations of Pomerance, Selfridge, and Wagstaff show that if $n < 25 \times 10^9$ is composite, it cannot be a strong pseudoprime for all five of 2, 3, 5, 7, 11. This leads to a simple primality test: Given an integer $n < 25 \times 10^9$, check that it passes the Strong Fermat Test for $b = 2, 3, 5, 7, 11$. If so, then it is prime.

In practice, if $n$ is composite, it is easy to find a $b$ such that $n$ is not a strong $b$-pseudoprime. Gary Miller proved under the assumption of the Generalized Riemann Hypothesis (an extension of the Riemann Hypothesis; see Section 5.7), that if $n$ is composite then there is a $b < 2(\ln n)^2$ such that $n$ is not a strong $b$-pseudoprime (the constant "2" in the formula is due to Eric Bach). Michael Rabin showed that there are at least $3(n-1)/4$ values of $b < n$ such that $n$ is not a strong $b$-pseudoprime. This leads to a *probabilistic* primality test, known as the **Miller–Rabin Primality Test**: Test $n$ using $m$ randomly chosen values of $b < n$. If $n$ is composite, the probability that it is a strong pseudoprime for one $b$ is at most $1/4$, so the probability that it passes all $m$ tests is at most $1/4^m$. Therefore, if $n$ passes all $m$ tests, then $n$ is prime with probability at least $1 - (1/4)^m$. We say that $n$ is *probably prime*. This is often how primes are chosen for cryptographic algorithms.

## 14.2.2    The Pocklington–Lehmer Primality Test

Suppose that $n$ passes the Strong Fermat Test for several values of $b$, so we suspect that $n$ is prime. If we want to *prove* that $n$ is actually prime, we can do so by partially factoring $n - 1$.

**Proposition 14.6.** *(Pocklington–Lehmer) Let* $n - 1 = FN$, *where* $\gcd(F, N) = 1$ *(where $F$ is for "factored" and $N$ is for "not necessarily* *factored"). Suppose that there exists b such that*

$$b^{n-1} \equiv 1 \pmod{n} \tag{14.1}$$

*and*

$$\gcd(b^{(n-1)/q} - 1, \ n) = 1 \tag{14.2}$$

*for all primes q dividing F. Then each prime factor p of n satisfies*
$p \equiv 1 \pmod{F}$. *Moreover, if $F > N$, then n is prime.*

*Proof.* Let $p$ divide $n$. Since $b^{n-1} \equiv 1 \pmod{p}$, Theorem 11.1 implies that $\text{ord}_p(b) \mid n - 1$. Write

$$n - 1 = k \cdot \text{ord}_p(b).$$

Let $q$ be one of the primes dividing $F$. If $q \mid k$, then

$$(n - 1)/q = (k/q)\text{ord}_p(b)$$

is a multiple of $\text{ord}_p(b)$. Therefore,

$$b^{(n-1)/q} = \left(b^{\text{ord}_p(b)}\right)^{k/q} \equiv (1)^{k/q} \equiv 1 \pmod{p}.$$

This contradicts the assumption that

$$\gcd(b^{(n-1)/q} - 1, \ n) = 1$$

since $p$ divides the gcd. Therefore, $q \nmid k$. This happens for each prime dividing $F$, so $\gcd(F, k) = 1$. Since

$$FN = n - 1 = k \cdot \text{ord}_p(b),$$

Proposition 2.16 implies that $F \mid \text{ord}_p(b)$. Theorem 11.1 says that $\text{ord}_p(b) \mid p - 1$, so $F \mid p - 1$, as claimed.

If $F > N$ then $F^2 > FN = n - 1$, so $F^2 \geq n$. If $F \mid p - 1$ then $p > F \geq \sqrt{n}$. Moreover, Proposition 2.7 says that if $n$ is composite then it has a prime factor $p \leq \sqrt{n}$. Since all prime factors are greater than $\sqrt{n}$, we see that $n$ cannot be composite, so $n$ must be prime.  $\square$

**Example.** Let's prove that $n = 9473$ is prime. We can factor

$$9472 = 2^8 \cdot 37$$

so we take $F = 2^8 = 256$ and $N = 37$. We compute $2^{(n-1)/2} \equiv 1 \pmod{n}$, so $b = 2$ does not satisfy the condition (14.2) of the Pocklington–Lehmer Test. However, $3^{n-1} \equiv 1 \pmod{n}$ and $3^{(n-1)/2} \equiv -1 \pmod{n}$, so

$$\gcd(3^{(n-1)/2} - 1, n) = \gcd(-2, n) = 1.$$

Therefore, $b = 3$ satisfies conditions (14.1) and (14.2) for $q = 2$. Since 2 is the only prime dividing $F$, the assumptions are satisfied. Moreover, $F > N$. Therefore $n$ is prime.

**Example.** Let's show that $n = 10^{22} + 9$ is prime. We can factor

$$n - 1 = \left(2^3 \cdot 3^2 \cdot 17 \cdot 8747 \cdot 11161\right) (83686681651),$$

so we let $F = 2^3 \cdot 3^2 \cdot 17 \cdot 8747 \cdot 11161$ and $N = 83686681651$. Calculations show that

$$7^{n-1} \equiv 1 \pmod{n}$$

and that

$$\gcd(7^{(n-1)/q} - 1, n) = 1$$

for $q = 2, 3, 17, 8747, 11161$. The assumptions of the Pocklington–Lehmer Test are met. Since $F > N$, we conclude that $n$ is prime.

If most of the factorization of $n - 1$ is not known (so that we can't say that $F > N$), the Pocklington–Lehmer Test does not prove $n$ is prime. In this case, more sophisticated methods are used. These are beyond the scope of this book.

---

**CHECK YOUR UNDERSTANDING:**

2. You want to use the Pocklington–Lehmer Test to prove 23 is prime. You calculate that $2^{22} \equiv 1 \pmod{23}$. What other calculation do you need to do to prove that 23 is prime? What are the values of $F$ and $N$ that you use?

---

## 14.2.3    The AKS Primality Test

For many years, there has been a search for primality testing algorithms that are fast, either in practice or in theory. On the theoretical front, in 2002 Agrawal, Kayal, and Saxena published what is known as the **AKS Primality Test**. We start with an integer $n > 1$ and do the following steps:

1. Check that $n$ is not a perfect power by computing $n^{1/k}$ for $2 \leq k \leq \log_2(n)$. If $n$ is a perfect power, then $n$ is composite.

2. Compute the smallest integer $r$ such that $\mathrm{ord}_r(n) > (\log_2(n))^2$.

3. If $\gcd(a, n) \neq 1, n$ for some $a$ with $2 \leq a \leq r$, then $n$ is composite.

4. If $n$ has passed test (3) and $n \leq r$, then $n$ is prime (because $n$ is relatively prime to all $a < n$).

5. If the algorithm has not stopped at steps (1) through (4), check that

$$(X + b)^n \equiv X^n + b \pmod{X^r - 1, n}$$

for all $b$ with $1 \leq b \leq \sqrt{\phi(r)} \log_2(n)$. If this is the case, $n$ is prime.

Step (5) requires some explanation. The congruence $(X + b)^n \equiv X^n + b$ $\pmod{X^r - 1, n}$ is a congruence involving polynomials. It means that when we take the difference $(X + b)^n - (X^n + b)$ and divide by $X^r - 1$, the remainder has all of its coefficients divisible by $n$.

The AKS Primality Test is not yet fast enough to beat the best current tests. However, it can be proved that, for integers with $m$ digits, the running time of the algorithm is bounded by a polynomial in $m$. In fact, on numbers of $m$ digits, the running time was first proved to be at most a constant times $m^{13}$ (actually, $m^{12}$ times a power of $\ln m$), and subsequent improvements have reduced the exponent. AKS is the first algorithm that works on all integers and is guaranteed to prove primality in polynomial time (that is, the running time is bounded by a polynomial in the number of digits). Let's talk about this a little.

1. There are algorithms for special numbers (for example, for

Fermat and Mersenne numbers). We'll talk about some of these in the next few pages. But these do not apply to arbitrary $n$.

2. There are algorithms that require unproved hypotheses. For example, as mentioned earlier, Miller showed, under the assumption of the Generalized Riemann Hypothesis, that if $n$ is composite then $n$ is not a strong $b$-pseudoprime for some $b < 2\,(\ln n)^2$. This means that we can test $n$ using each $b$ in that range and conclude that either $n$ is prime or the Generalized Riemann Hypothesis is false. It is generally expected that the GRH is true, so that Miller's result gives a primality test that runs in polynomial time.

3. The Miller–Rabin Test can be run in polynomial time, but it gives only a probabilistic answer, not a definitive answer.

4. Some of the best current algorithms are only slightly slower than polynomial time, but are faster than AKS in practice. How is this possible? For example, if one algorithm takes $2^m$ steps and a second algorithm takes $m^{100}$ steps, then the second is polynomial time, but the first is faster for $m \leq 996$.

**Example.** Let's prove that $n = 37$ is prime. To begin, we need to check $k$th roots of 37 for $k$ up to $\log_2 37$. Since $37^{1/2}$, $37^{1/3}$, $37^{1/5}$ are not integers, we know that 37 is not a perfect power. A calculation yields

$$\text{ord}_{29}(37) = 28 > (\log_2(37))^2 \approx 27.14,$$

so we have $r = 29$. We certainly have $\gcd(a, 37) = 1$ for $2 \leq a \leq r = 29$, so 37 passes step (3). We skip step (4) because $n = 37 > 29 = r$. Finally, we need to check that

$$(X + b)^{37} \equiv X^{37} + b \pmod{X^{29} - 1, 37}$$

for $1 \leq b \leq 27.6 = \sqrt{\phi(29)}\,\log_2(37)$. Let's do this for $b = 1$. We have

$$(X + 1)^{37} - (X^{37} + 1) = (X^{29} - 1)q(x) + r(X),$$

where $q(X)$ is a polynomial and

$$r(X) = 124403620 X^{28} + \cdots + 38608020$$

is the remainder. All of the coefficients of $r(X)$ are divisible by 37, so

$$(X + 1)^{37} \equiv X^{37} + 1 \pmod{X^{29} - 1, 37}.$$

We can do a similar calculation for each $b \leq 27$ and thus complete the proof that 37 is prime.

There is a simple reason that

$$(X + b)^{37} \equiv X^{37} + b \pmod{X^{29} - 1, 37}$$

for each $b$. By Lemma 8.3, each binomial coefficient $\binom{37}{j}$ is a multiple of 37, so

$$(X + b)^{37} = X^{37} + \binom{37}{1} X^{36} b + \binom{37}{2} X^{35} b^2 + \cdots + b^{37}$$

$$\equiv X^{37} + b^{37} \equiv X^{37} + b \pmod{37}.$$

Therefore, working mod 37 is sufficient; we didn't need to work mod $X^{29} - 1$. The same is true when $n$ is any prime. However, the key idea of the AKS Algorithm is that the use of mod $X^r - 1$ suffices to catch composites and it cuts down on the computations needed.

The justification that the AKS Algorithm works is slightly beyond the scope of this book.

## 14.2.4   Fermat Numbers

For numbers of special forms, there are often special primality tests that are more efficient than the general purpose ones that apply to all numbers. In this subsection and the next, we describe tests that apply to Fermat numbers and Mersenne numbers.

Recall that the $n$th Fermat number is

$$F_n = 2^{2^n} + 1.$$

There is an easy test for the primality of these numbers. It was proved by the French mathematician Pépin in 1877 using 5 in place of 3 in the following proposition. The next year, Proth pointed out that 3 could be used, as in the form stated here.

**Proposition 14.7.** *(Pépin's Test) Let $n \geq 1$. The Fermat number $F_n$ is prime if and only if*

$$3^{(F_n - 1)/2} \equiv -1 \pmod{F_n}.$$

*Proof.* Suppose $F_n$ is prime. Since $2^2 \equiv 1 \pmod 3$, we see that 2 raised to any even exponent is 1 mod 3. Therefore,

$$F_n = 2^{2^n} + 1 \equiv 1 + 1 \equiv 2 \pmod 3.$$

At this point, we need results from Chapter 13: Proposition 13.3 and Quadratic Reciprocity (Theorem 13.4) tell us that

$$3^{(F_n-1)/2} \equiv \left(\frac{3}{F_n}\right) = \left(\frac{F_n}{3}\right) = \left(\frac{2}{3}\right) \equiv -1 \pmod{F_n}.$$

Note that the first congruence is where we used the assumption that $F_n$ is prime.

Conversely, suppose that $3^{(F_n-1)/2} \equiv -1 \pmod{F_n}$. In the notation of the Pocklington–Lehmer Test (Proposition 14.6), we can write

$$F_n - 1 = 2^{2^n} = FN,$$

where $F = 2^{2^n}$ and $N = 1$. The only prime $q$ dividing $F$ is $q = 2$. Since

$$3^{(F_n-1)/2} \equiv -1 \pmod{F_n}$$

by hypothesis, we have

$$3^{F_n-1} \equiv 1 \pmod{F_n}$$

and

$$\gcd(3^{(F_n-1)/2} - 1, F_n) = \gcd(-2, F_n) = 1.$$

The Pocklington–Lehmer Test therefore implies that $F_n$ is prime.    □

**Example.** Let $n = 1$, so $F_1 = 2^{2^1} + 1 = 5$. We have $3^2 = 9 \equiv -1$ (mod 5), so 5 is prime.

**Example.** Let $n = 5$, so $F_5 = 2^{32} + 1 = 4294967297$. A computation yields

$$3^{(F_5-1)/2} \equiv 3^{2^{31}} \equiv 10324303 \pmod{4294967297}.$$

Therefore, $F_5$ is composite. We already know this from Section 2.9, where we gave the factorization of $F_5$. In the spirit of primality tests, the present result shows that $F_5$ is composite without factoring it.

## 14.2.5  Mersenne Numbers

Recall that when $p$ is prime, the $p$th Mersenne number is

$$M_p = 2^p - 1.$$

This sequence of numbers has yielded many large primes. In fact, at most times in history, the largest known prime has been a Mersenne prime, partly because there is a fast test for primality of these numbers.

**Proposition 14.8.** *(Lucas–Lehmer Test) Let $p$ be prime. Define a sequence by*

$$s_0 = 4, \quad s_{i+1} = s_i^2 - 2.$$

*Then*

$$M_p \text{ is prime} \iff s_{p-2} \equiv 0 \pmod{M_p}.$$

We prove this theorem in Chapter 19 using ideas from algebraic number theory. The Great Internet Mersenne Prime Search (see Section 2.9) uses the Lucas–Lehmer method to search for very large Mersenne primes.

**Example.** Let $p = 7$. Then $M_p = 2^7 - 1 = 127$. We calculate

$$s_0 = 4, \ s_1 = 14, \ s_2 = 194 \equiv 67, \ s_3 \equiv 42, \ s_4 \equiv 111, \ s_5 \equiv 0 \pmod{127}.$$

Therefore, $M_7$ is prime.

**Example.** Let $p = 11$. Then $M_{11} = 2047$. A calculation yields

$$s_9 \equiv 1736 \pmod{2047}.$$

Therefore, $2^{11} - 1$ is not prime. In fact, $2047 = 23 \times 89$.

---

# 14.3  Factorization

Suppose we have an integer $n$ and we know, because of some primality test, that $n$ is composite. How do we find the factors? In general, this is quite hard. In this section we look at four methods that are used in practice.

# 14.3.1    $x^2 \equiv y^2$

Fermat factorization relies on writing $n$ as a difference of two squares. A basic idea that underlies many factoring techniques is to generalize Fermat's idea and write a multiple of $n$ as a difference of two squares. In other words, find an $x$ and $y$ with $x^2 \equiv y^2 \pmod{n}$. Of course this is easy to do in a trivial way: just take $x = \pm y \pmod{n}$. But, as the next result shows, if we can obtain $x^2 \equiv y^2 \pmod{n}$ nontrivially, then we can factor $n$.

**Proposition 14.9.** *(Basic Factorization Principle) Let $n$ be a positive integer and let $x$ and $y$ be integers with*

$$x^2 \equiv y^2 \pmod{n} \quad but \quad x \not\equiv \pm y \pmod{n}.$$

*Then $\gcd(x - y, n)$ is a nontrivial factor of $n$ (that is, the gcd is not 1 or $n$).*

*Proof.* Of course, $\gcd(x - y, n)$ is a factor of $n$. We need to show that the gcd is not 1 or $n$. If $\gcd(x - y, n) = n$, then $n \mid x - y$, which means that $x \equiv y \pmod{n}$, contrary to assumption. Now suppose that $\gcd(x - y, n) = 1$. Since $x^2 \equiv y^2 \pmod{n}$,

$$(x - y)(x + y) = x^2 - y^2 \equiv 0 \pmod{n}.$$

But $\gcd(x - y, n) = 1$, so we can divide the congruence by $x - y$ and obtain $x + y \equiv 0 \pmod{n}$, which is the same as $x \equiv -y \pmod{n}$. This contradicts the assumptions. Therefore, the gcd is not 1 or $n$, so it is a nontrivial factor of $n$.                                                    $\square$

**Example.** We have $9^2 \equiv 2^2 \pmod{77}$, but $9 \not\equiv \pm 2 \pmod{77}$. Therefore,

$$\gcd(9 - 2, 77) = 7$$

is a nontrivial factor of 77.

Let's analyze why the example works. We have

$$9 \equiv 2 \pmod{7} \quad \text{and} \quad 9 \equiv -2 \pmod{11}.$$

Therefore,

$$9 - 2 \equiv 0 \pmod{7} \quad \text{but} \quad 9 - 2 \equiv -4 \pmod{11}.$$

This means that 7 is a common divisor of $9 - 2$ and 77, but 11 is not a common divisor of $9 - 2$ and 77. Therefore, $\gcd(9 - 2, 77)$ finds the factor 7.

An extension of these ideas shows that finding $x$ and $y$ as in Proposition 14.9 is equivalent to factoring $n$. For example, suppose $n = pq$ is a product of two distinct odd primes and we know the factors $p$ and $q$. Choose any number $y \not\equiv 0 \bmod p$ and $\bmod q$. The Chinese Remainder Theorem says that we can solve the simultaneous congruences

$$x \equiv +y \pmod{p} \quad \text{and} \quad x \equiv -y \pmod{q}.$$

Then $x \not\equiv \pm y \pmod{n}$. However, $x^2 \equiv y^2 \pmod{p}$ and $\pmod{q}$, so

$$x^2 \equiv y^2 \pmod{pq}.$$

This means that if we know $p$ and $q$ then we can produce $x$ and $y$ as in the proposition. For example, if $n = 7 \times 11$, then we can solve

$$x \equiv 1 \pmod{7} \quad \text{and} \quad x \equiv -1 \pmod{11}$$

and obtain $x = 43$. Then $43^2 \equiv 1^2 \pmod{77}$. We can use this to recover the factorization: $\gcd(43 - 1, 77) = 7$.

Now that we have Proposition 14.9, how do we find $x$ and $y$? We could compute a lot of squares mod $n$ until we get a match, but this might take a long time. Instead, we try to find combinations that yield squares.

**Example.** Let's factor $n = 3837523$. Suppose we have discovered, by some means, that

$$17847^2 \equiv 2^3 \cdot 5^3 \quad (\bmod n)$$
$$22111^2 \equiv 2^2 \cdot 3^7 \cdot 5^2 \cdot 7$$
$$32041^2 \equiv 2^{13} \cdot 5 \cdot 7^2$$
$$44153^2 \equiv 3 \cdot 5^2 \cdot 7^3.$$

If we multiply the first and third lines, we find

$$(17847 \cdot 32041)^2 \equiv (2^8 \cdot 5^2 \cdot 7)^2 \pmod{n}.$$

However,

$$17847 \cdot 32041 \equiv 2^8 \cdot 5^2 \cdot 7 \pmod{n},$$

so this does not help us. If we multiply the second and fourth lines, we obtain

$$(22111 \cdot 44153)^2 \equiv (2 \cdot 3^4 \cdot 5^2 \cdot 7^2)^2 \pmod{n}.$$

In this case, we're luckier:

$$22111 \cdot 44153 \not\equiv \pm 2 \cdot 3^4 \cdot 5^2 \cdot 7^2 \pmod{n}$$

and

$$\gcd(22111 \cdot 44153 - 2 \cdot 3^4 \cdot 5^2 \cdot 7^2, \, n) = 3511,$$

so we have a nontrivial factor of $n$. The other factor is $n/3511 = 1093$. We can express what we have done in matrix form:

$$\begin{matrix} 17847 \\ 22111 \\ 32041 \\ 44153 \end{matrix} \begin{bmatrix} 3 & 0 & 3 & 0 \\ 2 & 7 & 2 & 1 \\ 13 & 0 & 1 & 2 \\ 0 & 1 & 2 & 3 \end{bmatrix}.$$

The entries in the columns are the exponents of the primes 2, 3, 5, 7 in the factorizations coming from the labels of the rows. We want a sum of rows to have all of its entries even. For example, the sum of the first and third rows is $[16, 0, 4, 2]$, and the sum of the second and fourth rows is $[2, 8, 4, 4]$. These correspond to the relations $x^2 \equiv y^2 \pmod{n}$ that we found.

This example suggests that if we can find a method to produce squares congruent to products of small primes, we should be able to factor quickly. In 1931, D. H. Lehmer and R. E. Powers developed a method based on continued fractions that produces such squares. This continued fraction method was used to great advantage by Brillhart and Morrison in 1975 in their factorization of the seventh Fermat number. By using factor bases and linear algebra, they laid the groundwork for the quadratic sieve factorization method, which we'll discuss in Subsection 14.3.4. Moreover, theirs was one of the first factorization projects to use the massive computing power of modern computers.

Of course, before trying more complicated methods, it's good to do trial division by several small primes. If $n$ is not divisible by any small primes, then turn to more sophisticated techniques.

## 14.3.2   Factoring Pseudoprimes and Factoring Using RSA Exponents

Suppose $n$ is a $b$-pseudoprime, so $b^{n-1} \equiv 1 \pmod{n}$. Or, more generally, we might have an even exponent $m$ such that

$$b^m \equiv 1 \pmod{n}.$$

Often, this can be used to factor $n$. The method is very similar to the Strong Fermat Test.

**Algorithm 14.10.** *Suppose that we have an even exponent $m$ such that $b^m \equiv 1 \pmod{n}$. Write $m = 2^k s$ with $s$ odd and $k \geq 1$. Define*

$$b_0 \equiv b^s \pmod{n}$$
$$b_1 \equiv b_0^2 \pmod{n}$$
$$b_2 \equiv b_1^2 \pmod{n}$$
$$\vdots \qquad \vdots$$
$$b_k \equiv b_{k-1}^2 \pmod{n}.$$

*(Note that $b_k \equiv b^{2^k s} \equiv b^m \equiv 1 \pmod{n}$ by assumption.) Let $b_i \equiv 1$, with $i$ as small as possible. If $i \geq 1$, this means that $b_{i-1} \not\equiv 1 \pmod{n}$ and*

$$b_{i-1}^2 \equiv 1^2 \pmod{n}.$$

*If $b_{i-1} \not\equiv -1 \pmod{n}$, compute $\gcd(b_{i-1} - 1, n)$. This is a nontrivial factor of $n$.*

The justification for this algorithm is the Basic Factorization Principle (Proposition 14.9).

**Example.** Suppose we find that $2^{200} \equiv 1 \pmod{32817151}$. Write

$$200 = 2^3 \cdot 25.$$

Compute

$$b_0 \equiv 2^{25} \equiv 737281 \pmod{32817151}$$
$$b_1 \equiv b_0^2 \equiv 32800948$$
$$b_2 \equiv b_1^2 \equiv 1$$

(we don't need to compute $b_3$). We now have

$$\gcd(32800948 - 1, 32817151) = 4051,$$

which yields the factorization $32817151 = 4051 \times 8101$.

We can apply this idea to the RSA Algorithm. Let $n = pq$ be a product of two distinct primes. Suppose we know integers $e$ and $d$ such that $ed \equiv 1 \pmod{(p-1)(q-1)}$. In Algorithm 14.10, let $m = ed - 1$. Let $b \neq \pm 1, 0$ be any message. Then $b^{ed} \equiv b \pmod{n}$. If $\gcd(b, n) > 1$, we have a factor of $n$, so suppose $\gcd(b, n) = 1$. Then we can divide by $b$ and obtain $b^{ed-1} \equiv 1 \pmod{n}$. Therefore, the algorithm applies. If it fails to factor $n$, try a new $b$. We conclude that if we have some clever way of finding the decryption exponent $d$ in RSA, then we can factor $n$. Turning this around, we can say that if factorization is hard, then finding $d$ is hard.

### 14.3.3   Pollard's $p - 1$ Method

In 1974, John Pollard proposed what is known as **Pollard's $p - 1$ Method** to factor large integers. It was one of the standard factorization methods for the next 10 years, until it was superseded by quadratic methods and elliptic curve methods. In the following, we use $B!$ as the exponent. In practice, slightly different exponents are used.

Choose a number $b$ and a bound $B$. Compute $a \equiv b^{B!} \pmod{n}$ and then compute $\gcd(a - 1, n)$. If the gcd is 1, try a new value of $b$, try a larger $B$, or give up. If the gcd is $n$, use Algorithm 14.10. But if the gcd $\neq 1$ or $n$, it is a nontrivial factor of $n$.

**Example.** Let's factor $n = 3441959603$. Choose $b = 2$, let $B = 20$, and compute

$$2^{20!} = 2^{2432902008176640000} \equiv 1458887026 \pmod{n}.$$

Then

$$\gcd(1458887026 - 1, n) = 63361.$$

The factorization is $n = 63361 \times 54323$.

Let's analyze why the example works. If $p = 63361$ then $p - 1 = 2^7 \cdot 3^2 \cdot 5 \cdot 11$, which is a product of small primes. The number $B! = 20!$

contains more than 7 even numbers, so it's a multiple of $2^7$. It has more than 2 multiples of 3, so it's a multiple of $3^2$. It also contains 5 and 11. Therefore, $B!$ is a multiple of $p - 1$. Write $B! = (p-1)k$ for some $k$. Fermat's theorem tells us that

$$a \equiv 2^{B!} = \left(2^{p-1}\right)^k \equiv 1^k \equiv 1 \pmod{p}.$$

Therefore, $p$ divides $\gcd(a - 1, n)$. However, if $q = 54323$ then $q - 1 = 2 \cdot 157 \cdot 173$, so $B!$ is not a multiple of $q - 1$. Therefore, we do not expect $2^{B!} \pmod{q}$ to be anything special. In fact,

$$a \equiv 2^{B!} \equiv 42861 \pmod{q}.$$

Therefore, $q \nmid \gcd(a - 1, n)$, so this method does not pick up the factor $q$ of $n$.

When $p - 1$ divides $B!$, we have $b^{B!} \equiv 1 \pmod{p}$. Therefore, $\gcd(a - 1, n)$ includes $p$. What happens if the gcd is all of $n$? Let's look at an example.

**Example.** Let $n = 22198589711$. Let $b = 2$ and $B = 20$. Then

$$a = 2^{B!} \equiv 1 \pmod{n}.$$

But not all is lost. We can apply Algorithm 14.10. Write $B! = 2^{18} \cdot 9280784638125$ and then compute

$$b_0 \equiv 2^{9280784638125} \equiv 20518656667 \pmod{n}$$
$$b_1 \equiv b_0^2 \equiv 17303135189$$
$$b_2 \equiv b_1^2 \equiv 16792673782$$
$$b_3 \equiv b_2^2 \equiv 16081110901$$
$$b_4 \equiv b_3^2 \equiv 6773335884$$
$$b_5 \equiv b_4^2 \equiv 672323570$$
$$b_6 \equiv b_5^2 \equiv 1.$$

This yields

$$672323570^2 \equiv 1^2 \pmod{n}.$$

Then

$$\gcd(672323570 - 1, n) = 350351,$$

so $n = 22198589711 = 350351 \times 63361$. The fact that $a \equiv 2^{B!} \equiv 1$

(mod $n$) assured us that $b_j \equiv 1$ (mod $n$) for some $j \leq 18$. In the present example, we got $b_j \equiv 1$ early, namely when $j = 6$.

In the example, both $p - 1$ and $q - 1$ are products of primes less than 20, and 20! is a multiple of $p - 1$ and $q - 1$. Therefore, $2^{20!} \equiv 1$ (mod $p$) and (mod $q$). This is why the calculation of $\gcd(a - 1, n)$ yields $n$. The real difficulty is the opposite case. The method encounters problems when $p - 1$ has a large prime factor for each of the primes $p$ dividing $n$.

Of course, we could try making $B$ larger, but eventually the calculation of $b^{B!}$ (mod $n$) becomes too slow. We could try another $b$, but this usually doesn't help unless we get very lucky. In other words, the $p - 1$ method has limitations. In 1986, Hendrik Lenstra invented the elliptic curve factorization method, which uses the principles of the $p - 1$ method but has more flexibility in changing parameters after a failed attempt. It is very useful in finding prime factors of medium size, say around 20 to 25 digits.

---

**CHECK YOUR UNDERSTANDING:**

3. Suppose you know that $97^2 \equiv 5^2$ (mod 391). Use this information to factor 391.

4. Suppose you know that $119^2 \equiv 50$ (mod 14111) and $168^2 \equiv 2$ (mod 14111). Use this information to factor 14111.

---

## 14.3.4   The Quadratic Sieve

In 1981, Carl Pomerance discovered a method that substantially improved the search for squares congruent to products of small primes and thus invented the quadratic sieve. In this subsection, we present the algorithm and give an example of how it works.

We start with a composite integer $n$ that we want to factor and do the following procedure.

1. Let $s = \lceil \sqrt{n} \rceil =$ the smallest integer greater than or equal to $\sqrt{n}$. Form the polynomial

$$f(X) = (X + s)^2 - n.$$

Then $f(0) = s^2 - n$ is small in comparison to $n$. Moreover, if $0 \le j \le .4\sqrt{n} - 1$, then (since $s - 1 < \sqrt{n}$)

$$0 \le f(j) < \left(0.4\sqrt{n} - 1 + s\right)^2 - n$$
$$< \left(0.4\sqrt{n} + \sqrt{n}\right)^2 - n$$
$$= 1.96n - n$$
$$< n.$$

Therefore, for these $j$, if we compute $f(x) \pmod{n}$, we get $f(j)$ as the smallest positive residue. We are going to use a range of values $0 \le j \le A$ for some $A$ that is substantially smaller than $.4\sqrt{n}$. For each such $j$, we'll try to factor $f(j)$ using small primes.

2. Choose a bound $B$ and consider the set $\mathcal{B}$ of primes less than $B$. This set is called a **factor base**. Our goal is to factor numbers $f(j)$ using primes in $\mathcal{B}$. If $p \mid f(j)$ for some $j$, then $(j + s)^2 \equiv n \pmod{p}$, so the only primes $p$ that can occur have $n$ as a square mod $p$. Using the Legendre symbol and Quadratic Reciprocity from Chapter 13, we can decide whether $p$ satisfies this requirement. Discard the primes $p$ for which $n$ is not a square mod $p$ and obtain a new set $\mathcal{B}$. During this calculation, if we find a $p \in \mathcal{B}$ that divides $n$, we can stop because we have found a factor of $n$. Henceforth, assume that if $p \in \mathcal{B}$, then $p \nmid n$.

3. For each odd $p \in \mathcal{B}$, there are two solutions of $(X + s)^2 \equiv n \pmod{p}$, because $n$ is a square mod $p$. (If $p \mid n$, there is only one solution, but we have already excluded this case.) Call these solutions $x_{p,1}$ and $x_{p,2}$.

4. Make a list of values

$$\ln f(j) \quad \text{for} \quad 0 \le j \le A.$$

For each $p \in \mathcal{B}$, subtract $\ln p$ from the value of $\ln f(j)$ for each $j \equiv x_{p,1}$ or $x_{p,2} \pmod{p}$. If $f(j)$ is a product of distinct primes in $\mathcal{B}$, then the logarithm of each prime factor of $f(j)$ gets subtracted from $\ln f(j)$ and this entry in the list becomes 0.

5. For a few powers of each very small prime $p$ in $\mathcal{B}$, find the

roots $x_{p^i,1}$ and $x_{p^i,2}$ of $f(x) \equiv 0 \pmod{p^i}$. Subtract $\ln p$ from the values of $\ln f(j)$ for each $j \equiv x_{p^i,1}$ or $x_{p^i,2} \bmod p^i$ for each such $p$ and $i$.

6.  Look at the entries in the list that become small, say less than $\ln B$. Compute the factorizations of $f(j)$ for each of the corresponding values of $j$.

7.  Try to find products of combinations of these values that are squares. These yield relations of the form $x^2 \equiv y^2 \pmod{n}$. Compute $\gcd(x - y, n)$. If you're lucky, this yields a nontrivial factor of $n$. If not, find another relation $x^2 \equiv y^2 \pmod{n}$ and compute $\gcd(x - y, n)$ again.

**Example.** Let $n = 35947$. We start with a factor base of the primes less than 36, but then remove 5, 11, 13, 19, and 23 because $n$ is not a square mod these primes. Our factor base now has the six primes 2, 3, 7, 17, 29, 31. We have $s = \lceil \sqrt{n} \rceil = 190$. The polynomial is

$$f(X) = (X + s)^2 - n = (X + 190)^2 - 35947.$$

Make a list of $\ln f(j)$ for $0 \le j \le 100$. For brevity, we give only a few elements on the list (the other 94 elements on the list are represented by "$\cdots$").

| $j$ | 0 | 1 | 2 | 3 | 4 | $\cdots$ | 9 | $\cdots$ | 100 |
|---|---|---|---|---|---|---|---|---|---|
| $\ln f(j)$ | 5.03 | 6.28 | 6.82 | 7.17 | 7.43 | $\cdots$ | 8.20 | $\cdots$ | 10.78 |

The polynomial congruence $f(x) \equiv 0 \pmod{p}$ has solutions $x_{p,1}$ and $x_{p,2}$ as follows:

| $p$ | 2 | 3 | 7 | 17 | 29 | 31 |
|---|---|---|---|---|---|---|
| $x_{p,1}$ | 1 | 0 | 2 | 0 | 9 | 3 |
| $x_{p,2}$ |  | 1 | 3 | 11 | 17 | 20 |

Note that there is only one solution mod 2.

We subtract $\ln p$ from each $\ln f(j)$ for $j \equiv x_{p,1}$ or $x_{p,2} \pmod{p}$. For example, we subtract $\ln 7$ from $\ln f(j)$ for each $j \equiv 2 \pmod{7}$ and for each $j \equiv 3 \pmod{7}$. Therefore, we subtract $\ln 7$ from $\ln f(2)$, $\ln f(3)$,

In $f(9)$, $\ln f(10)$, and similarly up to $\ln f(100)$. We obtain the following:

| $j$ | 0 | 1 | 2 | 3 | 4 | $\cdots$ | 9 | $\cdots$ | 100 |
|---|---|---|---|---|---|---|---|---|---|
| $\ln f(j)$ | 5.03 | 6.28 | 6.82 | 7.17 | 7.43 | $\cdots$ | 8.20 | $\cdots$ | 10.78 |
| $\ln 2$ |  | .69 |  | .69 |  | $\cdots$ | .69 | $\cdots$ |  |
| $\ln 3$ | 1.10 | 1.10 |  | 1.10 | 1.10 | $\cdots$ | 1.10 | $\cdots$ | 1.10 |
| $\ln 7$ |  |  | 1.95 | 1.95 |  | $\cdots$ | 1.95 | $\cdots$ | 1.95 |
| $\ln 17$ | 2.83 |  |  |  |  | $\cdots$ |  | $\cdots$ |  |
| $\ln 29$ |  |  |  |  |  | $\cdots$ | 3.37 | $\cdots$ |  |
| $\ln 31$ |  |  |  | 3.43 |  | $\cdots$ |  | $\cdots$ |  |
|  | 1.10 | 4.49 | 4.87 | 0.00 | 6.33 | $\cdots$ | 1.09 | $\cdots$ | 7.73 |

For example, $f(3) = 2 \cdot 3 \cdot 7 \cdot 31$, so we have computed

$$\ln f(3) - \ln 2 - \ln 3 - \ln 7 - \ln 31 = 0.$$

On the other hand, $f(0) = 3^2 \cdot 17$, and we have computed

$$\ln f(0) - \ln 3 - \ln 17 = 1.10,$$

which is $\ln 3$. Moreover, $f(1) = 2 \cdot 3 \cdot 89$ and we have computed

$$\ln f(1) - \ln 2 - \ln 3 = 4.49,$$

which is $\ln 89$.

In a way, this process is similar to the Sieve of Eratosthenes (Section 2.4), where we crossed out the numbers in the congruence class 0 mod $p$ for several small primes $p$. See Exercise 30 for a similar idea.

When the end result is 0, as with $f(3)$, it means that we have the complete factorization. When the end result is small, as with $f(0)$, it means that $f(j)$ is divisible by a higher power of a prime in the factor base. When the end result is large, as with $f(1)$, it usually means that there is a large prime in the factorization of $f(j)$. Of course, it could mean that there is a large power of a prime in the factor base, but we're willing to miss a few cases in exchange for speed.

The entries 1.10 and 1.09 on the final line are less than or equal to $\ln 37$, which is the first prime larger than those in our factor base, so they must correspond to logarithms of numbers that can be factored using our factor base. We therefore compute

$$f(0) = 3^2 \cdot 17, \qquad f(9) = 2 \cdot 3^2 \cdot 7 \cdot 29.$$

We now have these two factorizations, along with $f(3) = 2 \cdot 3 \cdot 7 \cdot 31$. Overall, for $0 \le j \le 100$, there are nine values of $j$ that yield factorizations using only primes in the factor base. These values and the factorizations of $f(j)$ are

| $j$ | $f(j)$ |
|---|---|
| 0 | $3^2 \cdot 17$ |
| 3 | $2 \cdot 3 \cdot 7 \cdot 31$ |
| 9 | $2 \cdot 3^2 \cdot 7 \cdot 29$ |
| 17 | $2 \cdot 7 \cdot 17 \cdot 29$ |
| 34 | $3^3 \cdot 17 \cdot 31$ |
| 45 | $2 \cdot 3^4 \cdot 7 \cdot 17$ |
| 51 | $2 \cdot 3 \cdot 7 \cdot 17 \cdot 31$ |
| 79 | $2 \cdot 3^2 \cdot 7 \cdot 17^2$ |
| 96 | $3 \cdot 17 \cdot 29 \cdot 31$ |

(14.3)

The product of the factorizations for $j = 0, 9$, and $17$ yields

$$f(0) \cdot f(9) \cdot f(17) = \left(2 \cdot 3^2 \cdot 7 \cdot 17 \cdot 29\right)^2 = 62118^2.$$

Since

$$f(j) = (j + s)^2 - n \equiv (j + s)^2 \pmod{n},$$

we have (recall that $s = 190$)

$$\left((0 + s)(9 + s)(17 + s)\right)^2 \equiv 62118^2 \pmod{n}.$$

Now is the moment of truth. We have a relation $x^2 \equiv y^2 \pmod{n}$. If $x \equiv \pm y \pmod{n}$, we need to go back and find a new relation. This time, the truth is painful: we have

$$(0 + s)(9 + s)(17 + s) = 190 \cdot 199 \cdot 207$$
$$= 7826670$$
$$\equiv 62118 \pmod{35947}.$$

Back to the drawing board! We find another relation:

$$f(0) \cdot f(45) \cdot f(79) = \left(2 \cdot 3^4 \cdot 7 \cdot 17^2\right)^2 = 327726^2,$$

so

$$12010850^2 = \left((0+190)(45+190)(79+190)\right)^2$$
$$\equiv 32776^2 \pmod{35947}.$$

This time, we are lucky: $x \not\equiv \pm y \pmod{n}$, so we compute

$$\gcd(12010850 - 327726, 35947) = 349.$$

This yields

$$35947 = 349 \times 103.$$

The two factors are prime, so we have the prime factorization of $n$.

How did we figure out that $f(j)$ for $j = 0, 45,$ and $79$ can be combined to yield a square? Brillhart and Morrison pointed out that this is really a problem in linear algebra. From the table of factorizations of $f(j)$, form vectors from the exponents mod 2. For example, the factorization

$$f(79) = 2 \cdot 3^2 \cdot 7 \cdot 17^2 = 2^1 \cdot 3^2 \cdot 7^1 \cdot 17^2 \cdot 29^0 \cdot 31^0$$

gives the vector of exponents

$$(1, 2, 1, 2, 0, 0) \equiv (1, 0, 1, 0, 0, 0) \pmod 2.$$

Put the factorizations from (14.3) into a matrix, with each row corresponding to a value of $j$ and each column corresponding to a prime in the factor base:

$$\begin{bmatrix} 0 & 0 & 0 & 1 & 0 & 0 \\ 1 & 1 & 1 & 0 & 0 & 1 \\ 1 & 0 & 1 & 0 & 1 & 0 \\ 1 & 0 & 1 & 1 & 1 & 0 \\ 0 & 1 & 0 & 1 & 0 & 1 \\ 1 & 0 & 1 & 1 & 0 & 0 \\ 1 & 1 & 1 & 1 & 0 & 1 \\ 1 & 0 & 1 & 0 & 0 & 0 \\ 0 & 1 & 0 & 1 & 1 & 1 \end{bmatrix}.$$

Gaussian reduction mod 2 can be used to take combinations of rows and change the matrix to an upper triangular matrix. Since there are more rows than columns, there will be at least one row of all zeros at the bottom. This is a combination of rows that add up to zero mod 2. In our case, the sum of the first, sixth, and eighth rows is 0 mod 2.

These rows correspond to $j = 0$, 45, and 79. What this means is the sum of the exponents of each of the primes in $f(0) \cdot f(45) \cdot f(79)$ is even, which means the number is a square.

In practice, the algorithm is improved to deal with powers of primes in the factor base. For example, $f(x) \equiv 0 \pmod 9$ when $x \equiv 0$ or 7 mod 9, so an additional $\ln 3$ is subtracted from $f(0), f(7), f(9), f(16), \ldots, f(99)$. If we had done this, we would have obtained 0.00 at the bottom of the columns for $j = 0$ and $j = 9$.

There are several comments that need to be made about the algorithm.

1.  The logarithms of the primes were subtracted from $\ln f(j)$ rather than dividing $f(j)$ by $p$. This is because subtraction is much faster than division (which is why logarithms were invented). Subtracting $\ln p$ for every $j \equiv x_{p,1}$ or $x_{p,2}$ mod $p$ can then be done quickly. The naive method of trying to factor each $f(j)$ using the primes from the factor base requires many trial divisions. Most of the time, $f(j)$ contains a prime factor not in the factor base, so most values of $j$ are unsuccessful (our success rate was about 9%). Therefore, at first we are not as much interested in factoring $f(j)$ as we are in quickly discarding those $f(j)$ that have large prime factors. That's what is accomplished by the "sieving" method of subtracting logarithms of primes and keeping only those $j$ that yield small final values.

2.  There are two parts of the algorithm. This first part finds relations. This can be done by several computers working at the same time, each searching through different values of $j$. The second part of the algorithm is the linear algebra step. It is usually done on a large computer since it is much harder to break the calculations into pieces that small individual computers can do separately.

3.  The first notable success of the quadratic sieve was the factorization obtained in 1982 by Joseph Gerver:

$$n =$$
$$1767497181900566526866820090382275793007 6116201$$
$$= 1196349694438 26601 \times 286870274711101$$
$$\times 515009259868501.$$

The number $3^{225} - 1$ had been factored as a product of primes times this 47-digit number, and completing the factorization was a good testing ground. The factor base consisted of the first 999 primes $p$ for which $n$ is a square mod $p$, plus an entry for $\pm 1$ that allowed negative values $f(j)$ to be used. The computations were done for $-499,999,999 \leq j \leq 400,000,000$. This yielded 327 values of $j$ for which $f(j)$ factors as a product of primes in the factor base. This is not enough, but another trick was used. When the end value after subtracting the various $\ln p$ values was small, the factorization was computed and 25747 examples were found where the factorization used primes in the factor base and one additional slightly larger prime. There were 690 times where the extra prime for one $j$ matched the extra prime for another $j$. The product of these yielded a square times a product of primes in the factor base. These 690 additional relations, combined with the 327 relations, yielded 1017 relations. The exponents mod 2 were put into a $1017 \times 1000$ matrix. A relation among the rows yielded the factorization of $n$, as in our example.

4. The last notable success of the quadratic sieve, before it was surpassed by an improvement called the *number field sieve,* was the factorization of a 129-digit number in 1994 by a team led by Derek Atkins, Michael Graff, Arjen Lenstra, and Paul Leyland. They used approximately 1600 computers to find values of $j$ where $f(j)$ could be factored using primes from the factor base (they also allowed additional medium-sized primes, similar to what Gerver did). The factor base contained 524339 primes. They obtained a $569466 \times 524339$ matrix and found 205 examples of combinations of rows that added to 0 mod 2. The first three of these yielded examples of $x^2 \equiv y^2$ with $x \equiv \pm y$. With the fourth example, they got lucky and had $x \not\equiv \pm y$. This yielded the desired factorization. The 129-digit number was given in a *Scientific American* article in 1977 about the RSA Algorithm, which had recently been discovered. It was estimated that it would take $4 \times 10^{16}$ years to factor the number under technology available at that time. Since then, the algorithms and computers have improved significantly.

5. As with the index calculus (Section 11.5), the choice of the factor base is delicate. If the base is very large, it is easy to find squares mod $n$ that are products of primes in the factor base, but the linear algebra step becomes more difficult because the matrix is large. If the factor base is small, it is very hard to find squares mod $n$ that are products of small primes, but the linear algebra step is easier. The previous two remarks give examples of choices of factor bases that have been used in practice.

6. There is a series of challenge numbers published by RSA Laboratories. They are products of two primes and are designed to be benchmarks for progress in factoring and to demonstrate the security level of the RSA Algorithm. The most recent such challenge number to be factored is called RSA-768 (it has 768 binary bits, or 232 decimal digits). Its factorization was completed in 2009 using the number field sieve (a generalization of the quadratic sieve); this computation took almost three years.

---

# 14.4   Coin Flipping over the Telephone

Now that Alice and Bob have learned some new factorization techniques, they want to try a new method for flipping coins over the telephone. Here is how it goes.

1. Alice chooses two large primes, $p$ and $q$, and forms $n = pq$. She sends $n$ to Bob.

2. Bob chooses a random integer $r$ mod $n$ and computes $s \equiv r^2$ (mod $n$). He sends $s$ to Alice.

3. Alice knows $p$ and $q$, and computing square roots mod primes is straightforward (see Section 13.2). So Alice computes $a$ and $b$ with $a^2 \equiv s$ (mod $p$) and $b^2 \equiv s$ (mod $q$).

4. Alice, using the Chinese Remainder Theorem, computes $x_1$

and $x_2$ with

$$x_1 \equiv a \pmod{p}, \qquad x_1 \equiv b \pmod{q}$$

and

$$x_2 \equiv -a \pmod{p}, \qquad x_2 \equiv b \pmod{q}.$$

She randomly chooses $j = 1$ or 2 and sends $x_j$ to Bob.

5. If $x_j \not\equiv \pm r \pmod{n}$, Bob declares that Alice loses. If $x_j \equiv \pm r$ $\pmod{n}$, Alice wins.

6. If Bob says that Alice loses, Alice asks for $p$ and $q$. Bob computes

$$\gcd(x_j - r, n),$$

which equals $p$ or $q$, and sends the factors to Alice.

**Example.** Alice chooses $p = 31$ and $q = 41$, so $n = 1271$. Bob chooses $r = 55$ and computes $55^2 \equiv 483 \pmod{n}$. He sends 483 to Alice. Alice calculates $483 \equiv 18 \pmod{p}$ and $7^2 \equiv 18 \pmod{31}$. She also computes $483 \equiv 32 \pmod{q}$ and $27^2 \equiv 32 \pmod{41}$. Therefore, $a = 7$ and $b = 27$. She determines that

$$1216 \equiv 7 \pmod{31}, \qquad 1216 \equiv 27 \pmod{41}$$

and also that

$$396 \equiv -7 \pmod{31}, \qquad 396 \equiv 27 \pmod{41}.$$

Therefore, $x_1 = 1216$ and $x_2 = 396$. Alice chooses 396 and sends this to Bob. Bob sees that $396 \not\equiv \pm 55 \pmod{n}$, so he says he wins. Alice doesn't trust Bob, so she asks for $p$ and $q$. Bob computes

$$\gcd(396 - 55, 1271) = 31,$$

so he can factor $n$ and produce the factors. Note that if Alice chose 1216 instead, then Bob would calculate $\gcd(1216 - 55, 1271) = 1$, and therefore he would quickly have to find a method to factor $n$ if he wanted to cheat and claim that he wins.

Why does this work? Since $x_j \equiv a \pmod{p}$, we have

$$x_j^2 \equiv a^2 \equiv s \equiv r^2 \pmod{p}.$$

Similarly,
$$x_j^2 \equiv s \equiv r^2 \pmod{q},$$
so
$$x_j^2 \equiv r^2 \pmod{n}.$$

One of $x_1$ and $x_2$ is congruent to $\pm r \pmod{n}$ and the other is not. If $x_j \not\equiv \pm r \pmod{n}$, then $\gcd(x_j - r, n)$ gives a nontrivial factor of $n$, by the Basic Factorization Principle (Proposition 14.9). If $x_j \equiv \pm r \pmod{n}$, then the gcd gives only the trivial factorization. If $p$ and $q$ are chosen large enough to make factorization infeasible, then Bob is unable to cheat.

Can Alice cheat? One way is for her to send a random $x$, rather than $x_j$. Then it is very likely that $\gcd(x - r, n) = 1$, since the probability is only $1/p$ that a random $x - r$ is a multiple of $p$, and similarly for $q$. This means that Bob cannot factor using the gcd. If he suspects this is happening, he should calculate $x^2 \pmod{n}$ and check that it is the same as $r^2 \pmod{n}$. If not, he has caught Alice cheating.

# 14.5   Chapter Highlights

1. Fermat Factorization (Section 14.1)

2. Fermat Primality Test (Proposition 14.1)

3. Basic Factorization Principle (Proposition 14.9)

# 14.6   Problems

## 14.6.1   Exercises

**Section 14.1: Trial Division and Fermat Factorization**

1. Factor 2880 using trial division.
2. Factor 85680 using trial division.
3. Factor 1152 using trial division.

4. Factor 899 using Fermat Factorization.

5. Factor 391 using Fermat Factorization.

6. Factor 551 using Fermat Factorization.

7. Factor 621
   (a) by using Fermat Factorization
   (b) by using trial division.

8. Let $p$ and $q$ be primes with $p - q = t > 0$. Let $n = pq$.
   (a) In the Fermat Factorization method, if $n = x^2 - y^2$ with $x - y > 1$, show that $x = (p + q)/2$ and $y = (p - q)/2$.
   (b) Show that the Fermat method takes approximately $(p + q)/2 - \sqrt{pq}$ steps to factor $n$.
   (c) Show that

   $$\left(\frac{p+q}{2} - \sqrt{pq}\right)\left(\frac{p+q}{2} + \sqrt{pq}\right) = \frac{1}{4}t^2.$$

   (d) Assume that $t$ is much smaller than $p$, so that $(p + q)/2 \approx p$ and $\sqrt{pq} \approx p$. Show that

   $$\frac{p+q}{2} - \sqrt{pq} \approx \frac{t^2}{8p}.$$

   (e) Suppose $p$ and $q$ are 100-digit primes (so $p, q \approx 10^{99}$) and $p - q \approx 10^{80}$ (so the first 19 or 20 digits of $p$ and $q$ are the same). Show that the Fermat method takes approximately $10^{60}$ steps.
   (f) Suppose $p$ and $q$ are 100-digit primes (so $p, q \approx 10^{99}$) and $p - q \approx 10^{50}$ (so the first 49 or 50 digits of $p$ and $q$ are the same). Show that the Fermat method succeeds in very few steps.

## Section 14.2: Primality Testing

9. Use the Fermat Test with $b = 2$ to show that 15 is composite.

10. Using the facts that $2^{140} \equiv 67 \pmod{561}$ and $2^{280} \equiv 1 \pmod{561}$, show that 561 is composite (compare Exercise 24).

11. (a) Show that $2^{11} \equiv 1 \pmod{2047}$.
    (b) Use (a) to show that $2^{1023} \equiv 1 \pmod{2047}$.
    (c) Use (b) to show that 2047 is a strong 2-pseudoprime.

12. Show that 1729 is a Carmichael number.

13. Show that $41041 = 7 \cdot 11 \cdot 13 \cdot 41$ is a Carmichael number.

14. (a) Let $k \geq 1$ be such that $p = 6k + 1$, $q = 12k + 1$, and $r = 18k + 1$ are primes. Show that $n = pqr$ is a Carmichael number. (The case $k = 1$ gives $n = 1729$.)
    (b) Find six Carmichael numbers larger than 1729.

15. Suppose you use a pseudoprimality test to find primes for the RSA Algorithm. You choose a prime $p$ and a Carmichael number $r$, which you thought was prime because it passed several pseudoprimality tests. Assume that $\gcd(p, r) = 1$. You form the product $n = pr$ as your RSA modulus and then you choose $e$ and $d$ with $ed \equiv 1 \pmod{(p-1)(r-1)}$. Show that if $\gcd(m, n) = 1$, and $c \equiv m^e \pmod{n}$, then $m \equiv c^d \pmod{n}$.

16.  Use the Pocklington–Lehmer Primality Test with $n = 29$, $F = 7$, $N = 4$, and $b = 3$ to prove that 29 is prime.

17.  Use the Pocklington–Lehmer Primality Test with $n = 31$, $F = 6$, $N = 5$, and $b = 3$ to prove that 31 is prime.

18.  Use Pépin's Test to prove that $F_3$ and $F_4$ are prime.

19.  Let $n \geq 1$. Let $1 \leq k < 2^n$ be odd and let $m = k2^n + 1$. Suppose there is some integer $a$ such that

$$a^{(m-1)/2} \equiv -1 \mod m.$$

Show that $m$ is prime. (*Hint:* Look at the proof of Pépin's Test.)
This result is known as *Proth's theorem*.

## Section 14.3: Factorization

20.  Use the fact that $284^2 \equiv 123^2 \pmod{851}$ to factor 851.

21.  Use the fact that $333^2 \equiv 427^2 \pmod{893}$ to factor 893.

22.  Use the facts that $511^2 \equiv 300 \pmod{23711}$ and $1251^2 \equiv 75 \pmod{23711}$ to factor 23711.

23.  Use the facts that $937^2 \equiv 2 \cdot 3^2 \pmod{n}$ and $1666^2 \equiv 2 \cdot 7^2 \pmod{n}$ to factor $n = 28321$.

24.  Using the facts that $2^{140} \equiv 67 \pmod{561}$ and $2^{280} \equiv 1 \pmod{561}$, factor 561 (compare Exercise 10).

25.  (a) Show that 15 is a 4-pseudoprime.
     (b) Show that 15 is not a strong 4-pseudoprime.
     (c) Use (a) and (b) to factor 15 (of course, there is an easier way to factor 15).

26.  (a) Show that 35 is a 6-pseudoprime.
     (b) Show that 35 is not a strong 6-pseudoprime.
     (c) Use (a) and (b) to factor 35 (of course, there is an easier way to factor 35).

27.  Factor $n = 77$ by Pollard's $p - 1$ method by using $b = 2$ and $B = 3$.

28.  (a) Try to factor $n = 115$ by Pollard's $p - 1$ method by using $b = 2$ and $B = 3$. Why doesn't this work?
     (b) Try to factor $n = 115$ by Pollard's $p - 1$ method by using $b = 2$ and $B = 4$. (*Hint:* Compute $2^{24} \pmod{n}$ using successive squaring.)

29.  Try to factor $n = 91$ by Pollard's $p - 1$ method by using $b = 2$ and $B = 4$. (*Hint:* Compute $2^{24} \pmod{n}$ using successive squaring.) You should find that $2^{24} \equiv 1 \pmod{91}$. Use the method of Subsection 14.3.2 to factor $n$.

30.  Suppose we want to find all of the numbers up to 100 with the property that all of their prime factors are less than or equal to 10. (A number, all of whose prime factors are less than or equal to $B$, is called **B-smooth**, so we are looking for 10-smooth numbers.) Suppose we set up a table of $\ln j$ for $1 \leq j \leq 100$. Then, for each prime $p \leq 10$, we subtract $\ln p$ from $\ln j$ for each $j$ that is a multiple of $p$. Then we subtract $\ln p$ from each $j$ that is a multiple of $p^2$, and similarly for $p^3$, etc. Show that at the end of this process, the 10-smooth numbers are exactly those $j$ whose entry $\ln j$ has been reduced to 0.

# 14.6.2    Projects

1.  Let $n$ be a 2-pseudoprime, so $2^{n-1} \equiv 1 \pmod{n}$.
    (a) Let $M = 2^n - 1$. Show that $M - 1 = 2nk$ for some integer $k$.
    (b) Show that $2^{nk} \equiv 1 \pmod{M}$.
    (c) Show $nk$ is odd.
    (d) Show that $M$ is composite. (*Hint: n* is composite.)
    (e) Show that $M$ is a strong 2-pseudoprime.
    (f) Show that there are infinitely many strong 2-pseudoprimes.

2.  Let $n \geq 0$ and let $F_n = 2^{2^n} + 1$ be the $n$th Fermat number.
    (a) Show that

    $$2^{2^n} \equiv -1 \pmod{F_n}.$$

    (b) In the definition of a strong 2-pseudoprime, we have $F_n - 1 = 2^k \cdot s$
    with $k = 2^n$ and $s = 1$. Let $b_0 = 2$, $b_1 \equiv b_0^2 \pmod{F_n}$, etc. Show that

    $$b_n \equiv -1 \pmod{F_n}.$$

    (c) Show that $b_{n+1} \equiv 1 \pmod{F_n}$ and that $k \geq n + 1$. Conclude that
    $b_k \equiv 1 \pmod{F_n}$.
    (d) Show that every composite Fermat number is a strong 2-pseudoprime.
    (*Note:* This does not give a proof that there are infinitely many strong
    2-pseudoprimes. Such numbers are required to be composite and it has
    not been proved that infinitely many Fermat numbers are composite.)

3.  Suppose $n \equiv -1 \pmod{24}$ and we want to factor $n$ by the Fermat
    method, so we want to write $n = x^2 - y^2$.
    (a) Show that if $3 \mid y$ then $x^2 \equiv -1 \pmod{3}$, which is impossible.
    (b) Show that $2 \nmid y$.
    (c) Show that $y^2 \equiv 1 \pmod{24}$.
    (d) Show that $x \equiv 0 \pmod{12}$.
    Let $n = 68041439$. Then $n \equiv -1 \pmod{24}$, so let's try the method just
    described.
    (e) Let $x_1$ be the first multiple of 12 larger than $\sqrt{n}$. See whether $x_1^2 - n$
    is a square.
    (f) If $x_1^2 - n$ is not a square, let $x_2 = x_1 + 12$ and see whether $x_2^2 - n$ is a
    square. Continue in this way until you get a square. Use this information
    to factor $n$.
    *Remark:* D. H. Lehmer, one of the pioneers in factorization methods, used
    methods such as this to build factorization machines involving gears and
    bicycle chains (those were the days before computers).

# 14.6.3    Computer Explorations

1.  (a) There are seven 2-pseudoprimes less than 2000. Find them.
    (b) How many of the 2-pseudoprimes from (a) are strong 2-pseudoprimes?

2.  (a) Find a prime $p$ such that all of the prime factors of $p - 1$ are less than
    25.
    (b) Find a prime $q$ such that $q - 1$ is divisible by the prime 101.

(c) Let $n = pq$. Use the $p - 1$ method to factor $n$.

(d) Find a prime $r \neq p$ such that all of the prime factors of $r - 1$ are less than 25.

(e) Try to factor $m = pr$ by the $p - 1$ method. If the gcd is $m$, use Algorithm 14.10 to factor $m$.

3. Let $n = 4300736399$.

(a) Factor $n$ by the Fermat Factorization method.

(b) Try to factor $n$ by the $p - 1$ method. If you don't succeed, look at the factorization from part (a). Find the factorization of $p - 1$ for each prime factor $p$ of $n$. Explain why the $p - 1$ factorization method is expected to fail for this $n$.

(*Note:* Don't be misled by this example. Usually, Fermat Factorization is slower than other methods.)

4. Suppose you are testing integers up to 1000000 for primality. The product of the primes less than 20 is 9699690. A quick way to check for prime factors less than 20 is to compute $\gcd(n, 9699690)$.

(a) Count how many integers $n \leq 1000000$ satisfy $\gcd(n, 9699690) = 1$.

(b) Count how many primes are less than 1000000. How does this compare with the simple trial division test from (a)?

(c) Compute the product

$$\prod_{p<20} \left(1 - \frac{1}{p}\right)$$

$$= \left(1 - \frac{1}{2}\right)\left(1 - \frac{1}{3}\right)\left(1 - \frac{1}{5}\right) \cdots \left(1 - \frac{1}{19}\right),$$

where the product is over the primes less than 20. How does this relate to the number from (a)?

(d) The probability that an integer is divisible by a prime $p$ is $1/p$. Explain why the numbers in (a) and (c) are closely related.

## 14.6.4    Answers to "CHECK YOUR UNDERSTANDING"

1. (a) Look at $851 + 1^2$, $851 + 2^2$, $\ldots, 851 + 7^2 = 900 = 30^2$. This gives $851 = (30 + 7)(30 - 7) = 37 \cdot 23$.

(b) We have $\sqrt{851} \approx 29.2$, so we start with $x = 30$. This yields $30^2 - 851 = 49 = 7^2$, so we obtain $851 = (30 + 7)(30 - 7) = 37 \cdot 23$.

2. Let $n = 23$, so $n - 1 = 22 = 11 \cdot 2 = F \cdot N$. Let $b = 2$. We already know that $2^{n-1} \equiv 1 \pmod{n}$. The only prime dividing $F$ is $q = 11$, and $(n - 1)/q = 2$, so we need to compute $\gcd(2^2 - 1, 23) = 1$. The Pocklington–Lehmer Test says that 23 is prime.

3. Since $97 \not\equiv \pm 5 \pmod{391}$, the Basic Factorization Principle tells us to compute $\gcd(97 - 5, 391) = 23$, which yields the factorization $391 = 23 \cdot 17$.

4. We have $(119 \cdot 168)^2 \equiv 50 \cdot 2 \equiv 10^2 \pmod{14111}$. The Basic Factorization Principle tells us to compute $\gcd(119 \cdot 168 - 10, 14111) = 103$. Therefore, $14111 = 103 \cdot 137$.

# Chapter 15

# Geometry of Numbers

In 1889, Hermann Minkowski introduced a new method into number theory and started a subject called the *geometry of numbers*. Minkowski's theorem gives a simple geometric criterion for showing the existence of integers satisfying various conditions and is very useful in proving that certain Diophantine equations have solutions. In the following, we apply these methods to prove classical theorems about sums of squares and also to prove the existence of solutions to Pell's equation.

## 15.1 Volumes and Minkowski's Theorem

Often we want to find integers, not all 0, satisfying various inequalities. If we interpret the inequalities as describing a geometrical figure, we are asking for nonzero points with integer coordinates inside a region. Usually, the region is centered around the origin. How big does the region have to be in order to guarantee that there is a point? If we let the region bend around a lot, it is easy to get very large regions that contain no nonzero points. See Figure 15.1. Therefore, we need to impose some conditions. Let $\mathbb{R}^n$ denote $n$-dimensional space, so $\mathbb{R}^2$ is the usual plane and $\mathbb{R}^1$ is the real line.

**Definition 15.1.** *Let $B$ be a region in $\mathbb{R}^n$. We say that $B$ is* **centrally symmetric** *if*

$$(a_1, a_2, \ldots, a_n) \in B \implies (-a_1, -a_2, \ldots, -a_n) \in B.$$

**FIGURE 15.1**
**Centrally symmetric, not convex.     Centrally symmetric, convex.**

*We say that B is **convex** if, whenever P and Q are two points in B, the line segment $\overline{PQ}$ is contained in B.*

In other words, a centrally symmetric region is symmetric about the origin, and a convex region bulges outward. The region on the left in Figure 15.1 is centrally symmetric but not convex, since there are line segments whose endpoints are in the region but the line segments do not lie completely within the region. The region on the right in Figure 15.1 is both centrally symmetric and convex.

**Theorem 15.2.** *(Minkowski) Let B be a centrally symmetric convex region in $\mathbb{R}^n$ with vol$(B) > 2^n$, where vol$(B)$ denotes the n-dimensional volume. Then there exists a point in B with integer coordinates, not equal to the origin.*

*Proof.* Let $\frac{1}{2}B$ denote $B$ shrunk by a factor of $1/2$. This changes the volume by a factor of $(1/2)^n$, so we have vol$(\frac{1}{2}B) > 1$. For each point $P$ with integer coordinates (we'll call such a point an *integer point*), let $P + \frac{1}{2}B$ denote $\frac{1}{2}B$ translated by $P$, so it's centered about $P$.

We'll provide the technical details at the end of the proof, but the general idea is as follows: The sets $P + \frac{1}{2}B$ are filling up space at a rate of vol$(\frac{1}{2}B) > 1$ per integer point. But the volume of space is 1 per integer point. Therefore, there must be some overlap among the sets $P + \frac{1}{2}B$ as $P$ runs through the integer points. Let's assume this for the moment, so for some $P$ and $Q$ with $P \neq Q$, there is a point $T \in \mathbb{R}^n$ with

$$T \in \left(P + \frac{1}{2}B\right) \cap \left(Q + \frac{1}{2}B\right).$$

Write $T = P + \frac{1}{2}b_1 = Q + \frac{1}{2}b_2$, with $b_1, b_2 \in B$. Since $B$ is centrally symmetric, $-b_2 \in B$. Since $B$ is convex, the line segment from $b_1$ to

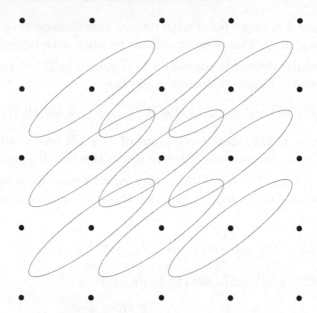

**FIGURE 15.2:** The sets $P + \frac{1}{2}B$.

$-b_2$ is contained in $B$. In particular, the midpoint of this segment is in $B$:

$$\frac{1}{2}(b_1 + (-b_2)) \in B.$$

But

$$\frac{1}{2}(b_1 - b_2) = (T - P) - (T - Q) = Q - P$$

is the difference of two points with integer coordinates, hence has integer coordinates. Therefore, $Q - P \in B$ is the desired point. This finishes the proof, except for the technical details.

We now need to justify the volume argument used above. First, suppose $B$ is unbounded. Let $r$ be a positive real number and let $B(r)$ be the intersection of $B$ with the ball of radius $r$ in $\mathbb{R}^n$:

$$B(r) = \{(x_1, x_2, \ldots, x_n) \in \mathbb{B} \mid x_1^2 + x_2^2 + \cdots + x_n^2 < r^2\}.$$

As $r \to \infty$, the sets $B(r)$ include more and more of $B$, and

$$\mathrm{vol}(B) = \lim_{r \to \infty} \mathrm{vol}(B(r)).$$

Since $2^n < \mathrm{vol}(B)$, we have $2^n < \mathrm{vol}(B(r))$ for some $r$. If we show that

$B(r)$ contains a nonzero point with integer coordinates, then this point is also a point of $B$. Therefore, it suffices to work with bounded sets $B$.

We now assume that $B$ is contained in a sphere in $\mathbb{R}^n$ of radius $r$, for some sufficiently large $r$. Consider a big box

$$R_N = \{(x_1, x_2, \ldots, x_n) \in \mathbb{R}^n \mid |x_i| \leq N \text{ for all } i\}.$$

If $P$ is a point in $R_N$, then every point of $P + \frac{1}{2}B$ has coordinates less than $N + \frac{1}{2}r$ in absolute value, hence lies in the box $R_{N+\frac{1}{2}r}$. There are $(2N+1)^n$ points $P \in R_N$ with integer coordinates. Each set $P + \frac{1}{2}B$ contributes $\text{vol}(\frac{1}{2}B)$ to $R_{N+\frac{1}{2}r}$. Assume these sets do not overlap. Then we must have

$$(2N+1)^n \, \text{vol}(\frac{1}{2}B) < \text{vol}(R_{N+\frac{1}{2}r}) = 2^n(N + \frac{1}{2}r)^n.$$

Divide by $(2N+1)^n$ and take the limit as $N \to \infty$:

$$1 < \text{vol}(\frac{1}{2}B) \leq \lim_{N \to \infty} \frac{2^n(N + \frac{1}{2}r)^n}{(2N+1)^n} = 1.$$

This is a contradiction, so we conclude that the sets overlap, as claimed. This completes the proof.                                                                    □

We need one more result on volumes. It requires the concept of determinant of a matrix. The reader who is not familiar with matrices and determinants can simply skip the proposition and look at the areas and volumes that we calculate as applications of the proposition. These are all that are needed in subsequent sections.

**Proposition 15.3.** *Let $B$ be a region in $\mathbb{R}^n$ described by inequalities of the form*

$$|f_1(x_1, x_2, \cdots, x_n)| < r_1,$$
$$|f_2(x_1, x_2, \cdots, x_n)| < r_2,$$
$$\cdots$$
$$|f_k(x_1, x_2, \cdots, x_n)| < r_k,$$

*where $f_1, \ldots, f_k$ are continuous functions and $r_1, \ldots, r_k$ are real numbers. Let*

$$A = \begin{pmatrix} a_{11} & a_{12} & \cdots & a_{1n} \\ \vdots & \vdots & \ddots & \vdots \\ a_{n1} & a_{n2} & \cdots & a_{nn} \end{pmatrix}$$

*be an invertible matrix with real entries. Let $B'$ be the region defined by inequalities*

$$|f_1(a_{11}x_1 + \cdots + a_{1n}x_n, \ldots, a_{n1}x_1 + \cdots + a_{nn}x_n)| < r_1$$
$$|f_2(a_{11}x_1 + \cdots + a_{1n}x_n, \ldots, a_{n1}x_1 + \cdots + a_{nn}x_n)| < r_2$$
$$\cdots \qquad\qquad \cdots$$
$$|f_k(a_{11}x_1 + \cdots + a_{1n}x_n, \ldots, a_{n1}x_1 + \cdots + a_{nn}x_n)| < r_k.$$

*Then*

$$vol(B) = |\det(A)| \; vol(B').$$

This formula is the basis for change of variables in multiple integrals and is discussed in many books on multivariable calculus.

The proposition has a lot of notation. Some examples should make clear what it says.

1. Let $B$ be the $2 \times 2$ square in $\mathbb{R}^2$ (= the plane) given by $|x_1| < 1$, $|x_2| < 1$. Suppose we want to find the area of the parallelogram $B'$ given by $|2x_1 - x_2| < 1$, $|3x_2| < 1$. Let $A = \begin{pmatrix} 2 & -1 \\ 0 & 3 \end{pmatrix}$. Then (since we're in the plane, we use area instead of volume)

$$4 = \text{area}(B) = |\det(A)| \, \text{area}(B') = 6 \, \text{area}(B').$$

Therefore, $\text{area}(B') = 4/6 = 2/3$.

2. More generally, the parallelogram

$$|ax_1 + bx_2| < r_1, \qquad |cx_1 + dx_2| < r_2$$

has area $4r_1 r_2/|ad - bc|$, since $\det \begin{pmatrix} a & b \\ c & d \end{pmatrix} = ad - bc$.

3. Here's an example that we'll need in the next section. Let $B$ be the disc $x_1^2 + x_2^2 < R^2$ in the plane. Let $p$ and $u$ be positive real numbers and let $A = \begin{pmatrix} p & u \\ 0 & 1 \end{pmatrix}$. Then $\det(A) = p$. Let $B'$ be the region defined by

$$(px_1 + ux_2)^2 + x_2^2 < R^2$$

(it's an ellipse, rotated so that its axes are not parallel to the coordinate axes). Then

$$\pi R^2 = \text{area}(B) = p \cdot \text{area}(B'),$$

so $\text{area}(B') = \pi R^2/p$.

4.  Finally, for an example that we'll need in Section 3, consider the ball $B$ in $\mathbb{R}^4$ defined by $x_1^2 + x_2^2 + x_3^2 + x_4^2 < R^2$. The volume of $B$ is $(1/2)\pi^2 R^4$ (this can be calculated by multiple integrals, but we'll take it as a known fact). Let $p, u, v$ be real numbers and let

$$A = \begin{pmatrix} u & v & p & 0 \\ v & -u & 0 & p \\ 1 & 0 & 0 & 0 \\ 0 & 1 & 0 & 0 \end{pmatrix}.$$

Then $\det(A) = p^2$. We find that the region in $\mathbb{R}^4$ given by

$$(ux_1 + vx_2 + px_3)^2 + (vx_1 - ux_2 + px_4)^2 + x_1^2 + x_2^2 < R^2$$

has volume $(1/2)\pi^2 R^4/p^2$.

---

**CHECK YOUR UNDERSTANDING:**

1. Give an example of a convex, centrally symmetric region in the plane with area greater than 3.9 but which does not contain any integer points other than $(0,0)$.

2. Give an example of a convex, centrally symmetric region in the plane with area less than .01 but which contains a point $(x, y) \neq (0,0)$ with integer coordinates.

---

# 15.2  Sums of Two Squares

Which integers can be written as the sum of two squares? As we'll see, the case of prime numbers is the key. The French mathematician

Albert Girard observed that every prime congruent to 1 mod 4 is a sum of two squares. Fermat, in a letter to Mersenne in 1640, said that he had a proof, and it is generally believed that this is the case. The first published proof is due to Euler in 1747. The proof we give is a simple consequence of Minkowski's theorem. In the Projects, we give a proof due to Zagier, and in Chapter 18 we give a proof using Gaussian integers.

**Theorem 15.4.** *(Fermat) Let $p$ be a prime with $p \equiv 1 \pmod{4}$. Then there exist integers $x$ and $y$ with $x^2 + y^2 = p$.*

*Proof.* We start with a lemma that gives a property that holds for primes $p \equiv 1 \pmod{4}$ but does not hold for primes $p \equiv 3 \pmod{4}$ (see Exercises 24 and 63 in Chapter 8). This lemma is also the Supplementary Law for $-1$ in Theorem 13.4.

**Lemma 15.5.** *Let $p$ be a prime with $p \equiv 1 \pmod{4}$. Then there exists $u$ with $u^2 \equiv -1 \pmod{p}$.*

*Proof.* Let $g$ be a primitive root mod $p$ and let $u \equiv g^{(p-1)/4} \pmod{p}$. (We need $p \equiv 1 \pmod{4}$ so that the exponent is an integer.) Then

$$u^2 \equiv g^{(p-1)/2} \equiv -1 \pmod{p}$$

by Proposition 11.6.                                                      □

Let $u$ be as in the lemma. Let $B$ be the set of points in $\mathbb{R}^2$ satisfying

$$(px_1 + ux_2)^2 + x_2^2 < 2p.$$

This is the interior of an ellipse. In the previous section, we showed that its area is $\pi(2p)/p = 2\pi$. Since $2\pi > 2^2$, Minkowski's Theorem says that there is a point $0 \neq (x, y)$ with integer coordinates in $B$. Since this point is in $B$,

$$0 < (px + uy)^2 + y^2 < 2p.$$

(The first inequality requires a little justification: If $(px + uy)^2 + y^2 = 0$ then $px + uy = 0$ and $y = 0$, from which it follows easily that $x = 0$, which is contrary to the condition that $(x, y) \neq (0, 0)$.) But

$$(px + uy)^2 + y^2 \equiv (uy)^2 + y^2 \equiv y^2(u^2 + 1) \equiv 0 \pmod{p}.$$

The only multiple of $p$ between 0 and $2p$ is $p$. Therefore, $(px+uy)^2+y^2 = p$, as desired. This proves Theorem 15.4.                              □

**Example.** Let $p = 13$. Since $5^2 \equiv -1 \pmod{13}$, we take $u = 5$. The elliptical region in the proof of Theorem 15.4 is

$$(13x_1 + 5x_2)^2 + x_2^2 < 26.$$

This is pictured in Figure 15.3.

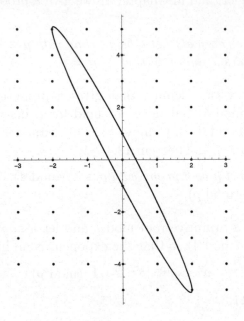

**FIGURE 15.3: The ellipse for $p = 13$.**

Note that there are four points inside the region: $(-1, 2)$, $(-1, 3)$, $(1, -2)$, $(1, -3)$. The first yields

$$13x_1 + 5x_2 = -3, \quad x_2 = 2, \quad (-3)^2 + 2^2 = 13.$$

The second yields

$$13x_1 + 5x_2 = 2, \quad x_2 = 3, \quad 2^2 + 3^2 = 13.$$

The other two points yield the relations $3^2 + (-2)^2 = 13$ and $(-2)^2 + (-3)^2 = 13$.

The observant reader might have noticed two points on the boundary of the region: $(-2, 5)$, $(2, -5)$. These correspond to $(-1)^2 + 5^2 = 26$ and $1^2 + (-5)^2 = 26$. This is why we needed to define the elliptical region $B$ with the strict inequality " $< 2p$" rather than with " $\leq 2p$."

We now discuss arbitrary positive integers. Suppose that $n$ and $m$ are integers such that

$$m = a^2 + b^2 \text{ and } n = c^2 + d^2.$$

Then

$$mn = (ac - bd)^2 + (ad + bc)^2. \tag{15.1}$$

Where does this identity come from? We'll discuss complex numbers in Chapter 18, but it's worth mentioning them here, too. The absolute value of the complex number $a + bi$ is $|a + bi| = \sqrt{a^2 + b^2}$, so we have

$$m = |a + bi|^2 \text{ and } n = |c + di|^2.$$

Therefore,

$$
\begin{aligned}
mn &= |a + bi|^2 |c + di|^2 \\
&= |(a + bi)(c + di)|^2 \\
&= |(ac - bd) + (ad + bc)i|^2 \\
&= (ac - bd)^2 + (ad + bc)^2.
\end{aligned}
$$

Of course, it's also possible to verify the formula (15.1) by algebraic calculation.

We conclude that if two numbers are sums of squares, then so is their product. This leads to the main result of this section.

**Theorem 15.6.** *Let $n$ be a positive integer. Then $n = x^2 + y^2$ for integers $x, y$ if and only if $n$ is a square times a product of primes $p \not\equiv 3 \pmod 4$. In other words, $n$ is a sum of two squares if and only if, in the prime factorization of $n$, every prime $p \equiv 3 \pmod 4$ has even exponent.*

For example, $65 = 5 \times 13$ and $130 = 2 \times 5 \times 13$ and $405 = 3^4 \times 5$ are sums of squares, while $35 = 5 \times 7$ and $189 = 3^3 \times 7$ are not sums of squares. We also include $49 = 7^2 + 0^2$ as a sum of two squares.

*Proof.* First, suppose that $n = u^2 v$ is a square times a number $v$ that is a product of primes $p \not\equiv 3 \pmod 4$. Each prime $p \not\equiv 3 \pmod 4$ is a sum of two squares: the odd primes are treated by Theorem 15.4 and the even prime is $2 = 1^2 + 1^2$. Therefore, Equation (15.1) implies that $v$, which is a product of primes that are sums of two squares, is a sum of two squares: $v = x_1^2 + y_1^2$. Therefore, $n = u^2 v = (ux_1)^2 + (uy_1)^2$ is a sum of two squares.

Conversely, suppose that the integer $n$ is a sum of two squares: $n = x^2 + y^2$. Let $p \equiv 3 \pmod 4$. We need to show that if $p^a$ is the power of $p$ in the factorization of $n$, then $a$ is even. Suppose $p^j$ is the largest power of $p$ that divides both $x$ and $y$. Write $x = p^j x_1$ and $y = p^j y_1$, where at least one of $x_1, y_1$ is not a multiple of $p$. Without loss of generality, we may assume that $x_1 \not\equiv 0 \pmod p$. We have

$$n = x^2 + y^2 = p^{2j}(x_1^2 + y_1^2),$$

so $p^{2j} \mid n$. Write $n = p^{2j} n_1$, with $n_1 = x_1^2 + y_1^2$. If $p \nmid n_1$, then we conclude that $p^{2j}$ is the power of $p$ in the factorization of $n$, and the exponent $2j$ is even. On the other hand, if $p \mid n_1$, then

$$x_1^2 + y_1^2 \equiv 0 \pmod p.$$

Since $x_1 \not\equiv 0 \pmod p$, we can write $(y_1/x_1)^2 \equiv -1 \pmod p$. As we saw in Exercise 24 in Chapter 8, or by Theorem 13.4, this is impossible when $p \equiv 3 \pmod 4$. Therefore, $p \nmid n_1$, so we conclude that $p$ occurs with even exponent in the factorization of $n$. This is true for every $p \equiv 3 \pmod 4$, so $n$ has the form claimed in the theorem.    $\square$

In Project 3, you get a chance to show that if an odd number $n \equiv 1 \pmod 4$ is prime then there is exactly one way (up to order) to write $n = x^2 + y^2$ with positive integers $x$, $y$. Moreover, if there are two ways to write an integer $n$ as a sum of two squares, we can use this information to factor $n$.

Observe that near the end of the proof, we proved the following: *If $p \equiv 3 \pmod 4$ and $x^2 + y^2 \equiv 0 \pmod p$, then $x \equiv y \equiv 0 \pmod p$.*

## 15.2.1    Algorithm for Writing $p \equiv 1 \pmod 4$ as a Sum of Two Squares

The proof of Theorem 15.4 does not readily yield the values of $x$ and $y$ such that $x^2 + y^2 = p$. Of course, we can simply try all values of $x < \sqrt{p}$ until $p - x^2$ is a square, but there is a faster way.

1.  Find a solution to $u^2 \equiv -1 \pmod p$. This may be done by

computing $x^{(p-1)/4}$ (mod $p$) for random values of $x$. If $x$ is a quadratic nonresidue mod $p$ then this yields the desired $u$. We may assume that $0 < u < p$. If $u > p/2$, change $u$ to $p - u$. Therefore, we may assume that $0 < u < p/2$.

2.  Apply the Euclidean Algorithm to $p$ and $u$. Let $r_k$ and $r_{k+1}$ be the first two remainders less than $\sqrt{p}$. Then $p = r_k^2 + r_{k+1}^2$.

For example, let $p = 557$. Then $(p - 1)/4 = 139$. We have $2^{139} \equiv 118$ (mod 557), and $118^2 \equiv -1$ (mod 557), so we let $u = 118$. (The preceding calculation shows that $2^{(p-1)/2} \equiv -1$ (mod $p$), so 2 is a quadratic nonresidue mod $p$ by Euler's Criterion (Proposition 13.3).) The Euclidean Algorithm is

$$557 = 4 \cdot 118 + 85$$
$$118 = 1 \cdot 85 + 33$$
$$85 = 2 \cdot 33 + 19$$
$$33 = 1 \cdot 19 + 14$$
$$19 = 1 \cdot 14 + 5$$
$$14 = 2 \cdot 5 + 4$$
$$5 = 1 \cdot 4 + 1$$
$$4 = 4 \cdot 1 + 0.$$

The first two remainders less than $\sqrt{557}$ are 19 and 14. We have $19^2 + 14^2 = 557$.

This algorithm is essentially due to Charles Hermite and Joseph Serret (independently) in 1848. Improvements were made by Henry Smith and John Brillhart. For the proof that the algorithm works, see F. W. Clarke, W. N. Everitt, L. L. Littlejohn, and S. J. R. Vorster, "H. J. S. Smith and the Fermat Two Squares Theorem," *American Mathematical Monthly* 106 (1999), no. 7, 652-665.

---

## CHECK YOUR UNDERSTANDING:

3. Using the fact that $22^2 \equiv -1$ (mod 97), write 97 as a sum of two squares.

# 15.3   Sums of Four Squares

The amazing fact that every positive integer is a sum of four squares was probably known to Diophantus, but the first proof is due to Lagrange in 1770.

**Theorem 15.7.** *(Lagrange) Every positive integer is a sum of four squares.*

*Proof.* As in the case of two squares, we first prove the theorem when $n = p$ is prime. We need an analogue of Lemma 15.5.

**Lemma 15.8.** *Let $p$ be prime. Then there are integers $u, v$ such that $u^2 + v^2 + 1 \equiv 0 \pmod{p}$.*

*Proof.* The case $p = 2$ is trivial (let $u = 1$ and $v = 0$), so assume $p$ is odd. There are $(p-1)/2$ nonzero squares mod $p$ by Exercise 34 in Chapter 6, or by Exercise 33 in Chapter 11. Since 0 is also a square, there are $1 + (p-1)/2 = (p+1)/2$ squares mod $p$. If we take each square and subtract it from $-1$, we get $(p+1)/2$ numbers of the form $-1 - v^2$ $\pmod{p}$. We have two sets: the squares with $(p+1)/2$ elements and the numbers $-1 - v^2$ with $(p+1)/2$ elements. There are only $p$ congruence classes mod $p$, and $(p+1)/2 + (p+1)/2 > p$. Therefore, the two sets must overlap. This means that $u^2 \equiv -1 - v^2 \pmod{p}$ for some $u, v$, which says that $u^2 + v^2 + 1 \equiv 0 \pmod{p}$. $\qquad \square$

**Example.** As an example of the lemma, let $p = 11$. The squares mod 11 are

$$\{0, 1, 4, 9, 5, 3\}$$

and the numbers of the formula $-1 - v^2$ are

$$\{-1, -2, -5, -10, -6, -4\} \equiv \{10, 9, 6, 1, 5, 7\} \text{ mod } 11.$$

Note that 1 is in both sets, which means that $1^2 \equiv -1 - 3^2$, or $1^2 + 3^2 + 1 \equiv 0 \pmod{11}$.

We now return to the proof of Lagrange's theorem. Let $u, v$ be as in the lemma. Let $B$ be the set of points in $\mathbb{R}^4$ satisfying

$$(ux_1 + vx_2 + px_3)^2 + (vx_1 - ux_2 + px_4)^2 + x_1^2 + x_2^2 < 2p.$$

As we calculated in Section 1, this is an ellipsoid with volume

$$\text{vol}(B) = (1/2)\pi^2(\sqrt{2p})^4/p^2 = 2\pi^2 > 2^4.$$

Since $B$ is centrally symmetric and convex, Minkowski's Theorem says that there is a point

$$(0,0,0,0) \neq (x_1, x_2, x_3, x_4) \in B$$

with integer coordinates. Then

$$0 < (ux_1 + vx_2 + px_3)^2 + (vx_1 - ux_2 + px_4)^2 + x_1^2 + x_2^2 < 2p.$$

But

$$(ux_1 + vx_2 + px_3)^2 + (vx_1 - ux_2 + px_4)^2 + x_1^2 + x_2^2$$
$$\equiv x_1^2(u^2 + v^2 + 1) + x_2^2(u^2 + v^2 + 1)$$
$$\equiv 0 \pmod{p}.$$

Since $p$ is the only multiple of $p$ between 0 and $2p$,

$$(ux_1 + vx_2 + px_3)^2 + (vx_1 - ux_2 + px_4)^2 + x_1^2 + x_2 r = p,$$

as desired. Therefore, every prime is a sum of four squares.

To prove that every positive integer is a sum of four squares, we need to extend Equation (15.1) to four squares. Here's the relevant identity:

$$(a^2 + b^2 + c^2 + d^2)(e^2 + f^2 + g^2 + h^2)$$
$$= (ae - bf - cg - dh)^2 + (af + be + ch - dg)^2$$
$$+ (ag + ce + df - bh)^2 + (ah + de + bg - cf)^2. \quad (15.2)$$

If you don't believe this, multiply it out and check it. To see how it can be deduced using quaternions, see Exercise 21. However, this identity was discovered before quaternions were discovered!

Every prime is a sum of four squares, and (15.2) says that products of sums of four squares are sums of four squares. Since every integer $n \geq 2$ is a product of primes, each such $n$ is a sum of four squares. Finally, $1 = 1^2 + 0^2 + 0^2 + 0^2$, so 1 is also a sum of four squares. This completes the proof.                                                                    $\square$

# 15.4    Pell's Equation

Fix a positive integer $d$ that is not a square. We want to find integers $x$ and $y$ such that

$$x^2 - dy^2 = 1.$$

This is called **Pell's Equation**. As discussed in the introductory chapter, this equation has a very long history, and it continues to attract attention in modern times.

Of course, there are the solutions $x = \pm 1, y = 0$, but we are looking for non-trivial solutions. Sometimes, this is easy. For $d = 63$, we have $8^2 - 63 \cdot 1^2 = 1$. But sometimes it is hard. For $d = 61$, the smallest solution is

$$1766319049^2 - 61 \cdot 226153980^2 = 1.$$

This solution was found by Bhāskara in the 12th century. We'll explain his method later in this section.

Although several mathematicians devised ways to find solutions, Lagrange was the first to publish a proof that solutions always exist (although it is likely that Fermat also knew how to prove this fact). In the following, we use Minkowski's theorem to prove solutions exist. Then we give Bhāskara's method for finding solutions. In Chapter 17, we show how continued fractions can be used to find solutions, and in Chapter 19, we interpret solutions in the context of algebraic number theory.

**Theorem 15.9.** *Let $d$ be a positive integer, not a square. Then*

$$x^2 - dy^2 = 1$$

*has a solution in integers $x, y$ with $x \neq \pm 1$.*

*Proof.* Let $A$ be a positive real number, and let $B_A$ be the parallelogram in the plane given by

$$|x - y\sqrt{d}| < A, \quad |x + y\sqrt{d}| < 3\sqrt{d}/A.$$

As we calculated at the end of Section 15.1,

$$\text{area}(B_A) = 6 > 2^2.$$

Since $B_A$ is convex and centrally symmetric, Minkowski's theorem implies that there is an integer point $(x_A, y_A) \neq (0,0)$ in $B_A$. Observe that such a point satisfies

$$|x_A^2 - dy_A^2| = |x_A - y_A\sqrt{d}| \cdot |x_A + y_A\sqrt{d}| < (A)(3\sqrt{d}/A) = 3\sqrt{d}, \quad (15.3)$$

which is independent of $A$.

Start with $A_1 = 1$. We get an integer point $(x_{A_1}, y_{A_1})$, which we'll call $(x_1, y_1)$ for simplicity of notation. Since $d$ is not a square, $\sqrt{d}$ is irrational, so $x_1 - y_1\sqrt{d} \neq 0$. Therefore, we can choose $A_2$ small enough that

$$A_2 < |x_1 - y_1\sqrt{d}|.$$

There is a point $(x_2, y_2) \neq (0,0)$ in $B_{A_2}$ with integer coordinates. Since

$$|x_2 - y_2\sqrt{d}| < A_2 < |x_1 - y_1\sqrt{d}|,$$

we must have $(x_2, y_2) \neq \pm(x_1, y_1)$.

Now choose $A_3$ small enough that $A_3 < |x_2 - y_2\sqrt{d}|$. We get a point $(x_3, y_3) \neq \pm(x_i, y_i)$ for $i = 1, 2$. Continuing in this way, we get distinct points $(x_j, y_j)$ for $j = 1, 2, 3, \ldots$ and each point satisfies

$$|x_j^2 - dy_j^2| < 3\sqrt{d},$$

by (15.3). Each number $x_1^2 - dy_1^2$ is a nonzero integer bounded by $3\sqrt{d}$. There are only finitely many such integers, so at least one integer, call it $N$, occurs infinitely often. This says that, for this $N$, we have infinitely many solutions to

$$x^2 - dy^2 = N.$$

We will use this fact to find integers $x$ and $y$ with $x^2 - dy^2 = 1$.

There are only finitely many choices for $x \pmod{N}$ and $y \pmod{N}$, so there must be two nonzero solutions $(u, v)$ and $(w, z)$ with $(u, v) \neq \pm(w, z)$ such that

$$u^2 - dv^2 = w^2 - dz^2 = N \text{ and } u \equiv w \pmod{N}, \quad v \equiv z \pmod{N}.$$
$$(15.4)$$

Using the fact that $w^2 - dz^2 = N$, we rationalize a denominator to get

$$\frac{u + v\sqrt{d}}{w + z\sqrt{d}} = \frac{u + v\sqrt{d}}{w + z\sqrt{d}} \frac{w - z\sqrt{d}}{w - z\sqrt{d}}$$
$$= \frac{(uw - vzd) + (vw - uz)\sqrt{d}}{N} \quad (15.5)$$
$$= x + y\sqrt{d},$$

where $x, y$ are rational numbers. By (15.4), we have

$$uw - vzd \equiv u^2 - v^2d \equiv 0 \pmod{N},$$
$$vw - uz \equiv vu - uv \equiv 0 \pmod{N}.$$

Therefore, $x = (uw - vzd)/N$ and $y = (vw - uz)/N$ are integers. Also,

$$
\begin{aligned}
x^2 - dy^2 &= \frac{1}{N^2}\left((uw - vzd)^2 - d(vw - uz)^2\right) \\
&= \frac{1}{N^2}\left(u^2w^2 + v^2z^2d^2 - dv^2w^2 - du^2z^2\right) \\
&= \frac{1}{N^2}(u^2 - dv^2)(w^2 - dz^2) \\
&= \frac{1}{N^2}(N)(N) \\
&= 1.
\end{aligned}
$$

Finally, we need to check that we didn't get a trivial solution. If $x = \pm 1$ and $y = 0$, then (15.5) implies that

$$\frac{u + v\sqrt{d}}{w + z\sqrt{d}} = \pm 1 + 0\sqrt{d} = \pm 1.$$

Therefore, $u + v\sqrt{d} = \pm(w + z\sqrt{d})$, which says that $(u, v) = \pm(w, z)$, contrary to the choices of $u, v, w, z$. Consequently, $(x, y)$ is not a trivial solution, so we have found a nontrivial solution. $\qquad\square$

**Corollary 15.10.** *Let $d$ be a positive integer, not a square. Then*

$$x^2 - dy^2 = 1 \tag{15.6}$$

*has infinitely many solutions in integers $x, y$.*

*Proof.* Let $(x_1, y_1)$ be a nontrivial solution. We may change the signs of $x_1$ and $y_1$ if necessary and assume that $x_1$ and $y_1$ are positive. Let $n$ be a positive integer. Define integers $x_n, y_n$ by

$$x_n + y_n\sqrt{d} = \left(x_1 + y_1\sqrt{d}\right)^n.$$

Then, changing $\sqrt{d}$ to $-\sqrt{d}$ yields

$$x_n - y_n\sqrt{d} = \left(x_1 - y_1\sqrt{d}\right)^n.$$

Therefore,

$$x_n^2 - dy_n^2 = (x_n + y_n\sqrt{d})(x_n - y_n\sqrt{d})$$
$$= \left(x_1 + y_1\sqrt{d}\right)^n \left(x_1 - y_1\sqrt{d}\right)^n$$
$$= \left((x_1 + y_1\sqrt{d})(x_1 - y_1\sqrt{d})\right)^n$$
$$= (x_1^2 - dy_1^2)^n$$
$$= 1^n = 1.$$

This means that $x_n, y_n$ is a solution of (15.6). Since $x_1, y_1, d$ are positive and

$$x_n = x_1^n + ndx_1^{n-2}y_1^2 + \cdots \geq n,$$

we see that $x_n \to \infty$ as $n \to \infty$. In particular, we get infinitely many different solutions, as desired.                                             $\square$

**Example:** When $d = 2$, we have $3^2 - 2 \cdot 2^2 = 1$, so we take $x_1 = 3$ and $y_1 = 2$. If $n = 2$, then

$$(3 + 2\sqrt{2})^2 = 17 + 12\sqrt{2},$$

and $17^2 - 2 \cdot 12^2 = 1$. If $n = 4$, we get

$$(3 + 2\sqrt{2})^4 = 577 + 408\sqrt{2},$$

and $577^2 - 2 \cdot 408^2 = 1$. These two solutions were known to the Indian mathematician Baudhāyana (around 800 B.C.E.), who used them to approximate $\sqrt{2}$:

$$\frac{577}{408} = 1.414215686\cdots, \qquad \sqrt{2} = 1.414213562\cdots.$$

## 15.4.1   Bhāskara's Chakravala Method

Suppose we have

$$x_1^2 - dy_1^2 = k_1 \quad \text{and} \quad x_2^2 - dy_2^2 = k_2.$$

We represent these equations by triples $(x_1, y_1, k_1)$ and $(x_2, y_2, k_2)$. Brahmagupta observed in the 7th century that these triples can be "multiplied" to get a triple $(x_3, y_3, k_3)$ with $k_3 = k_1 k_2$:

$$x_3^2 - dy_3^2 = k_1 k_2, \text{ where } x_3 = x_1 x_2 + dy_1 y_2 \text{ and } y_3 = x_1 y_2 + x_2 y_1.$$

This is simply the identity

$$(x_1 + y_1\sqrt{d})(x_2 + y_2\sqrt{d}) = x_3 + y_3\sqrt{d}.$$

Bhāskara's Chakravala Method (from *chakra*, which means *wheel* in Sanskrit) starts with an arbitrary triple $(a, b, k)$ with $a^2 - db^2 = k$. The goal is to obtain a triple with $k = 1$.

Let $e = \gcd(b, k)$. Suppose $e \neq 1$. If $p$ is a prime dividing $e$, then $p \mid b$ and $p \mid k$, so $p \mid k + db^2 = a^2$. Therefore, $p$ divides both $a$ and $b$, so $p^2 \mid a^2 - db^2 = k$ since $p$ divides both $a$ and $b$. This means that we can divide $a$ and $b$ by $p$ and divide $k$ by $p^2$ and thus obtain a smaller $k$. In this way, we can remove all common factors of $b$ and $k$ and thus assume that $\gcd(b, k) = 1$.

Multiply the triple $(a, b, k)$ by a triple $(m, 1, m^2 - d)$, where $m$ is yet to be chosen. This gives the triple $(am + bd, a + bm, k(m^2 - d))$, which, after dividing by $k^2$, yields

$$\left(\frac{am + bd}{|k|}\right)^2 - d\left(\frac{a + bm}{|k|}\right)^2 = \frac{m^2 - d}{k}$$

(we use $|k|$ for $x$ and $y$ to keep $x$ and $y$ positive; but we need to use $k$ on the right-hand side). Choose $m$ to be a positive integer such that $a + bm \equiv 0 \pmod{k}$ and such that $|m^2 - d|$ is as small as possible.

**Lemma 15.11.** *If $a + bm \equiv 0 \pmod{k}$, then $(m^2 - d)/k$ and $(am + bd)/k$ are integers.*

*Proof.* Since $bm \equiv -a \pmod{k}$ and $a^2 - db^2 = k$, we have

$$b^2(m^2 - d) = (bm)^2 - b^2d \equiv (-a)^2 - db^2 = k \equiv 0 \pmod{k}.$$

Since we are assuming that $\gcd(b, k) = 1$, we can divide by $b^2$ and get $m^2 - d \equiv 0 \pmod{k}$. Therefore, $(m^2 - d)/k$ is an integer.

Since $a \equiv -bm \pmod{k}$, we have

$$am + bd \equiv (-bm)m + bd \equiv -b(m^2 - d) \equiv 0 \pmod{k},$$

by what we just proved. Therefore, $(am + bd)/k$ is an integer.     □

The lemma says that we have a new integer triple. We keep repeating the process until we end up at a triple with $k = 1$ (Lagrange proved that we always eventually get to $k = 1$).

**Example.** Let $d = 19$. Start with the triple $(5, 1, 6)$, which corresponds to $5^2 - 19 \times 1^2 = 6$. Multiply by $(m, 1, m^2 - 19)$, with $m$ to be determined, and divide by $k = 6$ to get the triple

$$\left( \frac{5m + 19}{6}, \frac{5 + m}{6}, \frac{m^2 - 19}{6} \right).$$

We need $5 + m \equiv 0 \pmod 6$, so $m \equiv 1 \pmod 6$. The value of $m \equiv 1 \pmod 6$ that makes $|m^2 - 19|$ smallest is $m = 1$. This yields the triple

$$(4, \ 1, \ -3).$$

This corresponds to $4^2 - 19 \cdot 1^2 = -3$.

Now repeat the procedure with the triple $(4, 1, -3)$. Multiply by $(m, 1, m^2 - 19)$, with $m$ to be determined, and divide by $k = -3$ and $|k| = 3$. This yields

$$\left( \frac{4m + 19}{3}, \frac{4 + m}{3}, \frac{m^2 - 19}{-3} \right).$$

If $m \equiv 2 \pmod 3$ then $4 + m \equiv 0 \pmod 3$. The positive value of $m$ that makes $|m^2 - 19|$ smallest is $m = 5$. We get the new triple

$$(13, \ 3, \ -2),$$

corresponding to $13^2 - 61 \cdot 3^2 = -2$.

Continue the process with the triple $(13, 3, -2)$. Multiply by $(m, 1, m^2 - 19)$ and divide by 2 and $-2$ to get

$$\left( \frac{13m + 57}{2}, \frac{13 + 3m}{2}, \frac{m^2 - 19}{-2} \right).$$

If $m \equiv 1 \pmod 2$, then $13 + 3m \equiv 0 \pmod 2$. The value of $m$ that makes $|m^2 - 19|$ smallest is $m = 5$. We get the new triple

$$(61, \ 14, \ -3),$$

corresponding to $61^2 - 19 \cdot 14^2 = -3$.

Finally, we multiply $(61, 14, -3)$ by $(m, 1, m^2 - 19)$ and divide by 3 and $-3$ to get

$$\left( \frac{61m + 14 \cdot 19}{3}, \frac{61 + 14m}{3}, \frac{m^2 - 19}{-3} \right).$$

Choosing $m \equiv 1 \pmod 3$ makes $61 + 14m \equiv 0 \pmod 3$, and $m = 4$ makes $|m^2 - 19|$ smallest. This yields the triple

$$(170, \, 39, \, 1),$$

which corresponds to $170^2 - 19 \cdot 39^2 = 1$. We have therefore found the solution $(x, y) = (170, 39)$.

In Chapter 17, we'll use the closely related method of continued fractions to find solutions.

## 15.5   Chapter Highlights

1. Every prime $p \equiv 1 \pmod 4$ is a sum of two squares.

2. Every positive integer is a sum of four squares.

3. If $d > 0$ is not a square, then $x^2 - dy^2 = 1$ has nontrivial integer solutions.

## 15.6   Problems

### 15.6.1   Exercises

**Section 15.1: Volumes and Minkowski's Theorem**

1. Give an example of a rectangle $R$ that is not centrally symmetric and does not contain any points with integer coordinates (not even $(0,0)$) but with the area of $R$ larger than 1000.

2. Give an example of a region $R$ that is centrally symmetric, is not convex, does not contain any points with integer coordinates except $(0,0)$, and has area larger than 1000.

3.  Show that if the region bounded by a polygon is centrally symmetric, then the polygon has an even number of sides.

4.  Let $P$ be the parallelogram with vertices

    $$(-5, -1), (-3, -1), (3, 1), (5, 1).$$

    (a) Show that the area of $P$ is 4. (*Hint:* Area is base times height.)
    (b) Let $P'$ be the parallelogram with vertices

    $$(-4.99, -.99), (-3.01, -.99), (3.01, .99), (4.99, .99),$$

    so $P'$ lies slightly inside $P$. Show that the area of $P'$ is larger than 3.9 and the only point $(x, y)$ with integer coordinates that is inside $P'$ is $(0, 0)$.
    (c) The longest distance between two vertices of $P'$ is more than 10. Can you find a parallelogram $P''$ with area greater than 3.9 such that the only point $(x, y)$ with integer coordinates that is inside $P''$ is $(0, 0)$ and such that the longest distance between two vertices of $P''$ is more than 1000? This shows that even though our intuition for Minkowski's theorem might involve shapes such as squares, there are very elongated sets that also must be considered.

## Section 15.2: Sums of Two Squares

5.  Write 41 as a sum of two squares.

6.  Write 97 as a sum of two squares.

7.  Write 157 as a sum of two squares.

8.  Let $a$ and $b$ be nonzero integers. Show that when $\frac{a}{b} + \frac{b}{a}$ is expressed as a reduced fraction, its numerator contains no primes that are 3 mod 4.

9.  Determine which of the following numbers are sums of two squares:
    (a) $221 = 17 \times 13$, (b) $259 = 7 \times 37$, (c) $5040 = 2 \times 3 \times 4 \times 5 \times 6 \times 7$.

10. Write 65 as a sum of two squares in two different ways.

11. Use the facts that $13 = 2^2 + 3^2$ and $17 = 4^2 + 1^2$, plus Equation (15.1), to write $221 = 13 \cdot 17$ as a sum of two squares.

12. Use the facts that $29 = 2^2 + 5^2$ and $37 = 6^2 + 1^2$, plus Equation (15.1), to write $1073 = 29 \cdot 37$ as a sum of two squares.

13. Use the algorithm in Subsection 15.2.1, plus the fact that $546^2 \equiv -1$ (mod 1237), to write the prime 1237 as a sum of two squares.

14. Use the algorithm in Section 15.2.1, plus the fact that $17682^2 \equiv -1$ (mod 100049), to write the prime 100049 as a sum of two squares.

15. Let $p$ be a prime with $p \equiv 1$ or 3 (mod 8). Theorem 13.4 says that there exists $u$ with $u^2 \equiv -2$ (mod $p$).
    (a) Let $R$ be the elliptical region in the plane defined by $(px + uy)^2 + 2y^2 < 2p$. Show that the area of $R$ is $\sqrt{2}\pi$.
    (b) Show that there exist integers $x, y$ with $(x, y) \neq (0, 0)$ such that $0 < (px + uy)^2 + 2y^2 < 2p$.
    (c) Show that $(px + uy)^2 + 2y^2 \equiv 0$ (mod $p$).
    (d) Show that every prime $p \equiv 1$ or 3 (mod 8) can be written in the form $m^2 + 2n^2$ with integers $m$ and $n$.

16.   Let $n$ be a sum of two squares in two ways, so we have $n = a^2 + b^2 = c^2 + d^2$, with $a \geq b \geq 0$ and $c \geq d \geq 0$. Also, we may assume that $a > c$. Let $x = ad + bc$ and $y = ad - bc$. It can be shown (see Project 3) that $\gcd(x, n)$ and $\gcd(y, n)$ are nontrivial factors of $n$. Use the fact that $1000009 = 1000^2 + 3^2 = 972^2 + 235^2$ to factor $1000009$.

17.   Suppose that $n$ is not expressible as a sum of two squares of integers. Show that $n$ is not a sum of two squares of rational numbers.

## Section 15.3: Sums of Four Squares

18.   Write 23 as a sum of four squares.

19.   Write 123 as a sum of four squares.

20.   (a) Write 7 and 15 as sums of four squares.
      (b) Use Equation (15.2) to write $105 = 7 \cdot 15$ as a sum of four squares.
      (c) Write 105 as a sum of three squares.
      (d) Can 105 be written as a sum of two squares? Why or why not?

21.   The **quaternions** are expressions of the form $z = a + bi + cj + dk$, where $a, b, c, d$ are real numbers and $i, j, k$ satisfy the relations

$$i^2 = j^2 = k^2 = -1, \quad ij = k = -ji,$$
$$jk = i = -kj, \quad ki = j = -ik.$$

(a) Show that

$$(a_1 + b_1 i + c_1 j + d_1 k)(a_2 + b_2 i + c_2 j + d_2 k)$$
$$= (a_1 a_2 - b_1 b_2 - c_1 c_2 - d_1 d_2)$$
$$+ (a_1 b_2 + a_2 b_1 + c_1 d_2 - c_2 d_1)i$$
$$+ (a_1 c_2 + a_2 c_1 - b_1 d_2 + b_2 d_1)j$$
$$+ (a_1 d_2 + a_2 d_1 + b_1 c_2 - b_2 c_1)k.$$

(b) Define

$$\overline{a + bi + cj + dk} = a - bi - cj - dk$$

and

$$N(a + bi + cj + dk) = (a + bi + cj + dk)(\overline{a + bi + cj + dk}).$$

Show that

$$N(a + bi + cj + dk) = a^2 + b^2 + c^2 + d^2.$$

(c) It can be shown that if $z_1$ and $z_2$ are two quaternions, then $N(z_1 z_2) = N(z_1)N(z_2)$. Show that this relation is the same as Equation (15.2).

22.   Show that if $n \equiv 7 \pmod 8$, then $n$ is not a sum of 3 squares. (*Hint:* Lemma 5.7.)

23.   Show that every cube is congruent to 0, 1, or $-1$ mod 9. Deduce that there are infinitely many positive integers that are not sums of three cubes.

24.  (a) Express 23 and 239 as sums of 9 nonnegative cubes.
     (b) Show that 23 and 239 are not sums of 8 nonnegative cubes. (It has
     been shown that these are the only two numbers requiring 9 cubes.)
     (c) Using a computer, express 1290740 as a sum of six cubes. (It is conjec-
     tured that every number larger than 1290740 is a sum of five nonnegative
     cubes.)

25.  There is an integer $n \le 100$ that is a sum of 19 fourth powers but is not
     a sum of 18 fourth powers. Find it.

## Section 15.4: Pell's Equation

26.  Find integers $x, y$ with $x^2 - 5y^2 = 1$ and $y \ne 0$.

27.  Find integers $x, y$ with $x^2 - 7y^2 = 1$ and $y \ne 0$.

28.  We know that $(x, y) = (9, 4)$ is a solution of $x^2 - 5y^2 = 1$. Let $x_n + y_n\sqrt{5} = (9 + 4\sqrt{5})^n$. Explicitly calculate $(x_j, y_j)$ for $2 \le j \le 4$ and show that
     $x_j^2 - 5y_j^2 = 1$.

29.  (a) Suppose $x, y$ is an integer solution to $x^2 - 11y^2 = 1$ with $y \ne 0$. Show
     that $x \equiv \pm 1 \pmod{11}$.
     (b) Find integers $x, y$ with $x^2 - 11y^2 = 1$ and $y \ne 0$.

30.  Show that if $d$ is a positive integer that is divisible by a prime $p \equiv 3$
     (mod 4), then $x^2 - dy^2 = -1$ has no solutions. (*Hint:* Use part (b) of
     Theorem 13.4 or Exercise 24(a) of Chapter 8.)

31.  Let $d$ be a positive squarefree integer.
     (a) Show that if $d \equiv 1$ or $3$ or $7 \pmod 8$ and $x^2 - dy^2 = \pm 4$, then $x$ and
     $y$ are even. (*Hint:* An odd square is 1 mod 8.)
     (b) Show that if $d$ is even (and squarefree) and $x^2 - dy^2 = \pm 4$, then $x$
     and $y$ are even.
     (c) Suppose that $d$ is squarefree and $d \not\equiv 5 \pmod 8$. Show that if $x^2 - dy^2 = \pm 4$, then $x$ and $y$ are even.

32.  Let $d \equiv 5 \pmod 8$. The previous exercise says that this is the only case
     where $x^2 - dy^2 = \pm 4$ can possibly have a solution with $x$ and $y$ odd.
     Suppose $x, y$ is such a solution. Let

$$x_1 + y_1\sqrt{d} = \left(x + y\sqrt{d}\right)^3.$$

     Show that $x_1$ and $y_1$ are multiples of 8 and satisfy $x_1^2 - dy_1^2 = \pm 64$.
     (Therefore, $x_1/8$ and $y_1/8$ give integers $u, v$ with $u^2 - dv^2 = \pm 1$. This
     means that if we find a solution to $x^2 - dy^2 = \pm 4$ then we can use it
     to generate a solution to $u^2 - dv^2 = \pm 1$, and then square it to obtain a
     solution to $u^2 - dv^2 = +1$.)

33.  (a) Use one step of Bhāskara's Chakravala Method, starting with the
     triple $(8, 1, 3)$, to obtain a solution to $x^2 - 61y^2 = -4$.
     (b) Use the cubing procedure from the previous exercise to obtain a
     solution to $x^2 - 61y^2 = -1$.
     (c) Use squaring to produce a solution to $x^2 - 61y^2 = 1$.

## 15.6.2    Projects

1.  Edward Waring conjectured in 1770 that, for each positive integer $k$, there exists a number $g(k)$ such that every positive integer is a sum of $g(k)$ non-negative $k$th powers. For example, Lagrange's theorem says that for $k = 2$, every integer is a sum of $g(2) = 4$ squares. Waring's Conjecture was proved by Hilbert in 1909.
    (a) Show that if $n \equiv 7 \pmod 8$, then $n$ is not a sum of 3 squares.
    (b) Show that the cube of an integer is always congruent to 0, 1, or $-1$ mod 9. Deduce that there are infinitely many positive integers that are not sums of three non-negative cubes.
    (c) Express 23 and 239 as sums of 9 non-negative cubes.
    (d) Show that 23 is not a sum of 8 non-negative cubes. (It has been shown that 23 and 239 are the only two numbers requiring 9 cubes.)
    (e) Using a computer, express 1290740 as a sum of six cubes. (It is conjectured that every number larger than 1290740 is a sum of five non-negative cubes.)
    (f) There is an integer $n \leq 100$ that is a sum of 19 fourth powers but is not a sum of 18 fourth powers. Find it.

2.  Don Zagier's article "A One-Sentence Proof That Every Prime $p \equiv 1$ (mod 4) Is a Sum of Two Squares" (*American Mathematical Monthly* 97 (1990), p. 144) gives the following proof:
    The involution on the finite set $S = \{(x, y, z) \in \mathbb{N}^3 : x^2 + 4yz = p\}$ defined by

    $$(x, y, z) \mapsto \begin{cases} (x + 2z, z, y - x - z) & \text{if } x < y - z \\ (2y - x, y, x - y + z) & \text{if } y - z < x < 2y \\ (x - 2y, x - y + z, y) & \text{if } x > 2y \end{cases}$$

    has exactly one fixed point, so $|S|$ is odd and the involution defined by $(x, y, z) \mapsto (x, z, y)$ also has a fixed point. $\square$
    This project fills in the details.
    Let $p$ be an odd prime such that $p \equiv 1 \pmod 4$. Let $S$ be the set of triples $(x, y, z)$ of positive integers such that $x^2 + 4yz = p$.
    (a) Show that if $(x, y, z) \in S$ then $x \neq y - z$ and $x \neq 2y$.
    (b) Let

    $$A = \{(x, y, z) \in S \ | \ x < y - z\},$$
    $$B = \{(x, y, z) \in S \ | \ y - z < x < 2y\},$$
    $$C = \{(x, y, z) \in S \ | \ 2y < x\}.$$

    Let $f$ be the function on $S$ defined by

    $$f(x, y, z) = (x + 2z, z, y - z - x) \text{ if } (x, y, z) \in A$$
    $$f(x, y, z) = (2y - x, y, x - y + z) \text{ if } (x, y, z) \in B$$
    $$f(x, y, z) = (x - 2y, x - y + z, y) \text{ if } (x, y, z) \in C.$$

    Let $(x, y, z) \in S$ and let $f(x, y, z) = (x_1, y_1, z_1)$. Show that in all three cases, $x_1, y_1$, and $z_1$ are positive integers and that $x_1^2 + 4y_1z_1 = p$ (this means that $f(x, y, z) \in S$ for all $(x, y, z) \in S$).

(c) Show that $f$ maps all elements of $A$ into $C$, all elements of $B$ into $B$, and all elements of $C$ into $A$.

(d) A function $g$ defined on a set $T$ is called an *involution* of $T$ if $g(t) \in T$ and $g(g(t)) = t$ for all $t \in T$. Show that the function defined by $h((x, y, z)) = (x, z, y)$ is an involution of $S$.

(e) Show that $f$ (as defined above) is an involution of $S$.

(f) Show that if $T$ is a set with an odd number of elements and $g$ is an involution of $T$, then there exists an element $t \in T$ such that $g(t) = t$.

(g) Suppose $g$ is an involution of a finite set $T$ and that there exists exactly one $t \in T$ such that $g(t) = t$. Show that $T$ must have an odd number of elements.

(h) Show that there is exactly one triple $(x, y, z) \in S$ such that $f(x, y, z) = (x, y, z)$.

(i) Show that there are integers $u$ and $v$ such that $u^2 + v^2 = p$.

3.  Let $n$ be a sum of two squares in two ways, so we have $n = a^2 + b^2 = c^2 + d^2$ with $a \geq b \geq 0$ and $c \geq d \geq 0$. Also, we may assume that $a > c$. Let $x = ad + bc$ and $y = ad - bc$.

(a) Show that $xy \equiv 0 \pmod{n}$. (*Hint:* $a^2 \equiv -b^2$ and $d^2 \equiv -c^2 \pmod{n}$.)

(b) Show that $bc < dc < ad$ (you need to say why $c \neq 0$ and why $d \neq 0$). Conclude that $1 < y \leq x$.

(c) Show that $x^2 + (ac - bd)^2 = (a^2 + b^2)(c^2 + d^2)$.

(d) Show that $ac - bd > 0$. (*Hint:* Why can't we have both $a = b$ and $c = d$?)

(e) Show that $x < n$.

(f) We now have $1 < y \leq x < n$ and $xy \equiv 0 \pmod{n}$. Show that $\gcd(x, n)$ and $\gcd(y, n)$ are nontrivial factors of $n$.

(g) Show that if $n \equiv 1 \pmod 4$ is prime, then there is exactly one way (up to order and sign) of writing $n$ as a sum of two squares.

(h) Use the fact that $1000009 = 1000^2 + 3^2 = 972^2 + 235^2$ to factor $1000009$.

4.  An orchard is planted on the $xy$-plane by planting a tree at each point $(x, y)$ with integer coordinates such that $0 < x^2 + y^2 \leq 100$. The tree trunks each have radius $r$. You stand at the center of the orchard (since there is no tree at $(0, 0)$, this is possible). The goal is to show that if $r < 1/\sqrt{101}$ then you can see through the trees to the point $(10, 1)$, which is past the edge of the orchard, but if $r > 1/10$, then your view is blocked in all directions.

(a) In the diagram below (not drawn to scale), triangle $ABC$ is a right triangle, $AB$ has length 10, $BC$ has length 1, and $AD$ has length 1. The line $DE$ is perpendicular to $AC$. Show that triangle $ABC$ is similar to triangle $AED$.

(b) Show that the length of $DE$ is $1/\sqrt{101}$.

(c) Show that if $r < 1/\sqrt{101}$, then the line from $(0, 0)$ to $(10, 1)$ does not intersect any trees. That is, if $(x, y)$ has integer coordinates and $0 < x^2 + y^2 \leq 100$, then $(x, y)$ is farther than $r$ from the line.

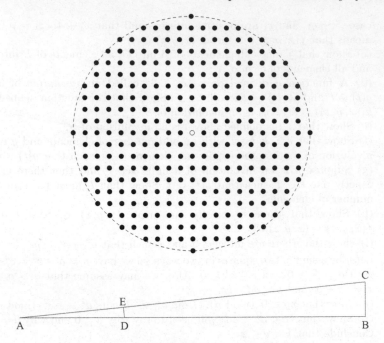

From now on, assume that the tree trunks have radius

$$r > 1/10.$$

Let $w$ be a real number satisfying

$$\frac{1}{10} < w < r \qquad \text{and} \qquad \frac{1}{10} < w < 1.$$

Let $L$ be any line through the origin. Form the rectangle $R$ with center

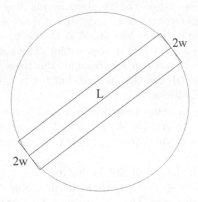

at the origin satisfying
(1) two sides are parallel to $L$, and

(2) the other two sides have length $2w$ and are tangent to the circle of radius 10 centered at the origin. (See the diagram.)

(d) Show that there is a point $(x, y) \neq (0, 0)$ with integer coordinates that is inside $R$.

(e) Let $L$ be as above and let $(x, y)$ be as in part (d). Show that $L$ intersects the circle of radius $r$ around $(x, y)$.

(f) There is a small technicality that must be dealt with: We must show that the point $(x, y)$ from part (d) is in the orchard; that is, we must show that $x^2 + y^2 \leq 100$: Show that $x^2 + y^2 < 101$ and explain why this implies that $x^2 + y^2 \leq 100$.

(g) Explain why you cannot see out of the orchard; that is, every line through the origin intersects a tree.

5.   The following gives a proof of Minkowski's theorem in the plane (a similar proof works in higher dimensions).

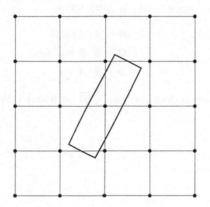

**FIGURE 15.4: A rectangle $B$.**

**FIGURE 15.5: Pieces of $B$.**

(a) Let $B$ be a convex centrally symmetric region in the plane with area larger than 4. Using the lines where one of the coordinates is an integer, cut the plane into unit squares. Finitely many of these squares contain pieces of $B$. Pile these squares containing pieces of $B$ on top of a unit square (do not rotate them). Show that there is some point in the unit square that has at least five points from inside $B$ above it.

(b) Show that there are five distinct points $(a_k, b_k)$ inside $R$, for $1 \leq k \leq 5$, such that $a_i - a_j$ and $b_i - b_j$ are integers for each $i, j$.

(c) Show that there are indices $i, j$ (with $i \neq j$) such that both $a_i - a_j$ and $b_i - b_j$ are even integers.

(d) Explain why $(-a_j, -b_j)$ is in $B$.

(e) Let $x = (a_i - a_j)/2$ and $y = (b_i - b_j)/2$. Show that $(x, y)$ is a point inside $B$ with integer coordinates.

## 15.6.3    Answers to "CHECK YOUR UNDERSTANDING"

1.  Take the interior of the square with vertices at $(\pm.9999, \pm.9999)$. This has area $4(.9999)^2 > 3.9$ and contains no integer point except the origin.

2.  Take the interior of the rectangle with vertices at $(\pm 1.1, \pm.001)$. This has area $.0044 < .01$ and contains the point $(1, 0)$.

3.  The Euclidean Algorithm for $\gcd(97, 22)$ is

$$97 = 4 \cdot 22 + 9$$
$$22 = 2 \cdot 9 + 4$$
$$9 = 2 \cdot 4 + 1.$$

The first two remainders less than $\sqrt{97}$ are 9 and 4. We have $97 = 9^2 + 4^2$.

# Chapter 16

---

# Arithmetic Functions

---

## 16.1 Perfect Numbers

Suppose that you are playing with sums of divisors of numbers (yes, mathematicians do things like this) and you discover the following two examples:

$$1 + 2 + 3 = 6$$
$$1 + 2 + 4 + 7 + 14 = 28.$$

What is so special about these? 1, 2, 4, 7, 14 are the positive divisors of 28 that are less than 28. Similarly, 1, 2, 3 are the divisors of 6 that are less than 6. Are there any other such examples? In fact, here is another:

$$1 + 2 + 4 + 8 + 16 + 31 + 62 + 124 + 248 = 496.$$

**Definition 16.1.** *A positive integer n is called* **perfect** *if n is the sum of the positive divisors of n that are less than n. It is called* **deficient** *if the sum of these divisors is less than n, and it is* **abundant** *if the sum of these divisors is greater than n.*

**Examples.** 6, 28, and 496 are perfect. However, 15 is deficient because $1 + 3 + 5 < 15$, and 12 is abundant because $1 + 2 + 3 + 4 + 6 > 12$.

**Definition 16.2.** *When n is a positive integer, $\sigma(n)$ is the sum of all positive divisors of n (including n).*

**Examples.** $\sigma(6) = 1 + 2 + 3 + 6 = 12$, $\quad \sigma(15) = 1 + 3 + 5 + 15 = 24$.

**Proposition 16.3.** *Let $p$ be prime and let $k \geq 1$. Then*

$$\sigma(p^k) = 1 + p + p^2 + \cdots + p^k = \frac{p^{k+1} - 1}{p - 1}.$$

*Proof.* The divisors of $p^k$ are $1, p, p^2, \ldots, p^k$, so the sum of these is $\sigma(p^k)$. The second equality follows from letting $x = p$ in the identity

$$x^{k+1} - 1 = (x - 1)(x^k + x^{k-1} + \cdots + x^2 + x + 1)$$

(if we divide both sides by $x - 1$ we have the formula for a geometric sum; see Appendix A).                                                    □

The advantage of using $\sigma(n)$ rather than the sum of the divisors less than $n$ is the following.

**Proposition 16.4.** *If $m$ and $n$ are positive integers with $\gcd(m, n) = 1$, then*

$$\sigma(mn) = \sigma(m) \cdot \sigma(n).$$

This property is called **multiplicativity** and will be proved in the next section (Corollary 16.11).

**Example.** $\sigma(3) = 1 + 3 = 4$ and $\sigma(5) = 1 + 5 = 6$. The proposition says that $\sigma(15) = \sigma(3)\sigma(5) = 4 \cdot 6 = 24$, which agrees with our previous calculation.

Since we are including $n$ in $\sigma(n)$, we have

$$n \text{ is perfect} \iff \sigma(n) = 2n.$$

We can use the multiplicativity of $\sigma(n)$ to produce all even perfect numbers.

**Theorem 16.5.** *Let $n$ be an even perfect number. Then there is a prime $p$ such that $2^p - 1$ is prime and*

$$n = 2^{p-1}(2^p - 1).$$

*Conversely, every $n$ of this form, with $p$ and $2^p - 1$ prime, is perfect.*

**Remark.** The formula that gives perfect numbers is in Euclid's *Elements* (Book IX, Proposition 36). The fact that all even perfect numbers have this form was proved by Euler.

*Proof.* Let's show that the formula yields perfect numbers. Since $2^p - 1$ is prime,

$$\sigma(2^p - 1) = 1 + (2^p - 1) = 2^p.$$

Also,

$$\sigma(2^{p-1}) = 1 + 2 + 4 + \cdots + 2^{p-1} = 2^p - 1$$

by Proposition 16.3. The multiplicativity of $\sigma$ (Proposition 16.4) says that

$$\sigma(n) = \sigma(2^{p-1}(2^p - 1)) = \sigma(2^{p-1}) \cdot \sigma(2^p - 1) = (2^p - 1)(2^p) = 2n,$$

so $n$ is perfect.

Now suppose $n$ is perfect and even. We want to show that $n$ has the desired form, namely, $n = 2^{p-1}(2^p - 1)$.

Write $n = 2^k m$ with $m$ odd and $k \geq 1$. Then

$$\sigma(n) = \sigma(2^k) \cdot \sigma(m).$$

As before, we have $\sigma(2^k) = 1 + 2 + 4 + \cdots + 2^k = 2^{k+1} - 1$. Although we don't know much about $m$, let's see what we can say. If $m = 1$, then $n = 2^k$ and $\sigma(n) = 2^{k+1} - 1 \neq 2n$. Therefore, $m > 1$, and we have at least two distinct divisors of $m$, namely, 1 and $m$. Write

$$\sigma(m) = 1 + m + s,$$

where $s \geq 0$ is the sum of the other divisors of $m$. The key step will be to show that $s = 0$.

Since $n$ is assumed to be perfect,

$$2^{k+1}m = 2n = \sigma(n) = \sigma(2^k) \cdot \sigma(m) = (2^{k+1} - 1)(1 + m + s).$$

When this is expanded and rearranged, we get

$$m = (2^{k+1} - 1)(s + 1). \tag{16.1}$$

Therefore, $s + 1$ is a divisor of $m$. Since $n$ is even, we have $k \geq 1$ and $2^{k+1} - 1 > 1$. Equation (16.1) implies that $m > s + 1$. If $s \neq 0$, then $s + 1$ cannot be 1, so $m > s + 1 > 1$ and we have found another divisor of $m$, namely, $s + 1$. This means that

$$\sigma(m) \geq 1 + m + (s + 1).$$

But $\sigma(m) = 1 + m + s$, so $1 + m + s \geq 1 + m + (s + 1)$, which is impossible. Therefore, we are forced to have $s = 0$. Since $s$ is the sum of the divisors of $m$ other than 1 and $m$, we conclude that $m$ has no other divisors. The only numbers with this property are 1 and primes. Since we already showed that $m > 1$, we must have $m$ prime. Also, Equation (16.1) implies that

$$m = (2^{k+1} - 1)(s + 1) = 2^{k+1} - 1.$$

Therefore,

$$n = 2^k m = 2^k(2^{k+1} - 1),$$

which shows that $n$ has the desired form. Since we showed that $m$ is prime, $2^{k+1} - 1$ is prime. As shown in Section 2.9, this means that $k+1$ is prime. This completes the proof. $\qquad\square$

Recall that primes of the form $M_p = 2^p - 1$ are called **Mersenne primes**. The first few Mersenne primes are

$$M_2 = 3, \quad M_3 = 7, \quad M_5 = 31, \quad M_7 = 127, \quad M_{13} = 8191.$$

The corresponding perfect numbers are

$$6, \quad 28, \quad 496, \quad 8128, \quad 33550336.$$

It is generally expected (but not yet proved) that there are infinitely many Mersenne primes, and therefore that there are infinitely many even perfect numbers. As of March 2017, the largest known Mersenne prime is $M_{74207281}$, and the corresponding perfect number is $2^{74207280}(2^{74207281} - 1)$.

What about odd perfect numbers? No odd perfect numbers are known, and it is conjectured that there are none. It has been proved that an odd perfect number must have at least 9 distinct prime factors. Results such as this lead most people to think that it is unlikely that odd perfect numbers exist. However, perhaps we should be slightly cautious. Most numbers have at least 9 prime factors, but we tend to meet only those with a few prime factors, so it's hard to have good intuition about such numbers.

---

**CHECK YOUR UNDERSTANDING:**

1. Show that 18 is an abundant number.

# 16.2   Multiplicative Functions

The relation $\sigma(mn) = \sigma(m) \cdot \sigma(n)$ when $\gcd(m, n) = 1$, which was very useful in the previous section, holds for several other functions. In this section, we develop the theory of such functions.

**Definition 16.6.** *A real-valued (or complex-valued) function $f(x)$ on the positive integers is called a* **multiplicative function** *if $f(mn) = f(m) \cdot f(n)$ for all pairs of positive integers $m, n$ with $\gcd(m, n) = 1$.*

The condition that $f(mn) = f(m)f(n)$ when $\gcd(m, n) = 1$ might seem strange. Why don't we require the multiplicativity for all $m, n$? In several number theoretic functions, complete multiplicativity (that is, no restriction on the gcd) does not occur. For example, Euler's $\phi$ function is multiplicative, but $\phi(4 \cdot 6) = 8 \neq 4 = \phi(4)\phi(6)$.

**Examples.** You've already seen some of these examples, while some others are new.

1.  The Euler $\phi$ function. For example,

    $$\phi(5) = 4, \quad \phi(7) = 6, \quad \phi(35) = 24 = \phi(5) \cdot \phi(7).$$

2.  The function $f(n) = n$ is multiplicative since $f(mn) = mn = f(m) \cdot f(n)$. We do not need the condition that $\gcd(m, n) = 1$ in this case.

3.  The function $f(n) = 1$ for all $n$ is multiplicative. Again, we do not need the condition that $\gcd(m, n) = 1$ in this case.

4.  The sum of divisors function, $\sigma(n)$, is multiplicative. This property was used in the previous section and will be proved in this section.

5.  Let $n$ be a positive integer and let $\tau(n)$ be the number of positive divisors of $n$ (that is, the number of integers $d$ such that $1 \leq d \leq n$ and $d \mid n$). For example,

    $$\tau(4) = 3, \quad \tau(75) = 6, \quad \tau(300) = 18.$$

    We'll show later (see Corollary 16.10) that $\tau$ is multiplicative.

6.  Define the Möbius function:

$$\mu(n) = \begin{cases} (-1)^r & \text{if } n \text{ is a product of } r \text{ distinct primes.} \\ 0 & \text{otherwise.} \end{cases}$$

For example,

$$\mu(15) = (-1)^2 = +1, \quad \mu(12) = 0,$$
$$\mu(2) = -1, \quad \mu(1) = (-1)^0 = +1.$$

Then $\mu$ is multiplicative. This is Exercise 20. Note that $\mu(n) = 0$ if $n$ is not squarefree.

Here is a very useful property of multiplicative functions.

**Proposition 16.7.** *If $f$ and $g$ are multiplicative functions and $f(p^j) = g(p^j)$ for all primes $p$ and all $j \geq 0$, then $f(n) = g(n)$ for all positive integers $n$.*

*Proof.* Let $n = p_1^{a_1} p_2^{a_2} \cdots p_r^{a_r}$ be the prime factorization of $n$. Since $\gcd(p_i^{a_i}, p_j^{a_j}) = 1$ if $i \neq j$, the multiplicativity of $f$ implies that

$$\begin{aligned} f(n) &= f(p_1^{a_1}) \cdot f(p_2^{a_2} \cdots p_r^{a_r}) \\ &= f(p_1^{a_1}) \cdot f(p_2^{a_2}) \cdot f(p_3^{a_3} \cdots p_r^{a_r}) \\ &= \cdots \\ &= f(p_1^{a_1}) \cdot f(p_2^{a_2}) \cdots f(p_r^{a_r}) \\ &= g(p_1^{a_1}) \cdot g(p_2^{a_2}) \cdots g(p_r^{a_r}) \\ &= g(p_1^{a_1} \cdot p_2^{a_2} \cdots p_r^{a_r}) \\ &= g(n). \end{aligned} \qquad \square$$

If we have one multiplicative function, we can produce another as follows.

**Proposition 16.8.** *Let $f$ be a multiplicative function and let*

$$g(n) = \sum_{d \mid n} f(d)$$

*(the sum is over the positive divisors of $n$). Then $g$ is multiplicative.*

*Proof.* We start with a lemma.

**Lemma 16.9.** *Let $m$ and $n$ be positive integers with $\gcd(m,n) = 1$. Let $d$ be a positive divisor of $mn$. Then $d$ has a unique decomposition $d = d_1 d_2$ with $d_1 \mid m$ and $d_2 \mid n$ (where $d_1, d_2$ are positive integers).*

**Example.** Let $m = 56 = 2^3 \cdot 7$ and $n = 75 = 3 \cdot 5^2$. Let $d = 70$. Then

$$70 \mid 4200 = 56 \cdot 75.$$

The lemma says that we can write $70 = 14 \cdot 5$, where $d_1 = 14 \mid 56$ and $d_2 = 5 \mid 75$.

*Proof.* Let $m = p_1^{a_1} p_2^{a_2} \cdots p_r^{a_r}$ and $n = q_1^{b_1} q_2^{b_2} \cdots q_s^{b_s}$ be the prime factorizations of $m$ and $n$. A divisor $d$ of $mn$ has the form

$$d = p_1^{a_1'} p_2^{a_2'} \cdots p_r^{a_r'} q_1^{b_1'} q_2^{b_2'} \cdots q_s^{b_s'}$$

with $0 \le a_i' \le a_i$ for all $i$ and $0 \le b_j' \le b_j$ for all $j$. Since we require that $d_1 \mid m$ and since $\gcd(m,n) = 1$, we have $\gcd(d_1, n) = 1$, so the primes $q_j$ cannot appear in $d_1$. Similarly, the primes $p_i$ cannot appear in $d_2$. Therefore, we have no choice and we must take

$$d_1 = p_1^{a_1'} p_2^{a_2'} \cdots p_r^{a_r'}, \qquad d_2 = q_1^{b_1'} q_2^{b_2'} \cdots q_s^{b_s'}.$$

Then $d = d_1 d_2$, with $d_1 \mid m$ and $d_2 \mid n$, as required.                    □

We now can prove Proposition 16.8. Let $g$ be as defined in the proposition and let $\gcd(m,n) = 1$. Then

$$g(mn) = \sum_{d \mid mn} f(d) = \sum_{d_1 \mid m,\, d_2 \mid n} f(d_1 d_2).$$

Because $d_1 \mid m$ and $d_2 \mid n$, and because $\gcd(m,n) = 1$, we have $\gcd(d_1, d_2) = 1$. The function $f$ is assumed to be multiplicative, so $f(d_1 d_2) = f(d_1)f(d_2)$. The sum becomes

$$\sum_{d_1 \mid m}\sum_{d_2 \mid n} f(d_1)f(d_2) = \left( \sum_{d_1 \mid m} f(d_1) \right)\left( \sum_{d_2 \mid n} f(d_2) \right) = g(m)g(n).$$

Therefore, $g$ is multiplicative.                                                □

**Corollary 16.10.** $\tau(n)$ *(the number of divisors of $n$) is multiplicative.*

*Proof.* In Proposition 16.7, let $f(n) = 1$ for all $n$. Then $f$ is certainly multiplicative. The function $g(n) = \sum_{d|n} 1$ is the sum of a set of 1's, one for each divisor of $n$. In other words, $g(n)$ is the number of divisors of $n$, which is $\tau(n)$. The proposition implies that $\tau(n)$ is multiplicative.    $\square$

When $n = p^k$, it's easy to evaluate $\tau(p^k)$. The divisors of $p^k$ are $1, p, p^2, \ldots, p^k$, so $\tau(p^k) = k + 1$. The multiplicativity of $\tau$ implies that

$$\tau(p_1^{a_1} p_2^{a_2} \cdots p_r^{a_r}) = (a_1 + 1)(a_2 + 1) \cdots (a_r + 1). \tag{16.2}$$

For example, $\tau(12) = \tau(2^2 \cdot 3) = (2 + 1)(1 + 1) = 6$.

**Corollary 16.11.** *$\sigma(n)$ (the sum of the divisors of $n$) is multiplicative.*

*Proof.* In Proposition 16.7, let $f(n) = n$, so

$$g(n) = \sum_{d|n} d = \sigma(n).$$

The proposition implies that $\sigma(n)$ is multiplicative.    $\square$

The proof of the next corollary shows how Proposition 16.7 can be used to show that two multiplicative functions are equal.

**Corollary 16.12.** *Let $n \geq 1$. Then*

$$\sum_{d|n} \phi(d) = n,$$

*where the sum is over all positive divisors of $n$.*

*Proof.* Let $f(n) = \phi(n)$, which is multiplicative by Proposition 8.6. Then $g(n) = \sum_{d|n} \phi(d)$ is multiplicative. Inspired by Proposition 16.7, we compute $g(p^j)$. The divisors of $p^j$ are $1, p, p^2, \ldots, p^j$. Therefore,

$$\begin{aligned}
g(p^j) &= \phi(1) + \phi(p) + \phi(p^2) + \cdots + \phi(p^j) \\
&= 1 + (p - 1) + (p^2 - p) + \cdots + (p^j - p^{j-1}) \\
&= p^j.
\end{aligned}$$

Let $h(n) = n$, which is multiplicative. Then $g(p^j) = h(p^j)$ for all primes $p$ and all $j \geq 0$. By Proposition 16.7, $g(n) = h(n)$ for all $n \geq 1$. This is the statement of the corollary.    $\square$

**Example.** Let $n = 12$. Then $\phi(1) + \phi(2) + \phi(3) + \phi(4) + \phi(6) + \phi(12) = 1 + 1 + 2 + 2 + 2 + 4 = 12$.

**Corollary 16.13.** *Let $n \geq 1$. Then*

$$\sum_{d|n} \mu(d) = \begin{cases} 1 & \text{if } n = 1 \\ 0 & \text{if } n > 1, \end{cases}$$

*where $\mu$ is the Möbius function defined in the examples preceding Proposition 16.7.*

*Proof.* Let $h(n) = 1$ if $n = 1$ and $h(n) = 0$ if $n > 1$. Then $h$ is multiplicative: $h(mn) = 0$ unless $m = n = 1$, and $h(m) \cdot h(n) = 0$ unless $m = n = 1$. Also, $h(1)h(1) = h(1 \cdot 1)$. In Proposition 16.8, let

$$f(n) = \mu(n) \text{ and } g(n) = \sum_{d|n} \mu(d).$$

When $n = p^j$ and $j \geq 1$, we have

$$\begin{aligned} g(p^j) &= \mu(1) + \mu(p) + \mu(p^2) + \cdots + \mu(p^j) \\ &= 1 - 1 + 0 + \cdots + 0 \\ &= 0 = h(p^j). \end{aligned}$$

Since $h(p^0) = 1$ and $g(p^0) = g(1) = 1$, we have proved that $g(p^j) = h(p^j)$ for all primes $p$ and all $j \geq 0$. Proposition 16.7 implies that $g(n) = h(n)$ for all $n \geq 1$. This is the statement that we were trying to prove. □

Corollary 16.13 allows us to solve for $f$ in terms of $g$ in Proposition 16.8.

**Proposition 16.14.** *(Möbius inversion) Suppose $f$ is any function (not necessarily multiplicative) and*

$$g(n) = \sum_{d|n} f(d).$$

*Then*

$$f(n) = \sum_{d|n} \mu(d)g\left(\frac{n}{d}\right).$$

*Proof.* The definition of $g$ says that

$$g\left(\frac{n}{d}\right) = \sum_{e|(n/d)} f(e),$$

which allows us to write

$$\sum_{d|n} \mu(d)g\left(\frac{n}{d}\right) = \sum_{d|n} \mu(d) \sum_{e|(n/d)} f(e).$$

In the right-hand side, we have pairs of integers $d$ and $e$ such that $d \mid n$ and $e \mid (n/d)$. This simply means that $de \mid n$. Notice that

$$d \mid n \text{ and } e \mid (n/d) \Longleftrightarrow e \mid n \text{ and } d \mid (n/e).$$

The sum becomes

$$\sum_{e|n} \sum_{d|(n/e)} \mu(d)f(e) = \sum_{e|n} \left( \sum_{d|(n/e)} \mu(d) \right) f(e) = f(n)$$

because Corollary 16.13 says that

$$\sum_{d|(n/e)} \mu(d) = \begin{cases} 1 \text{ if } n = e \\ 0 \text{ if } n > e, \end{cases}$$

which makes all the terms in the inner sum 0 except for the one with $n = e$. This completes the proof. $\qquad\square$

This proposition gives us some interesting formulas. Here is one for the $\tau$ function. Since

$$\tau(n) = \sum_{d|n} 1$$

Proposition 16.14 tells us that

$$1 = \sum_{d|n} \mu(d)\tau\left(\frac{n}{d}\right).$$

For example, let $n = 12$. Then

$$\mu(1)\tau(12)+\mu(2)\tau(6)+\mu(3)\tau(4)+\mu(4)\tau(3)+\mu(6)\tau(2)+\mu(12)\tau(1)$$
$$= 6 - 4 - 3 + 0 + 2 + 0 = 1.$$

If we apply Proposition 16.14 to the formula in Corollary 16.12, we recover a formula for $\phi(n)$. First, we have

$$\phi(n) = \sum_{d|n} \mu(d)\frac{n}{d}.$$

When $n = p^k$, this says that

$$\phi(p^k) = \mu(1)\frac{p^k}{1} + \mu(p)\frac{p^k}{p} + 0 + 0 + \cdots = p^k - p^{k-1} = p^k\left(1 - \frac{1}{p}\right).$$

The multiplicativity of $\phi$ yields the well-known formula

$$\phi(n) = \prod_{p|n} p^k\left(1 - \frac{1}{p}\right) = n\prod_{p|n}\left(1 - \frac{1}{p}\right).$$

---

**CHECK YOUR UNDERSTANDING:**

2. How many divisors does 60 have?
3. Evaluate $\sigma(40)$.

---

# 16.3    Chapter Highlights

1. $\phi$, $\sigma$, $\tau$, and $\mu$ are multiplicative.

2. Even perfect numbers have the form $2^{p-1}(2^p - 1)$, where $2^p - 1$ is prime.

---

# 16.4    Problems

## 16.3.1    Exercises

1. (a) List the divisors of 2, of 4, of 15, of 30, and of 60.
   (b) Compute $\sigma(n)$ for $n = 2$, 4, 15, 30, and 60.
   (c) Show that $\sigma(4)\sigma(15) = \sigma(60)$ but $\sigma(2)\sigma(30) \neq \sigma(60)$.

2. How many divisors does 1000 have?

3. What is the sum of the divisors of 20?

4. Find a value of $n$ that has exactly 32 divisors.

5. Show that $\phi(n) = n - 1$ if and only if $n$ is prime.

6. Show that $\sigma(n) = n + 1$ if and only if $n$ is prime.

7. Show that if $n = 2^p - 1$ is prime then $\sigma(\sigma(n)) = 2n + 1$.

8. Show that $\sigma(n) = 5$ has no solution.

9. Find four pairs $(m, n)$ of integers with $5 \le m < n$ such that $\phi(m) = \phi(n)$.

10. We know that $\phi(n) = \phi(n + 1)$ when $n = 3$. Find a value of $n \ge 4$ with $\phi(n) = \phi(n + 1)$.

11. Show that there is no integer $n$ with $\phi(n) = 14$.

12. The first few perfect numbers are 6, 28, 496, 8128.
    (a) Show that the last digit of an even perfect number is always 6 or 8.
    (b) Show by example that the last digits of the perfect numbers do not always alternate between 6 and 8.

13. Show that $n$ is perfect if and only if $2 = \sum_{d|n} \frac{1}{d}$ (for example, $2 = 1 + \frac{1}{2} + \frac{1}{3} + \frac{1}{6}$).

14. (a) Use the multiplicativity of $\tau(n)$ (= number of divisors) to show that $\tau(n)$ is odd if and only if $n$ is a square.
    (b) Pair a divisor $d$ of $n$ with $n/d$. Note that a divisor $d$ is paired with itself if and only if $n = d^2$. Deduce the result of part (a) directly from this observation.

15. There is a long row of doors labeled 1 through 10000. Originally, they are all closed. The first person opens all of them. The second person closes the 2nd, the 4th, the 6th, etc. The third person changes the status (that is, opens a closed door and closes an open door) of the 3rd, the 6th, the 9th, etc. This continues, with the $n$th person changing the status of the doors that are multiples of $n$ until the 10000th person changes the status of the 10000th door. When this is finished, which doors are open and which are closed?

16. Descartes thought he had found that $n = 3^2 \cdot 7^2 \cdot 11^2 \cdot 13^2 \cdot 22021$ is an odd perfect number. It appears that he thought 22021 is prime and calculated

$$\sigma(3^2)\sigma(7^2)\sigma(11^2)\sigma(13^2)(1 + 22021) = 2n.$$

Using the facts $\sigma(7^2) = 3 \cdot 19$ and $\sigma(13^2) = 183 = 3 \cdot 61$, factor 22021.

17. Let $\sigma(n) = \sum_{d|n} d$ be the sum of the divisors of $n$ and let $\phi(n)$ be the Euler phi function. Let $n = pq$ be the product of two distinct primes. Show that $\sigma(n) + \phi(n) = 2n + 2$.

18. Let $m$ and $n$ be positive integers.
    (a) Show that

$$\phi(m)\phi(n) = \phi(mn) \prod_{p|\gcd(m,n)} \left(1 - \frac{1}{p}\right).$$

    (b) Show that $\phi(m)\phi(n) \le \phi(mn)$, and that there is equality if and only if $\gcd(m, n) = 1$.

19. Let $\sigma(n) = \sum_{d|n} d$ be the sum of the divisors of $n$.
    (a) Let $p$ be an odd prime. Show that $\sigma(p^a)$ is odd if and only if $a$ is even.
    (b) Show that $\sigma(2^a)$ is always odd.
    (c) Show that $\sigma(n)$ is odd if and only if $n$ is a square or twice a square.
    (d) Show that a square cannot be a perfect number (as Lenstra remarked, this says that there are no "perfect squares").

20. Show that the Möbius function $\mu(n)$ is multiplicative.

21. (a) Let $\mu$ be the Möbius function. Let $g(n) = \sum_{d|n} \mu(d)^2$. Show that $g(n)$ is multiplicative.
    (b) Let $\omega(n)$ be the number of distinct prime factors dividing $n$ (for example, $\omega(60) = 3$ since 2, 3, 5 divide 60). Show that $2^{\omega(n)}$ is multiplicative.
    (c) Show that
    $$\sum_{d|n} \mu(d)^2 = 2^{\omega(n)}.$$

22. Let $n \geq 1$. Show that $\prod_{d|n} d = n^{\tau(n)/2}$.

23. Let $n \geq 12$ be a multiple of 6. Show that $n$ is abundant.

24. Assume that $m$ is perfect or abundant. Let $n \geq 2$ be an integer. Show that $mn$ is abundant.

25. We saw in the text that
    $$1 = \sum_{d|n} \mu(d)\tau\left(\frac{n}{d}\right).$$
    Suppose that $n = p_1 p_2 \cdots p_k$ is the product of $k$ distinct primes. Show that this formula is the same as
    $$1 = (2-1)^k = 2^k - \binom{k}{1}2^{k-1} + \binom{k}{2}2^{k-2} - \cdots + (-1)^k.$$

26. (a) Show that if $n > 1$ then
    $$\frac{\sigma(n)}{n} < \prod_{p|n} \frac{p}{p-1},$$
    where the product is over the distinct primes dividing $n$.
    (b) Show that if $n$ is an odd perfect number then $n$ has at least three distinct prime factors. (*Hint:* If there are two prime factors $p > q \geq 3$, then $\frac{p}{p-1}\frac{q}{q-1} \leq \frac{5}{4}\frac{3}{2} < 2$.)

27. Show that if $\sigma(m)$ is prime then $\tau(m)$ is prime. (*Hint:* If $(p^n - 1)/(p-1)$ is prime, then $n$ is prime.)

28. Show that if $2^n - 1$ is composite then $2^{n-1}(2^n - 1)$ is abundant.

29. Show that if $q = (3^k - 1)/2$ is prime and $n = 3^{k-1}(3^k - 1)/2$, then $\sigma(n) < 2n$.

# 16.4.2   Projects

1. Two distinct positive integers $a$ and $b$ are called an **amicable pair** if $b$ equals the sum of the divisors of $a$ that are less than $a$, and $a$ equals the

sum of the divisors of $b$ that are less than $b$.

(a) Show that 220 and 284 form an amicable pair.

(b) Show that $a$ and $b$ are amicable if and only if $\sigma(a) = \sigma(b) = a + b$.

(c) Suppose that $n > 1$ is an integer such that $p = 3 \cdot 2^{n-1} - 1$, $q = 3 \cdot 2^n - 1$, and $r = 9 \cdot 2^{2n-1} - 1$ are prime. Show that $a = 2^n pq$ and $b = 2^n r$ form an amicable pair. (This formula was discovered by Thabit ibn-Qurra almost 1200 years ago.)

(d) Show that $n = 2$ gives 220 and 284. Find two other amicable pairs given by this formula.

(e) Show that the amicable pair 1184 and 1210 (discovered by the 16-year-old student B. Paganini in 1866) is not given by the formula of part (c).

(f) Show that if $a$ and $b$ are given by the formula of (c), then $a + b$ is a multiple of 9.

(g) Show that the pair $a = 53151801712 = 2^4 \cdot 19 \cdot 6451 \cdot 27103$ and $b = 55270703248 = 2^4 \cdot 109 \cdot 307 \cdot 103231$ is amicable but $a + b$ is not a multiple of 9. (Of the millions of amicable pairs known, most satisfy $a + b \equiv 0 \pmod 9$.)

2.  This project gives another proof of the existence of primitive roots. Let $p$ be a prime.

    (a) Let $d \mid p - 1$ and let $b \not\equiv 0 \pmod p$. Show that if $\mathrm{ord}_p(b) = d$, then $\{1, b, b^2, \ldots, b^{d-1}\}$ gives all the solutions of $X^d - 1 \equiv 0 \pmod p$. (*Hint:* See Lemma 11.17.)

    (b) Let $d \mid p - 1$. Show that if $\mathrm{ord}_p(b) = d$, then there are exactly $\phi(d)$ values of $c \pmod p$ such that $\mathrm{ord}_p(c) = d$.

    (c) Let $N_d$ be the number of $c \pmod p$ with $\mathrm{ord}_p(c) = d$. Show that $N_d \leq \phi(d)$.

    (d) Show that if $d \mid p - 1$ then $N_d = \phi(d)$. (*Hint:* Use Corollary 16.12. What happens if $N_d < \phi(d)$ for some $d \mid p - 1$?)

    (e) Show that there are $\phi(p - 1)$ primitive roots mod $p$.

## 16.4.3   Computer Explorations

1.  Rosser and Schoenfeld showed that if $n \geq 3$ then

$$n/\phi(n) < 1.7811 \cdot \ln(\ln n) + \frac{5}{2\ln(\ln n)}$$

except for one value of $n$. Find that $n$. (*Hint:* Look at products of distinct small primes.)

2.  Carmichael conjectured that if $x$ and $m$ are positive integers with $\phi(x) = m$, then there is an integer $y \neq x$ with $\phi(y) = m$. In 1998, Ford showed that this statement is true for all $m < 10^{10^{10}}$. Verify that it is true for all $m \leq 100$. (*Note:* for some values of $m$, there might not be an $x$ with $\phi(x) = m$. The conjecture says that if there is one, then there is at least one more. The inequality of Computer Exploration 1 can be used to bound the largest $x$ with $\phi(x) \leq 100$.)

3.  Mertens conjectured that $|\sum_{n \leq x} \mu(n)| \leq \sqrt{x}$ for all $x \geq 1$. Odlyzko and

te Riele showed in 1985 that there are (very large, non-explicit) values of $x$ such that the conjecture is false. Show that it is true for all $x \leq 1000$.

4. A positive integer $n$ is called **highly composite** if $n$ has more divisors than any smaller number. That is, $\tau(n) > \tau(m)$ for all $m$ with $1 \leq m < n$. The first few highly composite numbers are 1, 2, 4, 6, 12, 24.

   (a) There are 20 highly composite numbers less than 10000. Can you find them?

   (b) Notice that the numbers $k!$ for $0 \leq k \leq 7$ are highly composite. Show that 8! and 9! are not highly composite.

5. A positive integer $n$ is called **highly abundant** if the sum of the divisors of $n$ is larger than the sum of the divisors of each smaller positive integer. That is, $\sigma(n) > \sigma(m)$ for all $m$ with $1 \leq m < n$. The first few highly abundant numbers are 1, 2, 3, 4, 6, 8, 10.

   (a) There are eight highly abundant numbers between 500 and 1000. Can you find them?

   (b) $n = 9! = 362880$ is not highly abundant. Find $m < n$ with $\sigma(m) > \sigma(n)$.

## 16.4.4   Answers to "CHECK YOUR UNDERSTANDING"

1. The divisors of 18 are 1, 2, 3, 6, 9, 18. Their sum is 39, which is larger than $36 = 2 \cdot 18$. Therefore, 18 is abundant.

2. By Equation (16.2), $\tau(60) = \tau(2^2 \cdot 3 \cdot 5) = (2+1)(1+1)(1+1) = 12$.

3. $\sigma(40) = \sigma(2^3)\sigma(5) = (1+2+4+8)(1+5) = 90$.

# Chapter 17

# Continued Fractions

Christiaan Huygens faced a challenge. In 1680, the Dutch astronomer was designing a mechanical planetarium for l'Académie Royale des Sciences in France. He calculated that in a 365-day year, the Earth goes through an angle of $359°45'40''31'''$ of arc, where $31'''$ denotes $31/60$ seconds. This is

$$(359 \cdot 60^3 + 45 \cdot 60^2 + 40 \cdot 60 + 31)/60 = 77708431/60$$

seconds of arc (note that this is slightly less than $360°$, which is why we need leap years). Saturn, which goes slower, goes through $2640858/60$ seconds of arc in a year. Therefore, in his planetarium, he needed gears in the ratio of

$$\frac{77708431}{2640858} \approx 29.42545$$

in order to make the relative motions of Earth and Saturn accurate. Since it was impossible to make a gear with 77708431 teeth, he approximated the ratio as

$$\frac{77708431}{2640858} \approx \frac{206}{7} \approx 29.42857.$$

Why did he choose this approximation, and how did he find it? Huygens wrote

$$\frac{77708431}{2640858} = 29 + \cfrac{1}{2 + \cfrac{1}{2 + \cfrac{1}{1 + \cfrac{1}{5 + \cfrac{1}{1 + \ddots}}}}}$$

(the fraction continues for 11 more terms). Expressions such as this are called **continued fractions**. Stopping before the 5 yields

$$\frac{206}{7} = 29 + \cfrac{1}{2 + \cfrac{1}{2 + \cfrac{1}{1}}},$$

which is the approximation that Huygens used.

In this chapter, we develop the theory of continued fractions, show how they give rational numbers approximating real numbers, and use them to find solutions of Pell's equation.

Since the proofs of some of the results are a little technical, we cover the highlights in Section 17.1 and defer the general theory and the proofs to subsequent sections. The last section of this chapter, where we prove that $e$ and $\pi$ are irrational, does not use continued fractions, but it is included in this chapter since it fits into the general theme of approximating real numbers by rational numbers.

# 17.1   Rational Approximations; Pell's Equation

Suppose we have a real number, for example $\pi = 3.14159265\cdots$, and we want to find a good approximation using rational numbers. We could use the rational number

$$\frac{314159}{100000},$$

but we can obtain a better approximation with

$$\frac{355}{113} = 3.14159292\cdots,$$

which also uses a much smaller denominator. Of course, we could use a denominator of 10000000 and obtain a better approximation, but we want to look at the size of the denominator when we consider how good an approximation is. Later in this chapter we'll show that if $x$ is a real number then there are integers $p$ and $q$ such that

$$\left| x - \frac{p}{q} \right| < \frac{1}{q^2}. \tag{17.1}$$

For example,

$$\left| \pi - \frac{22}{7} \right| \approx .00126 < \frac{1}{7^2}, \quad \text{and} \quad \left| \pi - \frac{355}{113} \right| \approx .00000027 < \frac{1}{113^2}.$$

However,

$$\left| \pi - \frac{314159}{100000} \right| \approx .00000265 > \frac{1}{100000^2},$$

so this is not as good an approximation.

But how do we find good approximations? The answer is continued fractions. Let's use $\pi$ as an example. The easiest approximation is $\pi \approx 3$. To refine this, we look at $\pi - 3 = .14159265 \cdots$. We need to approximate $.14159265 \cdots$. Since it is less than 1, we invert it to get

$$\frac{1}{.14159265 \cdots} = 7.0625133 \cdots = 7 + .0625133 \cdots$$

(calculations such as this one are easily done on a calculator). The integer part of this is 7 so we use

$$3 + \frac{1}{7} = \frac{22}{7}$$

as the next approximation. To continue the process, we need to find an approximation for $.0625133$, so we invert it and continue:

$$\frac{1}{.0625133 \cdots} = 15.99659 \cdots = 15 + .99659 \cdots.$$

We now have

$$\pi = 3 + \cfrac{1}{7 + \cfrac{1}{15.99659 \cdots}}$$

and the approximation

$$\pi \approx 3 + \cfrac{1}{7 + \cfrac{1}{15}} = \frac{333}{106}.$$

Then we approximate $.99659\cdots$ by inverting it to get

$$\frac{1}{.99659\cdots} = 1.0034\cdots.$$

We now have

$$\pi = 3 + \cfrac{1}{7 + \cfrac{1}{15 + \cfrac{1}{1.0034\cdots}}}$$

and the approximation

$$\pi \approx 3 + \cfrac{1}{7 + \cfrac{1}{15 + \cfrac{1}{1}}} = \frac{355}{113}.$$

We could continue:

$$\frac{1}{.0034\cdots} = 292.63\cdots.$$

The fact that 292 is relatively large is an indication that the previous step yields a very good approximation, namely 355/113. This will be explained in Corollary 17.7.

You might wonder why we didn't approximate 15.99659 by 16 rather than by 15. If we had done so, we would have obtained the *nearest integer continued fraction*. The theory for these is very similar to the theory we'll be describing for ordinary continued fractions. However, note that the algorithm we're using corrects itself: since 16 is better than 15, the next fraction we get is $15 + \frac{1}{1}$, which is 16. The nearest integer method skips a step here.

This process can be done for any real number $x$. Recall that $\lfloor x \rfloor$ denotes the largest integer less than or equal to $x$. Let $a_0 = \lfloor x \rfloor$ and $x_1 = 1/(x - a_0)$. For $k \geq 1$, let

$$a_k = \lfloor x_k \rfloor \quad \text{and} \quad x_{k+1} = 1/(x_k - a_k).$$

If $a_k = x_k$, which means that $x_k$ is an integer, we stop. In general, we obtain the continued fraction

$$[a_0, a_1, a_2, a_3, \ldots] \stackrel{\text{def}}{=} a_0 + \cfrac{1}{a_1 + \cfrac{1}{a_2 + \cfrac{1}{a_3 + \cfrac{}{\ddots}}}}$$

(the notation $[a_0, a_1, a_2, a_3, \ldots]$ saves a lot of paper). As we'll see later, when we stop after any $a_n$ we obtain rational approximations that always satisfy (17.1). Note that for each $k$ we have

$$x = [a_0, a_1, \ldots, a_{k-1}, x_k]. \tag{17.2}$$

If $x_k = [b_0, b_1, \ldots]$ then

$$x = [a_0, \ldots, a_{k-1}, b_0, b_1, \ldots].$$

**Technical point.** Suppose we compute the continued fraction of $x = -5.4321$. Then $a_0 = \lfloor x \rfloor = -6$, and we have $x = -6 + .5679 = -6 + 1/1.76087 \cdots$. Therefore, after the initial step, we are computing using only positive numbers. Therefore, throughout this chapter, we work only with continued fraction expansions of positive real numbers.

---

### CHECK YOUR UNDERSTANDING:

1. Compute the first three terms of the continued fraction of 2.357; that is, write $2.357 = [a_0, a_1, a_2, x_3]$.

---

## 17.1.1   Evaluating Continued Fractions

If you tried to evaluate

$$3 + \cfrac{1}{7 + \cfrac{1}{15 + \cfrac{1}{1}}} = 3 + \cfrac{1}{7 + \cfrac{1}{16}} = 3 + \cfrac{1}{\cfrac{113}{16}} = \frac{355}{113},$$

you probably found it rather annoying. And if you wanted to include one more term, namely the 292, you'd have to start all over. Fortunately, there is a faster way.

**Proposition 17.1.** *Let* $a_0, a_1, a_2, \ldots$ *be positive real numbers. Let*

$$p_{-2} = 0 \qquad p_{-1} = 1$$
$$q_{-2} = 1 \qquad q_{-1} = 0.$$

*For $k \geq 0$, define*

$$p_k = a_k p_{k-1} + p_{k-2}$$
$$q_k = a_k q_{k-1} + q_{k-2}.$$

(17.3)

*Then*

$$\frac{p_n}{q_n} = [a_0, a_1, a_2, \ldots, a_n]$$

*for each $n \geq 0$.*

The fraction $p_k/q_k$ is called the **$k$th convergent** of the continued fraction. We'll prove this proposition in Section 17.2. Although we're mostly interested in the situation where the $a_k$ are integers, we allow them to be real numbers here since we'll need that case in a few proofs. We could also allow the $a_k$ to be negative, as long as we don't end up dividing by 0, but we won't need this.

Rather than sorting out the mess of indices in the proposition, it's easier to calculate the numbers $p_k$ and $q_k$ as follows. Make a table with the $a_k$ in the top row and $p_{-2}, p_{-1}, q_{-2}, q_{-1}$ at the beginning, then fill in the $p_k$ and $q_k$:

|   |   | $a_0$ | $a_1$ | $a_2$ | $a_3$ | $a_4$ |
|---|---|-------|-------|-------|-------|-------|
| 0 | 1 | $p_0$ | $p_1$ | $p_2$ | $p_3$ | $p_4$ |
| 1 | 0 | $q_0$ | $q_1$ | $q_2$ | $q_3$ | $q_4$ |

For example, for $[3, 7, 15, 1]$, we start with

|   |   | 3 | 7 | 15 | 1 |
|---|---|---|---|----|---|
| 0 | 1 |   |   |    |   |
| 1 | 0 |   |   |    |   |

Note that we have filled in the initial values:

$$p_{-2} = 0, \ p_{-1} = 1, \ q_{-2} = 1, \ q_{-1} = 0.$$

Then we fill in the values in the table:

|   |   | 3 | 7  | 15  | 1   |
|---|---|---|----|-----|-----|
| 0 | 1 | 3 | 22 | 333 | 355 |
| 1 | 0 | 1 | 7  | 106 | 113 |

.

For example, to get the entry 333 that is below 15, multiply 15 times the entry 22 in the previous column and add to it the entry 3 from the column before that:

$$15 \times 22 + 3 = 333.$$

To get 355, multiply $1 \times 333$ and add 22. Similarly, to obtain $q_k = 113$ in the bottom row, multiply 1 (the $a_k$ at the top of the column) times 106 (which is $q_{k-1}$) and add 7 (which is $q_{k-2}$).

Note that we obtain all the intermediate approximations along the way. If we decide to include one more $a_k$, we can simply continue, rather than start over.

---

**CHECK YOUR UNDERSTANDING:**

2. Evaluate $[1, 3, 5, 7]$ as a rational number.

---

## 17.1.2    Pell's Equation

Now let's turn our attention to Pell's equation. We want to find integers $x$ and $y$ that are solutions to

$$x^2 - dy^2 = 1.$$

If we rewrite this as

$$\frac{x^2}{y^2} - d = \frac{1}{y^2},$$

we see that $x/y$ is approximately $\sqrt{d}$ when $y$ is large. With this in mind, let's compute the continued fraction for $\sqrt{d}$.

For example, if we let $d = 19$ and use the recursive definition of $a_k$, a computation yields the continued fraction

$$[4, 2, 1, 3, 1, 2, 8, 2, 1, 3, 1, 2, 8, 2, 1, 3, 1, 2, 8, \dots]$$

for $\sqrt{19}$. Note that, after the initial 4, it has a repeating pattern. In Section 17.5, we'll see that this is a general phenomenon of continued fractions of square roots of integers. To indicate the repetition, we write

$$[4, 2, 1, 3, 1, 2, 8, 2, 1, 3, 1, 2, 8, 2, 1, 3, 1, 2, 8, \dots] = [4, \overline{2, 1, 3, 1, 2, 8}].$$

Let's compute the numbers $p_k$ and $q_k$:

|   |   | 4 | 2 | 1 | 3 | 1 | 2 | 8 | 2 | 1 | 3 | 1 | 2 |
|---|---|---|---|----|----|----|-----|------|------|------|-------|-------|-------|
| 0 | 1 | 4 | 9 | 13 | 48 | 61 | 170 | 1421 | 3012 | 4433 | 16311 | 20744 | 57799 |
| 1 | 0 | 1 | 2 | 3  | 11 | 14 | 39  | 326  | 691  | 1017 | 3742  | 4759  | 13260 |

Since $a_6 = 8$ is relatively large, stopping with the preceding $a_k$ should give a good approximation (as with the 292 in the expansion for $\pi$, this will be made precise in Corollary 17.7). Therefore, we look at $170/39$ and compute

$$170^2 - 19 \cdot 39^2 = 1.$$

If we go to the next 8 (that is, the first column that is not computed in the table) and look at the preceding convergent, we have

$$57799^2 - 19 \cdot 13260^2 = 1.$$

This is not a coincidence.

**Theorem 17.2.** *Let $d$ be a positive integer, not a square. The continued fraction for $\sqrt{d}$ has the form*

$$[a_0, \overline{a_1, a_2, \dots, a_{n-1}, a_n}],$$

*where $a_n = 2a_0$. We have $a_k \leq a_0$ for $1 \leq k \leq n-1$. Let $m \geq 1$ and let $x = p_{mn-1}$ and $y = q_{mn-1}$. Then*

$$x^2 - dy^2 = \pm 1.$$

We'll prove the theorem in Section 17.5.

We now have an easy algorithm for solving $x^2 - dy^2 = 1$.

---

### Solving $x^2 - dy^2 = 1$

1. Compute $a_0 = \lfloor \sqrt{d} \rfloor$.
2. Compute the continued fraction $[a_0, a_1, a_2, \ldots]$ for $\sqrt{d}$. Stop when $a_n = 2a_0$.
3. Let $x_1 = p_{n-1}$ and $y_1 = q_{n-1}$.
4. If $x_1^2 - dy_1^2 = 1$, let $x = x_1$ and $y = y_1$ and stop.
5. If $x_1^2 - dy_1^2 = -1$, compute $x + y\sqrt{d} = (x_1 + y_1\sqrt{d})^2$. Then $x^2 - dy^2 = 1$.

---

**Example.** Let $d = 13$. Using a calculator we find that the continued fraction expansion for $\sqrt{13} = 3.60555\cdots$ is

$$[3, 1, 1, 1, 1, 6, \ldots].$$

Since 6 is twice the initial 3, we compute

$$[3, 1, 1, 1, 1] = \frac{18}{5}$$

and $18^2 - 13 \cdot 5^2 = -1$. Since this is $-1$, we compute

$$(18 + 5\sqrt{13})^2 = 18^2 + 2 \cdot 18 \cdot 5\sqrt{13} + 5^2 \cdot 13 = 649 + 180\sqrt{13}.$$

Then

$$649^2 - 13 \cdot 180^2 = 1.$$

The reason that squaring changes the $-1$ to $+1$ is that

$$x^2 - dy^2 = (x + y\sqrt{d})(x - y\sqrt{d}) = (x_1 + y_1\sqrt{d})^2 (x_1 - y_1\sqrt{d})^2$$
$$= \left( (x_1 + y_1\sqrt{d})(x_1 - y_1\sqrt{d}) \right)^2$$
$$= (x_1^2 - dy_1^2)^2 = (-1)^2 = 1.$$

The reason that we can stop when we get $a_n = 2a_0$ is that all the other $a_k$ are less than $2a_0$, so there is no danger of stopping at the wrong place. It is true that if we go a full period extra, we again get $a_{2n} = 2a_0$, and we get a solution of Pell's equation by stopping there. In fact, if we get $x_1^2 - dy_1^2 = -1$, then we can also obtain $x^2 - dy^2 = +1$ by continuing the continued fraction calculation. Then $x = p_{2n-1}$ and $y = q_{2n-1}$ give the same solution to Pell's equation as we get by squaring. But squaring is usually faster.

Occasionally, we stop as soon as we get started. For example, consider $d = 17$. The continued fraction is $[4, 8, 8, 8, \dots]$. Then we use only $a_0$, so $p_0 = 4$ and $q_0 = 1$, and

$$4^2 - 17 \cdot 1^2 = -1.$$

Squaring yields $x = 33$ and $y = 8$, and $33^2 - 17 \cdot 8^2 = 1$. A similar situation occurs for all $d$ of the form $n^2 + 1$.

---

**CHECK YOUR UNDERSTANDING:**

3. Use continued fractions to find a nontrivial solution of $x^2 - 11y^2 = 1$.

---

## 17.2    Basic Theory

This section develops the basic theory of continued fractions. We start with the fundamental recursive formulas. Then we show that the continued fraction of $x$ converges to $x$. Finally, we show that continued fractions give good rational approximations to real numbers.

We start by proving Proposition 17.1. Note that it is true for $n = 0$ since $p_0 = a_0 p_{-1} + p_{-2} = a_0$ and $q_0 = a_0 \cdot 0 + 1 = 1$.

We'll prove by induction on $n$ that

$$[a_0, a_1, \dots, a_{n-1}, x] = \frac{p_{n-1}x + p_{n-2}}{q_{n-1}x + q_{n-2}} \tag{17.4}$$

for all positive real numbers $x$.

Equation (17.4) is true for $n = 1$ since

$$[a_0, x] = a_0 + \frac{1}{x} = (a_0 x + 1)/x,$$

and $p_0x + p_{-1} = a_0x + 1$ and $q_0x + q_{-1} = x$. It's also true for $n = 0$ since $[a_0, \ldots, a_{n-1}, x]$ can be interpreted as $x$, and

$$x = \frac{1 \cdot x + 0}{0 \cdot x + 1} = \frac{p_{-1}x + p_{-2}}{q_{-1}x + q_{-2}}.$$

Now assume that (17.4) is true for $n = k$. We'll deduce it for $n = k+1$. We are assuming that

$$[a_0, a_1, \ldots, a_{k-1}, x] = \frac{p_{k-1}x + p_{k-2}}{q_{k-1}x + q_{k-2}}.$$

Let

$$x = a_k + 1/y,$$

for a positive real number $y$. Then

$$[a_0, a_1, \ldots, a_{k-1}, a_k + 1/y] = [a_0, a_1, \ldots, a_{k-1}, a_k, y]$$

(to see this, write out what these mean in terms of fractions). By the induction assumption,

$$[a_0, a_1, \ldots, a_{k-1}, a_k + 1/y] = \frac{p_{k-1}(a_k + 1/y) + p_{k-2}}{q_{k-1}(a_k + 1/y) + q_{k-2}}.$$

Multiply the numerator and denominator by $y$. The numerator becomes

$$p_{k-1}(a_ky + 1) + yp_{k-2} = (a_kp_{k-1} + p_{k-2})y + p_{k-1} = p_ky + p_{k-1}.$$

Similarly, the denominator becomes $q_ky + q_{k-1}$. Therefore,

$$[a_0, a_1, \ldots, a_{k-1}, a_k, y] = \frac{p_ky + p_{k-1}}{q_ky + q_{k-1}},$$

so (17.4) is true for $n = k + 1$. By induction, it is true for all $n \geq 0$. In (17.4), let $x = a_n$. Then

$$[a_0, a_1, \ldots, a_{n-1}, a_n] = \frac{p_{n-1}a_n + p_{n-2}}{q_{n-1}a_n + q_{n-2}} = \frac{p_n}{q_n}$$

for all $n \geq 0$.                                                          $\square$

**Corollary 17.3.** *For all* $n \geq 0$,

$$p_nq_{n-1} - q_np_{n-1} = (-1)^{n-1} \quad and \quad p_nq_{n-2} - q_np_{n-2} = (-1)^n a_n.$$

*Proof.* Let's prove the first equation. Once again, we'll use induction. The statement is true for $n = 0$ since

$$p_0 q_{-1} - q_0 p_{-1} = a_0 \cdot 0 - 1 \cdot 1 = -1 = (-1)^{0-1}.$$

The first equation (but not the second equation) also makes sense for $n = -1$, and it is true in this case:

$$p_{-1} q_{-2} - q_{-1} p_{-2} = 1 \cdot 1 - 0 \cdot 0 = 1 = (-1)^{-2}.$$

Assume the first equation is true for $n = k$, so

$$p_k q_{k-1} - q_k p_{k-1} = (-1)^{k-1}.$$

Then

$$
\begin{aligned}
p_{k+1} q_k - q_{k+1} p_k &= (a_{k+1} p_k + p_{k-1}) q_k - (a_{k+1} q_k + q_{k-1}) p_k \\
&= p_{k-1} q_k - q_{k-1} p_k \\
&= -(p_k q_{k-1} - q_k p_{k-1}) \\
&= -(-1)^{k-1} = (-1)^k.
\end{aligned}
$$

Therefore, the equation is true for $n = k + 1$. By induction, it is true for all $n$.

The second equation follows from the first one:

$$
\begin{aligned}
p_n q_{n-2} - q_n p_{n-2} &= (a_n p_{n-1} + p_{n-2}) q_{n-2} - (a_n q_{n-1} + q_{n-2}) p_{n-2} \\
&= a_n (p_{n-1} q_{n-2} - q_{n-1} p_{n-2}) \\
&= (-1)^{n-2} a_n,
\end{aligned}
$$

where we used the first equation of the proposition to get $(-1)^{n-2}$. Since $(-1)^{n-2} = (-1)^n$, we're done. $\qquad\square$

Dividing $p_n q_{n-1} - q_n p_{n-1} = (-1)^{n-1}$ by $q_{n-1} q_n$ yields

$$\frac{p_n}{q_n} - \frac{p_{n-1}}{q_{n-1}} = \frac{(-1)^{n-1}}{q_{n-1} q_n}.$$

This implies that

$$\frac{p_n}{q_n} > \frac{p_{n-1}}{q_{n-1}} \quad \text{if } n \text{ is odd} \tag{17.5}$$

and

$$\frac{p_n}{q_n} < \frac{p_{n-1}}{q_{n-1}} \text{ if } n \text{ is even.} \tag{17.6}$$

Similarly, the relation $p_n q_{n-2} - q_n p_{n-2} = (-1)^n a_n$ implies that

$$\frac{p_n}{q_n} < \frac{p_{n-2}}{q_{n-2}} \text{ if } n \text{ is odd} \tag{17.7}$$

and

$$\frac{p_n}{q_n} > \frac{p_{n-2}}{q_{n-2}} \text{ if } n \text{ is even.} \tag{17.8}$$

Therefore, we have

$$\frac{p_0}{q_0} < \frac{p_2}{q_2} < \frac{p_4}{q_4} < \cdots < \frac{p_5}{q_5} < \frac{p_3}{q_3} < \frac{p_1}{q_1}.$$

We point out that $p_{\text{even}}/q_{\text{even}} < p_{\text{odd}}/q_{\text{odd}}$ for every choice of even integer and odd integer. As an example, we'll show that $p_{12}/q_{12} < p_{51}/q_{51}$. To see this, note that

$$\frac{p_{12}}{q_{12}} < \frac{p_{52}}{q_{52}} < \frac{p_{51}}{q_{51}},$$

where the first inequality follows from (17.8) and the second inequality follows from (17.6). In general, we can always put convergents with high indices in between an even convergent and an odd convergent.

A standard fact about real numbers is that an increasing sequence of real numbers that is bounded above converges to a limit. Since the sequence of even convergents is increasing and bounded above by $p_1/q_1$, the limit

$$\lim_{m \to \infty} \frac{p_{2m}}{q_{2m}} = x_{\text{even}}$$

exists. Similarly, since the odd convergents are decreasing and bounded below by $p_0/q_0$,

$$\lim_{m \to \infty} \frac{p_{2m+1}}{q_{2m+1}} = x_{\text{odd}}$$

exists. But

$$\frac{p_{2n}}{q_{2n}} < x_{\text{even}} \le x_{\text{odd}} < \frac{p_{2n-1}}{q_{2n-1}} \tag{17.9}$$

for all $n$, so

$$|x_{\text{odd}} - x_{\text{even}}| < \left| \frac{p_n}{q_n} - \frac{p_{n-1}}{q_{n-1}} \right| = \frac{1}{q_{n-1}q_n}.$$

**Lemma 17.4.** *Assume that $a_0, a_1, a_2, \cdots$ are positive integers. As $n \to \infty$, we have $q_n \to \infty$.*

*Proof.* Let $n \geq 2$. Since $a_n \geq 1$,

$$q_n = a_n q_{n-1} + q_{n-2} \geq q_{n-1} + q_{n-2} = (a_{n-1} q_{n-2} + q_{n-3}) + q_{n-2} \geq 2q_{n-2}.$$

Since $q_0 = 1$, we find that $q_2 \geq 2$, $q_4 \geq 4$, $q_6 \geq 8$, and, in general, $q_{2m} \geq 2^m$. Therefore, $q_{2m} \to \infty$ as $m \to \infty$. Since $q_{2m+1} \geq q_{2m} \geq 2^m$, we also have $q_{2m+1} \to \infty$ as $m \to \infty$. Since the odd indices and the even indices yield sequences that go to $\infty$, we have $q_n \to \infty$.    □

Therefore, $1/(q_{n-1}q_n) \to 0$ as $n \to \infty$. This implies that $x_{\text{odd}} = x_{\text{even}}$. Call this limit $x_\infty$. We have proved the following.

**Proposition 17.5.** *Let $a_0, a_1, a_2, \ldots$ be a sequence of positive integers. Then*

$$\lim_{n \to \infty} [a_0, a_1, a_2, \ldots, a_n]$$

*exists.*

This proposition means that we can talk about continued fractions of the form $[a_0, a_1, a_2, \ldots]$ and regard them as representing real numbers.

There is a question that needs to be asked: If we start with a real number $x$ and form its continued fraction, we know that it converges, but what does it converge to? We'd like it to converge to $x$, but we don't know that yet. We'll now show that this is what occurs.

If some $x_k$ is an integer, the continued fraction stops after finitely many steps and $x = p_k/q_k$. In this case, we make the convention of saying that the continued fraction converges to $x$, even though it is really simply an equality. For the rest of this discussion, we assume that the continued fraction for $x$ does not terminate (as we'll see in the next section, this means that $x$ is irrational).

From Equations (17.2) and (17.4), we have

$$x = [a_0, a_1, \ldots, a_k, x_{k+1}] = \frac{p_k x_{k+1} + p_{k-1}}{q_k x_{k+1} + q_{k-1}}.$$

Therefore,

$$x - \frac{p_k}{q_k} = \frac{p_k x_{k+1} + p_{k-1}}{q_k x_{k+1} + q_{k-1}} - \frac{p_k}{q_k}$$

$$= \frac{q_k p_{k-1} - p_k q_{k-1}}{(q_k x_{k+1} + q_{k-1})q_k} \qquad (17.10)$$

$$= \frac{(-1)^k}{(q_k x_{k+1} + q_{k-1})q_k}.$$

Since $a_{k+1} \leq x_{k+1}$, it follows from (17.10) that

$$\left| x - \frac{p_k}{q_k} \right| = \frac{1}{(q_k x_{k+1} + q_{k-1})q_k} \leq \frac{1}{(q_k a_{k+1} + q_{k-1})q_k} = \frac{1}{q_{k+1} q_k}.$$

Since $q_k \to \infty$ as $k \to \infty$, we have proved the following:

**Theorem 17.6.** *Let $x$ be a real number. The continued fraction for $x$ converges to $x$. In fact,*

$$\left| x - \frac{p_k}{q_k} \right| \leq \frac{1}{q_k q_{k+1}}.$$

There is one more subtlety that must be addressed. See Project 4.
The following shows the effect that the numbers $a_n$ have.

**Corollary 17.7.** *If the continued fraction for $x$ is $[a_0, a_1, a_2, \dots]$, then*

$$\left| x - \frac{p_k}{q_k} \right| \leq \frac{1}{a_{k+1} q_k^2}.$$

*Proof.* Since $q_{k+1} = a_{k+1}q_k + q_{k-1}$, we have $q_{k+1} \geq a_{k+1}q_k$. Therefore,

$$\left| x - \frac{p_k}{q_k} \right| \leq \frac{1}{q_k q_{k+1}} \leq \frac{1}{a_{k+1} q_k^2}.$$

$\square$

The corollary explains why we obtained good approximations by stopping just before large values of $a_n$, because a large $a_n$ makes the right-hand side in the corollary small. For example, $22/7$ is a good approximation for $\pi$ because the next $a_n$ is 15. The approximation $355/113$ is good because the next $a_n$ is 292, which is quite large.

It is easy to check that no rational number with denominator less than 7 gives as good an approximation to $\pi$ as $22/7$ (see Exercise 8). The following result shows that continued fractions in general give best approximations with respect to the size of the denominator.

**Theorem 17.8.** *Let $x$ be an irrational number, let $a$ and $b$ be integers with $b > 0$, and let $p_n/q_n$ be the $n$th convergent of the continued fraction of $x$.*
*(a) If $|bx - a| < |q_n x - p_n|$, then $b \geq q_{n+1}$.*
*(b) If*

$$\left| x - \frac{a}{b} \right| \leq \left| x - \frac{p_n}{q_n} \right|,$$

*then $b \geq q_n$.*

*Proof.* We start by proving (a). Let

$$u = (-1)^{n+1} (q_{n+1}a - p_{n+1}b), \qquad v = (-1)^{n+1} (p_n b - q_n a).$$

Then

$$
\begin{aligned}
p_n u + p_{n+1} v &= (-1)^{n+1} (p_n q_{n+1} a - p_n p_{n+1} b + p_{n+1} p_n b - p_{n+1} q_n a) \\
&= (-1)^{n+1} (p_n q_{n+1} - p_{n+1} q_n) a \\
&= a
\end{aligned}
$$

by Corollary 17.3. Similarly,

$$
\begin{aligned}
q_n u + q_{n+1} v &= (-1)^{n+1} (q_n q_{n+1} a - q_n p_{n+1} b + q_{n+1} p_n b - q_{n+1} q_n a) \\
&= (-1)^{n+1} (p_n q_{n+1} - p_{n+1} q_n) b \\
&= b.
\end{aligned}
$$

Therefore,

$$
\begin{aligned}
|bx - a| &= |(q_n u + q_{n+1} v)x - (p_n u + p_{n+1} v)| \\
&= |u(q_n x - p_n) + v(q_{n+1} x - p_{n+1})|.
\end{aligned}
$$

Note that we cannot have both $u = 0$ and $v = 0$, since then $|bx - a| = 0$, which is impossible because $x$ is irrational.

If $v = 0$, then we have $|bx - a| = |u(q_n x - p_n)| \geq |q_n x - p_n|$, contrary to assumption. Therefore, $v \neq 0$.

Suppose now that $b < q_{n+1}$. Since $b = q_n u + q_{n+1} v$, we have $q_n u = b - q_{n+1} v$. If $v < 0$, then $q_n u > 0$, so $u > 0$. If $v > 0$, then

$$q_n u = b - q_{n+1} v \le b - q_{n+1} < 0,$$

so $u < 0$. Therefore, $u$ and $v$ have opposite signs, and $u \ne 0$.

From (17.9), and the fact that $x_{\text{odd}} = x_{\text{even}} = x$, we see that one of $q_n x - p_n$ and $q_{n+1} x - p_{n+1}$ is positive and the other is negative. Combining this with what we just proved, we find that

$$u(q_n x - p_n) \text{ and } v(q_{n+1} x - p_{n+1})$$

have the same sign. We now come to the crucial observation, and the reason that we've been worrying about the signs of various numbers. If $s$ and $t$ are real numbers and they have the same sign, then $|s + t| = |s| + |t|$. (*Proof:* If they are both positive, then the relation is simply $s + t = s + t$. If they are both negative, then $-|s+t| = s+t = -|s| - |t|$.) We therefore deduce that

$$
\begin{aligned}
|bx - a| &= |u(q_n x - p_n) + v(q_{n+1} x - p_{n+1})| \\
&= |u(q_n x - p_n)| + |v(q_{n+1} x - p_{n+1})| \\
&\ge |u(q_n x - p_n)| \\
&\ge |q_n x - p_n|,
\end{aligned}
$$

contrary to assumption. Therefore, we must have $b \ge q_{n+1}$. This proves (a).

We now prove (b). Assume that

$$\left| x - \frac{a}{b} \right| \le \left| x - \frac{p_n}{q_n} \right|.$$

If $b < q_n$, then we can multiply the inequalities and obtain

$$|bx - a| = b\left| x - \frac{a}{b} \right| < q_n \left| x - \frac{p_n}{q_n} \right| = |q_n x - p_n|,$$

so part (a) implies that $b \ge q_{n+1}$, contrary to assumption. Therefore, $b \ge q_n$. This completes the proof.  $\square$

# 17.3   Rational Numbers

Let's compute the continued fraction of 37/17:

$$\frac{37}{17} = 2 + \frac{1}{17/3} = 2 + \cfrac{1}{5 + \cfrac{1}{3/2}} = 2 + \cfrac{1}{5 + \cfrac{1}{1 + \cfrac{1}{2}}}.$$

Therefore,

$$\frac{37}{17} = [2, 5, 1, 2].$$

Now compare this with the Euclidean Algorithm for computing $\gcd(37, 17)$:

$$37 = 2 \cdot 17 + 3$$
$$17 = 5 \cdot 3 + 2$$
$$3 = 1 \cdot 2 + 1$$
$$2 = 2 \cdot 1 + 0.$$

We've done the same computations. The quotients in the Euclidean Algorithm are the values of $[a_0, a_1, \cdots]$ in the continued fraction.

Why is this happening? The Euclidean Algorithm starts with

$$a = q_1 b + r_1.$$

This is the same as

$$\left\lfloor \frac{a}{b} \right\rfloor = q_1 \quad \text{and} \quad \frac{a}{b} = q_1 + \frac{r_1}{b}.$$

The next step in the Euclidean Algorithm is

$$b = q_2 r_1 + r_2.$$

In the continued fraction algorithm, this is

$$\frac{b}{r_1} = q_2 + \frac{r_2}{r_1},$$

so we have

$$\frac{a}{b} = q_1 + \frac{1}{b/r_1} = q_1 + \frac{1}{q_2 + (r_2/r_1)}.$$

Continuing in this way, we eventually get to the last line of the Euclidean Algorithm:

$$r_{n-2} = q_n r_{n-1} + 0.$$

In the continued fraction, we have

$$\frac{a}{b} = [q_1, q_2, \cdots, q_{n-1}, (r_{n-1}/r_{n-2})] = [q_1, q_2, \cdots, q_{n-1}, q_n].$$

Since $q_n$ is an integer, the continued fraction calculation must stop; otherwise, we would divide by 0 in the next step.

Now let's look at the Extended Euclidean Algorithm in the case of $\gcd(37, 17)$:

|     | $x$  | $y$  |
| --- | ---- | ---- |
| 37  | 1    | 0    |
| 17  | 0    | 1    |
| 3   | 1    | $-2$ |
| 2   | $-5$ | 11   |
| 1   | 6    | $-13$ |
| 0   | $-17$ | 37.  |

Compare this with the evaluation of the convergents of the continued fraction $[2, 5, 1, 2]$:

|   |   | 2 | 5  | 1  | 2  |
| - | - | - | -- | -- | -- |
| 0 | 1 | 2 | 11 | 13 | 37 |
| 1 | 0 | 1 | 5  | 6  | 17 |

As we see, the numbers obtained, up to sign, are the same as those in the Extended Euclidean Algorithm. More precisely, if we number the row $a = bq_1 + r_1$ as the first row of the Euclidean Algorithm, then the last column of the $(k+2)$nd row contains $(-1)^k p_{k-1}$. To get the $(k+3)$rd row in the Extended Euclidean Algorithm, multiply the $(k+2)$nd row by the quotient, which is $a_k$, and subtract from the $(k+1)$st row. This calculation is

$$(-1)^{k-1}p_{k-2} - a_k(-1)^k p_{k-1} = (-1)^{k+1}p_k.$$

Multiply by $(-1)^{k-1}$ to get

$$p_{k-2} + a_k p_{k-1} = p_k,$$

which is the relation (17.3) used to calculate the values $p_k$. This is why the numbers match.

In conclusion, we see that the Euclidean Algorithm and the continued fraction algorithm for rational numbers are essentially the same. For future reference, we record one of our observations.

**Proposition 17.9.** *The continued fraction of x terminates if and only if x is rational.*

*Proof.* If the continued fraction terminates, it gives $x$ in terms of ratios of integers, so $x$ is rational. Conversely, if $x = a/b$ (with $\gcd(a, b) = 1$) is rational, then the Euclidean Algorithm for $\gcd(a, b)$ terminates with a 0 remainder. Correspondingly, the continued fraction algorithm encounters an integer $x_n$ and must terminate.                              □

# 17.4   Periodic Continued Fractions

This section determines which real numbers have purely periodic or eventually periodic continued fractions.

Let's calculate the continued fraction of $\sqrt{7}$. Rather than use the decimal expansion and a calculator, we'll do the calculations exactly:

$$\sqrt{7} = 2 + (\sqrt{7} - 2) = 2 + \cfrac{1}{\cfrac{\sqrt{7}+2}{3}} = [2, (\sqrt{7}+2)/3].$$

Since

$$\frac{\sqrt{7}+2}{3} = 1 + \frac{\sqrt{7}-1}{3} = 1 + \cfrac{1}{(\sqrt{7}+1)/2},$$

**Theorem 17.11.** *The continued fraction of a real number $x$ is purely periodic if and only if $x$ is a reduced quadratic irrational.*

*Proof.* First, assume that the continued fraction expansion is purely periodic, so we can write it in the form

$$x = [\overline{a_0, a_1, \ldots, a_n}].$$

In the notation of (17.2),

$$x = [a_0, a_1, \ldots, a_n, x]. \tag{17.11}$$

Let

$$\frac{p_k}{q_k} = [a_0, a_1, \ldots, a_k].$$

Then (17.11) and (17.4) imply that

$$x = [a_0, a_1, \ldots, a_n, x] = \frac{p_n x + p_{n-1}}{q_n x + q_{n-1}}.$$

Therefore,

$$q_n x^2 + (q_{n-1} - p_n)x - p_{n-1} = 0. \tag{17.12}$$

This says that $x$ is a root of a quadratic equation. Since the continued fraction of a rational stops and the continued fraction of $x$ does not stop, $x$ is irrational. Therefore, $x$ is a quadratic irrational.

We need to show that $x$ is reduced. First, $x > a_0 \geq 1$. Moreover, $x$ is a root of the equation

$$f(x) = q_n x^2 + (q_{n-1} - p_n)x - p_{n-1} = 0,$$

and $\overline{x}$ is the other root. We have $f(0) = -p_{n-1} < 0$ and

$$f(-1) = (q_n - q_{n-1}) + (p_n - p_{n-1}) > 0$$

since $q_n > q_{n-1}$ and $p_n > p_{n-1}$ (by (17.3)). By the Intermediate Value Theorem from calculus, $f(x)$ has a root between $-1$ and $0$. Since $x > 1$, this root must be $\overline{x}$, so $-1 < \overline{x} < 0$. Therefore, $x$ is reduced.

The harder part of the proof is showing that the expansion of a reduced quadratic irrational is purely periodic. The idea will be to show that all of the $x_n$ are reduced, next that there are only finitely many possibilities for the reduced numbers that can occur, so there must be a repetition, and finally that this repetition can be moved back to the beginning of the expansion.

**Lemma 17.12.** *Let $x$ be a quadratic irrational and let*

$$x = [a_0, a_1, \cdots, a_{n-1}, x_n]$$

*be the start of its continued fraction expansion. If $x_n$ is reduced, then $x_{n+1}$ is reduced.*

*Proof.* Let $a_n = \lfloor x_n \rfloor$. We have

$$x_{n+1} = 1/(x_n - a_n). \tag{17.13}$$

Since $0 < x_n - a_n < 1$, Equation (17.13) implies that $x_{n+1} > 1$. Taking conjugates in (17.13) yields

$$\overline{x_{n+1}} = 1/(\overline{x_n} - a_n).$$

But $\overline{x_n} - a_n < -a_n \le -1$, so $-1 < \overline{x_{n+1}} < 0$. Therefore, $x_{n+1}$ is reduced.                                                                 □

In the continued fraction for $\sqrt{7}$, we have that $x_1 = (\sqrt{7} + 2)/3$ is reduced. The lemma says that $x_2 = (\sqrt{7} + 1)/2$ is also reduced, and similarly for $x_3$ and $x_4$ (and for $x_5$, but it equals $x_1$).

The next lemma says that we can also work backwards through the continued fraction when we have reduced numbers.

**Lemma 17.13.** *Let $x$ and $y$ be quadratic irrationals and let*

$$x = [a_0, a_1, \cdots, a_{n-1}, x_n] \text{ and } y = [b_0, b_1, \cdots, b_{m-1}, y_m]$$

*be the starts of their continued fraction expansions. Suppose that $x_n = y_m$ and suppose that $x_{n-1}$ and $y_{m-1}$ are reduced. Then*

$$x_{n-1} = y_{m-1}.$$

*Proof.* Change $n$ to $n - 1$ and take conjugates in (17.13) to obtain $\overline{x_n} = 1/(\overline{x_{n-1}} - a_{n-1})$, hence

$$-1/\overline{x_n} = a_{n-1} - \overline{x_{n-1}}. \tag{17.14}$$

Since $0 < -\overline{x_{n-1}} < 1$ by the assumption that $x_{n-1}$ is reduced, (17.14) implies that

$$a_{n-1} < -1/\overline{x_n} < a_{n-1} + 1,$$

so $a_{n-1} = \lfloor -1/\overline{x_n} \rfloor$. Similarly, $b_{m-1} = \lfloor -1/\overline{y_m} \rfloor$. Since $x_n = y_m$, we have $a_{n-1} = b_{m-1}$. Therefore,

$$x_{n-1} = a_{n-1} + \frac{1}{x_n} = b_{m-1} + \frac{1}{y_m} = y_{m-1}.$$

This proves the lemma.                                                              □

We now need to investigate the nature of the numbers $x_n$ that occur during the computation of the continued fraction expansion of a quadratic irrational. If we start with $(a + b\sqrt{d})/c$, then

$$\frac{a + b\sqrt{d}}{c} = \frac{A \pm \sqrt{d_1}}{B},$$

where $A = ac$, $B = c^2$, and $d_1 = b^2 c^2 d$. Then $d_1 - A^2 = c^2(b^2 d - a^2)$ is a multiple of $B$. Therefore, by changing the signs of $A$ and $B$ if necessary, we will always assume that we start with

$$x = \frac{A + \sqrt{d}}{B} \quad \text{with integers } A, B, d \text{ such that } B \mid d - A^2.$$

Of course, we also assume that $d$ is not a square; otherwise, $x$ is rational and the continued fraction stops after finitely many terms. We can now trace the values of $x_n$ as we proceed through the computation of the continued fraction.

**Lemma 17.14.** *Let $d$ be a positive integer that is not a square and let*

$$x_k = \frac{A_k + \sqrt{d}}{B_k},$$

*where $A_k$ and $B_k$ are integers with $B_k \mid d - A_k^2$. Let $a_k$ be an integer and define*

$$A_{k+1} = a_k B_k - A_k \quad \text{and} \quad B_{k+1} = \frac{d - A_{k+1}^2}{B_k}.$$

*Then*

    *1. $A_{k+1}$ and $B_{k+1}$ are integers and $B_{k+1} \mid d - A_{k+1}^2$*

    *2. $1/(x_k - a_k) = (A_{k+1} + \sqrt{d})/B_{k+1}$.*

*Proof.* We have

$$
\begin{aligned}
x_k - a_k &= \frac{A_k + \sqrt{d}}{B_k} - a_k \\
&= \frac{(A_k - a_k B_k) + \sqrt{d}}{B_k} \\
&= \frac{\sqrt{d} - A_{k+1}}{B_k}.
\end{aligned}
$$

Therefore,

$$
\begin{aligned}
\frac{1}{x_k - a_k} &= \frac{B_k}{\sqrt{d} - A_{k+1}} \\
&= \frac{B_k}{\sqrt{d} - A_{k+1}} \frac{\sqrt{d} + A_{k+1}}{\sqrt{d} + A_{k+1}} \\
&= \frac{A_{k+1} + \sqrt{d}}{(d - A_{k+1}^2)/B_k} \\
&= \frac{A_{k+1} + \sqrt{d}}{B_{k+1}}.
\end{aligned}
$$

This proves (2).

Since $A_k$, $B_k$, and $a_k$ are integers, the definition of $A_{k+1}$ shows that $A_{k+1}$ is an integer. Also, working mod $B_k$, we have

$$
d - A_{k+1}^2 = d - (a_k B_k - A_k)^2 \equiv d - (0 - A_k)^2 = d - A_k^2 \equiv 0 \pmod{B_k}
$$

by assumption. Therefore, $B_k \mid d - A_{k+1}^2$, so $B_{k+1}$ is an integer. Since

$$
B_{k+1} B_k = d - A_{k+1}^2,
$$

we have $B_{k+1} \mid d - A_{k+1}^2$. This proves (1).      $\square$

If $x_k$ is reduced, we can say more.

**Lemma 17.15.** *Let $d$ be a positive integer that is not a square and let*

$$
x = \frac{A + \sqrt{d}}{B},
$$

*where $A, B$ are integers. If $x$ is reduced, then*

$$
B > 0 \quad and \quad A^2 < d.
$$

*Proof.* Since $x > 1$ and $\bar{x} < 0$, we have

$$0 < x - \bar{x} = 2\sqrt{d}/B.$$

Therefore, $B > 0$. Since $x\bar{x} < 0$ (because $x$ is positive and $\bar{x}$ is negative)

$$0 > \frac{A + \sqrt{d}}{B} \frac{A - \sqrt{d}}{B} = \frac{A^2 - d}{B^2},$$

so $A^2 < d$. $\qquad\qquad\qquad\square$

We can now finish the proof of Theorem 17.11. Since $x$ is assumed to be reduced, all $x_k$ are reduced, by Lemma 17.12. By Lemmas 17.14 and 17.15, each $x_k$ has the form

$$x_k = (A_k + \sqrt{d})/B_k,$$

with $|A_k| < \sqrt{d}$ and $B_k \mid d - A_k^2$. Therefore, there are only finitely many possibilities for $A_k$ and $B_k$, so there are only finitely many possibilities for $x_k$. Eventually, we have $x_i = x_j$ for some $i$ and $j$ with $i < j$. Lemma 17.13 with $x = y$ implies that $x_{i-1} = x_{j-1}$, and then $x_{i-2} = x_{j-2}$, and continuing until we find $x_0 = x_{j-i}$. But $x_0 = x$, so we have $x = x_{j-i}$. This means that the continued fraction repeats at the $(j - i)$th step, and the continued fraction is purely periodic. This proves Theorem 17.11. $\qquad\qquad\square$

## 17.4.2   Eventually Periodic Continued Fractions

Theorem 17.11 leads to the following important result.

**Theorem 17.16.** *(Lagrange) Let $x$ be a positive irrational real number. The continued fraction expansion of $x$ is eventually periodic if and only if $x$ is a quadratic irrational.*

*Proof.* Suppose first that the continued fraction of $x$ is eventually periodic, so

$$x = [a_0, a_1, \cdots, a_n, \overline{b_1, b_2, \cdots b_m}] = [a_0, a_1, \cdots, a_n, x_{n+1}]$$

with $x_{n+1} = \overline{[b_1, b_2, \cdots, b_m]}$. By Theorem 17.11, $x_{n+1}$ is a quadratic irrational. Equation (17.4) says that

$$x = \frac{p_n x_{n+1} + p_{n-1}}{q_n x_{n+1} + q_{n-1}}.$$

By rationalizing the denominator, we find that $x$ is a quadratic irrational.

The harder part is showing that a quadratic irrational has an eventually periodic expansion. By Theorem 17.11, it suffices to show that some $x_k$ is reduced. Once again, we need that

$$x = \frac{p_n x_{n+1} + p_{n-1}}{q_n x_{n+1} + q_{n-1}}.$$

This implies that $x(q_n x_{n+1} + q_{n-1}) = p_n x_{n+1} + p_{n-1}$, and when we isolate $x_{n+1}$ we get

$$x_{n+1} = \frac{p_{n-1} - x q_{n-1}}{x q_n - p_n} = -\frac{q_{n-1}}{q_n} \frac{x - (p_{n-1}/q_{n-1})}{x - (p_n/q_n)}.$$

Taking conjugates yields

$$\overline{x_{n+1}} = -\frac{q_{n-1}}{q_n} \frac{\overline{x} - (p_{n-1}/q_{n-1})}{\overline{x} - (p_n/q_n)}.$$

Let $n \to \infty$. Theorem 17.6 says that $p_n/q_n \to x$ and also $p_{n-1}/q_{n-1} \to x$. Therefore,

$$\lim_{n \to \infty} \frac{\overline{x} - (p_{n-1}/q_{n-1})}{\overline{x} - (p_n/q_n)} = \frac{\overline{x} - x}{\overline{x} - x} = 1,$$

so

$$\frac{\overline{x} - (p_{n-1}/q_{n-1})}{\overline{x} - (p_n/q_n)} > 0$$

for all sufficiently large $n$, say for $n \geq n_0$. Therefore,

$$\overline{x_{n+1}} = -\frac{q_{n-1}}{q_n} \frac{\overline{x} - (p_{n-1}/q_{n-1})}{\overline{x} - (p_n/q_n)} < 0$$

for $n \geq n_0$. Fix $n \geq n_0$. The definition of $x_{n+2}$ says that

$$x_{n+2} = 1/(x_{n+1} - a_{n+1}).$$

Since $\overline{x_{n+1}} < 0$,

$$\frac{1}{x_{n+2}} = \overline{x_{n+1}} - a_{n+1} < -1,$$

which implies that

$$-1 < \overline{x_{n+2}} < 0.$$

Since

$$x_{n+2} > a_{n+2} \geq 1,$$

we have proved that $x_{n+2}$ is reduced. Therefore, $x_{n+2}$ has a periodic continued fraction expansion. But this expansion is part of the expansion of $x$, so the continued expansion of $x$ is eventually periodic.    □

## 17.5    Square Roots of Integers

The goal of this section is to prove Theorem 17.2. Recall that it says the following:

*Let d be a positive integer, not a square. The continued fraction for $\sqrt{d}$ has the form*

$$[a_0, \overline{a_1, a_2, \ldots, a_{n-1}, a_n}],$$

*where $a_n = 2a_0$. We have $a_k \leq a_0$ for $1 \leq k < n$. Let $m \geq 1$ and let $x = p_{mn-1}$ and $y = q_{mn-1}$. Then*

$$x^2 - dy^2 = \pm 1.$$

*Proof.* The continued fraction computation starts as

$$\sqrt{d} = a_0 + (\sqrt{d} - a_0) = a_0 + \frac{1}{x_1}$$

$$= [a_0, x_1].$$

Since $0 < \sqrt{d} - a_0 < 1$,

$$x_1 = 1/(\sqrt{d} - a_0) > 1.$$

Also, $-\sqrt{d} - a_0 < -1$, so

$$-1 < \overline{x_1} = 1/(-\sqrt{d} - a_0) < 0.$$

Therefore, $x_1$ is reduced and its continued fraction expansion is purely periodic. This means that

$$\sqrt{d} = [a_0, \overline{a_1, a_2, \cdots, a_n}]$$

for some $a_1, a_2, \ldots, a_n$. Now let $z = a_0 + \sqrt{d}$. Then

$$z = 2a_0 + (\sqrt{d} - a_0) = 2a_0 + \frac{1}{x_1} = [2a_0, \overline{a_1, a_2, \cdots, a_n}].$$

But $z > 1$ and $-1 < \bar{z} = -\sqrt{d} + a_0 < 0$, so $z$ is reduced and its continued fraction expansion is purely periodic. Therefore, $a_n = 2a_0$ and $x_n = a_0 + \sqrt{d}$.

**Lemma 17.17.** *Let $\sqrt{d} = [a_0, a_1, \cdots, a_k, x_{k+1}]$, and let*

$$x_{k+1} = \frac{A_{k+1} + \sqrt{d}}{B_{k+1}}.$$

*Then*

$$p_k^2 - dq_k^2 = (-1)^{k-1} B_{k+1}.$$

*Proof.* From Equation (17.4),

$$\sqrt{d} = \frac{p_k x_{k+1} + p_{k-1}}{q_k x_{k+1} + q_{k-1}}.$$

Substitute

$$x_{k+1} = \frac{A_{k+1} + \sqrt{d}}{B_{k+1}}$$

and multiply numerator and denominator by $B_{k+1}$ to obtain

$$\sqrt{d} = \frac{p_k(A_{k+1} + \sqrt{d}) + p_{k-1} B_{k+1}}{q_k(A_{k+1} + \sqrt{d}) + q_{k-1} B_{k+1}}.$$

Cross-multiply to obtain

$$\sqrt{d}\,(q_k A_{k+1} + q_{k-1} B_{k+1} - p_k) = p_k A_{k+1} + p_{k-1} B_{k+1} - dq_k. \quad (17.15)$$

The left side is an integer times $\sqrt{d}$. The right side is an integer. If the integer on the left side is nonzero, then we can divide and conclude that $\sqrt{d}$ is rational, which is not the case. Therefore,

$$q_k A_{k+1} + q_{k-1} B_{k+1} - p_k = 0.$$

This implies that the right-hand side of (17.15) is 0:

$$p_k A_{k+1} + p_{k-1} B_{k+1} - dq_k = 0.$$

Multiply the first equation by $p_k$ and the second equation by $q_k$, then subtract. This yields

$$(p_k q_{k-1} - q_k p_{k-1}) B_{k+1} - (p_k^2 - dq_k^2) = 0.$$

Corollary 17.3 says that $p_k q_{k-1} - q_k p_{k-1} = (-1)^{k-1}$, so the last equation says that

$$(-1)^{k-1} B_{k+1} = p_k^2 - dq_k^2.$$

$\square$

Since $x = \sqrt{d} = (0 + \sqrt{d})/1$ satisfies the conditions of Lemma 17.14 with $A_0 = 0$ and $B_0 = 1$, we find that $A_j$ and $B_j$ are integers for all $j \geq 1$. Since $x_1$ is reduced, from Lemma 17.12 we know that $x_j$ is reduced for all $j \geq 1$. We then have $A_{k+1}^2 < d$ for all $k \geq 0$, by Lemma 17.15. Since $|A_{k+1}|$ is an integer less than $\sqrt{d}$, we must have $|A_{k+1}| \leq a_0 = \lfloor \sqrt{d} \rfloor$. Therefore,

$$B_{k+1} a_{k+1} < B_{k+1} x_{k+1} = A_{k+1} + \sqrt{d} \leq a_0 + \sqrt{d}.$$

Since $B_{k+1} a_{k+1}$ is an integer, $B_{k+1} a_{k+1} \leq a_0 + a_0 = 2a_0$, so

$$a_{k+1} \leq 2a_0 / B_{k+1}.$$

If $B_{k+1} \geq 2$, we have $a_{k+1} \leq a_0$.

If $B_{k+1} = 1$, which happens, for example, when $k + 1 = n$, then

$$p_k^2 - dq_k^2 = (-1)^{k-1} B_{k+1} = \pm 1$$

and

$$x_{k+1} = A_{k+1} + \sqrt{d}.$$

Since $x_{k+1}$ is reduced, $-1 < A_{k+1} - \sqrt{d} < 0$, which means that $\sqrt{d} - 1 < A_{k+1} < \sqrt{d}$, hence $A_{k+1} = \lfloor \sqrt{d} \rfloor = a_0$. Therefore, $x_{k+1} = a_0 + \sqrt{d}$ and

$$a_{k+1} = \lfloor x_k \rfloor = \lfloor a_0 + \sqrt{d} \rfloor = 2a_0.$$

We now have

$$x_{k+2} = 1/(x_{k+1} - a_{k+1}) = 1/(x_{k+1} - 2a_0) = 1/(\sqrt{d} - a_0) = x_1.$$

This means that the continued fraction for $\sqrt{d}$ repeats at $x_{k+2}$.

We have proved that $a_{k+1} \leq a_0$ when $B_{k+1} > 1$, and that $a_{k+1} = 2a_0$ and the continued fraction starts a repetition when $B_{k+1} = 1$. Moreover, when $B_{k+1} = 1$, we have $p_k^2 - dq_k^2 = \pm 1$. These are exactly the assertions of Theorem 17.2. $\qquad\square$

## 17.6   Some Irrational Numbers

Theorem 17.6 tells us that if $x$ is irrational then there are infinitely many fractions $p/q$ such that

$$\left| x - \frac{p}{q} \right| < \frac{1}{q^2}.$$

If $x$ is rational, this is not the case. Suppose $x = a/b$, where $a$ and $b$ are integers. Let $p/q$ be a rational number not equal to $a/b$. Since $aq - bp$ is a nonzero integer, $|aq - bp| \geq 1$. Therefore,

$$\left| \frac{a}{b} - \frac{p}{q} \right| = \left| \frac{aq - bp}{bq} \right| \geq \frac{1}{|bq|}.$$

If

$$\left| \frac{a}{b} - \frac{p}{q} \right| < \frac{1}{q^2},$$

then

$$\frac{1}{q^2} > \frac{1}{|bq|},$$

so $|q| < |b|$. Therefore, there are only finitely many such $q$, and consequently only finitely many such $p/q$. In other words, a rational number cannot be approximated closely by rational numbers other than itself.

By this reasoning, if a sequence of rational numbers converges very quickly to a real number $x$, then we have good rational approximations to $x$, which often allows us to conclude that $x$ is irrational. We give an example of this philosophy in the following.

**Theorem 17.18.** *The number $e = 2.71828\cdots$ is irrational.*

*Proof.* Recall the Taylor series

$$e^x = 1 + x + \frac{x^2}{2!} + \frac{x^3}{3!} + \cdots .$$

Setting $x = 1$ yields

$$e = 1 + 1 + \frac{1}{2!} + \frac{1}{3!} + \frac{1}{4!} + \frac{1}{5!} + \cdots .$$

Suppose $e = a/b$, where $a, b$ are positive integers. Since $e = 2.71828\cdots$, we know that $e$ is not an integer, so $b > 1$. We have

$$0 = b! \left( e - \frac{a}{b} \right)$$

$$= -b!\frac{a}{b} + b! \left( 1 + 1 + \frac{1}{2!} + \frac{1}{3!} + \cdots + \frac{1}{b!} \right)$$

$$+ b! \left( \frac{1}{(b+1)!} + \frac{1}{(b+2)!} + \cdots \right).$$

The quantity

$$-b!\frac{a}{b} + b! \left( 1 + 1 + \frac{1}{2!} + \frac{1}{3!} + \cdots + \frac{1}{b!} \right)$$

is an integer because $b!$ clears the denominators. Therefore,

$$N = b! \left( \frac{1}{(b+1)!} + \frac{1}{(b+2)!} + \cdots \right)$$

is also an integer since it plus an integer is 0. But

$$N = \frac{1}{b+1} + \frac{1}{(b+1)(b+2)} + \frac{1}{(b+1)(b+2)(b+3)} + \cdots$$

$$\leq \frac{1}{b+1} + \frac{1}{(b+1)(b+2)} + \frac{1}{(b+2)(b+3)} + \cdots$$

$$= \frac{1}{b+1} + \left( \frac{1}{b+1} - \frac{1}{b+2} \right) + \left( \frac{1}{b+2} - \frac{1}{b+3} \right) + \cdots$$

$$= \frac{1}{b+1} + \frac{1}{b+1} \quad \text{(the sum ``telescopes'' and the terms cancel)}$$

$$= \frac{2}{b+1}$$

$$< 1$$

since $b > 1$. Therefore, $N$ is an integer satisfying $0 < N < 1$, which is a contradiction. Therefore, $e \neq a/b$ and $e$ is irrational.    $\square$

Besides $e$, the other real number that occurs throughout mathematics is $\pi = 3.14159265 \cdots$. In 1761, Lambert proved it is irrational by using the continued fraction

$$\tan x = \cfrac{x}{1 - \cfrac{x^2}{3 - \cfrac{x^2}{5 - \cfrac{x^2}{7 - \cfrac{x^2}{\ddots}}}}}.$$

He showed that if $x$ is a nonzero rational, then the continued fraction equals an irrational number. But $\tan(\pi/4) = 1$ is rational, so $\pi/4$ must be irrational. The technical details of this proof are not difficult, and can be carried out by extending the theory developed in this chapter to more general continued fractions. Instead of doing that, we present a simpler proof due to Niven.

**Theorem 17.19.** $\pi$ and $\pi^2$ are irrational.

*Proof.* If $\pi$ is rational, then $\pi^2$ is rational, so it suffices to prove that $\pi^2$ is irrational. Suppose $\pi^2 = a/b$ for some positive integers $a$ and $b$. Choose $n$ large enough that

$$c_n = \frac{\pi a^n}{4^n n!} < 1. \tag{17.16}$$

Why is this possible? When $n \geq a/2$, we have

$$\frac{c_n}{c_{n-1}} = \frac{a}{4n} \leq \frac{1}{2},$$

so each $c_n$ is at most half of the preceding one. Therefore, $c_n \to 0$ as $n \to \infty$. In particular, $c_n < 1$ for large $n$.

Let

$$f(x) = x^n (1 - x)^n / n!.$$

**Lemma 17.20.** *For $n$ satisfying the inequality (17.16),*

$$0 < \int_0^1 \pi a^n f(x) \sin(\pi x) \, dx < 1.$$

*Proof.* We restrict our attention to $0 < x < 1$. In this range, $x(1-x)$ has a maximum value of $1/4$ and is positive. Therefore,

$$0 < f(x) = (1/n!)x^n(1-x)^n < (1/n!)(1/4^n).$$

Since $0 < \sin(\pi x) \le 1$ for $0 < x < 1$,

$$0 < \pi a^n f(x) \sin \pi x \le \pi a^n f(x) \le \pi a^n / (4^n n!) < 1$$

for $0 < x < 1$. It follows that

$$0 < \int_0^1 \pi a^n f(x) \sin(\pi x)\, dx < \int_0^1 1\, dx = 1.$$

$\square$

We'll now show that the integral is an integer, which will yield a contradiction. Let $f^{(j)}(x)$ denote the $j$th derivative of $f(x)$ (so $f^{(0)}(x) = f(x)$) and let $G(x) =$

$$b^n \left( \pi^{2n} f(x) - \pi^{2n-2} f^{(2)}(x) + \pi^{2n-4} f^{(4)}(x) - \cdots + (-1)^n f^{(2n)}(x) \right).$$

Since $f(x)$ is a polynomial of degree $2n$, we have $f^{(2n+2)}(x) = 0$, so

$$G''(x) + \pi^2 G(x) = b^n \left( \pi^{2n+2} f(x) + (-1)^n f^{(2n+2)}(x) \right) \qquad (17.17)$$

$$= b^n \pi^{2n+2} f(x) = \pi^2 a^n f(x).$$

**Lemma 17.21.**

$$\int \pi a^n f(x) \sin(\pi x)\, dx = \frac{1}{\pi} G'(x) \sin(\pi x) - G(x) \cos(\pi x) + C$$

*(where "+C" is the arbitrary constant of antidifferentiation).*

*Proof.* The formula can be obtained by integration by parts, but since we have the answer, it is easier to check it by differentiating. The derivative of the right-hand side is (use the product rule)

$$\frac{1}{\pi} G''(x) \sin(\pi x) + G'(x) \cos(\pi x) - G'(x) \cos(\pi x) + \pi G(x) \sin(\pi x)$$

$$= \frac{1}{\pi} G''(x) \sin(\pi x) + \pi G(x) \sin(\pi x)$$

$$= \frac{1}{\pi} \left( G''(x) + \pi^2 G(x) \right) \sin(\pi x)$$

$$= \pi a^n f(x) \sin(\pi x)$$

by (17.17). This proves the lemma.                                    $\square$

**Lemma 17.22.** $f^{(j)}(0)$ *and* $f^{(j)}(1)$ *are integers for all* $j \geq 0$.

*Proof.* The Binomial Theorem says that

$$(1-x)^n = 1 - \binom{n}{1}x + \binom{n}{2}x^2 - \cdots + (-1)^n x^n.$$

Therefore,

$$n!f(x) = x^n(1-x)^n = x^n - \binom{n}{1}x^{n+1} + \binom{n}{2}x^{n+2} - \cdots + (-1)^n x^{2n}.$$

If $j < n$ then $f^{(j)}(x)$ is a polynomial whose constant term is 0, so $f^{(j)}(0) = 0$. If $j = n + k \geq n$, then

$$n!f^{(n+k)}(0) = (n+k)!(-1)^k \binom{n}{k} = n!k! \binom{n+k}{n}(-1)^k \binom{n}{k},$$

which is $n!$ times an integer. Therefore, $f^{(n+k)}(0)$ is an integer.

We could do the same calculations with the derivatives at 1 by expanding $f(x)$ in powers of $(x-1)$, but it's more fun to use the following trick. Let

$$\begin{aligned} g(x) &= f(1-x) = (1-x)^n \left(1 - (1-x)\right)^n / n! \\ &= (1-x)^n (x)^n / n! \\ &= f(x). \end{aligned}$$

Differentiating the relation $f(x) = g(x) = f(1-x)$ yields

$$f^{(j)}(x) = g^{(j)}(x) = (-1)^j f^{(j)}(1-x),$$

so we have

$$f^{(j)}(1) = g^{(j)}(1) = (-1)^j f^{(j)}(1-1) = (-1)^j f^{(j)}(0),$$

which is an integer. This proves the lemma.  $\square$

The assumption that $\pi^2 = a/b$ implies that $G(0) =$

$$b^n \left( \pi^{2n} f(0) - \pi^{2n-2} f^{(2)}(0) + \pi^{2n-4} f^{(4)}(0) - \cdots + (-1)^n f^{(2n)}(0) \right)$$

$$= a^n f(0) - a^{n-1}b f^{(2)}(0) + a^{n-2}b^2 f^{(4)}(0) - \cdots + b^n(-1)^n f^{(2n)}(0).$$

This is an integer because every term is an integer. Similarly, $G(1)$ is an integer.

Lemma 17.21 implies that

$$\int_0^1 \pi a^n f(x) \sin(\pi x)\, dx = \frac{1}{\pi} G'(1)\sin(\pi) - G(1)\cos(\pi)$$

$$- \frac{1}{\pi} G'(0)\sin(0) + G(0)\cos(0)$$

$$= G(0) + G(1),$$

which is an integer. But Lemma 17.20 says that the value of the integral is strictly between 0 and 1. This is impossible, so we have a contradiction. Therefore, $\pi^2$ is irrational.                    □

It is interesting to note that both in the proof that $e$ is irrational and in the proof that $\pi^2$ is irrational, the contradiction was obtained by producing an integer between 0 and 1. This is a fairly common feature of irrationality proofs.

It is possible to make even stronger statements about $e$ and $\pi$. We say that a real (or complex) number is **algebraic** if it is a root of a polynomial

$$x^n + a_{n-1}x^{n-1} + \cdots + a_1 x + a_0$$

with rational coefficients. For example, $(3 + \sqrt{2})/4$ is algebraic since it is a root of $x^2 - (3/2)x + (7/16)$. Also, a rational number $a/b$ is the root of $x - (a/b)$, so rational numbers are algebraic. A number that is not algebraic is called **transcendental**. In 1873, Hermite proved that $e$ is transcendental, and in 1882, Lindemann proved that $\pi$ is transcendental. These proofs are beyond the scope of this book.

The fact that $\pi^2$ is irrational yields another proof of Euclid's theorem that there are infinitely many primes. As stated in Section 5.7,

$$\frac{\pi^2}{6} = \sum_{n=1}^{\infty} \frac{1}{n^2} = \prod_p \left(1 - p^{-2}\right)^{-1},$$

where the product is over all primes. If there were only finitely many primes, the product would be a rational number. Since $\pi^2$ is irrational, this is not the case. Therefore, there must be infinitely many primes.

# 17.7    Chapter Highlights

1.  Continued fractions give good rational approximations of real numbers (Theorem 17.6).

2.  Continued fractions yield solutions of Pell's equation (Theorem 17.2).

3.  The continued fraction of the square root of a positive integer has the form
    $$[a_0, \overline{a_1, a_2, \ldots, a_{n-1}, a_n}],$$
    where $a_n = 2a_0$ (Theorem 17.2).

# 17.8    Problems

## 17.8.1    Exercises

1.  Find the convergents $p_0/q_0$, $p_1/q_1$, $p_2/q_2$, and $p_3/q_3$ of the continued fraction expansion of 1.234567 and verify that
    $$\left| 1.234567 - \frac{p_n}{q_n} \right| < \frac{1}{q_n q_{n+1}}$$
    for $n = 0, 1, 2, 3$.

2.  Find $a, b$ less than 100 such that
    $$|.123456789 - (a/b)| < 10^{-8}.$$

3.  Find $a$ and $b$ less than 20 such that
    $$|.2357111317 - (a/b)| < .001.$$

4.  Find $a$ and $b$ with $0 < b < 20$ such that
    $$\left| \frac{3141}{5926} - \frac{a}{b} \right| < .001.$$

5.  Find $a$ and $b$ with $0 < b < 20$ such that
    $$\left| \frac{1357}{2468} - \frac{a}{b} \right| < .005.$$

6.  Evaluate $[1, 2, 3, 4]$ as a rational number.

7.  Evaluate $[4, 3, 2, 1]$ as a rational number.

8.  For each denominator $b$ with $1 \le b \le 6$, find the fraction $a/b$ that is closest to $\pi$. Show that none of these give as close an approximation to $\pi$ as $22/7$. (This provides an example for Theorem 17.8.)

9.  (a) Find the continued fraction for $\sqrt{7}$.
    (b) Find a nontrivial solution of $x^2 - 7y^2 = 1$.

10. (a) Find the continued fraction for $\sqrt{11}$.
    (b) Find a nontrivial solution of $x^2 - 11y^2 = 1$.

11. (a) Find the continued fraction for $\sqrt{15}$.
    (b) Find a nontrivial solution of $x^2 - 15y^2 = 1$.

12. (a) Find the continued fraction for $\sqrt{n^2 + 1}$, where $n \ge 1$ is an integer.
    (b) Find a nontrivial solution of $x^2 - (n^2 + 1)y^2 = 1$.

13. (a) Find the continued fraction for $\sqrt{n^2 - 1}$, where $n \ge 2$ is an integer.
    (b) Find a nontrivial solution of $x^2 - (n^2 - 1)y^2 = 1$.

14. (a) Find the continued fraction for $\sqrt{n^2 - 2}$, where $n \ge 3$ is an integer.
    (b) Find a nontrivial solution of $x^2 - (n^2 - 2)y^2 = 1$.

15. Evaluate $[\overline{1, 2, 3}]$ as a quadratic irrational.

16. Evaluate $[\overline{2, 1}]$ as a quadratic irrational.

17. Evaluate $[1, \overline{2}]$ as a quadratic irrational.

18. Evaluate $[2, \overline{4}]$ as a quadratic irrational.

19. Show that there are infinitely many Pythagorean triples $a, b, c$ such that $a + 1 = b$.

20. Show that there are infinitely many $n$ such that both $n$ and $2n + 1$ are squares.

## 17.8.2    Projects

1.  (a) Evaluate $[1, 1], [1, 1, 1], [1, 1, 1, 1]$ as rational numbers.
    (b) Let
    $$\frac{p_n}{q_n} = [1, 1, 1, \cdots, 1],$$
    where there are $n + 1$ ones (since the first 1 is $a_0$, the $(n + 1)$st is $a_n$). Use the recursion relations for $p_n$ and for $q_n$ to show that
    $$p_n = F_{n+2}, \qquad q_n = F_{n+1}.$$
    (c) Show that the infinite continued fraction $[\overline{1}] = [1, 1, 1, \cdots]$ equals the Golden Ratio $\phi = (1 + \sqrt{5})/2$.
    (d) Show that
    $$\lim_{n \to \infty} \frac{F_{n+2}}{F_{n+1}} = \phi.$$
    (This can also be deduced from the expressions for the Fibonacci numbers in terms of $\phi$.)

2.  Let $x$ be a real number. We used continued fractions to produce rational numbers $a/b$ with $|x - (a/b)| < 1/b^2$. Here we use a different method.

For a real number $y$, let $\{y\} = y - \lfloor y \rfloor$ be the fractional part of $y$. For example, $\{2.75\} = .75$ and $\{-3.2\} = .8$.

(a) Let $N$ be a positive integer. Consider the numbers

$$\{0\}, \{x\}, \{2x\}, \{3x\}, \ldots, \{Nx\}.$$

Show that there exist $i$ and $j$ with $0 \le i < j \le N$ such that

$$|\{jx\} - \{ix\}| < 1/N.$$

(*Hint:* Break up the interval from 0 to 1 into $N$ subintervals of length $1/N$ and explain why there must be some subinterval containing two numbers.)

(b) Use (a) to show that there exists an integer $b$ with $1 \le b \le N$, and an integer $a$ such that $|bx - a| < 1/N$.

(c) Let $a$ and $b$ be as in part (b). Show that $|x - (a/b)| < 1/b^2$.

3.  Let $x$ be a positive irrational number and suppose that $a$ and $b$ are positive integers with

$$\left| x - \frac{a}{b} \right| < \frac{1}{2b^2}.$$

We'll show that $a/b = p_k/q_k$ for some convergent of the continued fraction of $x$. Choose $k$ so that $q_k \le b < q_{k+1}$. By Theorem 17.8,

$$|bx - a| \ge |q_k x - p_k|.$$

(a) Show that

$$\left| x - \frac{p_k}{q_k} \right| < \frac{1}{2bq_k}.$$

(b) Show that if $a/b \ne p_k/q_k$ then

$$\frac{1}{bq_k} \le \left| \frac{a}{b} - \frac{p_k}{q_k} \right|.$$

(c) Show that if $a/b \ne p_k/q_k$ then

$$\frac{1}{bq_k} < \frac{1}{2bq_k} + \frac{1}{2b^2}.$$

(*Hint:* Write $a/b - p_k/q_k = (x - p_k/q_k) + (a/b - x)$ and use the fact that $|u + v| \le |u| + |v|$.)

(d) Use the inequality in (c) to show that $q_k > b$, which is a contradiction. Therefore, $a/b = p_k/q_k$.

4.  We know that if we start with a real number $x$ and find its continued fraction expansion, then the continued fraction converges to $x$. Suppose instead that we start with a continued fraction. By Proposition 17.5, we know that it converges to a real number $y$. If we compute the continued fraction of $y$, do we get the original continued fraction or a different one?

(a) Show that $[2, 5, 1] = [2, 6]$, so two continued fractions that terminate can equal the same rational number.

(b) More generally, suppose $a_k > 1$. Show that

$$[a_0, a_1, \ldots, a_{k-1}, a_k - 1, 1] = [a_0, a_1, \ldots, a_{k-1}, a_k].$$

Now suppose that $[a_0, a_1, a_2, \ldots] = [b_0, b_1, b_2, \ldots]$ (where $a_i$ and $b_i$ are positive integers) are two different continued fraction expansions of the same real number. Let $k$ be the smallest index such that $a_k \neq b_k$. Without loss of generality, assume that $a_k < b_k$.

(c) Show that $[a_k, a_{k+1}, \ldots] = [b_k, b_{k+1}, \ldots]$. (*Hint: $a_i = b_i$ for $0 \leq i < k$.*)

(d) Show that $[a_k, a_{k+1}, \ldots] \leq a_k + 1$, with equality if and only if the continued fraction terminates at $a_{k+1}$ and $a_{k+1} = 1$.

(e) Show that $[b_k, b_{k+1}, \ldots] \geq b_k$, with equality if and only if the continued fraction terminates at $b_k$.

(f) Show that $a_k + 1 = b_k$ and that $[a_0, a_1, a_2, \ldots]$ terminates at $a_{k+1} = 1$ and $[b_0, b_1, b_2, \ldots]$ terminates at $b_k$.

This shows that the continued fraction expansion is unique except for the case of rational numbers given in part (b).

## 17.8.3   Computer Explorations

1.  Find a nontrivial solution of $x^2 - 61y^2 = 1$.
2.  Find a nontrivial solution of $x^2 - 109y^2 = 1$.
3.  Calculate several terms of the continued fractions of $e$, $(e + 1)/(e - 1)$, and $(e^2 + 1)/(e^2 - 1)$ until you find a pattern. Calculate a few more terms and see if the pattern persists (if you have the correct pattern it should keep going).

## 17.8.4   Answers to "CHECK YOUR UNDERSTANDING"

1.  We have $\lfloor 2.357 \rfloor = 2$, so $a_0 = 2$ and $x_1 = 1/.357 = 2.8011 \cdots$. Since $\lfloor 2.8011 \rfloor = 2$, we have $a_1 = 2$ and $x_2 = 1/.8011 = 1.248 \cdots$. Next, $a_3 = 1$ and $x_3 = 1/.248 = 4.027 \cdots$. Therefore, $2.357 = [2, 2, 1, 4.027]$.

2.  Write the table

| | | 1 | 3 | 5 | 7 |
|---|---|---|---|---|---|
| 0 | 1 | 1 | 4 | 21 | 151 |
| 1 | 0 | 1 | 3 | 16 | 115 |

This yields the answer $151/115$.

3.  Calculate $\sqrt{11} = 3.31662 \cdots$. The continued fraction starts $[3, 3, 6, \ldots]$. Since $a_0 = 3$ and $6 = 2a_0$, we stop before the 6 and evaluate $[3, 3] = 3 + 1/3 = 10/3$. Then $10^2 - 11 \cdot 3^2 = 1$, so $(x, y) = (10, 3)$ is a solution.

# Chapter 18

# Gaussian Integers

When you try to solve a polynomial equation whose coefficients are integers, its solutions are often *not* integers. In order to solve $2x + 5 = 0$, $x^2 + 3x - 1 = 0$, or $x^2 + 1 = 0$, you have to go beyond the integers and work in the rational, real, or complex numbers. We encounter the same scenario in number theory, where we often find it natural to use irrational or complex numbers, even when trying to find integral solutions to equations. In this chapter, we look at numbers of the form $a + bi$, where $i = \sqrt{-1}$, and where $a$ and $b$ are integers. These numbers are called **Gaussian integers** and are usually denoted $\mathbb{Z}[i]$:

$$\mathbb{Z}[i] = \{a + bi \mid a, b \text{ are integers}\}$$

($\mathbb{Z}$ is a standard notation for the integers, from the German word *Zahlen*, meaning numbers). Gauss used $\mathbb{Z}[i]$ when he worked on Biquadratic Reciprocity, his generalization of Quadratic Reciprocity to 4th powers. The Gaussian integers have many properties in common with the integers and can be studied in their own right. However, in the last section of this chapter, we give three applications where they arise naturally when looking at problems involving the usual integers.

## 18.1 Complex Arithmetic

To start, it's worthwhile describing how to work with arbitrary complex numbers. Let $x$ and $y$ be real numbers. Then $z = x + yi$ is a complex number. Multiplying complex numbers is done in a way similar to multiplying linear polynomials, using the relation $i^2 = -1$. For example,

$$(3+2i)(5+8i) = 3 \cdot 5 + 2i \cdot 5 + 3 \cdot 8i + 2i \cdot 8i = 15 + 10i + 24i - 16 = -1 + 34i.$$

In general, the formula is

$$(x_1 + y_1 i)(x_2 + y_2 i) = (x_1 x_2 - y_1 y_2) + (x_1 y_2 + y_1 x_2)i.$$

The **complex conjugate** of $z$ is $\overline{z} = x - yi$. The **absolute value** of $z$ is

$$|z| = |x + yi| = \sqrt{x^2 + y^2}.$$

If $z = x + yi$ is represented by the point $(x, y)$ in the plane, then $|z|$ is the distance from $z$ to 0. A fact we'll need several times is that

$$z\overline{z} = (x + yi)(x - yi) = x^2 + y^2 = |z|^2.$$

Therefore,

$$z^{-1} = \frac{\overline{z}}{|z|^2} = \frac{x - yi}{x^2 + y^2}.$$

A basic property of the absolute value is that

$$|z_1 z_2| = |z_1||z_2|,$$

for any two complex numbers $z_1$, $z_2$. We ask you to prove this in Exercise 1.

Division is accomplished by rationalizing denominators:

$$\frac{x_1 + y_1 i}{x_2 + y_2 i} = \frac{x_1 + y_1 i}{x_2 + y_2 i} \frac{x_2 - y_2 i}{x_2 - y_2 i} = \frac{x_1 x_2 + y_1 y_2 + (y_1 x_2 - x_1 y_2)i}{x_2^2 + y_2^2}.$$

This can also be written as

$$\frac{z_1}{z_2} = \frac{z_1}{z_2} \frac{\overline{z_2}}{\overline{z_2}} = \frac{z_1 \overline{z_2}}{|z_2|^2}.$$

For example, since the conjugate of $2 - 3i$ is $2 + 3i$,

$$\frac{12 + 5i}{2 - 3i} = \frac{(12 + 5i)(2 + 3i)}{2^2 + 3^2} = \frac{9 + 46i}{13} = (9/13) + (46/13)i.$$

---

## CHECK YOUR UNDERSTANDING:

1. Show that $(23 - 2i)/(3 + 2i)$ is a Gaussian integer.

# 18.2    Gaussian Irreducibles

For number-theoretic applications, it is convenient to replace the absolute value with the square of the absolute value. Define the **norm** of $a + bi$ to be

$$N(a + bi) = a^2 + b^2 = (a + bi)(a - bi) = |a + bi|^2.$$

The advantage is that the norm of a Gaussian integer is an integer, and this isn't the case for the absolute value (for example, $|2 + 3i| = \sqrt{13}$). Moreover, squaring the relation $|z_1 z_2| = |z_1||z_2|$ yields

$$N\left((a + bi)(c + di)\right) = N(a + bi)\, N(c + di).$$

The following will be useful.

**Lemma 18.1.** *Let $\alpha \in \mathbb{Z}[i]$. The following are equivalent:*
*(a) $N(\alpha) = 1$*
*(b) $1/\alpha \in \mathbb{Z}[i]$*
*(c) $\alpha = \pm 1$ or $\pm i$.*

*Proof.* Write $\alpha = a + bi$. Then $N(a + bi) = a^2 + b^2$, which equals 1 if and only if either $(a, b) = (\pm 1, 0)$ or $(a, b) = (0, \pm 1)$. These correspond to $\alpha = \pm 1$ and $\pm i$. Therefore, (a) and (c) are equivalent. Suppose $1 = N(\alpha) = \alpha\bar{\alpha}$. Then $\bar{\alpha} = 1/\alpha$ and since $\bar{\alpha} \in \mathbb{Z}[i]$, (a) implies (b). Conversely, suppose $1/\alpha \in \mathbb{Z}[i]$. Let $\beta = 1/\alpha$. Then $1 = \alpha\beta$ implies that $1 = N(1) = N(\alpha)N(\beta)$. Since $N(\alpha)$ and $N(\beta)$ are positive integers whose product is 1, we must have $N(\alpha) = 1$, so (b) implies (a).    □

The numbers $\pm 1$ and $\pm i$ are called **units** of $\mathbb{Z}[i]$. This is standard terminology in ring theory for elements whose multiplicative inverses are in the set. (It might seem that the terms "unit" and "unit element" (for the number 1) could cause confusion, but the context should make the distinction clear.)

Units cause slight ambiguities in factorization. In the integers, we can factor $-12$ as $(-2)(2)(3)$, as $(2)(2)(-3)$, as $(-2)(-2)(-3)$, or as $-(2)(2)(3)$. Of course, these are all essentially the same factorization. In the Gaussian integers, there are four units, so the situation is even worse. For example, if $\alpha = \beta\gamma$, then $\alpha = (i\beta)(-i\gamma) = (-i\beta)(i\gamma) =$

$(-\beta)(-\gamma)$. To help us say that these are all essentially the same, we define the **associates** of $\alpha$ to be the numbers $u\alpha$, where $u$ is a unit. Every nonzero Gaussian integer has four associates. For example, the associates of $2 + 3i$ are

$$2 + 3i = (1)(2 + 3i), \quad -2 - 3i = (-1)(2 + 3i),$$
$$-3 + 2i = (i)(2 + 3i), \quad 3 - 2i = (-i)(2 + 3i).$$

We want to look at Gaussian integers that are the analogues of primes. In the integers, every number $n$ factors as $n = (-1)(-n)$, but we don't count that as a true factorization. Similarly, if $\alpha$ is a Gaussian integer, then $\alpha$ factors as $\alpha = (i)(-i\alpha)$. Again, this doesn't count as a true factorization. In other words, factoring off units does not count.

**Definition 18.2.** *A Gaussian integer $\alpha \neq 0$ is **irreducible** if $\alpha$ is not a unit and, whenever $\alpha = \beta\gamma$ is a factorization of $\alpha$ into Gaussian integers, one of the factors is a unit.*

In other words, irreducible elements are the ones that cannot be factored nontrivially.

**Proposition 18.3.** *If $N(\alpha) = p$ is prime, then $\alpha$ is irreducible.*

*Proof.* Suppose $\alpha = \beta\gamma$. Then $p = N(\alpha) = N(\beta)N(\gamma)$. Since $p$ is prime and $N(\beta)$ and $N(\gamma)$ are integers, we must have $N(\beta) = 1$ or $N(\gamma) = 1$. Lemma 18.1 implies that $\beta$ or $\gamma$ is a unit. Therefore, the only factorizations of $\alpha$ are trivial, so $\alpha$ is irreducible. $\qquad\square$

This gives a large supply of irreducibles. For example,

$$1 + i, \quad 2 - i, \quad 2 + 3i, \quad -6 + i$$

are irreducibles since their norms are primes. However, there are irreducibles whose norms are not prime. As the following result shows, 3 and 7, whose norms are 9 and 49, respectively, are irreducible in $\mathbb{Z}[i]$.

**Proposition 18.4.** *Let $p$ be a prime with $p \equiv 3 \pmod 4$. Then $p$ is irreducible in $\mathbb{Z}[i]$.*

*Proof.* Suppose $p = \beta\gamma$. Then

$$p^2 = N(p) = N(\beta)N(\gamma).$$

If neither $\beta$ nor $\gamma$ is a unit, then $N(\beta) > 1$ and $N(\gamma) > 1$. This implies that $N(\beta) = N(\gamma) = p$. Write $\beta = a + bi$. Then we have

$$p = a^2 + b^2.$$

But a square is congruent to 0 or 1 mod 4, so $a^2 + b^2 \equiv 0, 1, 2 \pmod{4}$. In particular, $a^2 + b^2 \neq p$. Therefore, $N(\beta) = p$ is impossible, so either $N(\beta) = 1$ or $N(\gamma) = 1$. Consequently, either $N(\beta) = 1$ or $N(\gamma) = 1$. Lemma 18.1 says that either $\beta$ or $\gamma$ must be a unit. This means that $p$ is irreducible.                                                                $\square$

In the proof, the main step was that primes congruent to 3 mod 4 cannot be written as sums of two squares. This fact is part of Theorem 15.6.

The proposition says that primes that are 3 mod 4 are still irreducible in $\mathbb{Z}[i]$. However, the primes 2 and 5 are no longer irreducible in $\mathbb{Z}[i]$ since $2 = -i(1 + i)^2$ and $5 = (2 + i)(2 - i)$.

Later, we'll prove the following.

**Proposition 18.5.** *The irreducible elements of $\mathbb{Z}[i]$ are the following and their associates:*
*(a) $1 + i$*
*(b) $p$, where $p$ is a prime with $p \equiv 3 \pmod{4}$*
*(c) $a + bi$ and $a - bi$, where $a^2 + b^2 = p$ with $p$ a prime such that $p \equiv 1$ (mod 4). The irreducibles $a + bi$ and $a - bi$ are not associates.*

The proof will be given in Subsection 18.5.1. The fact that primes that are 1 mod 4 are sums of two squares was proved in Section 15.2, but will also be proved in Subsection 18.5.1. If $p \equiv 1 \pmod{4}$, then $a$ and $b$ are unique. This is Exercise 11.

**Example.** The prime 29 is 1 mod 4, and $29 = 5^2 + 2^2$. This gives us two irreducibles: $5 + 2i$ and $5 - 2i$. We can also write $29 = 2^2 + 5^2$. This gives $2 + 5i$ and $2 - 5i$. These might look different from the previous two irreducibles, but

$$2 + 5i = i(5 - 2i) \text{ and } 2 - 5i = -i(5 + 2i),$$

so these new irreducibles are associates of the previous two irreducibles.

Our goal is to prove that every nonzero Gaussian integer is either a unit, an irreducible, or a product of irreducibles, and that this factorization is unique. The uniqueness needs a little explanation. We have

$$13 = (2 + 3i)(2 - 3i) = (3 + 2i)(3 - 2i).$$

Observe that

$$3 + 2i = i(2 - 3i) \text{ and } 3 - 2i = -i(2 + 3i),$$

so these irreducibles are associates. This is very much the analogue of factoring $15 = 3 \cdot 5 = (-3) \cdot (-5)$, which are regarded as essentially the same factorizations in the integers. In the integers, we make the convention that we always choose the primes to be positive, and we could make some similar type of choice in $\mathbb{Z}[i]$, but it is more standard to regard irreducibles as coming in sets of four associates: $\pm\alpha, \pm i\alpha$. Using any one of these in a factorization is just as good as using another one. We say that a factorization is unique if all other factorizations are obtained by multiplying irreducibles in the factorization by units.

When we proved that integers factor uniquely as products of primes (this is the "Fundamental Theorem of Arithmetic"), it was easy to show that every integer factors into a product of primes. Uniqueness was the difficult part and we used the Division Algorithm to show this. We encounter the same situation in the Gaussian integers: factorization is easy, uniqueness is more difficult. So, we begin with factorization into irreducibles as our first order of business.

**Proposition 18.6.** *Every nonzero Gaussian integer is either a unit, an irreducible, or a product of irreducibles.*

*Proof.* Suppose the proposition is false. Let $\alpha$ be a nonzero Gaussian integer that is not a unit, an irreducible, or a product of irreducibles, and assume that $N(\alpha)$ is as small as possible among such counterexamples. Since $\alpha$ is not a unit or an irreducible, $\alpha = \beta\gamma$ with neither $\beta$ nor $\gamma$ a unit. Therefore, $N(\beta) > 1$ and $N(\gamma) > 1$. But $N(\alpha) = N(\beta)N(\gamma)$, so $N(\beta) < N(\alpha)$ and $N(\gamma) < N(\alpha)$. By the minimality of $N(\alpha)$, both $\beta$ and $\gamma$ must be irreducibles or factor as products of irreducibles. Therefore, $\alpha = \beta\gamma$ is a product of irreducibles. This contradiction shows that no counterexamples exist, so the proposition must be true. $\square$

In Section 18.4, we show that the factorization is unique. The main step will be showing that if an irreducible divides a product, then it divides one of the factors. For the integers, the proof uses the Division Algorithm, and we similarly use a division algorithm for the Gaussian integers.

---

**CHECK YOUR UNDERSTANDING:**

2. Show that $6 + i$ is irreducible.
3. Find the associates of $1 + 4i$.
4. Explain why the two factorizations $17 = (4 + i)(4 - i)$ and $17 = (1 + 4i)(1 - 4i)$ are essentially the same factorization.

---

# 18.3    The Division Algorithm

Perhaps the most important fact about the integers is that we can do division with remainder: Given an integer $a$ and an integer $b > 0$, there are integers $q$ and $r$ such that

$$a = bq + r \text{ and } 0 \leq r < b.$$

A similar property is true for Gaussian integers. When working with integers, we use the absolute value to tell us the size of a number. For Gaussian integers, it's the norm that plays an analogous role.

**Theorem 18.7.** *Let $\alpha$ and $\beta$ be Gaussian integers with $\beta \neq 0$. Then there exist Gaussian integers $\eta$ and $\rho$ such that*

$$\alpha = \beta\eta + \rho \text{ and } 0 \leq N(\rho) < N(\beta).$$

For example, if $\alpha = 23 - 9i$ and $\beta = 3 + 2i$, then

$$23 - 9i = (3 + 2i)(4 - 5i) + (1 - 2i),$$

and $N(1 - 2i) < N(3 + 2i)$.

*Proof.* Let $\delta = \alpha/\beta$. Then $\delta$ is a complex number and the set of points $z$

in the complex plane satisfying $|z - \delta| < 1$ is a disc of radius 1 centered at $\delta$. Every point inside a unit square is a distance at most $1/\sqrt{2} < 1$ from a corner. Therefore, there is a point $\eta = a + bi$ with $a, b$ integers, inside this disc:

$$|\eta - \delta| < 1.$$

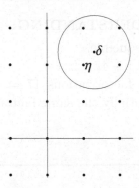

This says that

$$\left| \eta - \frac{\alpha}{\beta} \right| < 1.$$

Multiplication by $|\beta|$ yields

$$|\alpha - \beta\eta| < |\beta|.$$

Let $\rho = \alpha - \beta\eta$. Then $\rho$ is a Gaussian integer since $\alpha, \beta, \eta$ are Gaussian integers, and $|\rho| < |\beta|$. Therefore,

$$N(\rho) = |\rho|^2 < |\beta|^2 = N(\beta).$$

Since $\alpha = \beta\eta + \rho$, this completes the proof. $\qquad\square$

Although it is rare that an explicit calculation is done with the Division Algorithm, here is how it can be done in practice. First, let's consider the case where $\beta$ is an integer. For example, let $\alpha = 191 + 239i$ and $\beta = 34$. Then

$$191 + 239i = 34(6 + 7i) + (-13 + i).$$

Note that we used the Division Algorithm for the integers on the real part and the imaginary part. However, in order to make the remainder as small as possible, we allowed negative remainders and wrote $191 = 34 \cdot 6 - 13$ instead of $191 = 34 \cdot 5 + 21$, since $-13$ is smaller in absolute value than $21$. Since $N(-13 + i) = 170 < N(34) = 1156$, this is the desired division with remainder.

Now let's consider the general situation. Suppose $\alpha = 52 - 7i$ and $\beta = 3 - 5i$. We want

$$52 - 7i = (3 - 5i)\eta + \rho.$$

Since it is easier to work with integers, we multiply by $3 + 5i$ to obtain

$$191 + 239i = (52 - 7i)(3 + 5i) = 34\eta + (3 + 5i)\rho.$$

This is the division problem we just solved:

$$191 + 239i = 34(6 + 7i) + (-13 + i).$$

Now divide by $3 + 5i$:

$$52 - 7i = (3 - 5i)(6 + 7i) + (-13 + i)/(3 + 5i).$$

The first two divisions were easy since $191 + 239i$ and $34$ were originally obtained as multiples of $3 + 5i$. Moreover, since $191 + 239i = (52 - 7i)(3 + 5i)$ and $34 = (3 - 5i)(3 + 5i)$ are multiples of $3 + 5i$, we know that $-13 + i$ must also be a multiple of $3 + 5i$. In fact, it is:

$$\rho = \frac{-13 + i}{3 + 5i} = \frac{-13 + i}{3 + 5i}\frac{3 - 5i}{3 - 5i} = \frac{-34 + 68i}{34} = -1 + 2i.$$

Therefore, we obtain the desired division with remainder:

$$52 - 7i = (3 - 5i)(6 + 7i) + (-1 + 2i).$$

Observe that $N(-1 + 2i) < N(3 - 5i)$, so $\rho$ is a suitable remainder.

In general, to solve $\alpha = \beta\eta + \rho$, multiply by $\overline{\beta}$ and let $B = \beta\overline{\beta}$. Divide $\alpha\overline{\beta}$ by the integer $B$, and make the coordinates of the remainder as small as possible in absolute value. This gives $\alpha\overline{\beta} = B\eta + \rho'$. Then divide by $\overline{\beta}$ to get $\alpha = \beta\eta + \rho$, with $\rho = \rho'/\overline{\beta}$.

# 18.4   Unique Factorization

Let $\alpha$ and $\beta$ be Gaussian integers. We say that $\alpha$ **divides** $\beta$ (notation: $\alpha \mid \beta$) if there exists a Gaussian integer $\gamma$ such that $\alpha\gamma = \beta$. For example,

$$2 + i \mid -1 + 7i \quad \text{since} \quad -1 + 7i = (2 + i)(1 + 3i)$$

and

$$2 + i \mid 6 + 3i \quad \text{since} \quad 6 + 3i = 3(2 + i).$$

If $\alpha \mid \beta$ and $u$ is a unit, then $u\alpha \mid \beta$. This is seen as follows: $\alpha \mid \beta$ implies that there exists $\gamma$ with $\beta = \alpha\gamma$. Therefore, $\beta = (u\alpha)(u^{-1}\gamma)$. Since $u$ is a unit, $u^{-1}$ is a Gaussian integer, so $u^{-1}\gamma$ is a Gaussian integer. Therefore, $u\alpha$ times a Gaussian integer equals $\beta$, so $u\alpha \mid \beta$. Therefore, we can say that $\alpha \mid \beta$ if and only if some associate of $\alpha$ divides $\beta$.

An important concept for the usual integers is the greatest common divisor and it is useful to define an analogous concept for the Gaussian integers. We could define greatest common divisors to be those common divisors with largest norm. Instead, we give the following definition, which also works in situations where we don't have something like the norm to help us decide whether one number is larger than another.

**Definition 18.8.** *Let $\alpha$ and $\beta$ be Gaussian integers and assume that at least one of them is nonzero. We say that $\gamma$ is a **greatest common divisor** of $\alpha$ and $\beta$ if*
*(a) $\gamma \mid \alpha$ and $\gamma \mid \beta$; and*
*(b) whenever $\delta \mid \alpha$ and $\delta \mid \beta$ then $\delta \mid \gamma$.*

Note that if $\gamma$ is a greatest common divisor of $\alpha$ and $\beta$, then every associate of $\gamma$ is also a gcd. This follows from the above remark that one number divides another if and only if this divisibility holds for an associate of the number.

When the gcd is defined as in Definition 18.8, it is not obvious that it exists. The following takes care of this, along with proving the analogue of Theorem 2.12.

**Theorem 18.9.** *Let $\alpha$ and $\beta$ be Gaussian integers and assume that at* *least one of them is nonzero. Then*

*(a)* $\gamma = \gcd(\alpha, \beta)$ *exists.*

*(b) If* $\gamma'$ *is another gcd of* $\alpha$ *and* $\beta$, *then* $\gamma'$ *is an associate of* $\gamma$.

*(c) There exist* $x, y \in \mathbb{Z}[i]$ *such that* $\alpha x + \beta y = \gamma$.

*(d) If* $\delta$ *is a common divisor of* $\alpha$ *and* $\beta$, *then* $N(\delta) \leq N(\gamma)$.

*(e) If* $\delta$ *is a common divisor of* $\alpha$ *and* $\beta$, *and* $N(\delta) = N(\gamma)$, *then* $\delta$ *is also a gcd of* $\alpha$ *and* $\beta$.

*Proof.* It is possible to develop the Euclidean Algorithm and the Extended Euclidean Algorithm to find gcd's, but a non-constructive approach, similar to the proof of Theorem 2.12, suffices for our purposes.

We begin by proving (a): $\gamma = \gcd(\alpha, \beta)$ exists. Let $I$ be the set of all linear combinations of $\alpha$ and $\beta$:

$$I = \{\alpha x + \beta y \mid x, y \in \mathbb{Z}[i]\}.$$

Then $\alpha = 1 \cdot \alpha + 0 \cdot \beta$ and $\beta = 0 \cdot \alpha + 1 \cdot \beta$ are in $I$, so $I$ contains at least one nonzero element. Let

$$\gamma = \alpha x_0 + \beta y_0$$

be a nonzero element of $I$ with smallest possible norm: $0 < N(\gamma) \leq N(s)$ for all nonzero $s \in I$. We claim that $\gamma$ is a gcd of $\alpha$ and $\beta$.

By the Division Algorithm, we can divide $\alpha$ by $\gamma$ and get

$$\alpha = \gamma \eta + \rho$$

with $N(\rho) < N(\gamma)$. Since

$$\rho = \alpha - \gamma\eta = \alpha - (\alpha x_0 + \beta y_0)\eta = (1 - \eta x_0)\alpha + (-\eta y_0)\beta,$$

we have $\rho \in I$. But $N(\rho) < N(\gamma)$ and $N(\gamma)$ is the minimum value for norms of nonzero elements of $I$, so we must have $\rho = 0$. This means that $\alpha = \gamma\eta$, so $\gamma \mid \alpha$. Similarly, $\gamma \mid \beta$. This proves that $\gamma$ is a common divisor.

Suppose $\delta$ is a common divisor of $\alpha$ and $\beta$, so $\alpha = \eta_1 \delta$ and $\beta = \eta_2 \delta$, for some $\eta_1, \eta_2 \in \mathbb{Z}[i]$. Then

$$\gamma = \alpha x_0 + \beta y_0 = \eta_1 \delta x_0 + \eta_2 \delta y_0 = (\eta_1 x_0 + \eta_2 y_0)\delta,$$

so $\delta \mid \gamma$. This shows that $\gamma$ satisfies the conditions for being a gcd, so that gcd's exist. This completes the proof of part (a). In addition, we have also proved part (c) of the theorem for this $\gamma$ since $\gamma = \alpha x_0 + \beta y_0$.

We now prove part (b) and part (c) for *all* gcd's. Suppose that $\gamma'$ is another gcd. Then $\gamma'$ is a common divisor, so $\gamma' \mid \gamma$, so we can write $\gamma' \mu_1 = \gamma$ for some $\mu_1$. Reversing the roles of $\gamma$ and $\gamma'$, we find that $\gamma \mid \gamma'$, so we can write $\gamma \mu_2 = \gamma'$. Therefore, $\gamma \mu_2 \mu_1 = \gamma$, so $\mu_2 \mu_1 = 1$. This implies that $N(\mu_2) N(\mu_1) = N(1) = 1$, so $N(\mu_1) = N(\mu_2) = 1$. By Lemma 18.1, $\mu_2 = \pm 1$ or $\pm i$. This proves part (b) of the theorem. Also, $\gamma' = \gamma \mu_2 = \alpha(\mu_2 x_0) + \beta(\mu_2 x_0)$, so (c) holds for all gcd's.

The only remaining parts are (d) and (e). These are relatively straightforward: If $\delta$ is a common divisor of $\alpha$ and $\beta$, then the definition of gcd says that $\delta \mid \gamma$. This implies that $\delta \eta = \gamma$ for some $\eta$, which implies that $N(\delta) N(\eta) = N(\gamma)$. Therefore $N(\delta) \leq N(\gamma)$. This proves (d). If $N(\delta) = N(\gamma)$ then $N(\eta) = 1$. By Lemma 18.1, $\eta$ is a unit, so $\gamma = \eta \delta$ and $\delta$ are associates and both are gcd's. This proves (e).                □

**Remark.** Parts (c) and (d) show that the greatest common divisor is the largest in terms of the norm, so the definition we gave for gcd gives the same answer as what might seem to be the more natural definition.

We can now prove a crucial property of irreducibles. The proof is the same as the one with integers.

**Corollary 18.10.** *Let $\pi$ be an irreducible in $\mathbb{Z}[i]$, and let $\alpha$ and $\beta$ be Gaussian integers. If $\pi \mid \alpha\beta$, then $\pi \mid \alpha$ or $\pi \mid \beta$.*

*Proof.* If $\pi \mid \alpha$, we're done, so let's assume that $\pi \nmid \alpha$. Let $\gamma = \gcd(\alpha, \pi)$. Then $\gamma \mid \pi$. Since $\pi$ is irreducible, $\gamma$ (or an associate) equals either 1 or $\pi$. But $\gamma \neq \pi$ because $\gamma \mid \alpha$ and $\pi \nmid \alpha$. Therefore, $\gamma$ is a unit. By part (c) of Theorem 18.9, there exist $x_1, y_1 \in \mathbb{Z}[i]$ such that $\alpha x_1 + \pi y_1 = \gamma$. Since $\gamma$ is a unit, we can multiply both sides of the equation by $\gamma^{-1}$. Let $x = \gamma^{-1} x_1$ and $y = \gamma^{-1} y_1$, so we have

$$\alpha x + \pi y = 1.$$

Multiply by $\beta$ to obtain

$$\alpha\beta x + \pi\beta y = \beta.$$

Since $\pi \mid \alpha\beta$ and $\pi \mid \pi\beta y$, we have $\pi \mid \beta$.                □

An irreducible that satisfies the property of Corollary 18.10 is usually called *prime* in ring theory. So we have proved that irreducibles in $\mathbb{Z}[i]$ are primes in $\mathbb{Z}[i]$. However, since we will be working with both the integers and $\mathbb{Z}[i]$, to avoid confusion we will use "prime" for primes in the integers and "irreducible" for the irreducible elements of $\mathbb{Z}[i]$.

As in the case of integers, we immediately deduce the following.

**Corollary 18.11.** *Let $\pi$ be an irreducible in $\mathbb{Z}[i]$, let $m \geq 2$, and let $\alpha_1, \alpha_2, \ldots, \alpha_m$ be Gaussian integers. If $\pi \mid \alpha_1 \alpha_2 \cdots \alpha_m$, then $\pi \mid \alpha_j$ for some $j$.*

*Proof.* We proceed by induction on $m$, the number of factors. By Corollary 18.10, if $\pi | \alpha_1 \alpha_2$, then $\pi | \alpha_1$ or $\pi | \alpha_2$, so Corollary 18.11 is true for $m = 2$.

Assume that it is true for $m = k \geq 2$. We'll show that it's true for $m = k + 1$. Suppose

$$\pi \mid \alpha_1 \alpha_2 \cdots \alpha_{k+1} = (\alpha_1 \alpha_2 \cdots \alpha_k)\, \alpha_{k+1}.$$

Corollary 18.10 says that

$$\pi \mid \alpha_1 \alpha_2 \cdots \alpha_k \quad \text{or} \quad \pi \mid \alpha_{k+1}.$$

If $\pi \mid \alpha_{k+1}$, we're done. If $\pi \mid \alpha_1 \alpha_2 \cdots \alpha_k$, then the induction assumption says that $\pi \mid \alpha_j$ for some $j$. Therefore, the statement is true for $m = k + 1$. By induction, it is true for all $m$. $\qquad\square$

As with the integers, we can now prove the uniqueness of factorization into irreducibles in $\mathbb{Z}[i]$.

**Theorem 18.12.** *Every nonzero Gaussian integer is either a unit, an irreducible, or a product of irreducibles. This factorization is unique up to the order of the factors and multiplication of irreducibles by units.*

*Proof.* We showed in Proposition 18.6 that factorizations exist. It remains to show uniqueness. Assume that there is a Gaussian integer that can be written as a product of irreducibles in more than one way. Among these, let $\alpha$ have the smallest norm. Write

$$\alpha = \pi_1 \pi_2 \cdots \pi_r = \pi_1' \pi_2' \cdots \pi_s', \tag{18.1}$$

where each $\pi_j$ and $\pi_j'$ is irreducible. Since $\pi_1$ divides the left-hand side, it has to divide the right-hand side as well. From Corollary 18.11, we see that $\pi_1 \mid \pi'_j$ for some $j$. After rearranging terms, we can assume that $\pi'_j = \pi'_1$. Since $\pi'_1$ is irreducible and $\pi_1 \mid \pi'_1$, we must have that $\pi_1$ and $\pi'_1$ differ by a unit, otherwise we would have a nontrivial factorization of $\pi'_1$. Write $\pi'_1 = u\pi_1$ for some unit $u$ and write Equation (18.1) as

$$\alpha = \pi_1\pi_2 \cdots \pi_r = u\pi_1\pi'_2 \cdots \pi'_s.$$

Divide both sides by $\pi_1$. This gives us

$$\mu = \pi_2\pi_3 \cdots \pi_r = \pi''_2\pi'_3 \cdots \pi'_s$$

for some integer $\mu$, where $\pi''_2 = u\pi'_2$. Since the two factorizations of $\alpha$ were different, these must be two different factorizations of $\mu$. Because $\mu = \alpha/\pi_1$, we know that $N(\mu) < N(\alpha)$. This contradicts our assumption that $\alpha$ has the smallest norm among these numbers that can be factored as a product of irreducibles in more than one way. Therefore, our initial assumption on non-unique factorization was incorrect and consequently every integer can be factored into a product of irreducibles in exactly one way.                                    $\square$

Suppose you have a Gaussian integer and you want to factor it into irreducibles. For example, suppose we want to factor $8+i$. First, compute the norm and factor it into primes:

$$(8 + i)(8 - i) = N(8 + i) = 8^2 + 1^2 = 65 = 5 \times 13.$$

Then factor each prime into irreducibles in $\mathbb{Z}[i]$ to obtain

$$(8 + i)(8 - i) = 5 \times 13 = (2 + i)(2 - i)(3 + 2i)(3 - 2i),$$

where we used the facts that $5 = 2^2 + 1^2$ and $13 = 3^2 + 2^2$ to obtain the irreducibles on the right. We now know, by uniqueness of factorization, that the factorization of $8 + i$ comes from the irreducibles in this equation. Try a combination:

$$(2 + i)(3 + 2i) = 4 + 7i.$$

This didn't work, so try another combination:

$$(2 + i)(3 - 2i) = 8 - i.$$

This almost works. Take complex conjugates to get

$$(2 - i)(3 + 2i) = 8 + i,$$

which is the desired factorization.

Here is another example. Let's factor $-6+27i$ into irreducibles. Immediately, we can factor off the 3, which is irreducible in $\mathbb{Z}[i]$, by Proposition 18.4. We are left with $-2 + 9i$. Its norm is

$$(-2 + 9i)(-2 - 9i) = N(-2 + 9i) = 85 = 5 \times 17$$
$$= (2 + i)(2 - i)(4 + i)(4 - i).$$

The combination

$$(2 + i)(4 - i) = 9 + 2i$$

is almost what we want. In fact, since $i(9 + 2i) = -2 + 9i$, we have

$$-2 + 9i = i(2 + i)(4 - i),$$

so

$$-6 + 27i = i(2 + i)(4 - i)(3).$$

If we don't like the unit $i$ in the factorization, we could choose $-1 + 2i$ as an irreducible and write

$$-6 + 27i = (-1 + 2i)(4 - i)(3).$$

**Remark.** Sometimes it is convenient to collect all the associates of an irreducible together in a factorization. For example, instead of writing $2 = (1 + i)(1 - i)$, we can write $2 = -i(1 + i)^2$. As in this example, we sometimes cannot absorb the unit into the powers of irreducibles, so we write

$$\alpha = u p_1^{a_1} \pi_2^{a_2} \cdots \pi_k^{a_k},$$

where $u$ is a unit. We see this phenomenon in the integers, too. We can write $-36 = (-2)(2)(3)(3)$, which is a product of irreducibles. But we can also write this as $-36 = -2^2 3^2$, and the unit $-1$ is not absorbed into the squares of primes.

## 18.5    Applications

The Gaussian integers are interesting in their own right, but in this section we show how they can be used to derive information about integers.

### 18.5.1    Sums of Two Squares

In Section 15.2, we determined which integers are sums of two squares. The key step was showing that primes congruent to 1 mod 4 are sums of two squares. Since $N(a + bi) = a^2 + b^2$, it is not surprising that Gaussian integers can be used to prove this fact.

**Theorem 18.13.** *Let $p$ be a prime with $p \equiv 1$ (mod 4). Then there are integers $a$ and $b$ with $p = a^2 + b^2$.*

*Proof.* Let $g$ be a primitive root mod $p$ and let $x \equiv g^{(p-1)/4}$ (mod $p$). Since $p \equiv 1$ (mod 4), the exponent $(p - 1)/4$ is an integer, so this definition of $x$ makes sense. By Proposition 11.6,

$$x^2 \equiv g^{(p-1)/2} \equiv -1 \pmod{p}.$$

This means that $p \mid x^2 + 1$. Now, let's work in $\mathbb{Z}[i]$. Since $x^2 + 1 = (x + i)(x - i)$,

$$p \mid (x + i)(x - i).$$

If $p$ is irreducible, then Proposition 18.10 tells us that $p \mid x + i$ or $p \mid x - i$. Suppose $p \mid x + i$. Then there exists $c + di \in \mathbb{Z}[i]$ such that

$$p(c + di) = x + i.$$

The coefficients of $i$ tell us that $pd = 1$, which is impossible since $d$ is an integer and $p \geq 5$. Therefore, $p \nmid x + i$. Similarly $p \nmid (x - i)$. Therefore, $p$ is not irreducible, so it factors: $p = \alpha\beta$, with $N(\alpha) < N(p)$ and $N(\beta) < N(p)$. Since

$$p^2 = N(p) = N(\alpha)N(\beta),$$

the only choice is $N(\alpha) = N(\beta) = p$. Write $\alpha = a + bi$. Then

$$p = N(\alpha) = a^2 + b^2,$$

as desired.                                                                    □

**Corollary 18.14.** *Let $n$ be a positive integer. Suppose that in the prime factorization of $n$, the primes that are 3 mod 4 occur with even exponents. Then $n$ is expressible as a sum of two squares.*

*Proof.* The assumptions say that we can write $n$ as a square times a product of primes that are not 3 mod 4. That is, $n = s^2 m$, where $m = p_1 p_2 \cdots p_r$ is a product of primes, each of which either is equal to 2 or is 1 mod 4. Theorem 18.13 and the fact that $2 = 1^2 + 1^2$ say that the primes in the factorization of $m$ are sums of two squares. Write $p_j = a_j^2 + b_j^2$. Let

$$(a_1 + b_1 i)(a_2 + b_2 i) \cdots (a_r + b_r i) = a + bi.$$

Then

$$m = p_1 p_2 \cdots p_r = N(a_1 + b_1 i) N(a_2 + b_2 i) \cdots N(a_r + b_r i)$$
$$= N(a + bi) = a^2 + b^2.$$

Therefore, $n = s^2 m = (sa)^2 + (sb)^2$ is a sum of two squares. $\square$

The converse of Corollary 18.14 is also true. See Theorem 15.6.

We now return to Proposition 18.5 and present its proof. Recall that this proposition characterizes the irreducible elements in $\mathbb{Z}[i]$.

*Proof of Proposition 18.5:* Let $\pi$ be an irreducible in $\mathbb{Z}[i]$ and let

$$n = N(\pi) = \pi\overline{\pi}.$$

Write the prime factorization of $n$ in integers as

$$n = p_1^{a_1} p_2^{a_2} \cdots p_r^{a_r}.$$

Since $\pi \mid n$, Corollary 18.11 implies that $\pi \mid p_j$ for some $j$. Write $p_j = \pi\alpha$ for some $\alpha$. Then

$$p_j^2 = N(p_j) = N(\pi)N(\alpha).$$

Since $\pi$ is not a unit, $N(\pi) \neq 1$, so $N(\pi) = p_j$ or $p_j^2$.

If $N(\pi) = p_j^2$, then $N(\alpha) = 1$, so $\alpha$ is a unit. This means that $\pi = (1/\alpha)p_j$ is a unit times $p_j$. If $p_j \not\equiv 3 \pmod 4$ then $p_j$ is a sum of two squares, by Theorem 18.13. But $p_j = a^2 + b^2$ implies $p_j = (a+bi)(a-bi)$, so $p_j$ is not irreducible, so it cannot equal $\pi$. Therefore, we must have

$p_j \equiv 3 \pmod 4$. Such primes are irreducible in $\mathbb{Z}[i]$, by Proposition 18.4.

If $N(\pi) = p_j$, write $\pi = a + bi$. Then $a^2 + b^2 = N(\pi) = p_j$, so $p_j = 2$ or $p_j \equiv 1 \pmod 4$. Therefore, $\pi$ has the form given in part (c) of the proposition.

Finally, we need to show that when $p$ is odd, $a + bi$ and $a - bi$ are not associates. The associates of $a + bi$ are $a + bi$, $-a - bi$, $-b + ai$, $b - ai$. If $a - bi$ equals one of the first two, then $b = 0$ or $a = 0$, so $p = a^2$ or $b^2$, which is impossible. If $a - bi = -b + ai$ or $b - ai$, then $a = \pm b$. Then $p = a^2 + b^2 = 2a^2$. Since $p$ is odd, this is impossible.

Note that when $p = 2 = 1^2 + 1^2$, we have $1 - i = -i(1 + i)$, so $1 + i$ and $1 - i$ are associates. This is why $1 + i$ is singled out in the statement of the proposition. This completes the proof of Proposition 18.5. $\square$

**Remark.** When we write an integer as a sum of two squares, there are many ways to write the same result by changing signs and order. For example, if $n = a^2 + b^2$, then

$$n = a^2 + b^2 = (-a)^2 + (-b)^2 = (-b)^2 + (a)^2 = (b)^2 + (-a)^2$$
$$= (a)^2 + (-b)^2 = (-a)^2 + (b)^2 = (-b)^2 + (-a)^2 = (b)^2 + (a)^2.$$

These correspond to the norms of

$$a + bi, \ -a - bi = -(a + bi), \ -b + ai = i(a + bi), \ b - ai = -i(a + bi)$$

and their complex conjugates. In other words, if we obtain a sum of two squares from $N(\alpha)$, then we obtain the trivial modifications of this sum by looking at the associates of $\alpha$ and their complex conjugates.

**Remark.** Implicit in the proof of Corollary 18.14 is a way to write a composite as a sum of squares in two distinct ways. For example, let $n = 65 = 5 \times 13$. Then

$$5 = (2 + i)(2 - i) \text{ and } 13 = (2 + 3i)(2 - 3i).$$

Two ways to combine these irreducibles are

$$(2 + i)(2 + 3i) = 1 + 8i \quad \text{and} \quad (2 + i)(2 - 3i) = 7 - 4i.$$

These yield
$$65 = 1^2 + 8^2 = 7^2 + 4^2.$$

The other combinations of irreducibles yield these two equations up to sign and order.

## 18.5.2    Pythagorean Triples

In Theorem 5.5, we gave a formula for primitive Pythagorean triples. Here we derive the same formula by using Gaussian integers. Suppose

$$a^2 + b^2 = c^2,$$

where $a, b, c$ are positive integers that are pairwise relatively prime. The relative primality implies that two of $a, b, c$ are odd, which means that at least one of $a$ and $b$ is odd. A square is 0 or 1 mod 4. If both $a$ and $b$ are odd, then $a^2 + b^2 \equiv 2 \pmod 4$, and this cannot be congruent to $c^2$. Therefore, one of $a, b$ is odd and the other is even. Write

$$c^2 = (a + bi)(a - bi).$$

Assume for the moment that $a + bi$ is a unit times a square. We get that

$$a + bi = u(m + ni)^2 = u\left((m^2 - n^2) + 2mni\right),$$

where $u$ is a unit. The four choices of $u$ just permute $m^2 - n^2$ and $2mn$ and change their signs. Therefore, up to sign and order,

$$a = m^2 - n^2 \text{ and } b = 2mn.$$

Finally, $c^2 = a^2 + b^2 = (m^2 - n^2)^2 + (2mn)^2 = (m^2 + n^2)^2$, so

$$c = m^2 + n^2.$$

These are the formulas of Theorem 5.5.

We now need to justify that $a + bi$ is a unit times a square. First, we show that $a+bi$ and $a-bi$ are relatively prime. Let $\delta = \gcd(a+bi, a-bi)$. Then $\delta \mid a + bi$ and $\delta \mid a - bi$ yields

$$\delta \mid 2a = (a + bi) + (a - bi) \quad \text{and} \quad \delta \mid 2b = i(a - bi) - i(a + bi).$$

Since $\gcd(a, b) = 1$ in the integers, there are integers $x$ and $y$ such that $ax + by = 1$. Therefore, $2ax + 2by = 2$. Since $\delta$ divides both summands, $\delta \mid 2$. Since $2 = -i(1 + i)^2$, and $1 + i$ is irreducible, we have $\delta = 1, 1+i$, or $(1 + i)^2$ (remember that $\delta$ can always be multiplied by a unit). In particular, if $\delta \neq 1$, then $1 + i$ divides $\delta$, hence $1 + i$ divides $a + bi$.

**Lemma 18.15.** *If* $1 + i \mid a + bi$, *then* $a + b \equiv 0 \pmod 2$.

*Proof.* If $1 + i \mid a + bi$, then $a + bi = (1+i)(u+vi) = (u-v) + (u+v)i$ for some integers $u, v$, Therefore, $a + b = (u - v) + (u + v) = 2u \equiv 0$ (mod 2). □

The lemma says that $a + b$ is even. However, one of $a$ is even and the other is odd, so $a + b$ is odd. This contradiction implies that $\delta = 1$, so $a + bi$ and $a - bi$ are relatively prime.

We now need the analogue of Lemma 5.6.

**Lemma 18.16.** *Let $k \geq 2$, let $\alpha$ and $\beta$ be relatively prime Gaussian integers such that $\alpha\beta = \eta^k$ for some Gaussian integer $\eta$. Then each of $\alpha$ and $\beta$ is a unit times a kth power of a Gaussian integer.*

*Proof.* Factor $\eta$ into powers of irreducibles:

$$\eta = u\pi_1^{x_1}\pi_2^{x_2}\pi_3^{x_3} \cdots$$

with integers $x_j \geq 0$, and where $u$ is a unit. Then

$$\alpha\beta = \eta^k = u^k \pi_1^{kx_1} \pi_2^{kx_2} p_3^{kx_3} \cdots .$$

Let $\pi$ be an irreducible occurring in the factorization of $\alpha$ into irreducibles; let's say $\pi^c$ is the exact power of $\pi$ in $\alpha$. Since $\gcd(\alpha, \beta) = 1$, the irreducible $\pi$ does not occur in the factorization of $\beta$. Therefore, $\pi^c$ is the power of $\pi$ occurring in $\alpha\beta$. Since the factorization of $\alpha\beta = \eta^k$ is unique, we must have $\pi^c = v\pi_j^{kx_j}$ for some $j$ and some unit $v$, and $c = kx_j$. This means that every irreducible in the factorization of $\alpha$ occurs with exponent that is a multiple of $k$, and this implies that $\alpha$ is a unit times a kth power. By exactly the same reasoning, $\beta$ is a unit times a kth power. □

The lemma tells us that $a + bi$ is a unit times a square, as claimed. This finishes the derivation of the formula for Pythagorean triples.

## 18.5.3   $y^2 = x^3 - 1$

**Proposition 18.17.** *The only integer solutions to $y^2 = x^3 - 1$ are $x = 1$ and $y = 0$.*

*Proof.* Suppose $y^2 = x^3 - 1$. If $x$ is even then $x^3 \equiv 0 \pmod 4$, so $y^2 \equiv -1 \pmod 4$, which is impossible. Therefore, $x$ is odd and $y^2 = x^3 - 1$ is even, so $y$ is even. Rewrite the equation as

$$x^3 = y^2 + 1 = (y+i)(y-i).$$

We claim that $\gcd(y+i, y-i) = 1$. If we assume this to be true, then Lemma 18.16 says that $y + i$ is a unit times a cube:

$$y + i = u(a+bi)^3.$$

But $u^4 = 1$ for every unit in $\mathbb{Z}[i]$, so $u = (u^{-1})^3$. This says that $u$ is a cube, so we can absorb it into $(a+bi)^3$ and thus assume that

$$y + i = (a+bi)^3 = (a^3 - 3ab^2) + (3a^2b - b^3)i.$$

This implies that

$$1 = 3a^2b - b^3 = b(3a^2 - b^2).$$

Therefore, $b \mid 1$, so $b = \pm 1$, which implies that $\pm 1 = 3a^2 - b^2 = 3a^2 - 1$. Therefore,

$$3a^2 = 0 \text{ or } 2.$$

Since $a$ is an integer, the only solution to this is $a = 0$. We now have $y + i = (a+bi)^3 = (0 \pm i)^3 = \mp i$, so $y = 0$, and therefore $x = 1$.

It remains to show that $\gcd(y+i, y-i) = 1$. Suppose $\pi$ is an irreducible dividing $y + i$ and $y - i$. Then

$$\pi \mid 2 = i(y-i) - i(y+i).$$

Since $2 = -i(1+i)^2$ is the factorization of 2 into irreducibles, we must have $\pi = 1+i$ (we can multiply $\pi$ by a unit if necessary). But $\pi \mid y+i$, so $1 + i \mid y + i$. By Lemma 18.15, $y + 1$ is even. But $y$ is even, so we have a contradiction. Therefore, $\pi$ cannot exist, which means that $\gcd(y+i, y-i) = 1$, as claimed. This completes the proof.    □

The similar-looking equation $y^2 = x^3 + 1$ has the trivial solutions $x = 0, y = \pm 1$, but it also has the nontrivial solutions $x = 2$, $y = \pm 3$. This solution is special. Catalan conjectured in 1844 that 8 and 9 are the only two powers of positive integers (with exponents greater than 1) that differ by 1. This was proved in 2002 by Preda Mihăilescu.

## 18.6   Chapter Highlights

1. Every nonzero Gaussian integer is either a unit, an irreducible, or a product of irreducibles, and this factorization into irreducibles is unique.

2. Every prime $p \equiv 1 \pmod 4$ is a sum of two squares.

## 18.7   Problems

### 18.7.1   Exercises

1. (a) Prove that
$$(x_1 x_2 - y_1 y_2)^2 + (x_1 y_2 + y_1 x_2)^2 = (x_1^2 + y_1^2)(x_2^2 + y_2^2).$$
   (b) Use (a) to prove that if $z_1$ and $z_2$ are complex numbers, then $|z_1 z_2| = |z_1||z_2|$.

2. List the associates of $3 + 5i$.

3. List the associates of $1 + i$.

4. Factor $4 + 6i$ into a product of irreducibles.

5. Factor $7 + 6i$ into a product of irreducibles.

6. Does 9 divide $(7 + i)(3 + 4i)$?

7. Prove the following or show that it is false by giving a counterexample: If $\alpha$ and $\beta$ are Gaussian integers with $N(\alpha) \mid N(\beta)$, then $\alpha \mid \beta$.

8. If $\alpha$, $\beta$, and $\gamma$ are Gaussian integers, solve $\alpha\beta\gamma = \alpha + \beta + \gamma = 1$.

9. (a) Factor 85 into a product of irreducibles in the Gaussian integers.
   (b) Use the result of (a) to write 85 as a sum of two squares in two ways.

10. (a) Let $n$ be an integer and let $\alpha$ be a Gaussian integer. Show that if $\alpha \mid n$ then $\bar{\alpha} \mid n$.
    (b) Show that no power $(2 + 3i)^k = n$, with $k \neq 0$, is an integer. (*Hint:* Unique factorization in $\mathbb{Z}[i]$.)

11. Suppose $p \equiv 1 \pmod 4$ is prime and $p = a^2 + b^2 = c^2 + d^2$.
    (a) Show that $a + bi \mid c + di$ or $a + bi \mid c - di$.
    (b) By changing the sign of $d$ if necessary, assume that $a + bi \mid c + di$. Write $(a + bi)\gamma = c + di$ for some $\gamma$. Show that $\gamma$ is a unit.
    (c) Show that $(c, d) = (a, b), (-a, -b), (-b, a),$ or $(b, -a)$.
    This says that up to order and sign, the decomposition of $p$ as a sum of two squares is unique. For a different proof, see Project 3 in Chapter 15.

12.  (a) Let $\alpha$ be a nonzero Gaussian integer. Show that the four associates of $\alpha$ are the vertices of a square in the complex plane.

(b) Show that every nonzero Gaussian integer has an associate in the first quadrant or on the positive $x$-axis.

## 18.7.2   Projects

1.  (a) Let $a + bi$ be a Gaussian integer and let $n$ be a positive integer. Show that we can write $a = q_1 n + r_1$ and $b = q_2 n + r_2$, where $|r_1| \leq n/2$ and $|r_2| \leq n/2$.

(b) In the notation of (a), show that $N(r_1 + r_2 i) < N(n)$.

(c) Let $\alpha$ and $\beta$ be Gaussian integers with $\beta \neq 0$. Let $n = \beta \overline{\beta}$ and $a + bi = \alpha \overline{\beta}$, and let $r_1$ and $r_2$ be as in (a). Show that $\overline{\beta} \mid r_1 + r_2 i$.

(d) Write $(r_1 + r_2 i)/\overline{\beta} = \rho$ and let $\eta = q_1 + q_2 i$. Show that $\alpha = \beta \eta + \rho$ and $N(\rho) < N(\beta)$.

This gives a "non-geometric" proof of the Division Algorithm for Gaussian integers.

## 18.7.3   Computer Explorations

1.  Count how many non-associated Gaussian irreducibles have norm less than or equal to $x$ for $x = 10000$, $x = 1000000$, and $x = 100000000$. Compare this to $x/\ln(x)$. (The Prime Number Theorem for Gaussian integers says that the count divided by $x/\ln(x)$ goes to 1 as $x \to \infty$.)

## 18.7.4   Answers to "CHECK YOUR UNDERSTANDING"

1.  Rationalize the denominator:

$$\frac{23 - 2i}{3 + 2i} = \frac{23 - 2i}{3 + 2i} \frac{3 - 2i}{3 - 2i} = \frac{65 - 52i}{13} = 5 - 4i,$$

which is a Gaussian integer.

2.  $N(6 + i) = 37$, which is prime, so Proposition 18.3 says that $6 + i$ is irreducible.

3.  The associates of $1 + 4i$ are obtained by multiplying by units:

$$(1 + 4i)(1) = 1 + 4i, \quad (1 + 4i)(i) = -4 + i,$$
$$(1 + 4i)(-1) = -1 - 4i, \quad (1 + 4i)(-i) = 4 - i.$$

4.  Since $4 - i = (1 + 4i)(-i)$ and $4 + i = (1 - 4i)(i)$, the irreducibles in one factorization are associates of the irreducibles in the other factorization.

# Chapter 19

# Algebraic Integers

In Chapter 18, we studied the Gaussian integers and saw how they can be used to answer questions about the usual integers. In the present chapter, we consider the more general situation of quadratic fields and algebraic integers. These also have interesting applications, and we include a few. But we also encounter a situation where there is no unique factorization into irreducibles and show how this causes difficulties. The material in this chapter is the beginning of the subject called algebraic number theory, which started in the 1800s and is still an active area of research.

## 19.1   Quadratic Fields and Algebraic Integers

Let $d \neq 1$ be a squarefree integer, positive or negative. For example, $d = 5$, $d = -1$, and $d = -3$ are possibilities. Let

$$\mathbb{Q}(\sqrt{d}) = \{a + b\sqrt{d} \mid a, b \text{ are rational numbers}\}.$$

For example, $2 - 7\sqrt{5}$ and $\frac{1}{2} + \frac{3}{11}\sqrt{5}$ are elements of $\mathbb{Q}(\sqrt{5})$. We can add and subtract numbers in $\mathbb{Q}(\sqrt{d})$ and obtain numbers in $\mathbb{Q}(\sqrt{d})$ as answers. We can also multiply. For example,

$$\left(2 - 7\sqrt{5}\right)\left(4 + 3\sqrt{5}\right) = 2 \cdot 4 + 2 \cdot 3\sqrt{5} - 7\sqrt{5} \cdot 4 - 7 \cdot 3 \cdot 5 = -97 - 22\sqrt{5}.$$

The technique of rationalizing the denominator lets us divide:

$$\frac{3 + 4\sqrt{5}}{1 - 2\sqrt{5}} = \frac{3 + 4\sqrt{5}}{1 - 2\sqrt{5}} \frac{1 + 2\sqrt{5}}{1 + 2\sqrt{5}} = \frac{43 + 10\sqrt{5}}{-19} = -\frac{43}{19} - \frac{10}{19}\sqrt{5}.$$

Note that when $a$ or $b$ is nonzero and we divide by $a + b\sqrt{d}$, we are not dividing by 0 because $d$ is not a square. The set $\mathbb{Q}(\sqrt{d})$ is called a **quadratic field** and it has properties similar to the rational numbers; namely, we can add, subtract, multiply, and divide (by nonzero numbers) and obtain another element in the set.

The integers are a subset of the rational numbers and we can add, subtract, and multiply integers and obtain integers as answers. We cannot always divide, and this led to the interesting theory we've developed about divisibility, congruences, and prime numbers. Similarly, we can add, subtract, and multiply Gaussian integers, but not necessarily divide. We would like to have a subset with these properties when we work with $\mathbb{Q}(\sqrt{d})$ so that we can develop results analogous to those we proved about the integers. This will be supplied by the algebraic integers.

An **algebraic integer** is a complex number $\alpha$ that is a root of a polynomial of the form

$$X^n + a_{n-1}X^n + \cdots + a_1X + a_0,$$

for some $n \geq 1$, where each $a_i$ is an integer. Note that the coefficient of $X^n$ is 1. For example, 7 is an algebraic integer because it is a root of $X - 7$. Also, $2 + 5i$ is an algebraic integer because it is a root of $X^2 - 4X + 29$. However, $1/2$ is not an algebraic integer because Theorem 5.4 tells us that if a rational number is a root of a polynomial with integer coefficients, then its denominator must divide the coefficient of the highest power of $X$. Since we have assumed that the coefficient of $X^n$ in our polynomial is 1, the denominator must divide 1. Therefore, $1/2$ cannot be a root of such a polynomial. In this way, we see that integers are the only rational numbers that are algebraic integers.

We can write down many more algebraic integers. For example,

$$\frac{3 + 7\sqrt{5}}{2}$$

is an algebraic integer because it is a root of the polynomial $X^2 - 3X - 59$. Also, $\sqrt[3]{2}$ is a root of $X^3 - 2$, so it is an algebraic integer.

Let's determine which elements of $\mathbb{Q}(\sqrt{d})$ are algebraic integers. The form of the answer depends on $d \pmod 4$.

**Proposition 19.1.** *Let $d \neq 1$ be a squarefree integer. The algebraic integers in $\mathbb{Q}(\sqrt{d})$ are the numbers of the form*

$$a + b\sqrt{d} \text{ with } a, b \text{ integers, if } d \equiv 2 \text{ or } 3 \pmod 4$$

$$a + b\frac{1 + \sqrt{d}}{2} \text{ with } a, b \text{ integers, if } d \equiv 1 \pmod 4.$$

The standard notation is

$$\mathbb{Z}[\sqrt{d}] = \{a + b\sqrt{d} \mid a \text{ and } b \text{ are integers}\}$$

$$\mathbb{Z}[(1 + \sqrt{d})/2] = \{a + b\frac{1 + \sqrt{d}}{2} \mid a \text{ and } b \text{ are integers}\}.$$

*Proof.* Let $\alpha = u + v\sqrt{d}$, where $u$ and $v$ are rational numbers. Then $\alpha$ is a root of the polynomial $X^2 - 2uX + (u^2 - dv^2)$. You can check this with either the quadratic formula or direct substitution.

First, let's show that the numbers in the statement of the proposition are actually algebraic integers. If $\alpha$ has the form $a + b\sqrt{d}$, where $a, b$ are integers, then $\alpha$ is a root of $X^2 - 2aX + (a^2 - db^2)$, and the coefficients of this polynomial are integers. Therefore, $\alpha$ is an algebraic integer.

Now suppose that $d \equiv 1 \pmod 4$ and $\alpha = a + b(1 + \sqrt{d})/2$. Then $\alpha = u + v\sqrt{d}$ with $u = a + \frac{1}{2}b$ and $v = \frac{1}{2}b$, so $\alpha$ is a root of

$$X^2 - (2a + b)X + \left(a^2 + ab + b^2\frac{1 - d}{4}\right).$$

Since $d \equiv 1 \pmod 4$, we see that $(1 - d)/4$ is an integer and therefore the coefficients of the polynomial are integers. Therefore, $\alpha$ is an algebraic integer.

We now want to show that all algebraic integers in $\mathbb{Q}(\sqrt{d})$ have the form stated in the proposition. Suppose $\alpha = u + v\sqrt{d}$, with rational numbers $u$ and $v$, and that $X^2 - 2uX + (u^2 - dv^2)$ has integer coefficients. Since $2u$ and $u^2 - dv^2$ are integers, so is

$$(2u)^2 - 4(u^2 - dv^2) = 4dv^2 = (2v)^2d.$$

Now we'll show that $2v$ is an integer by showing that it can't have any prime factors in its denominator. So, assume that $2v$ has a prime $p$ in its denominator. Then $(2v)^2$ has $p^2$ in its denominator. But $d$ is assumed to be squarefree, so it cannot cancel $p^2$. Since $(2v)^2 d$ is an integer, it cannot have $p$ in its denominator, and we conclude that there is no such $p$. Therefore, the denominator of $2v$ has no prime factors, so $2v$ is an integer.

Write $2u = x$ and $2v = y$, where $x$ and $y$ are integers. Then

$$x^2 - dy^2 = 4(u^2 - dv^2) \equiv 0 \pmod{4}.$$

Suppose $d \equiv 2, 3 \pmod{4}$. Since squares are congruent to 0 or 1 mod 4, it follows that $x^2 - dy^2 \equiv 0 \pmod{4}$ only if $x$ and $y$ are even. Therefore, $u = x/2$ and $v = y/2$ are integers, so $\alpha = u + v\sqrt{d}$ has the desired form.

Suppose that $d \equiv 1 \pmod{4}$. In this case, $x^2 - dy^2 \equiv 0 \pmod{4}$ implies that $x$ and $y$ are either both odd or both even. In other words, $x \equiv y \pmod{2}$. Write $x = y + 2a$ for some integer $a$, and let $b = y$. Then

$$\alpha = u + v\sqrt{d} = \frac{1}{2}\left(x + y\sqrt{d}\right) = \frac{1}{2}\left(b + 2a + b\sqrt{d}\right) = a + b\frac{1 + \sqrt{d}}{2}.$$

Therefore, $\alpha$ has the desired form.

This completes the proof, except for a technicality that we need to address. The definition of algebraic integer requires only that $\alpha$ be the root of *some* polynomial with integer coefficients. We have used only the quadratic polynomial. The following lemma resolves this problem.

**Lemma 19.2.** *Let* $f(X) = X^n + a_{n-1}X^{n-1} + \cdots + a_1 X + a_0$ *be a polynomial with integer coefficients. Suppose* $f(X)$ *factors as*

$$f(X) = g(X) \cdot h(X),$$

*where* $g(X) = X^\ell + b_{\ell-1}X^{\ell-1} + \cdots + b_1 X + b_0$ *and* $h(X) = X^m + c_{m-1}X^{m-1} + \cdots + c_1 X + c_0$ *are polynomials with rational coefficients. Then* $g(X)$ *and* $h(X)$ *have integer coefficients.*

*Proof.* Let $p$ be a prime and let $p^a$ be the highest power of $p$ occurring in the denominator of the $b_i$'s. Among the coefficients for which $p^a$ divides the denominator, let $b_i$ have the highest index $i$. This means that $p^a$ does not divide the denominator of $b_k$ for $k > i$. It is possible that $p$ does not divide the denominator of any coefficient. In this case $a = 0$ and $i = n$.

Similarly, let $p^b$ be the highest power of $p$ dividing the denominator of some $c_j$ and let $j$ be the highest index for which this occurs.

In the product $f(X) = g(X)h(X)$, we have

$$a_{i+j} = b_0 c_{i+j} + b_1 c_{i+j-1} + \cdots + b_i c_j + b_{i+1} c_{j-1} + \cdots + b_{i+j} c_0.$$

The term $b_i c_j$ has $p^{a+b}$ in its denominator. This is not the case for the other terms. For example, since $i + 1 > i$, $b_{i+1}$ does not have $p^a$ in its denominator. Since $p^b$ is the largest power of $p$ in a denominator for $g(X)$, the coefficient $c_{j-1}$ does not have more than $p^b$ in its denominator. Therefore, $b_{i+1}c_{j-1}$ does not have $p^{a+b}$ in its denominator. The same reasoning applies to the other terms, so there is nothing to cancel the $p^{a+b}$ in the denominator of $b_i c_j$. Therefore, $a_{i+j}$ has $p^{a+b}$ in its denominator. But $a_{i+j}$ is assumed to be an integer, so $a + b = 0$. This means that $a = b = 0$, so no coefficient of $g(X)$ or $h(X)$ has a $p$ in its denominator. Since $p$ was arbitrary, there are no primes in the denominators of the coefficients, which means they are integers. This proves the lemma.                                                                    $\square$

We can now finish the proof of the proposition. Let $\alpha = u + v\sqrt{d}$ be an algebraic integer in $\mathbb{Q}(\sqrt{d})$. Then $\alpha$ is a root of a polynomial $f(X)$ with integer coefficients (where the coefficient of the highest degree term is 1). If $v = 0$, then $\alpha = u$ is a rational number. The Rational Root Theorem (Theorem 5.4) tells us that $\alpha$ is an integer. Now suppose that $v \neq 0$. If $f(u+v\sqrt{d}) = 0$, then $f(u-v\sqrt{d}) = 0$. Therefore, $X-(u+v\sqrt{d})$ and $X - (u - v\sqrt{d})$ are factors of $f(X)$, so their product, which is

$$X^2 - 2uX + (u^2 - dv^2),$$

is a factor of $f(X)$. The lemma implies that $X^2 - 2uX + (u^2 - dv^2)$ has integer coefficients. Therefore, using the polynomial of degree 2, rather than an arbitrary degree polynomial, was sufficient. This completes the proof of Proposition 19.1.                                                           $\square$

It is easy to see that $\mathbb{Z}[\sqrt{d}]$ is closed under addition, subtraction, and multiplication. This is also the case for $\mathbb{Z}[\frac{1+\sqrt{d}}{2}]$, but the fact that it's closed under multiplication is more subtle: Let $\beta = (1 + \sqrt{d})/2$. Then $\beta^2 = \beta + (d - 1)/4$. Therefore,

$$\begin{aligned}
(a + b\beta)(x + y\beta) &= ax + (bx + ay)\beta + by\beta^2 \\
&= ax + (bx + ay)\beta + (by)(\beta + (d-1)/4) \\
&= ax + by(d - 1)/4 + (bx + ay + by)\beta.
\end{aligned}$$

Since $d \equiv 1 \pmod 4$, this last expression is in $\mathbb{Z}[\beta]$. Therefore, $\mathbb{Z}[(1 + \sqrt{d})/2]$ is closed under multiplication.

A (non-empty) set of numbers that is closed under addition, subtraction, and multiplication is called a **ring**, so we have verified that $\mathbb{Z}[(1 + \sqrt{d})/2]$ (when $d \equiv 1 \mod 4$) and $\mathbb{Z}[\sqrt{d}]$ are rings.

---

**CHECK YOUR UNDERSTANDING:**

1. Show that $1 + 3\sqrt{2}$ and $\frac{5+\sqrt{13}}{2}$ are algebraic integers by exhibiting them as roots of explicit polynomials.
2. Show that $(-7 + 7\sqrt{-5})/(1 + 2\sqrt{-5})$ is an algebraic integer.

---

# 19.2   Units

In the integers, the numbers $\pm 1$ play a special role because they are the only integers that have multiplicative inverses that are also integers. In the Gaussian integers, $\pm 1$ and $\pm i$ have inverses in $\mathbb{Z}[i]$. In this section, we look at quadratic algebraic integers with this property. When $d > 0$, this will give us another way of looking at Pell's equation.

**Definition 19.3.** *Let $R = \mathbb{Z}[(1 + \sqrt{d})/2]$ (when $d \equiv 1 \mod 4$) or $\mathbb{Z}[\sqrt{d}]$. A **unit** of $R$ is a number $u \in R$ such that $1/u \in R$. In other words, $u$ is a unit if there exists $v \in R$ such that $uv = 1$.*

For the Gaussian integers, the norm played a major role in determining the units. The same situation happens here. For $\mathbb{Q}(\sqrt{d})$, define the **norm** to be

$$N(a + b\sqrt{d}) = a^2 - db^2 = (a + b\sqrt{d})(a - b\sqrt{d}).$$

For example,

$$N(3 + 4\sqrt{7}) = 3^2 - 7 \cdot 4^2 = -103,$$
$$N(2 + 3i) = N(2 + 3\sqrt{-1}) = 2^2 + 3^2 = 13,$$
$$N\left(3 + \frac{1 + \sqrt{5}}{2}\right) = N\left(\frac{7}{2} + \frac{1}{2}\sqrt{5}\right) = \left(\frac{7}{2}\right)^2 - 5\left(\frac{1}{2}\right)^2 = 11.$$

We see from the second example that the present norm agrees with the norm defined for the Gaussian integers in Chapter 18.

Before proceeding, we record two useful facts about the norm.

**Lemma 19.4.** *The norm of an algebraic integer in $\mathbb{Q}(\sqrt{d})$ is an integer.*

*Proof.* If $\alpha = a + b\sqrt{d}$ with integers $a$ and $b$, then $N(\alpha) = a^2 - db^2$, which is an integer. If $\alpha = a + b(1 + \sqrt{d})/2$ (and $d \equiv 1 \pmod 4$), then

$$N(\alpha) = \left(a + \frac{1}{2}b\right)^2 - \frac{1}{4}db^2 = a^2 + ab + b^2\frac{1-d}{4},$$

which is an integer because $d \equiv 1 \pmod 4$.                                                   $\square$

**Lemma 19.5.** *Let $\alpha, \beta \in \mathbb{Q}(\sqrt{d})$. Then $N(\alpha\beta) = N(\alpha)N(\beta)$.*

*Proof.* If $\alpha = a + b\sqrt{d}$, let $\overline{\alpha} = a - b\sqrt{d}$. If $\beta = x + y\sqrt{d}$, then

$$\overline{\alpha\beta} = \overline{(ax + byd) + (ay + bx)\sqrt{d}} = (ax + byd) - (ay + bx)\sqrt{d}$$

and

$$\overline{\alpha}\overline{\beta} = (a - b\sqrt{d})(x - y\sqrt{d}) = (ax + byd) - (ay + bx)\sqrt{d}.$$

Therefore, $\overline{\alpha}\overline{\beta} = \overline{\alpha\beta}$, so

$$N(\alpha\beta) = \alpha\beta\overline{\alpha\beta} = \alpha\beta\overline{\alpha}\overline{\beta} = \alpha\overline{\alpha}\beta\overline{\beta} = N(\alpha)N(\beta).$$                $\square$

We now have the machinery to determine which quadratic algebraic integers are units. The answer depends on whether $d$ is positive or negative.

**Proposition 19.6.** *Let $d \neq 1$ be squarefree and let $R$ be the algebraic integers in $\mathbb{Q}(\sqrt{d})$ (as given in Proposition 19.1). The units of $R$ are the elements of norm $\pm 1$.*
*(a) If $d < 0$ and $d \neq -1$ or $-3$, then the only units of $R$ are $\pm 1$.*
*(b) If $d = -1$, the units of $R = \mathbb{Z}[i]$ are $\pm 1$ and $\pm i$.*
*(c) If $d = -3$, the units of $R = \mathbb{Z}[(1 + \sqrt{-3})/2]$ are $\pm 1$, $\pm(1 + \sqrt{-3})/2$,*

*and* $\pm(1 - \sqrt{-3})/2$.

*(d) If $d > 0$, there are infinitely many units in $R$:*

*If $d \equiv 2$ or $3$ (mod 4) they are given by numbers of the form*

$a + b\sqrt{d}$, *where $a$ and $b$ are integers with $a^2 - db^2 = \pm 1$.*

*If $d \equiv 1$ (mod 4) they are given by numbers of the form*

$$a + b\frac{1 + \sqrt{d}}{2}, \text{ where } a \text{ and } b \text{ are integers}$$

*with $(2a + b)^2 - db^2 = \pm 4$.*

*Proof.* If $u$ is a unit, then there exists $v$ with $uv = 1$. Taking norms, we find that $N(u)N(v) = N(1) = 1$. Since $N(u)$ and $N(v)$ are integers, we have $N(u) = \pm 1$. Conversely, $N(u) = \pm 1$ means that $u\bar{u} = \pm 1$ (in the notation of the proof of Lemma 19.5), so $(u)(\pm\bar{u}) = 1$. Therefore, $\pm\bar{u}$ is the inverse of $u$. If $R = \mathbb{Z}[\sqrt{d}]$, then it is clear that $\bar{u} \in R$. But if $R = \mathbb{Z}[(1 + \sqrt{d})/2]$, we need to check this: Let $\alpha = a + b\frac{1+\sqrt{d}}{2}$. Then

$$\bar{\alpha} = a + b\frac{1 - \sqrt{d}}{2} = (a + b) - b\frac{1 + \sqrt{d}}{2} \in \mathbb{Z}[(1 + \sqrt{d})/2].$$

Therefore, in both cases, if $N(u) = \pm 1$, the inverse of $u$ is in $R$. We conclude that the units of $R$ are exactly those elements with norm $\pm 1$.

If $d = -1$, we need to solve $a^2 + b^2 = 1$. The only solutions are $(a, b) = (\pm 1, 0), (0, \pm 1)$, which correspond to $\pm 1$ and $\pm i$.

If $d = -3$, we need to solve $a^2 + ab + b^2 = \pm 1$. Multiplying by 4 and completing the square yields

$$\pm 4 = 4a^2 + 4ab + 4b^2 = (2a + b)^2 + 3b^2.$$

The only integer solutions are $(2a+b, b) = (\pm 1, \pm 1)$ and $(\pm 2, 0)$. These yield

$$(a, b) = (0, 1), (-1, 1), (0, -1), (1, -1), (1, 0), (-1, 0),$$

which correspond to

$$\frac{1 + \sqrt{-3}}{2}, \quad \frac{-1 + \sqrt{-3}}{2}, \quad \frac{-1 - \sqrt{-3}}{2}, \quad \frac{1 - \sqrt{-3}}{2}, \quad 1, \quad -1.$$

Suppose $d < 0$ and $d \neq -1$ or $-3$. If $d \equiv 2, 3$ (mod 4), then we need to solve $a^2 + |d|b^2 = \pm 1$. If $b \neq 0$, then $a^2 + |d|b^2 \geq |d|b^2 \geq |d| \geq 2$.

Therefore, we must have $b = 0$, so $a^2 = \pm 1$. This yields $a = \pm 1$, so the only units are $\pm 1$. If $d \equiv 1 \pmod 4$, we need to solve $a^2 + ab + b^2(1 - d)/4 = \pm 1$. As in the case where $d = -3$, we multiply by 4 and complete the square to obtain

$$\pm 4 = 4a^2 + 4ab + (1 - d)b^2 = (2a + b)^2 + (-d)b^2.$$

Since $d \neq -3$ and $d \equiv 1 \pmod 4$, we have $d \leq -7$. If $b \neq 0$, we have $(2a + b)^2 + (-d)b^2 \geq (-d)b^2 \geq 7$. Therefore, we must have $b = 0$. Therefore, $\pm 4 = 4a^2$, and $a = \pm 1$. The only units are $\pm 1$.

Now suppose $d > 0$. We showed in Chapters 15 and 17 that Pell's equation, $a^2 - db^2 = \pm 1$, has infinitely many nontrivial solutions. Each solution $a, b$ yields a unit $a + b\sqrt{d}$.

If $d \equiv 1 \pmod 4$, then there could be more units (see the Example below for $d = 5$). We have $N(a + b\frac{1+\sqrt{d}}{2}) = \pm 1$ if and only if $N((2a + b) + b\sqrt{d}) = \pm 4$, which happens if and only if $(2a+b)^2 - db^2 = \pm 4$. In the case where $b$ is even, we can divide by 4 and get $(a + \frac{1}{2}b)^2 - d(\frac{1}{2}b)^2 = \pm 1$. This case yields units of the form $x + y\sqrt{d}$ where $x$ and $y$ are integers. However, when $b$ is odd, we get units that have the 2 in the denominator. (However, the cube of such a unit has the form $x + y\sqrt{d}$; see Exercise 32 in Chapter 15.) □

**Example.** Let $d = 7$. Pell's equation $x^2 - 7y^2 = 1$ has the solution $(x, y) = (8, 3)$. Therefore, $8 + 3\sqrt{7}$ is a unit of $\mathbb{Z}[\sqrt{7}]$. The inverse of $8 + 3\sqrt{7}$ is $8 - 3\sqrt{7}$ since

$$(8 + 3\sqrt{7})(8 - 3\sqrt{7}) = 8^2 - 7 \cdot 3^3 = 1$$

(this is just the calculation that $N(8 + 3\sqrt{7}) = 1$).

**Example.** Let $d = 5$. Since $x^2 - 5y^2 = 1$ has the solution $(x, y) = (9, 4)$, we obtain the unit

$$u_1 = 9 + 4\sqrt{5} = 5 + 8\frac{1 + \sqrt{5}}{2}.$$

We can also use Pell's equation with $-1$. The solution $2^2 - 5 \cdot 1^2 = -1$ yields

$$u_2 = 2 + \sqrt{5} = 1 + 2\frac{1 + \sqrt{5}}{2}.$$

We can also look directly at the equation $x^2 - 5y^2 = \pm 4$ and obtain the solution $x = 1, y = 1$. Solving $2a + b = 1$, $b = 1$ yields the solution $(a, b) = (0, 1)$, which yields the unit

$$\phi = \frac{1 + \sqrt{5}}{2}.$$

Another solution is $x = 3, y = 1$, which yields $a = 1, b = 1$ and the unit

$$\phi^2 = 1 + \frac{1 + \sqrt{5}}{2}.$$

A straightforward calculation shows that

$$u_2 = \phi^3, \qquad u_1 = \phi^6.$$

In fact, it can be shown that all of the units of $\mathbb{Z}[(1 + \sqrt{5})/2]$ are of the form $\pm\phi^n$, where $n$ is an integer (possibly negative). The number $\phi = (1+\sqrt{5})/2$ is often called the **Golden Ratio**. It is closely connected with the Fibonacci numbers (see Appendix A).

---

**CHECK YOUR UNDERSTANDING**

3. Find a unit $u \neq \pm 1$ in $\mathbb{Z}[\sqrt{11}]$.

---

# 19.3   $\mathbb{Z}[\sqrt{-2}\,]$

$\mathbb{Z}[\sqrt{-2}]$ has many properties similar to those of the Gaussian integers. In particular, there is a division algorithm, and this implies that there is unique factorization into irreducibles. In this section, we sketch the proofs and then give an application.

Recall that

$$N(a + b\sqrt{-2}) = a^2 + 2b^2 = \left|a + b\sqrt{-2}\,\right|^2 = (a + b\sqrt{-2})\overline{(a + b\sqrt{-2})}.$$

As in the Gaussian integers, the norm tells us the size of a number.

When we proved the Division Algorithm for the Gaussian integers (Theorem 18.7), we looked at a square in the plane whose vertices were of the form $s + ti, s + 1 + ti, s + (t+1)i$, and $s + 1 + (t+1)i$. As we'll see in the following proof, the essential geometric figure is now a rectangle.

**Proposition 19.7.** *Let $\alpha$ and $\beta$ be elements of $\mathbb{Z}[\sqrt{-2}\,]$ with $\beta \neq 0$. Then there exist $\eta$ and $\rho$ in $\mathbb{Z}[\sqrt{-2}\,]$ such that*

$$\alpha = \beta\eta + \rho \text{ and } 0 \leq N(\rho) < N(\beta).$$

*Proof.* Let $\delta = \alpha/\beta$, which is a complex number. Write $\delta = x + y\sqrt{-2}$. Let $s$ and $t$ be integers with

$$s \leq x \leq s+1, \qquad t \leq y \leq t+1.$$

Then $\delta$ is in the rectangle with corners at $s + t\sqrt{-2}$, $s+1+t\sqrt{-2}$, $s + (t+1)\sqrt{-2}$, $s+1+(t+1)\sqrt{-2}$. Every point inside this rectangle is a

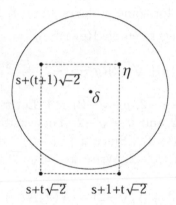

distance at most $\sqrt{3}/2 < 1$ from a corner. Therefore, there is a point $\eta = a + b\sqrt{-2}$ with $a, b$ integers ($a = s$ or $s+1$ and $b = t$ or $t+1$) such that

$$|\delta - \eta| \leq \sqrt{3}/2 < 1.$$

This says that

$$\left| \frac{\alpha}{\beta} - \eta \right| < 1.$$

Multiplication by $|\beta|$ yields

$$|\alpha - \beta\eta| < |\beta|.$$

Let $\rho = \alpha - \beta\eta$. Then $\rho \in \mathbb{Z}[\sqrt{-2}\,]$ since $\alpha, \beta, \eta$ are in $\mathbb{Z}[\sqrt{-2}\,]$, and we have $|\rho| < |\beta|$, so $N(\rho) < N(\beta)$. Since $\alpha = \beta\eta + \rho$, this completes the proof. $\qquad\square$

The Division Algorithm yields the following. We omit the proof since it is almost identical to the proof of Theorem 18.12.

**Theorem 19.8.** *Every nonzero element of $\mathbb{Z}[\sqrt{-2}]$ is either $\pm 1$, an irreducible, or a product of irreducibles. This factorization into irreducibles is unique up to the order of the factors and multiplication of irreducibles by $\pm 1$.*

Let's use this theorem to find all integer solutions of $y^2 = x^3 - 2$. We start with some observations about the integers $x$ and $y$. If $x$ or $y$ is even, so is the other. But then $y^2 \equiv 0 \pmod 4$ and $x^3 \equiv 0 \pmod 4$, so we obtain $0 \equiv 0 - 2 \pmod 4$, which is a contradiction. Therefore, $x$ and $y$ are odd. Let $d = \gcd(x, y)$, so $d$ must be odd. Then $d \mid x^3 - y^2 = 2$, which means $d = 1$ or $2$. Since $d$ is odd, $d = 1$. By Theorem 2.12, we can write $sx + ty = 1$ for some integers $s$ and $t$.

Now the algebraic integers enter. Rewrite $y^2 = x^3 - 2$ as

$$(y + \sqrt{-2})(y - \sqrt{-2}) = x^3.$$

We claim that $\gcd(y + \sqrt{-2}, y - \sqrt{-2}) = 1$. Let $\pi$ be an irreducible that divides both $y + \sqrt{-2}$ and $y - \sqrt{-2}$. Then $\pi$ divides their difference, which is $2\sqrt{-2} = -(\sqrt{-2})^3$. Since $\pi \mid (\sqrt{-2})^3$, the analogue of Corollary 18.11 says that $\pi \mid \sqrt{-2}$. Write $\pi\alpha = \sqrt{-2}$ for some $\alpha \in \mathbb{Z}[\sqrt{-2}]$. Then

$$2 = N(\sqrt{-2}) = N(\pi)N(\alpha).$$

But $\pi$ is an irreducible, so it is not a unit. Therefore, $N(\pi) \neq 1$, and we must have $N(\pi) = 2$ and $N(\alpha) = 1$ (note that norms for $\mathbb{Z}[\sqrt{-2}]$ are non-negative). Therefore, $\alpha$ is a unit, so it equals $\pm 1$. This means that $\pi = \pm\sqrt{-2}$. We can always change $\pi$ by a unit, so we can assume that $\pi = \sqrt{-2}$. Since $\pi = \sqrt{-2}$ is assumed to divide $y + \sqrt{-2}$, we also have $\pi \mid y$. In addition, $(y + \sqrt{-2}) \mid x^3$, so $\pi \mid x^3$. This implies that $\pi \mid x$. But we know that there are integers $s$ and $t$ with $sx + ty = 1$. Since $\pi$ divides both $x$ and $y$, we have $\pi \mid 1$, which is impossible because $\pi$ is not a unit. This contradiction shows that $\pi$ does not exist, so $\gcd(y + \sqrt{-2}, y - \sqrt{-2}) = 1$. We now need the following.

**Lemma 19.9.** *Let $k \geq 2$, and let $\alpha$ and $\beta$ be relatively prime elements of $\mathbb{Z}[\sqrt{-2}]$ such that $\alpha\beta = \eta^k$ for some $\eta \in \mathbb{Z}[\sqrt{-2}]$. Then each of $\alpha$ and $\beta$ is $\pm 1$ times a $k$th power of an element of $\mathbb{Z}[\sqrt{-2}]$.*

We omit the proof, which is essentially the same as the proof of Lemma 18.16.

From the lemma, we conclude that

$$y + \sqrt{-2} = \pm \left(a + b\sqrt{-2}\right)^3 = \pm \left((a^3 - 6ab^2) + (3a^2b - 2b^3)\sqrt{-2}\right).$$

The coefficients of $\sqrt{-2}$ yield

$$1 = \pm b(3a^2 - 2b^2),$$

so $b = \pm 1$ and $3a^2 - 2b^2 = \pm 1$. Therefore, $3a^2 - 2 = \pm 1$. The only solutions are $a = \pm 1$. Since $y = \pm(a^3 - 6ab^2) = \pm(a^3 - 6a)$, we have $y = \pm 5$. We conclude that the only solutions of $y^2 = x^3 - 2$ are $(x, y) = (3, 5)$ and $(3, -5)$.

**Historical Note:** After his death, Fermat's notes and letters were published by his son. One of them read "Can one find in whole numbers a square different from 25, which when increased by 2, becomes a cube?" (*Oeuvres de Fermat*, Volume 3, Page 269, Gauthier–Villars, Paris, 1896) Euler provided the answer to Fermat's question (*Elements of Algebra*, Part II, Section 193, Springer-Verlag, New York, 1984) by assuming that $\mathbb{Z}[\sqrt{-2}]$ has unique factorization. The above is what he did, expressed in modern language.

---

# 19.4   $\mathbb{Z}[\sqrt{3}]$

In the previous section, we saw that $\mathbb{Z}[\sqrt{-2}]$ behaves very much like the Gaussian integers. In the present section, we look at $\mathbb{Z}[\sqrt{3}]$, which has different features. For example, there are infinitely many units.

There is a division algorithm for $\mathbb{Z}[\sqrt{3}]$, but we need to modify the norm so that it takes only non-negative values. Define

$$\overline{N}(a + b\sqrt{3}) = |N(a + b\sqrt{3})| = |(a + b\sqrt{3})\overline{(a + b\sqrt{3})}| = |a^2 - 3b^2|,$$

where $\overline{a + b\sqrt{3}} = a - b\sqrt{3}$.

**Proposition 19.10.** *Let $\alpha$ and $\beta$ be elements of $\mathbb{Z}[\sqrt{3}]$ with $\beta \neq 0$. Then there exist $\eta$ and $\rho$ in $\mathbb{Z}[\sqrt{3}]$ such that*

$$\alpha = \beta\eta + \rho \text{ and } 0 \leq \overline{N}(\rho) < \overline{N}(\beta).$$

*Proof.* Let $\delta = \alpha/\beta$. Write $\delta = x + y\sqrt{3}$. Let $s$ and $t$ be integers with

$$|s - x| \leq \frac{1}{2}, \qquad |t - y| \leq \frac{1}{2}.$$

Then

$$\overline{N}(\delta - (s + t\sqrt{3})) = \overline{N}\left(x - s + (y - t)\sqrt{3}\right)$$
$$= |(x - s)^2 - 3(y - t)^2|.$$

Since $(x - s)^2 \leq (1/2)^2$ and $(y - t)^2 \leq (1/2)^2$,

$$-3/4 \leq -3(y - t)^2 \leq (x - s)^2 - 3(y - t)^2 \leq (x - s)^2 \leq 1/4.$$

In particular, $|(x - s)^2 - 3(y - t)^2| \leq 3/4$, so

$$\overline{N}(\delta - (s + t\sqrt{3})) < 1. \tag{19.1}$$

Let $\eta = s + t\sqrt{3}$. Then, since $\delta = \alpha/\beta$, Equation (19.1) says that

$$\overline{N}\left(\frac{\alpha}{\beta} - \eta\right) < 1.$$

Multiplication by $\overline{N}(\beta)$ yields

$$\overline{N}(\alpha - \beta\eta) < \overline{N}(\beta).$$

Let $\rho = \alpha - \beta\eta$. Then $\rho \in \mathbb{Z}[\sqrt{3}]$ since $\alpha, \beta, \eta$ are in $\mathbb{Z}[\sqrt{3}]$, and we have $\overline{N}(\rho) < \overline{N}(\beta)$. Since $\alpha = \beta\eta + \rho$, this completes the proof. □

The Division Algorithm again yields the following. We omit the proof since it is almost identical to the proof of Theorem 18.12.

**Theorem 19.11.** *Every nonzero element of $\mathbb{Z}[\sqrt{3}]$ is either a unit, an irreducible, or a product of irreducibles. This factorization into irreducibles is unique up to the order of the factors and multiplication of irreducibles by units.*

Because there are infinitely many units in $\mathbb{Z}[\sqrt{3}]$, they play a significant role in any problem that involves factoring. Fortunately, we can determine all of them explicitly. The number $2 + \sqrt{3}$ is a unit in $\mathbb{Z}[\sqrt{3}]$. Its inverse is $2 - \sqrt{3}$. Up to sign, all units of $\mathbb{Z}[\sqrt{3}]$ are powers of these numbers.

**Proposition 19.12.** *The units of $\mathbb{Z}[\sqrt{3}]$ are of the form $\pm(2 + \sqrt{3})^n$, where $n$ is an integer (possibly negative).*

*Proof.* Let $\epsilon = 2 + \sqrt{3}$. Suppose there is a unit $u = a + b\sqrt{3}$ with $1 < u < \epsilon$. Since $u$ is a unit, we have $|u\bar{u}| = |N(u)| = 1$, so

$$|\bar{u}| = 1/|u| < 1.$$

Therefore,

$$|2a| = |u + \bar{u}| \le |u| + |\bar{u}| < \epsilon + 1 = 3 + \sqrt{3} \approx 4.7,$$

so $a = 0, \pm1, \pm2$. Using the fact that $a^2 - 3b^2 = \pm1$, we can try the various possibilities for $a$ and $b$ and we find that none of these yield units $u$ with $1 < u < \epsilon$. Therefore, $\epsilon$ is the smallest unit larger than 1.

Now suppose there exists a unit $v$ that is not of the form $\pm\epsilon^n$. By changing the sign of $v$, if necessary, we may assume that $v > 0$. Let $m$ be the integer (possibly negative) such that

$$\epsilon^m < v < \epsilon^{m+1}.$$

Let $u = v\epsilon^{-m}$. Then $u$ is a unit (products and quotients of units are units) and

$$1 < u < \epsilon.$$

As we showed above, this is impossible. Therefore, $v$ does not exist, so every unit is of the desired form. $\qquad\square$

Here is an example of how units play a role. Let's try to find all integer solutions of $y^2 = x^3 + 3$. Rewrite the equation as

$$(y + \sqrt{3})(y - \sqrt{3}) = x^3.$$

As in the previous section, the factors $y + \sqrt{3}$ and $y - \sqrt{3}$ are relatively prime, and we conclude that $y + \sqrt{3}$ is a unit times a cube:

$$y + \sqrt{3} = \pm(2 + \sqrt{3})^n(a + b\sqrt{3})^3.$$

If we ignore the unit (that is, let $n = 0$), then we have

$$y + \sqrt{3} = \pm\left((a^3 + 9ab^2) + (3a^2b + 3b^3)\sqrt{3}\right).$$

Equating the coefficients of $\sqrt{3}$ yields

$$1 = \pm 3(a^2 b + b^3),$$

which is impossible. We might be tempted to conclude that $y^2 = x^3 + 3$ has no solutions. But $(x, y) = (1, 2)$ is a solution. To find it, we need to include units. Let $n = 1$. Then we have

$$y + \sqrt{3} = \pm(2 + \sqrt{3})(a + b\sqrt{3})^3.$$

Equating the coefficients of $\sqrt{3}$ yields

$$1 = \pm(a^3 + 9ab^2 + 6a^2 b + 6b^3).$$

This is much more complicated than previous equations we have encountered. We can see that $a = \pm 1$, $b = 0$ are solutions, and these yield $(x, y) = (1, \pm 2)$. Showing that these are the only solutions is much harder, and we omit the details.

Generally, working with $\mathbb{Q}(\sqrt{d})$ is easier when $d < 0$ than when $d > 0$. The existence of infinitely many units is one of the major reasons.

## 19.4.1   The Lucas–Lehmer Test

In Subsection 14.2.5, we stated the Lucas–Lehmer Test for primality of Mersenne numbers. Using $\mathbb{Z}[\sqrt{3}]$, we can now prove this criterion. Recall that when $p$ is prime, the $p$th Mersenne number is

$$M_p = 2^p - 1.$$

**Proposition 19.13.** *(Lucas–Lehmer Test) Let $p$ be an odd prime. Define a sequence by*

$$s_0 = 4, \quad s_{i+1} = s_i^2 - 2.$$

*Then*

$$M_p \text{ is prime} \iff s_{p-2} \equiv 0 \pmod{M_p}.$$

*Proof.* We start with a lemma that relates $s_n$ to $\mathbb{Z}[\sqrt{3}]$.

**Lemma 19.14.** *For $n \geq 0$,*

$$s_n = (2 + \sqrt{3})^{2^n} + (2 - \sqrt{3})^{2^n}.$$

*Proof.* We'll prove this by induction. We have

$$(2+\sqrt{3})^{2^0} + (2-\sqrt{3})^{2^0} = (2+\sqrt{3}) + (2-\sqrt{3}) = 4 = s_0,$$

so the lemma is true for $n = 0$. Assume that it is true for $n = k$, so

$$s_k = (2+\sqrt{3})^{2^k} + (2-\sqrt{3})^{2^k}.$$

Then, using the relation $(2+\sqrt{3})(2-\sqrt{3}) = 1$, we obtain

$$s_{k+1} = s_k^2 - 2 = \left((2+\sqrt{3})^{2^k} + (2-\sqrt{3})^{2^k}\right)^2 - 2$$

$$= \left((2+\sqrt{3})^{2^k}\right)^2 + 2\left((2+\sqrt{3})(2-\sqrt{3})\right)^{2^k}$$

$$+ \left((2-\sqrt{3})^{2^k}\right)^2 - 2$$

$$= (2+\sqrt{3})^{2^{k+1}} + 2 \cdot 1^{2^k} + (2-\sqrt{3})^{2^{k+1}} - 2$$

$$= (2+\sqrt{3})^{2^{k+1}} + (2-\sqrt{3})^{2^{k+1}}.$$

Therefore, the formula is true for $n = k+1$. By induction, it is true for all $n \geq 0$.                                                              □

Assume $q = M_p = 2^p - 1$ is prime. We want to show that $q \mid s_{p-2}$. We start by observing that $q \equiv 7 \pmod 8$. By Proposition 13.3,

$$2^{(q-1)/2} \equiv \left(\frac{2}{q}\right) = 1 \pmod q.$$

Therefore,

$$2^{2^{p-1}} = 2^{(q+1)/2} = 2 \cdot 2^{(q-1)/2} \equiv 2 \pmod q. \tag{19.2}$$

Also, since $q \equiv 3 \pmod 4$ and

$$q = 2^p - 1 \equiv (-1)^p - 1 = -2 \equiv 1 \pmod 3,$$

quadratic reciprocity tells us that

$$3^{(q-1)/2} \equiv \left(\frac{3}{q}\right) = -\left(\frac{q}{3}\right) = -\left(\frac{1}{3}\right) = -1 \pmod q. \tag{19.3}$$

We now need to work with congruences mod $q$ in $\mathbb{Z}[\sqrt{3}]$. We say that

$a + b\sqrt{3} \equiv c + d\sqrt{3} \pmod{q}$ if $a \equiv c$ and $b \equiv d \pmod{q}$. Lemma 8.3 says that $q$ divides all of the binomial coefficients $\binom{q}{i}$ for $1 \le i \le q-1$, so

$$
\begin{aligned}
(1 + \sqrt{3})^q \\
= 1 + \binom{q}{1}\sqrt{3} + \binom{q}{2}(\sqrt{3})^2 + \cdots + \binom{q}{q-1}(\sqrt{3})^{q-1} + (\sqrt{3})^q \\
\equiv 1 + (\sqrt{3})^q \\
\equiv 1 + 3^{(q-1)/2}\sqrt{3} \\
\equiv 1 - \sqrt{3} \quad \pmod{q},
\end{aligned}
$$

where we used (19.3) in the last step. Since $2(2 + \sqrt{3}) = (1 + \sqrt{3})^2$, we have

$$
\begin{aligned}
2^{2^{p-1}}(2 + \sqrt{3})^{2^{p-1}} = (1 + \sqrt{3})^{2^p} \\
= (1 + \sqrt{3})^{q+1} \\
= (1 + \sqrt{3})(1 + \sqrt{3})^q \\
\equiv (1 + \sqrt{3})(1 - \sqrt{3}) = -2 \quad \pmod{q}.
\end{aligned}
$$

From (19.2), we obtain

$$
2(2 + \sqrt{3})^{2^{p-1}} \equiv -2 \pmod{q}. \tag{19.4}
$$

Since $q$ is odd, we can divide this congruence by 2 and obtain

$$
-1 \equiv (2 + \sqrt{3})^{2^{p-1}} \pmod{q}. \tag{19.5}
$$

Multiply this by $(2 - \sqrt{3})^{2^{p-2}}$. The relation $(2 + \sqrt{3})(2 - \sqrt{3}) = 1$, along with properties of exponents, yields

$$
\begin{aligned}
-1 \cdot (2 - \sqrt{3})^{2^{p-2}} &\equiv (2 - \sqrt{3})^{2^{p-2}}(2 + \sqrt{3})^{2^{p-1}} \\
&= (2 - \sqrt{3})^{2^{p-2}}(2 + \sqrt{3})^{2^{p-2}}(2 + \sqrt{3})^{2^{p-2}} \\
&= 1^{2^{p-2}}(2 + \sqrt{3})^{2^{p-2}} \\
&= (2 + \sqrt{3})^{2^{p-2}} \quad \pmod{q}.
\end{aligned}
$$

Therefore,

$$
s_{p-2} = (2 + \sqrt{3})^{2^{p-2}} + (2 - \sqrt{3})^{2^{p-2}} \equiv 0 \pmod{q}.
$$

Therefore, $q = M_p$ divides $s_{p-2}$.

Conversely, assume that $M_p$ divides $s_{p-2}$. Suppose that $M_p$ is not prime. By Proposition 2.7, there exists a prime

$$\ell \le \sqrt{M_p} < \sqrt{2^p}$$

with $\ell \mid M_p$.

Every number in $\mathbb{Z}[\sqrt{3}]$ is congruent mod $\ell$ to one of the numbers $a + b\sqrt{3}$ with $0 \le a, b \le \ell - 1$. Therefore, there are $\ell^2$ congruence classes for $\mathbb{Z}[\sqrt{3}]$ mod $\ell$. Consider the numbers

$$1, \quad 2 + \sqrt{3}, \quad (2 + \sqrt{3})^2, \quad \ldots, (2 + \sqrt{3})^{\ell^2}.$$

There are $\ell^2 + 1$ numbers, so two of them must be congruent mod $\ell$:

$$(2 + \sqrt{3})^i \equiv (2 + \sqrt{3})^j \pmod{\ell},$$

with $0 \le i < j \le \ell^2$. Multiply both sides of this congruence by $(2 - \sqrt{3})^i = (2 + \sqrt{3})^{-i}$. This yields

$$1 \equiv (2 + \sqrt{3})^k \pmod{\ell}, \tag{19.6}$$

with $k = j - i$ satisfying $0 < k \le \ell^2$. Define $\mathrm{ord}_\ell(2 + \sqrt{3})$ to be the smallest positive $k$ satisfying (19.6). We have proved that

$$\mathrm{ord}_\ell(2 + \sqrt{3}) \le \ell^2. \tag{19.7}$$

Since $M_p \mid s_{p-2}$ and $\ell \mid M_p$, we have $\ell \mid s_{p-2}$. Therefore,

$$(2 + \sqrt{3})^{2^{p-2}} + (2 - \sqrt{3})^{2^{p-2}} = s_{p-2} \equiv 0 \pmod{\ell}.$$

Multiply this congruence by $(2 + \sqrt{3})^{2^{p-2}}$ and again use the relation $(2 + \sqrt{3})(2 - \sqrt{3}) = 1$ to obtain

$$0 \equiv (2 + \sqrt{3})^{2^{p-2}} \left( (2 + \sqrt{3})^{2^{p-2}} + (2 - \sqrt{3})^{2^{p-2}} \right)$$

$$= (2 + \sqrt{3})^{2^{p-1}} + 1^{2^{p-2}} \pmod{q},$$

so

$$(2 + \sqrt{3})^{2^{p-1}} \equiv -1 \pmod{q}. \tag{19.8}$$

Squaring yields

$$(2 + \sqrt{3})^{2^p} \equiv 1 \pmod{q}. \tag{19.9}$$

The analogue of Theorem 11.1 holds:

$$(2 + \sqrt{3})^m \equiv 1 \pmod{\ell} \iff \operatorname{ord}_\ell(2 + \sqrt{3}) \mid m.$$

We omit the proof since it is the same as for Theorem 11.1. From (19.9), we find that $\operatorname{ord}_\ell(2 + \sqrt{3})$ divides $2^p$, so it is a power of 2. From (19.8), we see that $\operatorname{ord}_\ell(2 + \sqrt{3})$ does not divide $2^{p-1}$, so we must have

$$\operatorname{ord}_\ell(2 + \sqrt{3}) = 2^p > \ell^2.$$

This contradicts (19.7). The assumption that $M_p$ is not prime is what led to this contradiction, so we conclude that $M_p$ is prime. This completes the proof. □

## 19.5   Non-Unique Factorization

We have seen that $\mathbb{Z}[i]$, $\mathbb{Z}[\sqrt{-2}]$, and $\mathbb{Z}[\sqrt{3}]$ have unique factorization. This is not always the case. Consider $\mathbb{Z}[\sqrt{-5}]$. We can write

$$6 = 2 \cdot 3 = (1 + \sqrt{-5})(1 - \sqrt{-5}).$$

We claim that these are distinct factorizations.

**Proposition 19.15.** *The numbers* 2, 3, $1 + \sqrt{-5}$, *and* $1 - \sqrt{-5}$ *are irreducible in* $\mathbb{Z}[\sqrt{-5}]$. *Moreover, they are not associates; that is, they do not differ by multiplication by units of* $\mathbb{Z}[\sqrt{-5}]$.

*Proof.* Let's start with 2. We'll use the norm: $N(a + b\sqrt{-5}) = a^2 + 5b^2$. Suppose $2 = \alpha\beta$, where neither $\alpha$ nor $\beta$ is a unit. Then

$$4 = N(2) = N(\alpha)N(\beta).$$

Since $\alpha$ and $\beta$ are not units, $N(\alpha) \neq 1$ and $N(\beta) \neq 1$. Therefore, $N(\alpha) = 2$. If $\alpha = a + b\sqrt{-5}$, this means that $2 = a^2 + 5b^2$, which is impossible. Therefore, 2 does not factor.

The proof that 3 is irreducible is similar. It reduces to showing that $a^2 + 5b^2 = 3$ has no integer solutions, which is easily checked.

If $1 + \sqrt{-5}$ factors as $\alpha\beta$, then

$$6 = N(1 + \sqrt{-5}) = N(\alpha)N(\beta),$$

so $N(\alpha) = 2$ or 3. Both of these are impossible. Similarly, $1 - \sqrt{-5}$ is irreducible.

The only units in $\mathbb{Z}[\sqrt{-5}]$ are $\pm 1$ by Proposition 19.6. Therefore, 2, 3, $1+\sqrt{-5}$, and $1-\sqrt{-5}$ are not associates. This completes the proof.  $\square$

It was proved (independently) by Alan Baker, Kurt Heegner, and Harold Stark that there are only nine $\mathbb{Q}(\sqrt{d})$ with $d < 0$ whose rings of algebraic integers have unique factorization:

$$d = -1, -2, -3, -7, -11, -19, -43, -67, -163.$$

On the other hand, there are many $d > 0$ such that the ring of algebraic integers in $\mathbb{Q}(\sqrt{d})$ has unique factorization, and it is conjectured that there are infinitely many.

Let's give one example where non-unique factorization causes problems. Suppose we want to find all integer solutions to $y^2 = x^3 - 61$. We have considered this type of equation several times (Sections 18.5, 19.3, and 19.4), and the procedure might be familiar by now. Write

$$(y + \sqrt{-61})(y - \sqrt{-61}) = x^3.$$

The factors $y + \sqrt{-61}$ and $y - \sqrt{-61}$ can be shown to be relatively prime. If we have unique factorization, we obtain

$$y + \sqrt{-61} = \pm(a + b\sqrt{-61})^3.$$

Equating the coefficients of $\sqrt{-61}$ yields

$$1 = \pm(3a^2 b - 61b^3) = \pm b(3a^2 - 61b^2).$$

Therefore, $b \mid 1$, so $b = \pm 1$. This yields $3a^2 - 61 = \pm 1$, which has no solutions. Unfortunately, $(x, y) = (5, 8)$ is a solution of $y^2 = x^3 - 61$. What went wrong?

The ring $\mathbb{Z}[\sqrt{-61}]$ does not have unique factorization. For example,

$$125 = (8 + \sqrt{-61})(8 - \sqrt{-61}) = 5^3.$$

The techniques used to prove Proposition 19.15 show that $8+\sqrt{-61}$, $8-$

$\sqrt{-61}$, and 5 are irreducibles in $\mathbb{Z}[\sqrt{-61}]$, and they are not associates. Therefore, 125 has two distinct factorizations, and this is exactly what caused the problem. In fact, $8 + \sqrt{-61}$ and $8 - \sqrt{-61}$ are two relatively prime non-cubes whose product is a cube, so the analogue of Lemma 19.9 fails.

In order to work with situations where there is not unique factorization, Kummer and Dedekind in the 1800s developed the theory of ideals, a basic concept in algebraic number theory. This led to abstract ring theory and has had a substantial influence on modern mathematics.

# 19.6   Chapter Highlights

1. The determination of the algebraic integers in $\mathbb{Q}(\sqrt{d})$.

2. The only algebraic integers in $\mathbb{Q}(\sqrt{d})$ that are units are $\pm 1$ when $d < -3$. When $d > 0$, the units correspond to solutions of Pell's equation (Proposition 19.6).

# 19.7   Problems

## 19.7.1   Exercises

1. Prove that the product of two units is a unit.
2. Show that 7 and $\sqrt{-5}$ are irreducible in $\mathbb{Z}[\sqrt{-5}]$.
3. (a) Show that $a^2 + ab - b^2 \equiv 0 \pmod 2$ implies that $a \equiv b \equiv 0 \pmod 2$.
   (b) Show that $a^2 + ab - b^2 = \pm 2$ has no solutions.
   (c) Show that 2 is irreducible in $\mathbb{Z}[(1 + \sqrt{5})/2]$.
4. Show that $\sqrt{5}$ is irreducible in $\mathbb{Z}[(1 + \sqrt{5})/2]$.
5. (a) Let $\phi = (1 + \sqrt{5})/2$. Show that $\phi$, $1 + \phi$, $1 + 2\phi$, and $2 + 3\phi$ are units in $\mathbb{Z}[\phi]$.
   (b) Let $F_n$ be the $n$th Fibonacci number (see Appendix A for the definition of Fibonacci numbers). Show that

$$F_{n-1} + F_n \phi$$

   is a unit of $\mathbb{Z}[\phi]$ for each $n \geq 1$. (*Hint:* Use induction. Assume the statement is true for $n = k$. Then multiply by $\phi$.)

6.  (a) Show that 2, 7, $1 + \sqrt{-13}$, $1 - \sqrt{-13}$ are irreducible in $\mathbb{Z}[\sqrt{-13}]$.
    (b) Show that 14 does not uniquely factor into irreducibles in $\mathbb{Z}[\sqrt{-13}]$.
    (c) Generalize this to find examples of non-unique factorization in $\mathbb{Z}[\sqrt{-n}]$ whenever $n > 4$ is squarefree and satisfies $n \equiv 1 \pmod{4}$.

7.  (a) Show that $\sqrt{2} - 1$ is a unit in $\mathbb{Z}[\sqrt{2}]$.
    (b) Let $a + b\sqrt{2} \in \mathbb{Z}[\sqrt{2}]$. Show that

$$\left(\sqrt{2} - 1\right)\left(a + b\sqrt{2}\right) = (2b - a) + (a - b)\sqrt{2}.$$

    (c) Use norms to show that

$$a^2 - 2b^2 = -\left((2b - a)^2 - 2(a - b)^2\right)$$

    (of course, this can also be shown by explicit calculation).
    Look at the proofs in Chapter 5 that $\sqrt{2}$ is irrational and note that the expressions are similar. We'll now use these techniques to prove that $\sqrt{3}$ is irrational. Suppose $a, b$ are integers with $a/b = \sqrt{3}$, with $b > 0$ as small as possible. Multiply $a + b\sqrt{3}$ by the unit $2 - \sqrt{3}$ of $\mathbb{Z}[\sqrt{3}]$ to get $(2a - 3b) + (2b - a)\sqrt{3}$.
    (d) Show that $0 < 2b - a < b$.
    (e) Show that $(2a - 3b)^2 = 3(2b - a)^2$. Deduce that $\sqrt{3}$ is irrational.

8.  Let $\mathbb{Z}[\sqrt{-3}] = \{a + b\sqrt{-3} \mid a, b \text{ are integers}\}$. This is not the the full set of algebraic integers in $\mathbb{Q}(\sqrt{-3})$.
    (a) Give an algebraic integer in $\mathbb{Q}(\sqrt{-3})$ that is not in $\mathbb{Z}[\sqrt{-3}]$.
    (b) Show that the units of $\mathbb{Z}[\sqrt{-3}]$ are $\pm 1$.
    (c) Show that $(1 + \sqrt{-3})(1 - \sqrt{-3}) = 2 \cdot 2$ and that $1 \pm \sqrt{-3}$ and 2 are not associates in $\mathbb{Z}[\sqrt{-3}]$ but are associates in $\mathbb{Z}[\frac{1+\sqrt{-3}}{2}]$.
    Part (c) shows that the lack of unique factorization in $\mathbb{Z}[\sqrt{-3}]$ can be explained by the "missing element" $(1 + \sqrt{-3})/2$.

9.  Assume that the algebraic integers in $\mathbb{Q}(\sqrt{d})$ have the unique factorization property; namely, every nonzero element is either a unit, an irreducible, or a product of irreducibles, and this factorization into irreducibles is unique. Let $\pi$ be an irreducible. Show that if $\pi \mid \alpha\beta$ then $\pi \mid \alpha$ or $\pi \mid \beta$.

## 19.7.2    Projects

1.  We'll determine all the irreducibles in $\mathbb{Z}[\sqrt{-2}]$. The proofs follow closely the proofs for $\mathbb{Z}[i]$ (see Subsection 18.5.1).
    (a) Show that $\sqrt{-2}$ is irreducible in $\mathbb{Z}[\sqrt{-2}]$.
    (b) Let $p$ be an odd prime. Show that $-2$ is a square mod $p$ if and only if $p \equiv 1$ or 3 (mod 8). (This part requires Theorem 13.4(b,c).)
    (c) Let $p$ be a prime with $p \equiv 5$ or 7 (mod 8). Show that $x^2 + 2y^2 = p$ has no solutions. Deduce that $p$ is irreducible in $\mathbb{Z}[\sqrt{-2}]$.
    (d) Let $p$ be a prime with $p \equiv 1$ or 3 (mod 8). Show that there exists an integer $x$ such that
$$p \mid (x + \sqrt{-2})(x - \sqrt{-2})$$

in $\mathbb{Z}[\sqrt{-2}]$. (*Hint:* Use part (b).)

(e) Let $p$ be a prime with $p \equiv 1$ or $3 \pmod{8}$. Show that $p$ is not irreducible in $\mathbb{Z}[\sqrt{-2}]$ and that it factors as the product of two irreducibles.

(f) Let $p$ be a prime with $p \equiv 1$ or $3 \pmod{8}$. Show that $p$ can be written in the form $a^2 + 2b^2$.

(g) Show that the irreducibles from (a), (c), and (e) give all irreducibles in $\mathbb{Z}[\sqrt{-2}]$. (*Hint:* Look at the proof of the analogous fact for $\mathbb{Z}[i]$.)

2. Let's find all solutions of $x^2 + 2y^2 = z^2$.

(a) Show that $x$ is even if and only if $z$ is even, and show that if $x$ and $z$ are even then $y$ is even. This says that we can divide $x$, $y$, $z$ by their gcd and therefore assume that $x, y, z$ are pairwise relatively prime and that $x$ and $z$ are odd.

(b) Let $n$ be an integer. Suppose $\sqrt{-2} \mid n$ in $\mathbb{Z}[\sqrt{-2}]$. Show that $n$ is even. (*Hint:* Either write it out explicitly, or use norms.)

(c) Show that if $\sqrt{-2} \mid x + y\sqrt{-2}$ in $\mathbb{Z}[\sqrt{-2}]$, then $x$ is even.

(d) Let $\pi \neq \pm\sqrt{-2}$ be an irreducible in $\mathbb{Z}[\sqrt{-2}]$. Suppose that $\pi \mid 2x$ and $\pi \mid 2y\sqrt{-2}$. Show that $\pi \mid x$ and $\pi \mid y$.

(e) Show that an irreducible cannot divide two relatively prime integers. (*Hint:* Use Theorem 2.12.)

(f) Show that $x + \sqrt{-2}y$ and $x - \sqrt{-2}y$ are relatively prime in $\mathbb{Z}[\sqrt{-2}]$.

(g) Show that $x + \sqrt{-2}y = \pm\alpha^2$ for some $\alpha \in \mathbb{Z}[\sqrt{-2}]$.

(h) Show that there are integers $a$ and $b$ such that

$$x = \pm(a^2 - 2b^2), \quad y = \pm 2ab, \quad z = \pm(a^2 + 2b^2).$$

(This is the same answer that we obtained in Project 3 in Chapter 5.)

3. Assume that the algebraic integers in $\mathbb{Q}(\sqrt{d})$ have the unique factorization property; namely, every nonzero element is either a unit, an irreducible, or a product of irreducibles, and this factorization into irreducibles is unique. Let $\pi$ be an irreducible. Let $p$ be a prime dividing $N(\pi)$ and let $\gamma$ be an irreducible dividing $p$.

(a) Show that $N(\gamma) = \pm p$ or $\pm p^2$.

(b) Show that if $N(\gamma) = \pm p^2$ then $p = \gamma u$ for some unit $u$.

(c) Show that if $N(\gamma) = \pm p$ then $\gamma$ is an associate of either $\pi$ or $\bar{\pi}$. (*Hint:* Use Exercise 9.)

(d) Show that if $\pi$ is an irreducible that is not an associate of a prime number, then, up to sign, $N(\pi)$ is a prime number.

4. Let $\alpha = a + b\frac{1+\sqrt{-163}}{2}$, where $a$ and $b$ are integers with $b \neq 0$.

(a) Show that $N(\alpha) \geq 41$.

(b) Let $\gamma \in \mathbb{Z}[\frac{1+\sqrt{-163}}{2}]$. Suppose that $\gamma$ is not divisible by any integer other than $\pm 1$ and suppose that $1 < N(\gamma) < 41^2$. Show that $\gamma$ is irreducible.

(c) Show that $N(n + \frac{1+\sqrt{-163}}{2}) < 41^2$ for $-40 \leq n \leq 39$.

(d) It is known (but it's a deep fact) that $\mathbb{Z}[\frac{1+\sqrt{-163}}{2}]$ has the unique factorization property. Assuming this, show that $n^2 + n + 41$ is prime for $-40 \leq n \leq 39$. (*Hint:* Use Project 3.)

# 19.7.3    Answers to "CHECK YOUR UNDERSTANDING"

1.  $1 + 3\sqrt{2}$ is a root of $x^2 - 2x - 17$ and $\frac{5+\sqrt{13}}{2}$ is a root of $x^2 - 5x + 3$.

2.  Rationalize the denominator:

$$\frac{-7 + 7\sqrt{-5}}{1 + 2\sqrt{-5}} \frac{1 - 2\sqrt{-5}}{1 - 2\sqrt{-5}} = \frac{63 - 21\sqrt{-5}}{21} = 3 - \sqrt{-5},$$

which is an algebraic integer.

3.  Solve Pell's equation: $x^2 - 11y^2 = \pm 1$. Using either continued fractions or "trial and error" yields $(x, y) = (10, 3)$. Therefore, $u = 10 + 3\sqrt{11}$ is a unit.

# Chapter 20

## The Distribution of Primes

We know that there are infinitely many primes. This was easily proved using only the fact that every integer greater than 1 has a prime factor. In this chapter, we'll continue the theme from Section 5.7 and obtain quantitative statements about how primes are distributed.

In the first section, we consider gaps between consecutive primes and prove *Bertrand's Postulate*, which says that if $x > 1$ then there is always a prime between $x$ and $2x$.

In the second section, we prove a weak form of the Prime Number Theorem, showing that the number of primes less than $x$ is approximately $x/\ln(x)$.

## 20.1   Bertrand's Postulate

> *Chebyshev said it, but I'll say it again;*
> *There's always a prime between $n$ and $2n$.*
> – N.J. Fine

In 1845, the French mathematician Joseph Bertrand, in connection with his work on permutation groups, conjectured that for every $n > 1$ there is always a prime between $n$ and $2n$. This was proved a few years later by Pafnuty Chebyshev. The proof we give is due to Paul Erdős.

**Theorem 20.1.** *Let $x \geq 1$ be a real number. Then there is a prime number $p$ satisfying $x < p \leq 2x$.*

*Proof.* The main part of the proof is the case when $x$ is an integer. The passage to real numbers $x$ is fairly straightforward.

Let $n$ be an integer. We need a way to isolate the prime numbers between $n$ and $2n$. The trick is to use the binomial coefficient

$$\binom{2n}{n} = \frac{(2n)!}{n!\,n!},$$

which is an integer (see Appendix A).

**Key Observation:** All of the prime factors of the integer $\binom{2n}{n}$ are $\leq 2n$ and all primes $p$ with $n < p \leq 2n$ divide $\binom{2n}{n}$.

This statement is essential to everything in the proof. It says that the binomial coefficient captures all of the primes that we're looking for. Difficulties arise because it might capture a few extra primes. The technical parts of the proof are aimed at showing that these extras do not cause any trouble. You might ask, "Why don't we use $(2n)!$ since it's simpler and also captures the relevant primes?" The advantage of $\binom{2n}{n}$ is that the $(n!)^2$ in the denominator cancels out a lot of the primes less than $n$, so most of what remains is exactly what we need — the primes from $n$ to $2n$.

Why is the Key Observation true? First note that $(2n)!$ is a product of integers up to $2n$, so each of the primes $p \leq 2n$ divides this product, and these are the only primes dividing the product. The primes $p$ with $n < p \leq 2n$ do not appear in the factorization of $n!$, so they are not canceled by the $(n!)^2$ in the denominator. Therefore, they remain in the factorization of $\binom{2n}{n}$.

Let's look at an example with $n = 11$:

$$\binom{22}{11} = \frac{22!}{(11!)^2} = 705432 = 2^3 \cdot 3 \cdot 7 \cdot 13 \cdot 17 \cdot 19.$$

There are four 5's in 22!, coming from 5, 10, 15, and 20, but they are all canceled out by the multiples of 5 in $(11!)^2$. Similarly, the 11's cancel out. All but one 3 and all but one 7 cancel, and there is a small power of 2 remaining. The total contribution from the primes up through $n = 11$ is

$$2^3 \cdot 3 \cdot 7 = 168,$$

which is much less than 705432, the value of the binomial coefficient. Therefore, the binomial coefficient must have prime factors larger than

11. But 705432 is constructed by starting with a product of numbers up to 22, so its prime factors must be less than 22. We conclude that there must be primes between 11 and 22. Looking at the factorization, we see the primes 13, 17, and 19.

The proof of the theorem follows a similar line of reasoning. Namely, we will show that the contribution to $\binom{2n}{n}$ from the primes up to $n$ is too small to account for the size of the number, so there must be additional prime factors larger than $n$. These prime factors are less than $2n$, so there must be primes between $n$ and $2n$.

We need five lemmas. We state them now, show how they are used to prove the theorem, and then prove them.

**Lemma 20.2.** *Let $n \geq 1$. Then*

$$\frac{1}{2n} \cdot 4^n \leq \binom{2n}{n} \leq \frac{1}{2} \cdot 4^n.$$

This lemma shows that the binomial coefficient is very large, and we'll soon see that the primes $p \leq n$ do not contribute enough to make it as large as it is.

**Lemma 20.3.** *Let $n \geq 3$. If $p$ is a prime with $(2/3)n < p \leq n$ then*

$$p \nmid \binom{2n}{n}.$$

This lemma is an important step. It shows that a lot of primes $p \leq n$ cannot contribute anything to the prime factorization of the binomial coefficient.

**Lemma 20.4.** *Let $n \geq 1$ and let $p$ be a prime. Let $p^a$ be a power of $p$ dividing $\binom{2n}{n}$. Then $p^a \leq 2n$.*

This lemma might look obvious at first. For example, take $a = 1$. The lemma says that if $p$ divides the binomial coefficient then $p \leq 2n$. Yes, this is obvious. But now consider $a = 2$, so we are supposing that $p^2$ divides the binomial coefficient. There are many multiples of $p$ in $(2n)!$ and some of them are canceled by multiples of $p$ in the denominator. The lemma says that at most one factor of $p$ is not canceled unless $p^2 \leq 2n$. In particular, if $p > \sqrt{2n}$, then $p$ can divide $\binom{2n}{n}$ but $p^2$ cannot. This allows us to keep the medium-sized primes under control.

**Lemma 20.5.** *Let $n \geq 1$. Then*

$$\prod_{p \leq n} p \leq 4^n,$$

*where the product is over the prime numbers $p \leq n$.*

This lemma can be improved. The Prime Number Theorem can be used to prove that the product of the primes less than $n$ is approximately $e^n$. However, the slightly worse, but more explicit, estimate of the lemma is all that we need.

**Lemma 20.6.** *If $n \geq 468$, then*

$$4^{n/3} > (2n)^{1+\sqrt{2n}}.$$

This final lemma is a technical result that is needed to finish the proof.

*Proof of Theorem.* Let's use the lemmas to prove the theorem. It's much easier to work with integers, so we start by taking $n \geq 1$ to be an integer and proving that there is a prime $p$ with $n < p \leq 2n$.

We suppose that $n$ is a positive integer such that there is no prime $p$ with $n < p \leq 2n$. Our aim is to obtain a contradiction.

All of the prime factors of $\binom{2n}{n}$ are less than or equal to $2n$, but we're assuming that there are no primes between $n$ and $2n$, so all the primes factors of $\binom{2n}{n}$ are less than or equal to $n$. The second lemma says that there are no prime factors between $2n/3$ and $n$, so all prime factors satisfy $p \leq 2n/3$.

Suppose $p > \sqrt{2n}$. Then $p^2 > 2n$, so the case $a \geq 2$ of the third lemma cannot occur for such a $p$. This means that the prime factorization of $\binom{2n}{n}$ contains $p$ to at most the first power. Therefore, the contribution to the prime factorization that comes from the primes with $\sqrt{2n} < p \leq 2n/3$ is at most

$$\prod_{\sqrt{2n} < p \leq 2n/3} p \leq \prod_{p \leq 2n/3} p \leq 4^{2n/3},$$

by the fourth lemma.

The remaining primes in the prime factorization of $\binom{2n}{n}$ are less than or equal to $\sqrt{2n}$. There are at most $\sqrt{2n}$ such primes (yes, this is a rather crude estimate, but it is both easy and sufficient). By the third lemma, the power $p^a$ of a prime dividing $\binom{2n}{n}$ satisfies $p^a \leq 2n$, so the contribution of each prime power to the factorization is at most $2n$. This means we are considering at most $\sqrt{2n}$ prime powers in the factorization and each is at most $2n$. Therefore, the primes up to $\sqrt{2n}$ contribute at most

$$(2n)^{\sqrt{2n}}$$

to $\binom{2n}{n}$.

Putting everything together, we find that

$$\frac{4^n}{2n} \leq \binom{2n}{n} \leq \left((2n)^{\sqrt{2n}}\right) 4^{2n/3}.$$

(The first inequality is from the first lemma. The two factors in the last expression are from the primes less than $\sqrt{2n}$ and from the primes between $\sqrt{2n}$ and $2n/3$.) The inequality can be rearranged:

$$4^{n/3} \leq (2n)^{1+\sqrt{2n}}.$$

But this contradicts the fifth lemma if $n \geq 468$. Therefore, if $n \geq 468$, there is a prime $p$ with $n < p \leq 2n$.

What about $n \leq 467$? Surely, Bertrand must have checked small values of $n$ before making his conjecture. That's what we'll do, too. Consider the following sequence of primes:

$$2, \quad 3, \quad 5, \quad 7, \quad 13, \quad 23, \quad 43, \quad 83, \quad 163, \quad 317, \quad 631.$$

Each prime on this list is less than twice the one before it. Let $3 \leq n \leq 467$. Then there are two successive primes $p, q$ on the list with $p < n \leq q$. If $n < q$ then

$$n < q < 2p < 2n,$$

so there is a prime between $n$ and $2n$. If $n = q$ then $n \neq 631$ because $n \leq 467$. Let $r$ be the next prime after $q$ on the list. Then $r$ is less than $2q = 2n$, so we have

$$n = q < r < 2n.$$

We have shown that when $3 \leq n \leq 467$, there is always a prime between $n$ and $2n$. There are only two possibilities left: $n = 1$ and $n = 2$. But $1 < 2 \leq 2$ and $2 < 3 \leq 4$, so the theorem is also true for these two values of $n$. That's it. There are no more values of $n$, so we have proved that there is always a prime $p$ with $n < p \leq 2n$ when $n \geq 1$.

Finally, we prove the theorem for real values of $x$. Suppose $x \geq 1$ and there is no prime $p$ with $x < p \leq 2x$. Let $n$ be the largest integer with $n \leq x$. We now know that there is a prime $p$ with $n < p \leq 2n$, so

$$x < n + 1 \leq p \leq 2n \leq 2x.$$

Therefore, there is a prime between $x$ and $2x$, so we are done.    □

We now prove the lemmas.

**Proof of Lemma 20.2:** We'll proceed by induction on $n$. First, let's check the lemma for $n = 1$:

$$\frac{4}{2 \cdot 1} = 2 \leq 2 = \binom{2}{1} \leq \frac{1}{2} \cdot 4^1.$$

Assume that we know that the first inequality of the lemma is true for $n = k$, so

$$\frac{4^k}{2k} \leq \binom{2k}{k}.$$

Then

$$\binom{2(k+1)}{k+1} = \frac{(2k+2)(2k+1)}{(k+1)^2} \binom{2k}{k} = \frac{2(2k+1)}{(k+1)} \binom{2k}{k}$$

$$\geq \frac{2(2k+1)}{(k+1)} \frac{4^k}{2k} = \frac{2k+1}{2k} \frac{4^{k+1}}{2(k+1)}$$

$$\geq \frac{4^{k+1}}{2(k+1)}.$$

Therefore, the first inequality of the lemma is also true for $n = k + 1$. By induction, it is true for all $n$.

The proof of the second inequality is similar. The case $n = 1$ was easily checked to be true. Suppose that the inequality is true for $n = k$, so

$$\binom{2k}{k} \leq \frac{1}{2} \cdot 4^k.$$

Then

$$\binom{2(k+1)}{k+1} = \frac{(2k+2)(2k+1)}{(k+1)^2}\binom{2k}{k}$$

$$\leq \frac{(2k+2)(2k+2)}{(k+1)^2}\binom{2k}{k} = 4\binom{2k}{k}$$

$$\leq 4 \cdot \frac{1}{2} \cdot 4^k = \frac{1}{2} \cdot 4^{k+1}.$$

Therefore, the second inequality of the lemma is also true for $n = k+1$. By induction, it is true for all $n$. $\quad\square$

**Proof of Lemma 20.3:** Since $n \geq 3$, and $2n/3 < p$, we have $2 < p$, so $p$ is odd. If $2n/3 < p \leq n$ then $2p \leq 2n$ and $3p > 2n$, so the multiples of $p$ in $(2n)!$ are $p$ and $2p$. Therefore, $p^2$ is the exact power of $p$ dividing $(2n)!$ (fortunately, $p \neq 2$; otherwise, the 2 in $2p$ would yield an extra factor of 2). Since $p \leq n$ and $n < 4n/3 < 2p$, the only multiple of $p$ in $n!$ is $p$ itself. Therefore, $p^1$ is the power of $p$ in $n!$, and $p^2$ is the power of $p$ in $(n!)^2$. The $p^2$ in the numerator of $(2n)!/(n!)^2$ is canceled by the $p^2$ in the denominator, so $p$ does not divide $\binom{2n}{n}$. $\quad\square$

**Proof of Lemma 20.4:** For a real number $x$, recall that $\lfloor x \rfloor$ is the largest integer less than or equal to $x$. We claim that

$$\lfloor 2x \rfloor - 2\lfloor x \rfloor = 0 \text{ or } 1.$$

To prove this, write $x = m + r$, where $m$ is an integer and $0 \leq r < 1$. Then $\lfloor x \rfloor = m$ and $\lfloor 2x \rfloor = 2m + \delta$, where $\delta = 0$ if $0 \leq r < 1/2$ and $\delta = 1$ if $1/2 \leq r < 1$. Therefore,

$$\lfloor 2x \rfloor - 2\lfloor x \rfloor = 2m + \delta - 2m = \delta \in \{0, 1\},$$

as claimed.

Let $p^c$ satisfy $p^c \leq 2n < p^{c+1}$. By Theorem 5.9, the power of $p$ dividing $n!$ is $p^b$, where

$$b = \left\lfloor \frac{n}{p} \right\rfloor + \left\lfloor \frac{n}{p^2} \right\rfloor + \left\lfloor \frac{n}{p^3} \right\rfloor + \cdots + \left\lfloor \frac{n}{p^c} \right\rfloor,$$

and the power of $p$ dividing $(2n)!$ is this expression with $2n$ in place of

$n$. Therefore, the power of $p$ in $\binom{2n}{n}$ is $p^d$ with

$$d = \left\lfloor \frac{2n}{p} \right\rfloor + \left\lfloor \frac{2n}{p^2} \right\rfloor + \cdots + \left\lfloor \frac{2n}{p^c} \right\rfloor - 2\left( \left\lfloor \frac{n}{p} \right\rfloor + \left\lfloor \frac{n}{p^2} \right\rfloor + \cdots + \left\lfloor \frac{n}{p^c} \right\rfloor \right)$$

$$= \left( \left\lfloor \frac{2n}{p} \right\rfloor - 2\left\lfloor \frac{n}{p} \right\rfloor \right) + \left( \left\lfloor \frac{2n}{p^2} \right\rfloor - 2\left\lfloor \frac{n}{p^2} \right\rfloor \right) + \cdots + \left( \left\lfloor \frac{2n}{p^c} \right\rfloor - 2\left\lfloor \frac{n}{p^c} \right\rfloor \right)$$

$$\leq c,$$

since each term in the sum is 0 or 1. Therefore, $d \leq c$, so $p^d \leq p^c \leq 2n$. Since $p^a$ is a power of $p$ dividing $\binom{2n}{n}$ and $p^d$ is the largest such power, we have $p^a \leq p^d \leq 2n$. $\qquad\square$

**Proof of Lemma 20.5:** Before getting to the main part of the proof, we need one more binomial coefficient:

$$2\binom{2m+1}{m} = \frac{2(2m+1)!}{m!\,(m+1)!} = \frac{(2m+2)(2m+1)!}{(m+1)m!\,(m+1)!}$$

$$= \frac{(2m+2)!}{(m+1)!\,(m+1)!} = \binom{2m+2}{m+1} \leq \frac{1}{2} \cdot 4^{m+1},$$

by Lemma 20.2. Therefore,

$$\binom{2m+1}{m} \leq 4^m.$$

If $p$ is a prime with $m+2 \leq p \leq 2m+1$, then $p$ divides $(2m+1)!$ but it divides neither $m!$ nor $(m+1)!$. Therefore, $p$ divides the numerator of the expression for $\binom{2m+1}{m}$ but is not canceled out by the denominator, so $p$ divides $\binom{2m+1}{m}$. This means that the product of all primes from $m+2$ to $2m+1$ divides $\binom{2m+1}{m}$, so

$$\prod_{m+2 \leq p \leq 2m+1} p \leq \binom{2m+1}{m} \leq 4^m \qquad\qquad (20.1)$$

(the product is over primes $p$ in the interval $[m+2,\ 2m+1]$).

Now we can prove the lemma. It is certainly true for $n = 1$ (the empty product is defined to be 1, which is less than 4). It is also true for $n = 2$. Assume now that $n \geq 3$ and that the inequality of the lemma is true for all integers $j \leq n$. We'll prove it is then true for $n + 1$.

First, suppose that $n + 1$ is even. Then $n + 1$ is not prime ($n + 1 \geq 4$, so $n + 1 \neq 2$), so

$$\prod_{p \leq n+1} p = \prod_{p \leq n} p \leq 4^n < 4^{n+1},$$

where we have used the induction assumption that the inequality holds for $n$.

Now suppose $n + 1 = 2m + 1$ is odd. Then (20.1) and the induction assumption yield

$$\prod_{p \leq n+1} p = \prod_{p \leq m+1} p \prod_{m+2 \leq p \leq 2m+1} p \leq 4^{m+1} \cdot 4^m = 4^{2m+1} = 4^{n+1}.$$

Therefore, the inequality holds for $n + 1$ in both cases. By induction, the inequality holds for all $n$.  □

**Proof of Lemma 20.6:** Let $f(x) = \ln(x)/x$ for $x > 0$. Then $f'(x) < 0$ when $x > e$, so $f(x)$ is decreasing for $x > e$. Let

$$g(x) = e^{f(x)} = e^{\ln(x) \cdot \frac{1}{x}} = x^{1/x}.$$

Then $g(x)$ is also decreasing for $x > e$.

Suppose $n \geq 500$. Then

$$(2n)^{\sqrt{2/n}} = \left( (\sqrt{2n})^{1/\sqrt{2n}} \right)^4 = g(\sqrt{2n})^4 \leq g(\sqrt{1000})^4 < 1.55,$$

and

$$(2n)^{1/n} = \left( (2n)^{1/(2n)} \right)^2 \leq g(1000)^2 < 1.014.$$

Therefore,

$$(2n)^{1+\sqrt{2n}} = \left( (2n)^{1/n} (2n)^{\sqrt{2/n}} \right)^n < (1.014 \times 1.55)^n$$

$$< \left( 4^{1/3} \right)^n = 4^{n/3},$$

so the lemma is true for $n \geq 500$. A computer calculation shows that the inequality is also true for $468 \leq n \leq 500$.  □

## 20.2   Chebyshev's Approximate Prime Number Theorem

We know that there are infinitely many primes, but suppose we ask for more quantitative information. For example, there are 78498 primes less than $10^6$ and 50847534 primes less than $10^9$. Is there a formula that tells us the general situation? Let

$$\pi(x) = \text{ number of primes } p \le x.$$

We want a simple continuous function that gives a good approximation to $\pi(x)$. Around 1800, Legendre considered this question and conjectured that

$$\pi(x) \approx \frac{x}{\ln(x) - 1.08366}.$$

About the same time, Gauss, in his private notes, conjectured a similar formula, and a few years later, Dirichlet also proposed what turns out to be a better approximation:

$$\pi(x) \approx \text{li}(x) = \int_0^x \frac{dt}{\ln t} = \int_0^2 \frac{dt}{\ln t} + \int_2^x \frac{dt}{\ln t} = 1.045 + \int_2^x \frac{dt}{\ln t}$$

where the first integral on the right is improper at both 0 and 1, but converges (the Cauchy principal value must be used at 1). It can be shown that

$$\lim_{x \to \infty} \frac{\text{li}(x)}{x / \ln x} = 1,$$

(see Exercise 4) so the approximations to $\pi(x)$ by $\text{li}(x)$ and by $x / \ln x$ are of approximately the same size.

Around 1850, Chebyshev made the first real progress, proving that there exist constants $c_1$ and $c_2$ such that

$$c_1 \frac{x}{\ln x} \le \pi(x) \le c_2 \frac{x}{\ln x}$$

for all $x$ larger than a specific bound. Chebyshev used

$$c_1 = \ln\left(\frac{2^{\frac{1}{2}} 3^{\frac{1}{3}} 5^{\frac{1}{5}}}{30^{\frac{1}{30}}}\right) \approx .921 \text{ and } c_2 = \frac{6c_1}{5} \approx 1.106.$$

In 1859, Riemann made a big advance toward understanding $\pi(x)$ by relating $\pi(x)$ to the zeros of the zeta function $\zeta(s)$ (see Section 5.7). In particular, he showed that if all zeros of $\zeta(s)$ with $0 \leq \operatorname{Re}(s) \leq 1$ (where $\operatorname{Re}(s)$ denotes the real part of the complex number $s$) have $\operatorname{Re}(s) = 1/2$, then $\operatorname{li}(x)$ gives a very good approximation to $\pi(x)$. Finally, in 1896, Hadamard and de la Valée-Poussin independently proved that $\zeta(s) \neq 0$ when $\operatorname{Re}(s) = 1$, which suffices to prove the Prime Number Theorem.

**Theorem 20.7.** *(Prime Number Theorem)*

$$\lim_{x \to \infty} \frac{\pi(x)}{x/\ln x} = 1.$$

The fact that complex numbers were used to prove a theorem about integers bothered several people who thought that every theorem should have a proof in the context in which the theorem is stated. Eventually, in 1949, Selberg and Erdős discovered "elementary" proofs of the Prime Number Theorem. These proofs are elementary in the sense that they avoid any real or complex analysis, but they are in many ways more technical and harder to understand than the "non-elementary" proofs. We don't give a proof of the Prime Number Theorem in this book. (For proofs, see almost any book on analytic number theory.) Instead, we content ourselves with the following version of Chebyshev's result.

**Theorem 20.8.** *Let $x \geq 3$. Then*

$$\frac{1}{2} \cdot \frac{x}{\ln x} \leq \pi(x) \leq 3 \cdot \frac{x}{\ln x}.$$

*Proof.* The theorem consists of two inequalities. Although they look similar, they tell us different things. The first inequality tells us that there are lots of primes. Where are we going to find them? In the proof of Bertrand's Postulate in the previous section, we found primes in intervals of the form $[x, 2x]$. A slight modification of these techniques finds enough primes for our purposes. The second inequality gives an upper bound on the number of primes up to $x$. Since we know from the proof of Bertrand's Postulate how to trap the primes between $n$ and $2n$, we just need to use the fact that our trap, namely the binomial coefficient $\binom{2n}{n}$, is not very large, which shows that the number of primes we trap is not very large.

Let's start with the upper bound. Let $n \geq 1$ be an integer. There are $\pi(2n) - \pi(n)$ primes $p$ with $n < p \leq 2n$. It follows that

$$n^{\pi(2n) - \pi(n)} \leq \prod_{n < p \leq 2n} p$$

(each factor on the right is greater than $n$ and there are $\pi(2n) - \pi(n)$ factors). By the "Key Observation" in the proof of Bertrand's Postulate (Section 20.1), all the primes between $n$ and $2n$ divide $\binom{2n}{n}$, so

$$\prod_{n < p \leq 2n} p \leq \binom{2n}{n} \leq \frac{4^n}{2} < 4^n,$$

where the middle inequality is Lemma 20.2. We now have

$$n^{\pi(2n) - \pi(n)} < 4^n.$$

Take logarithms of both sides and divide by $\ln n$ to obtain

$$\pi(2n) - \pi(n) \leq \frac{n \ln 4}{\ln n}.$$

This is really close; you can see the $n / \ln n$ on the right. But it's going to take some technical work to get the final answer.

Let $n = 2^j$. Then we have

$$\pi(2^{j+1}) - \pi(2^j) \leq \frac{2^j \ln 4}{j \ln 2} = \frac{2^{j+1}}{j}. \tag{20.2}$$

We claim that

$$\pi(2^k) \leq 4.3 \frac{2^{k-1}}{k - 1}$$

for all integers $k \geq 2$.

We'll prove the claim by induction. It's true for $2 \leq k \leq 16$ by straightforward computation:

$$\pi(2^2) = 2, \quad \pi(2^3) = 4, \quad \pi(2^4) = 6, \quad \dots, \quad \pi(2^{16}) = 6542,$$

and all of these satisfy the inequality. Now suppose that $m \geq 16$ and that the claim is true for $k = m$. Let's prove it for $k = m + 1$. Using the inequality (20.2) with $j = m$, we have

$$\pi(2^{m+1}) = \left(\pi(2^{m+1}) - \pi(2^m)\right) + \pi(2^m)$$
$$\leq \frac{2^{m+1}}{m} + 4.3\frac{2^{m-1}}{m-1}$$
$$= \left(2 + 4.3\frac{m}{2(m-1)}\right)\frac{2^m}{m}$$
$$\leq 4.3\frac{2^m}{m},$$

since a straightforward calculation shows that $4.3m/2(m-1) \leq 2.3$ when $m \geq 16$. Therefore, the claim is true for $k = m+1$. By induction, the claim is true for all $k \geq 2$.

Let $x \geq 4$ be a real number. Choose $k$ such that $4 \leq 2^{k-1} \leq x < 2^k$. Since $\pi(x)$ is nondecreasing and $x/\ln(x)$ is increasing for $x > e$,

$$\pi(x) \leq \pi(2^k) \leq 4.3\frac{2^{k-1}}{k-1} = 4.3\ln(2)\frac{2^{k-1}}{\ln(2^{k-1})}$$
$$\leq 4.3\ln(2)\frac{x}{\ln x} \leq 3\frac{x}{\ln x}.$$

This is the desired upper bound.

Now we prove the lower bound. Let $x \geq 20$. Choose an integer $n$ with $2n \leq x < 2n + 2$. Let

$$\binom{2n}{n} = \prod p^{a_p}$$

be the prime factorization of the binomial coefficient. From Lemma 20.4, $p^{a_p} \leq 2n$ for each prime, which implies that

$$a_p \leq \frac{\ln(2n)}{\ln p}.$$

Taking logarithms of the prime factorization yields

$$\ln\binom{2n}{n} = \sum_{p|\binom{2n}{n}} a_p \ln p \leq \sum_{p|\binom{2n}{n}} \frac{\ln(2n)}{\ln p}\ln p \leq \pi(2n)\ln(2n),$$

since there are at most $\pi(2n)$ primes dividing $\binom{2n}{n}$. This tells us that

$$\ln\binom{2n}{n} \leq \pi(2n)\ln(2n),$$

while Lemma 20.2 implies that

$$n \ln 4 - \ln(2n) \leq \ln \binom{2n}{n}.$$

Putting these two inequalities together and dividing by $\ln(2n)$ yields

$$\frac{n \ln 4}{\ln(2n)} - 1 \leq \pi(2n) \leq \pi(x),$$

which says that

$$\pi(x) \geq \ln 2 \frac{2n}{\ln(2n)} - 1.$$

We're almost done. We just need to change the $2n$ to $x$, get rid of the $-1$, and keep track of what happens to the inequality. We put the necessary work in a lemma.

**Lemma 20.9.** *If $n \geq 11$, then*

$$\frac{2n \ln 2}{\ln(2n)} - 1 \geq \frac{1}{2} \frac{2n + 2}{\ln(2n + 2)}.$$

*Proof.* Since

$$\frac{d}{dt} \frac{t}{\ln t} = \frac{-1 + \ln t}{(\ln t)^2} \leq \frac{\ln t}{(\ln t)^2} = \frac{1}{\ln t} \leq \frac{1}{\ln x}$$

for $t \geq x > 1$,

$$\frac{x + 2}{\ln(x + 2)} - \frac{x}{\ln x} = \int_x^{x+2} \frac{-1 + \ln t}{(\ln t)^2} dt \leq \int_x^{x+2} \frac{1}{\ln x} dt = \frac{2}{\ln x}.$$

The last integral is for a constant function (that is, independent of $t$) over an interval of length 2. Let $x = 2n$ and rearrange the terms to obtain

$$\frac{2n}{\ln(2n)} \geq \frac{2n + 2}{\ln(2n + 2)} - \frac{2}{\ln(2n)}.$$

Multiply by $\ln 2$ and subtract 1 to obtain

$$\ln(2) \frac{2n}{\ln(2n)} - 1 \geq \ln(2) \left( \frac{2n + 2}{\ln(2n + 2)} - \frac{2}{\ln(2n)} \right) - 1$$

$$\geq \frac{1}{2} \frac{2n + 2}{\ln(2n + 2)} + f(2n),$$

where
$$f(x) = (\ln(2) - .5)\frac{x+2}{\ln(x+2)} - \frac{2\ln 2}{\ln x} - 1.$$

Since $f(x)$ is an increasing function for $x > 1$, and $f(22) = .0101\cdots >$ 0, we have $f(2n) > 0$ for $n \geq 11$. This yields the lemma. $\qquad\square$

Returning to the main proof, when $x \geq 22$ we use the lemma and obtain

$$\pi(x) \geq \frac{1}{2}\frac{2n+2}{\ln(2n+2)} \geq \frac{1}{2}\frac{x}{\ln x},$$

where the second inequality follows because $x/\ln x$ is an increasing function for $x > e$.

Don't forget $3 \leq x < 22$. The theorem is not as interesting for small values of $x$ because we can count the primes, but we can still show that it is true. We have

$$\frac{1}{2}\frac{x}{\ln x} \leq \frac{1}{2}\frac{8}{\ln 8} < 1.95 < 2 \leq \pi(x) \text{ for } 3 \leq x \leq 8$$

$$\frac{1}{2}\frac{x}{\ln x} \leq \frac{1}{2}\frac{22}{\ln 22} < 3.56 < 4 \leq \pi(x) \text{ for } 8 \leq x \leq 22.$$

This completes the proof of the lower bound for all $x \geq 3$. $\qquad\square$

---

## 20.3   Chapter Highlights

1.  Bertrand's Postulate: When $x > 1$, there is always a prime between $x$ and $2x$.

2.  Chebyshev's Approximate Prime Number Theorem:

$$\frac{1}{2} \cdot \frac{x}{\ln x} \leq \pi(x) \leq 3 \cdot \frac{x}{\ln x}$$

when $x \geq 3$.

## 20.4   Problems

### 20.4.1   Exercises

1.   Let $n > 1$. Show that $n!$ cannot be a perfect $k$th power with $k \geq 2$ (*Hint:* Use Bertrand's Postulate to get a prime that divides $n!$ only once.)

2.   (a) Show that if $n > 1$ then $1 + \frac{1}{2} + \frac{1}{3} + \frac{1}{4} + \cdots + \frac{1}{n}$ is not an integer. (*Hint:* Let $2^k$ be the highest power of 2 less than or equal to $n$. Show that $1/2^k$ is the only term whose denominator is a multiple of $2^k$, and therefore no other term can cancel this power of 2.)
     (b) Show that if $n > 1$ then $1 + \frac{1}{2} + \frac{1}{3} + \frac{1}{4} + \cdots + \frac{1}{n}$ is not an integer. (*Hint:* If $n \geq 4$, Bertrand's Postulate says there is a prime $p$ with $n/2 < p \leq n$. Show that no other term cancels this $p$ in the denominator.)

3.   Let $n \geq 2$ be an integer. Show that it is impossible to arrange the numbers $1, 2, 3, \ldots, n^2$ in a square array such that the product of the numbers in each column is the same for each column. In other words, there are no "multiplicative magic squares." (*Hint:* Use Bertrand's Postulate to find a prime $p$ that occurs but such that no other multiple of $p$ occurs.)

4.   Use L'Hôpital's Rule to show that

$$\lim_{x \to \infty} \frac{\mathrm{li}(x)}{x/\ln x} = 1.$$

5.   (a) Suppose you know that

$$.921 \frac{x}{\ln x} < \pi(x) < 1.106 \frac{x}{\ln x} \tag{20.3}$$

     for all $x \geq 10^6$. Use this to show that $\pi(2x) - \pi(x) > 0$ for $x \geq 10^6$. This shows that strengthenings of Theorem 20.8 can be used to prove Bertrand's Postulate.
     (b) Show that (20.3) implies that $\pi(2x) - \pi(x) > 10^6$ for all sufficiently large $x$. (Bertrand's Postulate says that there is always at least one prime between $x$ and $2x$. This says that, eventually, there are at least $10^6$ primes between $x$ and $2x$.)

6.   Let $n \geq 2$ be a positive integer and let $p$ be the smallest prime greater than $n$.
     (a) Show that $(p-1)/2 \leq n$.
     (b) Show that if $p \geq 7$ then $p - 1 \mid n!$ (*Hint:* If $p \geq 7$, then $2 \neq (p-1)/2$.)
     (c) (Booker, Granville) Show that $p$ is the smallest prime dividing

$$(n!)^{n!} - 1.$$

     (*Hint:* If $n \geq 5$ then $p \geq 7$, so you can use part (b). Then check the cases $2 \leq n \leq 4$ individually.)
     (*Remark:* This can be regarded as an "improvement" of Euclid's proof that there infinitely many primes. That proof produces new primes at

each step, but it does not produce the primes in order. The present proof shows that if $n$ is a prime then the smallest prime dividing $(n!)^{n!} - 1$ is the next prime.)

## 20.4.2    Projects

1.  The number $n = 30$ has the interesting property (let's call it *prime finding*) that whenever $1 < m \le n$ and $\gcd(m, n) = 1$, then $m$ is prime (these $m$ are 7, 11, 13, 17, 19, 23, 29).
    (a) Show that $n = 18$ also has this prime finding property.
    (b) In this part and in (c) through (f), assume that $n$ has the prime finding property. Let $p$ be a prime with $p^2 \le n$. Show that $p \mid n$.
    (c) Let $p_1 = 2, p_2 = 3, p_3 = 5, \ldots$ be the primes. Let $p_r \le \sqrt{n} < p_{r+1}$. Show that

    $$p_1 p_2 \cdots p_r < p_{r+1}^2.$$

    (d) Use Bertrand's Postulate to show that $p_{r+1} < 2p_r$ and $p_{r+1} < 4p_{r-1}$.
    (e) Show that $p_1 p_2 \cdots p_{r-2} < 8$.
    (f) Show that $r \le 4$ and conclude that $n < 121$.
    (g) Find all $n > 1$ with the prime finding property (there are nine of them; 2 is the smallest and 30 is the largest).
2.  Let $p_n$ be the $n$th prime, so $\pi(p_n) = n$. In Section 20.2, we proved that there are constants $A$ and $B$ such that $Ax/\ln x \le \pi(x) \le Bx/\ln x$.
    (a) Show that $Ap_n/\ln p_n \le n \le Bp_n/\ln p_n$.
    (b) Using the fact that $\ln(\ln x)/\ln x \to 0$ as $x \to \infty$, show that

    $$0.9 \le \frac{\ln n}{\ln p_n} \le 1.1$$

    for all sufficiently large $n$. (*Note:* 0.9 could be replaced by any number less than 1, and 1.1 could be replaced by any number greater than 1.)
    (c) Show that

    $$\frac{1}{1.1B} n \ln n \le p_n \le \frac{1}{.9A} n \ln n$$

    for all sufficiently large $n$.
    *Remark:* Barkley Rosser proved in 1937 that $p_n > n \ln n$ for all $n$. It follows from the prime number theorem, essentially by the calculations in (a), (b), and (c), that $p_n/(n \ln n) \to 1$ as $n \to \infty$.
    (d) Show that

    $$\sum_{n=1}^{\infty} \frac{1}{n \ln n}$$

    diverges.
    (e) Use (c) and (d) to show that

    $$\sum_{n=1}^{\infty} \frac{1}{p_n}$$

    diverges.

## 20.4.3   Computer Explorations

1.  A *twin prime pair* is a pair $p, p+2$ such that both $p$ and $p+2$ are prime.
    Let $\pi_2(x)$ be number of primes $p \le x$ such that $p + 2$ is also prime. It is
    conjectured that
    $$\pi_2(x) \approx 1.32 \frac{x}{(\ln x)^2}. \qquad (20.4)$$

    (a) Count pairs for $x = 10^4$, $10^5$, $10^6$, $10^7$ and compute the ratio
    $\pi_2(x)/(x/(\ln x)^2)$. Notice that the ratio is larger than 1.32 but is decreas-
    ing toward that value. (You might want to use one of the pseudoprimality
    tests to check large integers for primality; they are accurate enough to
    give reasonably good counts.)
    (b) A generalization of Bertrand's Postulate could say that for $x \ge 7$ there
    is always a twin prime pair between $x$ and $2x$. Check this for $x \le 10^5$.
    (*Hint:* You don't need to check every $x$. Use the trick at the end of the
    proof of Bertrand's Postulate.)
    (c) Suppose that
    $$\frac{x}{(\ln x)^2} \le \pi_2(x) \le \frac{1.9x}{(\ln x)^2}$$

    for all sufficiently large $x$ (for example, all $x > 10^6$). (*Note:* This has
    not been proved, but part (a) indicates that it is probably true.) Show
    that $\pi_2(2x) > \pi_2(x)$ for all sufficiently large $x$ (that is, for something like
    $x > 10^{12}$).
    The computational evidence is very strong in favor of the conjecture that
    there are infinitely many twin prime pairs, but this conjecture is still
    unproved.

2.  The estimates in Theorem 20.8 are rather imprecise. Compute the ratio

    $$\frac{\pi(x)}{x/\ln x}$$

    for $x = 10^k$ for $1 \le k \le 7$ to see how close this ratio if to 1 (the Prime
    Number Theorem says that the ratio converges to 1 as $x \to \infty$).

# Chapter 21

# Epilogue: Fermat's Last Theorem

## 21.1  Introduction

In the late 1630s, Pierre de Fermat was reading Diophantus's *Arithmetica*, which had recently received a new translation from Greek into Latin by Bachet. The *Arithmetica* consists of hundreds of algebraic problems covering linear, quadratic, and higher degree equations and uses algebraic techniques to find their solutions. Fermat must have been intrigued by Problem 8 from Book II, which asks how "to divide a square number into two squares." Diophantus starts with a square, for example 16, and tries to write it in the form $x^2 + y^2$ with nonzero rational numbers $x$ and $y$. To do so, he starts with an "obvious" solution, namely $(x, y) = (4, 0)$, and uses it to find a nontrivial solution. Take, for example, a line with slope $-2$ through $(4, 0)$, described parametrically by

$$x = 4 - t, \quad y = 2t$$

for a parameter $t$. Substitute this into $x^2 + y^2 = 16$:

$$16 = (4 - t)^2 + (2t)^2 = 16 - 8t + 5t^2.$$

This says that $8t = 5t^2$. The solution $t = 0$ corresponds to the point $(4, 0)$ that we already have. The solution $t = 8/5$ yields $(x, y) = (12/5,\ 16/5)$:

$$16 = \left(\frac{16}{5}\right)^2 + \left(\frac{12}{5}\right)^2.$$

This method that he used to find this solution is essentially the same as the one described in Section 5.4. Other values of the slope yield other solutions to the problem, and it is clear that any square can be decomposed into a sum of two squares in this way, since there is always an "obvious" solution to use as a starting point.

Fermat wrote the following in the margin of his copy of the *Arithmetica*:

> On the other hand, it is impossible to separate a cube into two cubes, or a biquadrate into two biquadrates, or generally any power except a square into two powers with the same exponent. I have discovered a truly marvelous proof of this, which however the margin is not large enough to contain.

Fermat is asserting that the equation

$$x^n + y^n = z^n \tag{21.1}$$

has no solutions if $x, y$, and $z$ are nonzero integers and $n > 2$.

Fermat never provided a proof of (21.1), except for the case $n = 4$. Near the end of his life, he made a list of his major accomplishments in number theory and mentioned only the cases $n = 3$ and $n = 4$ of (21.1).

After Fermat's death, many of his notes and letters were published by his son Samuel. Eventually all of Fermat's assertions and claims were resolved except (21.1), so this became known as Fermat's Last Theorem, even though it was neither the last thing he did nor was it a theorem. Throughout the 18th and 19th centuries, many mathematicians tried to prove Fermat's Last Theorem. Although none succeeded, progress was made. For example, Euler gave a proof for $n = 3$ and Dirichlet and Legendre showed in 1825 that (21.1) has no solution for $n = 5$.

There is a slight simplification of the problem. Suppose that you want to prove that the sum of two tenth powers cannot be a tenth power. Since a tenth power is automatically a fifth power, it suffices to prove that the sum of two fifth powers cannot be a fifth power. Because every integer $n \geq 3$ is a multiple either of 4 or of an odd prime, it suffices to prove Fermat's Last Theorem for $n = 4$ and for odd primes. Since Fermat already proved it for $n = 4$, only odd prime exponents need to be considered.

In the mid-nineteenth century, the great German mathematician Ernst Kummer began the development of modern algebraic number theory.

Part of his work gave criteria that guarantee the validity of Fermat's Last Theorem for a given exponent $n$. (If Kummer's criteria are not satisfied, it doesn't mean that Fermat's Last Theorem is false for that exponent, only that his method does not apply in that situation.) Using these ideas, Kummer proved Fermat's Last Theorem for all exponents less than 100. With extensions of these techniques and the use of computers, the list of exponents was expanded. By 1993 it was known that Fermat's Last Theorem is true for exponents up to 4000000.

In the 1980s, techniques from algebraic geometry (which studies, for example, curves described by polynomial equations) started playing a major role in the work on Fermat's Last Theorem. In 1983, Gerd Faltings proved the Mordell Conjecture, which had been one of the main unsolved problems in number theory. A small piece of this conjecture is relevant to Fermat's Last Theorem, namely

$$\text{for each } n \geq 4, x^n + y^n = z^n \text{ has only a } \textit{finite} \text{ number}$$
$$\text{of solutions with } \gcd(x, y, z) = 1$$

(if $\gcd(x, y, z) = d > 1$, then $x$, $y$, and $z$ can be divided by $d$ to get a "primitive" solution). Of course, this is not good enough since we want the *exact* number of solutions. Note that there are always a few solutions: if $n$ is odd there are primitive solutions $\pm(1, 0, 1)$, $\pm(0, 1, 1)$, and $\pm(-1, 1, 0)$. If $n$ is even, there are $(\pm 1, 0, \pm 1)$ and $(0, \pm 1, \pm 1)$. The problem is to show that these are the only ones.

Faltings's work made clear that the ideas of algebraic geometry have implications for Fermat's Last Theorem. Then in 1985, Gerhard Frey suggested that the Modularity Conjecture (see below) in the theory of elliptic curves could be used in proving Fermat's Last Theorem. In 1986, Ken Ribet showed that Frey's idea is correct by proving that the Modularity Conjecture implies Fermat's Last Theorem. Finally, in 1994, Andrew Wiles, with the assistance of Richard Taylor, proved the Modularity Conjecture and therefore completed the proof of Fermat's Last Theorem.

## 21.2  Elliptic Curves

We now give a brief discussion of elliptic curves and show how they are used in the proof of Fermat's Last Theorem.

An **elliptic curve** is a curve of the form

$$y^2 = x^3 + Ax^2 + Bx + C \quad \text{where } A, B, C \text{ are rational numbers.}$$

(There is a technical requirement that the cubic polynomial does not have multiple roots, but this will not concern us here.) For example, $y^2 = x^3 + 36$ gives an elliptic curve. Two of the points on it are $(4, 10)$ and $(0, 6)$. A method going back to Diophantus uses these two points to produce a third point. Draw the line through $(4, 10)$ and $(0, 6)$. It has equation $y = x + 6$. We know that this line intersects the curve in our two points. Let's see if it intersects anywhere else. To find the intersections, substitute $x + 6$ for $y$ to get

$$(x + 6)^2 = x^3 + 36.$$

This rearranges to

$$x^3 - x^2 - 12x = 0.$$

Therefore, we want to solve $x^3 - x^2 - 12x = 0$. Since we know that $x = 0$ and $x = 4$ correspond to intersections, these values of $x$ yield solutions. So we have a cubic equation where we know two of the roots, and we can factor:

$$x^3 - x^2 - 12x = x(x - 4)(x + 3).$$

We now have the third root, which is $x = -3$. This gives $y = x + 6 = 3$, so we obtain the new point $(-3, 3)$ on the curve. For technical reasons, we reflect this point across the $x$-axis and obtain the point $(-3, -3)$, and regard this as the point produced by $(4, 10)$ and $(0, 6)$. This process can be continued. Take the points $(4, 10)$ and $(-3, -3)$. The line through them is

$$y = \frac{13}{7}x + \frac{18}{7}.$$

A calculation shows that this line intersects the curve $y^2 = x^3 + 36$ in the points

$$(4, 10), \quad (-3, -3), \quad \left(\frac{120}{49}, \frac{2442}{343}\right).$$

After reflecting this third point across the $x$-axis, we say that the point $(120/49, -2442/343)$ is produced by $(4, 10)$ and $(-3, -3)$. (This process is usually called the **addition law** for points on elliptic curves.)

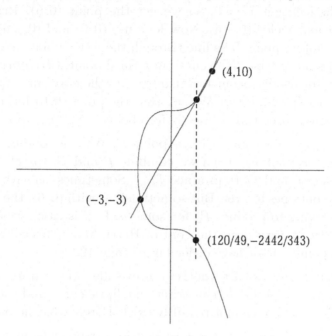

**FIGURE 21.1: The addition law on $y^2 = x^3 + 36$.**

We can also use $(4, 10)$ and $(4, 10)$ (these are two points that happen to be equal). For the line through these points, we use the tangent line to the curve at $(4, 10)$. By implicit differentiation, we have $2yy' = 3x^2$. Substituting $(x, y) = (4, 10)$ yields $20y' = 48$, so $y' = 12/5$, which is the slope of the tangent line. The equation of the tangent line is

$$y = \frac{12}{5}x + \frac{2}{5}.$$

A calculation shows that it intersects the curve $y^2 = x^3 + 36$ at $(4, 10)$ and at

$$\left( \frac{-56}{25}, \frac{-622}{125} \right).$$

Reflecting this point across the $x$-axis gives the point produced by $(4, 10)$ and $(4, 10)$. We could now use $(4, 10)$ and this new point to produce another point, and then use $(4, 10)$ and this point to produce another point, etc. It can be shown that this process can go on forever.

Let's use another point. Start with $(0, 6)$ and find the point produced by $(0, 6)$ and $(0, 6)$. The tangent line is horizontal: $y = 6$. When we intersect it with the curve, we obtain the cubic equation $36 = x^3 + 36$. This has the solution $x = 0$, so we get the point $(0, 6)$. Reflecting across the $x$-axis yields $(0, -6)$. Now let's use $(0, 6)$ and $(0, -6)$ to try to produce another point. The line through these two points is vertical: $x = 0$. It does not intersect the curve in a third point. (*Technical point:* If we work in projective geometry, it intersects the curve in a "point at infinity.") Therefore, we stop. We say that the point $(0, 6)$ has order 3, because we used $(0, 6)$ three times before being forced to stop.

In general, suppose we start with a point $P$. We can combine $P$ and $P$ to produce a point $P_2$. Then we combine $P$ and $P_2$ to get $P_3$, and continue, using $P$ and $P_k$ to produce $P_{k+1}$. Sometimes, as with $(4, 10)$, this process continues forever. But sometimes, as with $(0, 6)$, the process stops while trying to produce $P_n$ for some $n$. In this case, we say that $P$ has order $n$. A very deep theorem of Barry Mazur from the 1970s says that a point cannot have order larger than 12.

Let's return to Fermat's Last Theorem. Frey's idea was to use a solution to Fermat's Last Theorem to construct an elliptic curve and then show that this elliptic curve cannot possibly exist. Here's what he did.

As pointed out above, it suffices to consider exponents in Fermat's Last Theorem that are odd primes $q$. In fact, we can assume that $q > 4000000$, since it was already proved for smaller exponents. Assume that Fermat's Last Theorem is *false* so there are non-zero integers $a, b$, and $c$ with $a^q + b^q = c^q$. Now, use these three integers to construct the elliptic curve

$$E_{Frey} : y^2 = x(x - a^q)(x + b^q). \tag{21.2}$$

This curve has properties that contradict what, at the time, was a conjectured property of elliptic curves. So, the $a, b$, and $c$ used to construct this curve should not exist and Fermat's Last Theorem must be true.

The conjectured property that Frey relied on is known as the Modularity Conjecture (also known as the Taniyama Conjecture, the Shimura–Taniyama Conjecture, and the Taniyama–Weil Conjecture; anyway, it's now a theorem), which asserts that all elliptic curves have a property know as modularity. We'll discuss what modularity means, but first we'll explain what's so special about $E_{Frey}$.

Every cubic polynomial has three roots (double roots are counted twice). Call them $r_1, r_2$, and $r_3$. The discriminant of a cubic whose leading coefficient is 1 is defined to be

$$\Delta = (r_3 - r_1)^2 (r_3 - r_2)^2 (r_2 - r_1)^2.$$

As with quadratic polynomials, the discriminant tells whether there are complex roots of the polynomial and whether there are repeated roots. In the case of $E_{Frey}$, the discriminant has a special form. The roots of the right-hand side of (21.2) are

$$r_1 = 0, \quad r_2 = a^q \text{ and } r_3 = -b^q.$$

Therefore,

$$\Delta = (-b^q - 0)^2 (-b^q - a^q)^2 (a^q - 0)^2 = b^{2q} c^{2q} a^{2q}$$

where we used the fact that $a^q + b^q = c^q$ to change $-b^q - a^q$ to $-c^q$. This tells us that

$$\Delta = (abc)^{2q}.$$

It's this special property of the discriminant being a perfect power that makes $E_{Frey}$ so special. Experience shows that if the discriminant is a perfect $q$th power, then the curve should have properties similar to a curve that has a point of order $q$. But Mazur showed that there are no curves that have points of order larger than 12, so $E_{Frey}$ should act like a curve that does not exist! Therefore, we suspect that $E_{Frey}$ does not exist. But we just wrote down the equation for $E_{Frey}$. How can this be? Recall that we can write the equation of $E_{Frey}$ only when there is a counterexample to Fermat's Last Theorem. The conclusion is that there should be no counterexamples, and therefore Fermat's Last Theorem is true.

The problem is how to make this argument precise. Frey suggested that the Modularity Conjecture could be used, and Ribet proved that this is the case.

## 21.3   Modularity

We now have to explain what modularity means. The actual definition is beyond the scope of this book, so instead we give some examples.

We begin by looking at the elliptic curve

$$y^2 = x^3 + x.$$

Let's reduce this equation mod various primes. We start with 3:

$$y^2 \equiv x^3 + x \pmod 3.$$

Since we're working mod 3, there are only three possibilities for the values of $x$. Here's a chart that shows what $y$ can be for each value of $x$.

| $x$ | $x^3 + x$ | $x^3 + x \pmod 3$ | $y$ |
|---|---|---|---|
| 0 | 0 | 0 | 0 |
| 1 | 2 | 2 | - |
| 2 | 10 | 1 | 1, 2 |

The "-" indicates that $y^2 \equiv 2 \pmod 3$ has no solutions. We see that there are three points on the curve mod 3:

$$(0,0), (2,1), (2,2).$$

Now let's look at look at $E : y^2 \equiv x^3 + x \pmod 5$. The following shows that there are three points on the curve mod 5: $(0,0), (2,0), (3,0)$.

| $x$ | $x^3 + x$ | $x^3 + x \pmod 5$ | $y$ |
|---|---|---|---|
| 0 | 0 | 0 | 0 |
| 1 | 2 | 2 | - |
| 2 | 10 | 0 | 0 |
| 3 | 30 | 0 | 0 |
| 4 | 68 | 3 | - |

If $p$ is any prime, we can define $N_p$ to be the number of solutions to $y^2 \equiv x^3 + x \pmod p$ for any $p$. We have calculated that $N_3 = 3$ and $N_5 = 3$. Here's a chart that shows the value of $N_p$ for all odd primes up to 31.

| $p$ | 3 | 5 | 7 | 11 | 13 | 17 | 19 | 23 | 29 | 31 |
|-----|---|---|---|----|----|----|----|----|----|----|
| $N_p$ | 3 | 3 | 7 | 11 | 19 | 15 | 19 | 23 | 19 | 31 |

These values might look random, but there is a formula for $N_p$. Let's look at the table more closely.

$$N_p = p \text{ if } p = 3, 7, 11, 19, 23, 31.$$

So we have a formula for half the primes: $N_p = p$ if $p \equiv 3 \pmod 4$. If $p \equiv 1 \pmod 4$, the situation is more complicated. Observe that

$$5 = 1^2 + 2^2 \text{ and } N_5 = 5 - 2 \cdot 1$$
$$13 = 3^2 + 2^2 \text{ and } N_{13} = 13 + 2 \cdot 3$$
$$17 = 1^2 + 4^2 \text{ and } N_{17} = 17 - 2 \cdot 1$$
$$29 = 5^2 + 2^2 \text{ and } N_{29} = 29 - 2 \cdot 5.$$

There is a pattern: If $p \equiv 1 \pmod 4$, write $p = a^2 + b^2$ with $b$ even and $a \equiv 1 \pmod 4$. Then

$$N_p = p - 2 \cdot a.$$

For example, $13 = (-3)^2 + 2^2$ and $N_{13} = 13 - 2(-3) = 19$. We now have a formula for $N_p$ for all odd primes. The proof that this formula is correct is due to Gauss and is beyond the scope of this book, so we omit it.

Let's consider another curve: $y^2 = x^3 - 4x^2 + 16$. Here is a table of values of $N_p$.

| $p$ | 3 | 5 | 7 | 11 | 13 | 17 | 19 | 23 | 29 | 31 |
|-----|---|---|---|----|----|----|----|----|----|----|
| $N_p$ | 4 | 4 | 9 | 12 | 9 | 19 | 19 | 24 | 29 | 24 |

The pattern here is much harder to figure out: Consider the infinite product

$$x \prod_{n=1}^{\infty} (1 - x^n)^2 (1 - x^{11n})^2 = x - 2x^2 - x^3 + 2x^4 + x^5 + 2x^6 - 2x^7 + \cdots.$$

Let $p \neq 11$ be an odd prime (there is a technicality for 11) and let $a_p$ be the coefficient of $x^p$. Then $N_p = p - a_p$. For example, the coefficient of $x^7$ is $-2$ and $N_7 = 7 - (-2) = 9$.

The numbers $N_p$ tell us information about the curve. For example, $y^2 = x^3 - 4x^2 + 16$ has a point of order 5, namely $(0, 4)$, and this is seen in the fact that $N_p \equiv -1 \pmod 5$ for all $p \neq 11$ (if we included the "point at infinity" in our count, then we would have that the number of points mod $p$ is a multiple of 5). When we said that $E_{Frey}$ should act like a curve with a point of order $q$, a consequence is that we should have $N_p \equiv -1 \pmod q$ for most primes $p$. But how do we get hold of the numbers $N_p$?

We say that an elliptic curve is **modular** if the numbers $a_p$ satisfy a pattern (this can be made much more precise using the theory of modular forms, which is well beyond the scope of this book). This pattern might be fairly simple, as in the case of $y^2 = x^3 + x$, or it might be complicated, as in the case of $y^2 = x^3 - 4x^2 + 16$. But there is a pattern. This allows us to get information about the numbers $N_p$. What Ribet did was to show, under the assumption that $E_{Frey}$ is modular, that the numbers $N_p$ behave in a way that is impossible. Therefore, if it can be shown that $E_{Frey}$ is modular, then it cannot exist, which proves Fermat's Last Theorem.

Finally, in 1994, Andrew Wiles, with the assistance of Richard Taylor, proved that all elliptic curves of a certain type are modular. Since $E_{Frey}$ (if it exists) is in this set, the proof of Fermat's Last Theorem is complete. In 2000, Christophe Breuil, Brian Conrad, Fred Diamond, and Richard Taylor extended the techniques used by Taylor and Wiles and showed that all elliptic curves are modular.

# Appendix A

## Supplementary Topics

This appendix contains some topics that are used in a few places in the main text. The sections of the appendix may be used for reference when they are needed. However, the topics covered are very basic ideas of mathematics, so the reader who is unfamiliar with them will profit from learning them regardless of their relations to the rest of the book.

## A.1   What Is a Proof?

If you are studying chemistry and read that the boiling point of salt water changes as the concentration of salt changes, you might try an experiment to see if this is really true. How do you proceed? You use a thermometer to measure the water temperature and you calculate the concentration of the salt in the water. After performing this experiment several times, you conclude that the statement is, in fact, correct.

If you are studying mathematics and read that the sum of two even integers is always an even integer, you might also try an experiment to see if this is really true. How do you proceed in this case? You can write down two lists of even numbers, take a number from each list, and add them up. If you do this correctly, can you conclude that the sum of two even numbers is always even? If you suspect that this does not completely justify the statement, you're on to something. This is not how mathematics works. Although mathematicians can do experiments (often with computers) to try to decide if something seems to be true, they do not use experiments to conclude that things are true. Instead, mathematicians *prove* things. What is a proof? Why do we prove things? How is a proof different from experimental verification? We'll answer these questions in the next few paragraphs.

Simply put, we prove things to convince others (and ourselves) that a statement we're asserting to be true really is true. We start off with an initial assumption or hypothesis and use a chain of logical deductions to arrive at what we want to prove. The validity of our proof rests on our correct use of logic, not on the consistency of experimental data. Let's see how this works as we prove that the sum of two even integers is an even integer.

To begin, coming up with several examples

$$4 + 12 = 16, \quad 36 + 20 = 56, \quad 128 + 416 = 544$$

does not constitute a proof because it's impossible to write down the sum of every pair of even integers. Even if you calculated the sum of 1000 different pairs of even integers, it's possible that the next pair you'd try would contradict what you believe to be true. Instead, you need to come up with some general method that handles all possibilities. In order to do this, you need to begin with a clear and precise definition of even integers. (You can't prove something about even integers until you have defined them.) This is the motivation behind the following definition.

**Definition A.1.** *An integer $n$ is an* **even** *integer if $n$ is a multiple of 2. In other words, there is an integer $k$ with $n = 2k$.*

For example, 26 is even because $26 = 2 \cdot 13$, and 100 is even because $100 = 2 \cdot 50$.

Now that we've defined even, let's take another look at what we want to prove, namely that the sum of two even integers is also an even integer. When trying to prove something, it's essential to recognize what your assumptions are and be absolutely certain what it is you want to prove. In this case, we're assuming that we have two even integers and we want to prove that their sum is even. Sometimes it helps to link our assumption with what we want to prove in an "if, then" sentence, called a conditional statement. Here's what a conditional statement looks like in this example.

**If** <u>you have two even integers</u>, **then** <u>their sum is even</u>.

The underlined clause that follows the "if" is called the **hypothesis**, and the underlined clause that follows the "then" is called the **conclusion**. The hypothesis is what you assume; the conclusion is what you want to prove.

Now, let's begin the proof. We start off with our even integers: we'll call them $m$ and $n$. (Giving the numbers names such as $m$ and $n$ makes it much easier to work with them. Don't skip this step.) We want to prove that $m + n$ is also even. The best way to begin is to make use of the hypotheses. Since both $m$ and $n$ are even, we'll write down the only thing we know about even integers, namely their definition. Since $m$ is even, we know that $m = 2k_1$ for some integer $k_1$, and since $n$ is even, $n = 2k_2$ for some integer $k_2$. What do we do now? Since we want to prove something about the sum of $m$ and $n$, we'll add $m$ to $n$ and see what happens. We see that $m + n = 2k_1 + 2k_2$. Our goal is to show that $m + n$ is even, which means (by the definition) that we want to express $m + n$ as a multiple of 2. If fact, this is what we've done:

$$m + n = 2k_1 + 2k_2 = 2(k_1 + k_2).$$

This shows that $m + n$ is twice the integer $k_1 + k_2$, which tells us that $m + n$ is even.

This proof is in many ways a model for how you should think of any proof you try to do. First, make sure you have an unambiguous definition of all your terms. Next, write what you want to prove as a conditional statement, making sure you understand the hypothesis and conclusion. Then, translate the hypothesis into a mathematical statement. Finally, develop a strategy that will allow you to go from the hypothesis to the conclusion. Although many of the proofs in this book are more complex than the present example, the method that we just outlined will always apply.

When you finish reading a proof (yes, you should always try to read the proofs; that's how to learn how concepts fit together and how to do proofs), it is very good practice to look back and see where the hypotheses were used. In our example, the hypothesis was that $m$ and $n$ are even. How was this used? It allowed us to write $m = 2k_1$ and $n = 2k_2$, which is what got us started. If a proof is not using one of the hypotheses, it is likely that something is wrong (or that the hypothesis is not needed).

No discussion of how to prove things is complete without saying how to disprove things. Suppose you see the statement that all integers are less than 100. That's absurd, you say. For example, 123 is an integer and it is not less than 100. You have disproved the statement. Whenever you want to disprove a statement, all you need to do is find a counterexample. One counterexample suffices. This is in complete contrast to most proofs, where an example is not the same as a proof.[1]

For practice, let's try to prove or disprove the following statements. For each, you should first try a few examples ("experiments") to get a feel for what is being said. If you find a counterexample, you are done, since you have disproved the statement. If your examples seem to indicate that the statement is true, try to give a proof. We'll give answers after the statements, but you should try each before looking at the solutions.

**Prove or Disprove:**

1.  The sum of two odd integers is always even. (*Note:* First, you need a definition of *odd*; we say that $n$ is odd if there is an integer $k$ such that $n = 2k + 1$.)

2.  The product of two even integers is always a multiple of 4.

3.  Every multiple of 3 is odd.

4.  Every multiple of 6 is even.

5.  Every odd number larger than 1 is a prime number.

**Solutions:**

1.  Some examples show that this seems to be true: $5 + 9 = 14$, $21 + 111 = 132$, $1 + 3 = 4$. Now let's prove it. First, write the statement as a conditional statement: If $m$ and $n$ are odd integers, then $m + n$ is even. The hypothesis is that $m$ and $n$ are odd. To write this mathematically, use the definition: $m = 2k_1 + 1$ for some integer $k_1$, and $n = 2k_2 + 1$ for some integer $k_2$. Our goal is to say something about the sum $m + n$,

---

[1]An exception: If the statement you want to prove is an existence statement, then an example might suffice. For example, if the statement says that there exists a cube that is two more than a square, then the example $3^3 = 5^2 + 2$ shows that the statement is true.

so we compute

$$m + n = (2k_1 + 1) + (2k_2 + 1) = 2k_1 + 2k_2 + 2.$$

What do we do now? Look at the conclusion. It says that $m+n$ is even, which means that we need to be able to express $m+n$ as a multiple of 2. Our expression for $m + n$ lets us do this:

$$m + n = 2k_1 + 2k_2 + 2 = 2(k_1 + k_2 + 1).$$

Therefore, $m + n$ is even, which is what we wanted to prove.

2.  Try some examples: $2 \cdot 6 = 12$, $8 \cdot 6 = 48$, $10 \cdot 10 = 100$. The products are multiples of 4, so the statement seems to be true. Let's prove it. The hypothesis is that we have two even integers. Let's call them $m$ and $n$. Since $m$ and $n$ are even, we can write $m = 2k_1$ and $n = 2k_2$ for some integers $k_1$ and $k_2$. The conclusion says something about the product $mn$, so we write

    $$mn = (2k_1)(2k_2) = 4(k_1 k_2).$$

    This says that $mn$ is a multiple of 4, so we have completed the proof.

3.  Let's try some examples: $3 \cdot 1 = 3$, $3 \cdot 2 = 6$, $3 \cdot 3 = 9$. Two out of three; not bad. In most sporting events, winning two out of three is awesome. But in math, one counterexample to a statement is enough to disprove the statement. Since $2 \cdot 3 = 6$ shows that there is a multiple of 3 that is even, the statement that all multiples of 3 are odd is false. So we have disproved the statement.

4.  Try some examples: $6 \cdot 1 = 6$, $6 \cdot 2 = 12$, $6 \cdot 3 = 18$. All of these are even, so the statement looks correct. Let's try to prove it. A conditional form of the statement is "If $n$ is an integer, then $6n$ is even." The hypothesis is that $n$ is an integer. There's not much that we can say about $n$, so let's look at what we're trying to do. The conclusion says something about $6n$, so let's look at this number. We're trying to prove that $6n$ is even, which means that we need to express it as a multiple of 2. In fact, this is easy:

    $$6n = 2(3n).$$

    Therefore, $6n$ is a multiple of 2, so $6n$ is even. This proves the statement.

5.  Let's look at some odd numbers: 3, 5, 7, 9, 11, 13. The numbers 3, 5, 11, and 13 are primes, but 9 is 3·3, so it is not prime. Since the odd number 9 is not prime, we have a counterexample, which means the statement is false.

Throughout this book you will see proofs and wonder how in the world anyone could have thought of them. Don't be disheartened. They might represent hours, or years, of work by many people. We don't see their mistakes and false starts — we see only the final successes. Andrew Wiles, who proved Fermat's Last Theorem (see Chapter 1 and Chapter 21), explained it as follows:

"Perhaps I could best describe my experience of doing mathematics in terms of entering a dark mansion. One goes into the first room, and it's dark, completely dark. One stumbles around bumping into the furniture, and gradually, you learn where each piece of furniture is, and finally, after six months or so, you find the light switch. You turn it on, and suddenly, it's all illuminated. You can see exactly where you were." *(PBS NOVA Broadcast, October 28, 1997)*

Although none of the proofs you will be asked to do in this book should take six months, don't be afraid of stumbling around and making mistakes. Learning what doesn't work is sometimes as important as learning what does work.

We'll wrap up this introductory section with a brief discussion of some of the terminology concerning proofs you'll be seeing throughout this book. A **proposition** is a statement that we'll be able to prove. A **theorem** is an extremely important proposition, and is usually a highlight of the topic under consideration. If something is called a theorem, you should make a special effort to remember its statement and to understand what it says. A **lemma** is a statement that is used to prove a proposition or theorem. Often, a lemma is singled out because it is useful and interesting in its own right, but is not considered to be as important as a proposition or a theorem. (Admittedly, this can be somewhat subjective. One person's lemma can, at times, be someone else's proposition.) A **corollary** is a result that is an easy consequence of a proposition or theorem.

A statement that says "$A$ is true if and only if $B$ is true" is a combination of two statements: "If $A$ is true then $B$ is true" and "If $B$ is true then $A$ is true." To prove such a statement, you assume $A$ is true and deduce that $B$ is true. Then you assume $B$ is true and deduce that $A$ is true.

Often, "$A$ if and only if $B$" is written as "$A \Leftrightarrow B$." Similarly, "$A \Rightarrow B$" means that "$A$ implies $B$."

Sometimes you will see a statement saying that $A$, $B$, and $C$ are equivalent. This is the same as saying that $A$ is true if and only if $B$ is true, $B$ is true if and only if $C$ is true, and $A$ is true if and only if $C$ is true. However, it is not necessary to prove all six implications. Instead, prove that if $A$ is true then $B$ is true, then prove that if $B$ is true then $C$ is true, and finally prove that if $C$ is true then $A$ is true. This suffices to prove all of the necessary implications and is much more efficient. For example, since $A$ implies $B$ and $B$ implies $C$, we automatically conclude that $A$ implies $C$ without proving it directly.

## A.1.1   Proof by Contradiction

Sometimes, the easiest way to prove a statement is to suppose the statement is false and deduce a false consequence. This is known as *proof by contradiction*. Classically, it was called *reductio ad absurdum* (Latin for "reduction to absurdity").

In the following, we give three examples of proof by contradiction. In each case, proof by contradiction arises naturally because the conclusion is a negative statement ("there is no such number" and "the number is not rational"). The proof proceeds by assuming the positive statement ("there is such a number" or "the number is rational") and showing that this leads to a contradiction, so the negative statement must be true.

Recall that a rational number is a number that can be expressed as the ratio of two integers; for example, $2/3$, $-15/2$, $71 = 71/1$, and $0 = 0/1$ are rational. An irrational number is a number that is not rational. We show in Chapter 5 that $\sqrt{2}$ and $\sqrt{3}$ are examples of irrational numbers.

**I.** *Prove that there is no largest integer.*

*Proof.* If someone told you that there exists a largest integer, you would probably say something like this: "That's ridiculous. Adding one to any integer gives you a larger integer. So there can't be a largest one."

This reasoning is essentially the proof of the statement: Suppose that the statement to be proved is false. Then there exists a largest integer $n$. Let $m = n + 1$. Then $m$ is an integer and $m > n$, which contradicts the assumption that $n$ was the largest integer. So, no largest integer can exist. This completes the proof.                                           □

**II.** *Prove that there is no smallest positive rational number. (Recall that "positive" means greater than 0, and does not include 0.)*

*Proof.* Let's suppose the statement is false. Then there is a smallest positive rational number. Call it $r$. Since $r$ is rational, we can write $r = a/b$, where $a$ and $b$ are integers. Then $r/2 = a/(2b)$ is positive, rational, and smaller than $r$, which contradicts the assumption that $r$ is the smallest positive rational number. Therefore, the assumption that there is a smallest positive rational number has led to a contradiction. The only possibility that remains is that there is no smallest positive rational number.                                                                 □

**III.** *Prove that a rational number plus an irrational number is irrational.*

*Proof.* Let's translate the statement into symbols: It says that if $x$ is rational and $y$ is irrational, then $z = x + y$ is irrational. We need to show that $z$ is irrational, which means that $z$ cannot be written in the form $z = c/d$ with integers $c$ and $d$. Because we're saying that something is *not* rational, it's very natural to use a proof by contradiction. So let's assume that $z$ is rational, which means that $z = c/d$ for some integers $c$ and $d$. Moreover, one of the hypotheses is that $x$ is rational, so $x$ can be written in the form $x = a/b$, where $a$ and $b$ are integers. Therefore, $x + y = z$ tells us that $y = z - x$, which we can write as

$$y = \frac{c}{d} - \frac{a}{b} = \frac{bc - ad}{bd}.$$

But $bc - ad$ and $bc$ are integers, so $y$ can be expressed as the ratio of two integers. This contradicts the assumption that $y$ is irrational. Therefore, the assumption that $z$ is rational led to the false consequence that $y$ is rational. Therefore, we conclude that $z$ must be irrational.          □

A consequence of what we just proved is that $3 - \sqrt{2}$ is irrational. How do we show this? As mentioned earlier, we show in Chapter 5 that $\sqrt{2}$ is irrational. Suppose $3 - \sqrt{2}$ is rational. Then $3 = (3 - \sqrt{2}) + \sqrt{2}$ is a rational plus an irrational, which is irrational by what we just proved. But 3 is rational, so we have a contradiction. Therefore, our assumption that $3 - \sqrt{2}$ is rational must be false, so $3 - \sqrt{2}$ is irrational.

For the record, we now note that an irrational plus an irrational can be either rational or irrational. For example, $\sqrt{2} + \sqrt{2} = 2\sqrt{2}$ is an example where the sum of two irrational numbers is irrational. On the other hand, $\sqrt{2} + (3 - \sqrt{2}) = 3$ is an example where the sum of two irrationals is rational.

When we have an "if–then" statement such as "if $A$ is true then $B$ is true," its *contrapositive* is "if $B$ is false then $A$ is false." A statement is true if and only if its contrapositive is true. A proof by contradiction is really just proving the contrapositive of the statement in question.

*Warning:* Proof by contradiction is a very useful tool, but do not overuse it. It is not uncommon to see something like the following: You are asked to prove that $x^2 - x = 12$ has a solution that is an integer. You write, "Suppose it does not have such a solution. Let $x = 4$. Then $4^2 - 4 = 12$, so the equation has a solution. Contradiction. Therefore, the equation has an integral solution." What's wrong? Technically, nothing. But what you've really done is say, "Suppose this does not have a solution. But it does! Let $x = 4$. Since $4^2 - 4 = 12$, the equation has a solution that is an integer." It makes much more sense to state directly that $x = 4$ is a solution. The moral is, don't use proof by contradiction when a direct proof is more straightforward.

*Another Warning:* When proving something, and you are not doing proof by contradiction, make sure you start with the hypothesis and end with the conclusion. For example, suppose you are asked to prove that if $x \geq 10$ then $2x + 7 \geq 27$. Many people will write the following steps:

$$2x + 7 \geq 27$$
$$2x \geq 20$$
$$x \geq 10.$$

Since a correct statement is deduced, it is claimed that the statement $2x+7 \geq 10$ must be true. What's wrong? The hypothesis and conclusion were reversed. The proof showed that if $2x + 7 \geq 27$ is assumed then it can be concluded that $x \geq 10$. This is the reverse of what is being asked. The correct argument is the following:

$$x \geq 10$$
$$2x \geq 20$$
$$2x + 7 \geq 27.$$

Note that this starts with the hypothesis $x \geq 10$ and ends with the conclusion $2x \geq 27$.

## A.2   Geometric Series

Certain types of expressions occur frequently in mathematics. One of the most common is a **geometric series**, which is a sum of the form $a + ar + ar^2 + \cdots + ar^{n-1}$. Such sums can be evaluated quite easily:

**Proposition A.2.** *Let $a$ and $r$ be real numbers with $r \neq 1$, and let $n \geq 1$. Then*

$$a + ar + ar^2 + \cdots + ar^{n-1} = \frac{ar^n - a}{r - 1}.$$

*Proof.* Let $S$ be the sum. Then

$$rS = ar + ar^2 + ar^3 + \cdots + ar^n$$
$$= -a + a + ar + ar^2 + \cdots + ar^{n-1} + ar^n$$

(*a* was subtracted from and added to the previous expression)

$$= -a + S + ar^n.$$

Therefore, $(r-1)S = -a + ar^n$. Dividing by $r-1$ yields the formula.  $\square$

This formula occurs in various contexts. For example, let $a = 1$ and $r = x$. Then

$$1 + x + x^2 + \cdots + x^{n-1} = \frac{x^n - 1}{x - 1}.$$

Multiplying by $x - 1$ yields the factorization

$$x^n - 1 = (x - 1)(x^{n-1} + x^{n-2} + \cdots + x + 1).$$

For example, when $n = 2$, this says $x^2 - 1 = (x - 1)(x + 1)$.

We can also consider an infinite geometric series.

**Proposition A.3.** *Let $a$ and $r$ be real numbers with $|r| < 1$. Then*

$$a + ar + ar^2 + ar^3 + \cdots = \frac{a}{1 - r}.$$

*Proof.* In Proposition A.2, let $n \to \infty$. Since $|r| < 1$, we know that $r^n \to 0$. Therefore,

$$a + ar + ar^2 + ar^3 + \cdots = \lim_{n \to \infty} \left( a + ar + ar^2 + \cdots + ar^{n-1} \right)$$

$$= \lim_{n \to \infty} \frac{ar^n - a}{r - 1}$$

$$= \frac{0 - a}{r - 1}$$

$$= \frac{a}{1 - r}.$$

$\square$

**Example.** An example of this proposition is given by repeating decimals:

$$0.42424242\ldots = .42 + .42(.01)^1 + .42(.01)^2 + .42(.01)^3 + \cdots$$

$$= \frac{.42}{1 - .01} = \frac{42}{99} = \frac{14}{33},$$

since $a = .42$ is the first term and $r = .01$ is the ratio between successive terms.

Another example is

$$1 + \frac{1}{2} + \frac{1}{4} + \frac{1}{8} + \cdots = \frac{1}{1 - \frac{1}{2}} = 2.$$

---

**CHECK YOUR UNDERSTANDING**

1. Show that $1 + 2 + 4 + \cdots + 2^{n-1} = 2^n - 1$ for all $n \geq 1$.
2. Interpret $.3333 \cdots$ as a geometric series and use this to show that it equals $1/3$.

---

# A.3    Mathematical Induction

Mathematicians like to make educated guesses and then prove that their guesses are correct. For example, if we add up even integers, we get the following:

$$2 + 4 = 6 = 2 \cdot 3$$

$$2 + 4 + 6 = 12 = 3 \cdot 4$$

$$2 + 4 + 6 + 8 = 20 = 4 \cdot 5.$$

Looking at this pattern, we make an educated guess (or conjecture) that

$$2 + 4 + 6 + \cdots + 2n = n(n+1). \tag{A.1}$$

We could then check this for several more values of $n$. For example, for $n = 10$, we have

$$2 + 4 + 6 + 8 + 10 + 12 + 14 + 16 + 18 + 20 = 110 = 10 \cdot 11, \tag{A.2}$$

so it works in this case. Now suppose we want to check it for $n = 11$. We don't have to go back and add up $2 + 4 + \cdots + 20 + 22$. Instead, we already know that Equation (A.2) is true, so we add 22 and get

$$2 + 4 + \cdots + 20 + 22 = 110 + 22 = 132 = 11 \cdot 12.$$

This says that Equation (A.1) holds for $n = 11$. In this way, if we know the equation holds for one value of $n$, we can verify it for the next value of $n$ without recomputing the whole sum. Rather, we can simply add on one more term. Let's try this in general. Suppose we know that Equation (A.1) holds for $n = k$. This means that

$$2 + 4 + 6 + \cdots + 2k = k(k+1).$$

Let's try to deduce that it holds for the next value of $n$, namely $k + 1$. Add $2(k+1)$ to obtain

$$(2 + 4 + 6 + \cdots + 2k) + 2(k+1) = k(k+1) + 2(k+1)$$
$$= (k+1)(k+2).$$

This is Equation (A.1) for $n = k + 1$. We have shown that if we know the equation is true for one value of $n$, then it is true for the next value of $n$.

All we need now is a place to start: Let $n = 1$. The sum on the left in (A.1) has one term, namely 2, and the right side is $1(1 + 1) = 2$, so (A.1) is true for $n = 1$. We have just shown that it is true for the next value of $n$, which is 2. And then it is true for the next value of $n$, which is 3. This keeps going, and we find that (A.1) is true for all $n \geq 1$.

What we just did is called **mathematical induction**. The general principle is the following.

If we want to prove some statement is true for all positive integers, we need to show two things:

1. The statement is true for $n = 1$.

2. If the statement is true for $n = k$ then it is true for $n = k+1$.

We then conclude that the statement is true for all positive integers $n$.

Mathematical induction is often compared to proving that you can climb to the top of a ladder. You prove that you can get onto the ladder. This is the $n = 1$ case. Then you show that you can always get from one rung to the next. This is the $n = k$ to $n = k + 1$ case. Therefore, you can climb every rung of the ladder.

Let's look at another example. Consider sums of odd numbers:

$$1 = 1^2$$
$$1 + 3 = 2^2$$
$$1 + 3 + 5 = 3^2$$
$$1 + 3 + 5 + 7 = 4^2.$$

We guess that

$$1 + 3 + 5 + \cdots + (2n - 1) = n^2. \tag{A.3}$$

We have already seen that (A.3) is true for $n = 1$. Now let's assume it's true for $n = k$, so

$$1 + 3 + 5 + \cdots + (2k - 1) = k^2.$$

Add $2k + 1$ to obtain

$$(1 + 3 + 5 + \cdots + (2k - 1)) + (2k + 1) = k^2 + (2k + 1)$$
$$= (k + 1)^2.$$

This is (A.3) for $n = k + 1$, so we have proved that if (A.3) is true for $n = k$ then it is also true for $n = k + 1$. Mathematical induction tells us that (A.3) is true for all positive integers $n$.

As another example, let's prove that

$$n + 1 \leq 2^n$$

for all $n \geq 1$. The statement is certainly true for $n = 1$. Now assume that it is true for $n = k$, so

$$k + 1 \leq 2^k.$$

Adding 1 to each side, we obtain

$$\begin{aligned} k + 2 &\leq 2^k + 1 \\ &\leq 2^k + 2^k \\ &\leq 2^{k+1}, \end{aligned}$$

which is the statement for $n = k+1$. Therefore, if the inequality is true for $n = k$ then it is true for $n = k + 1$. Mathematical induction tells us that the inequality is true for all $n \geq 1$.

**Remark.** Sometimes the induction does not start at $n = 1$. Suppose we want to prove that $10n + 9 \leq 2^n$ for all $n \geq 7$. We then take the base case to be $n = 7$ (instead of $n = 1$) and then check that the inequality is true for this case. We then show, as in the preceding calculation, that if the inequality is true for $n = k$ with $k \geq 7$ then it is true for $n = k + 1$:

$$10k + 9 \leq 2^k \implies 10(k + 1) + 9 \leq 2^k + 10 \leq 2^k + 2^k = 2^{k+1}.$$

The last inequality used the fact that $10 \leq 2^k$ when $k \geq 7$. (In fact, this step would have worked for $k \geq 4$, but showing "true for 4" $\implies$ "true for 5" is worthless when the statement is false for 4.) We conclude that $10n + 9 \leq 2^n$ for all $n \geq 7$.

There is another form of induction that is useful. **Strong induction** assumes that a statement is true for *all* $n \leq k$ and proves it for $n = k+1$. For example, suppose we want to prove that every integer $n > 1$ is a prime or a product of primes. Assume that we have proved this for all $n \leq k$. Now consider $k+1$. If $k+1$ is prime, we're done. Otherwise, $k+1$ factors as $k + 1 = ab$ with $1 < a, b < k + 1$. The induction hypothesis says that $a$ and $b$ are either prime or products of primes. Therefore, $k + 1 = ab$ is a prime or a product of primes.

Closely related to strong induction is the **Well-Ordering Principle**, which states that a non-empty set of positive integers has a smallest element. For example, let's prove again that every integer $n > 1$ is either prime or a product of primes. Let $S$ be the set of integers $n > 1$ that are neither prime nor a product of primes. Suppose that $S$ is non-empty. The Well-Ordering Principle says that $S$ has a smallest element. Call it $s$. The elements of $S$ are not prime, so $s$ factors as $s = ab$ with $1 < a, b < s$. Since $s$ is the smallest element of $S$, both $a$ and $b$ are prime or products of primes. Therefore, $s = ab$ is a product of primes, contradicting the fact that $s$ is in $S$. This contradiction shows that $S$ must be empty, which is exactly the statement that every integer $n > 1$ is either prime or a product of primes. As we see, this proof is very close to the one using strong induction. It is often a matter of personal choice whether to use strong induction or the Well-Ordering Principle.

The Well-Ordering Principle can also be used to give an alternate proof of the induction example we began this section with. We want to prove that Equation (A.1)

$$2 + 4 + 6 + \cdots + 2n = n(n+1).$$

is true for all $n$. We begin by letting $S$ be the set of integers $n \geq 1$ for which

$$2 + 4 + 6 + \cdots + 2n \neq n(n+1).$$

Suppose that $S$ is non-empty. Then the Well-Ordering Principle says that $S$ has a smallest element, which we'll call $m$. Since we know that (A.1) is true for $n = 1$, we know that $m$ must be larger than 1. Since $m > 1$, we know that $m - 1$ is still a positive integer. Moreover, $m - 1 < m$. So $m - 1$ is a positive integer smaller than the smallest element of $S$, so (A.1) must be true for $m - 1$. This means that

$$2 + 4 + \cdots + 2(m-1) = (m-1)(m-1+1) = (m-1)m.$$

Adding $2m$ to both sides of this equation, we get

$$2 + 4 + \cdots + 2(m-1) + 2m = (m-1)m + 2m$$
$$= m^2 - m + 2m = m^2 + m = m(m+1).$$

But this shows that (A.1) is true for $m$, so it is impossible for $S$ to be non-empty. Therefore,

$$2 + 4 + \cdots + 2n = n(n+1)$$

must be true for all $n \geq 1$.

Finally, here's an interesting example of strong induction. We'll play a game. A move in this game is to divide a pile of markers into two piles, say of size $a$ and $b$. Our score for that move is $ab$. Suppose we start with a pile of 12 markers. We make moves by choosing a pile and breaking it into two smaller piles and adding the score to our total. We continue until all the piles are size 1. The claim is that no matter how we play our total score will be 66. For example, suppose the game proceeds as follows:

$$
\begin{aligned}
12 &= 5+7 \quad (\text{score } = 5 \cdot 7 = 35) \\
&= 2+3+7 \quad (\text{score } = 2 \cdot 3 = 6) \\
&= 2+3+5+2 \quad (\text{score } = 5 \cdot 2 = 10) \\
&= 2+3+5+1+1 \quad (\text{score } = 1 \cdot 1 = 1) \\
&= 1+1+3+5+1+1 \quad (\text{score } = 1 \cdot 1 = 1) \\
&= 1+1+2+1+5+1+1 \quad (\text{score } = 1 \cdot 2 = 2) \\
&= 1+1+1+1+1+5+1+1 \quad (\text{score } = 1 \cdot 1 = 1) \\
&= 1+1+1+1+1+2+3+1+1 \quad (\text{score } = 2 \cdot 3 = 6) \\
&= 1+1+1+1+1+1+1+3+1+1 \quad (\text{score } = 1 \cdot 1 = 1) \\
&= 1+1+1+1+1+1+1+2+1+1+1 \quad (\text{score } = 2 \cdot 1 = 2) \\
&= 1+1+1+1+1+1+1+1+1+1+1+1 \quad (\text{score } = 1 \cdot 1 = 1).
\end{aligned}
$$

The total score is

$$
35+6+10+1+1+2+1+6+1+2+1 = 66.
$$

Proving this always yields 66 by trying all cases would be painful. However, it is much easier to consider the general statement: If we start with a pile of $n$ markers, our total score will be $\frac{1}{2}n(n-1)$. This is certainly true for $n = 1$ since we cannot make any moves so our total score is 0. Let's assume the statement is true for all $n \leq k$ and try to prove it for $n = k+1$.

The first move splits the pile of $k+1$ markers into two piles, say of sizes $a$ and $b$, with $a + b = k + 1$. This gives us $ab$ points. After that, the game really breaks into two subgames, one for the markers in the $a$ pile and one for the markers in the $b$ pile. By the induction assumption, the $a$ pile earns us $\frac{1}{2}a(a-1)$ total points and the $b$ pile earns us $\frac{1}{2}b(b-1)$

total points. The total number of points we earn is the $ab$ for the first move plus these points:

$$ab + \frac{1}{2}a(a-1) + \frac{1}{2}b(b-1) = \frac{1}{2}\left(a^2 + 2ab + b^2 - a - b\right)$$
$$= \frac{1}{2}\left(a+b\right)\left(a+b-1\right)$$
$$= \frac{1}{2}(k+1)(k).$$

Therefore, the claim holds for $n = k+1$. Strong induction tells us that the claim holds for all $n \geq 1$. In particular, if we start with $n = 12$ markers, we will always earn $\frac{1}{2}(12)(11) = 66$ points.

---

## CHECK YOUR UNDERSTANDING

3. Use induction to prove that $1 + 2 + 4 + \cdots + 2^{n-1} = 2^n - 1$ for all $n \geq 1$.

---

# A.4    Pascal's Triangle and the Binomial Theorem

Consider the following triangular array of numbers:

$$
\begin{array}{ccccccccc}
 & & & & 1 & & 1 & & \\
 & & & 1 & & 2 & & 1 & \\
 & & 1 & & 3 & & 3 & & 1 \\
 & 1 & & 4 & & 6 & & 4 & & 1 \\
1 & & 5 & & 10 & & 10 & & 5 & & 1 \\
\end{array}
$$

$$
\begin{array}{ccccccccccccc}
1 & & 6 & & 15 & & 20 & & 15 & & 6 & & 1 \;.
\end{array}
$$

Each number is obtained by adding together the two numbers above it in the preceding row. For example, the 15 that is the third entry in the

6th row is the sum $5 + 10$ of the numbers above it in the 5th row. This array is often called **Pascal's Triangle**, after the great French mathematician and philosopher Blaise Pascal. The triangle did not originate with Pascal; it had been known for hundreds of years in many countries around the world. But Pascal organized its properties in his book *Traité du triangle arithmétique* (1653) and used them to develop probability theory and analyze gambling strategies.

In this section we define binomial coefficients and show that they are the entries in Pascal's triangle.

We all learned in high school that

$$(a + b)^2 = a^2 + 2ab + b^2,$$

and you might remember that

$$(a + b)^3 = a^3 + 3a^2b + 3ab^2 + b^3.$$

There are many times when we need to calculate $(a+b)^n$ for an arbitrary positive integer $n$ and the *Binomial Theorem* allows us to do just that.

In order to preview what the Binomial Theorem says, we calculate $(a + b)^n$ for $n = 1, 2, 3, 4, 5$, and 6:

$$(a + b)^1 = a + b$$

$$(a + b)^2 = a^2 + 2ab + b^2$$

$$(a + b)^3 = a^3 + 3a^2b + 3ab^2 + b^3$$

$$(a + b)^4 = a^4 + 4a^3b + 6a^2b^2 + 4ab^3 + b^4$$

$$(a + b)^5 = a^5 + 5a^4b + 10a^3b^2 + 10a^2b^3 + 5ab^4 + b^5$$

$$(a + b)^6 = a^6 + 6a^5b + 15a^4b^2 + 20a^3b^3 + 15a^2b^4 + 10ab^5 + b^6.$$

The first thing to notice is that when calculating $(a + b)^n$, each term is of the form $a^ib^j$ with $i + j = n$. The coefficient of $a^ib^j$ is the $j$th entry in the $n$th row of Pascal's Triangle (where the initial 1 is counted as the 0th element of the row).

Although this is nice, we still want a general theorem that gives a formula for the value of each coefficient in the expansion of $(a + b)^n$. For this we need the binomial coefficients:

**Definition A.4.** *Let $m$ and $r$ be integers with $0 \leq r \leq m$. Then*

$$\binom{m}{r} = \frac{m!}{r!(m - r)!}$$

*is called a* **binomial coefficient**.

If you have a set with $m$ elements, $\binom{m}{r}$ is the number of subsets with $r$ elements. In other words, it is the number of ways to choose a subset of $r$ marbles from a set of $m$ marbles. For this reason, $\binom{m}{r}$ is often read as "$m$ choose $r$."

If we recall that $0! = 1$, we see that

$$\binom{4}{0} = 1, \quad \binom{4}{1} = 4, \quad \binom{4}{2} = 6, \quad \binom{4}{3} = 4, \quad \text{and} \quad \binom{4}{4} = 1$$

and these are exactly the numbers that occur in the fourth line of Pascal's Triangle. We'll prove that the binomial coefficients always tell us the coefficients of $(a + b)^n$ in a moment, but first we need a lemma.

In Pascal's Triangle, each number is the sum of the two integers that are above it. The same property holds for binomial coefficients.

**Proposition A.5.** *Let $m$ and $r$ be integers with $1 \leq r \leq m$. Then*

$$\binom{m}{r - 1} + \binom{m}{r} = \binom{m + 1}{r}.$$

*Proof.* Starting with the left-hand side, we have

$$\binom{m}{r-1} + \binom{m}{r}$$

$$= \frac{m!}{(r-1)!(m+1-r)!} + \frac{m!}{r!(m-r)!}$$

$$= \frac{m!r}{r(r-1)!(m+1-r)!} + \frac{m!(m+1-r)}{r!(m-r)!(m+1-r)!}$$

$$= \frac{m!}{r!(m+1-r)!}\left((r) + (m+1-r)\right)$$

$$= \frac{m!(m+1)}{r!(m+1-r)!}$$

$$= \binom{m+1}{r}.$$

This proves the proposition.               □

We can now write the first six rows of Pascal's Triangle as follows:

$$\binom{1}{0} \quad \binom{1}{1}$$

$$\binom{2}{0} \quad \binom{2}{1} \quad \binom{2}{2}$$

$$\binom{3}{0} \quad \binom{3}{1} \quad \binom{3}{2} \quad \binom{3}{3}$$

$$\binom{4}{0} \quad \binom{4}{1} \quad \binom{4}{2} \quad \binom{4}{3} \quad \binom{4}{4}$$

$$\binom{5}{0} \quad \binom{5}{1} \quad \binom{5}{2} \quad \binom{5}{3} \quad \binom{5}{4} \quad \binom{5}{5}$$

$$\binom{6}{0} \quad \binom{6}{1} \quad \binom{6}{2} \quad \binom{6}{3} \quad \binom{6}{4} \quad \binom{6}{5} \quad \binom{6}{6}.$$

The proposition says, for example, that $\binom{3}{1} + \binom{3}{2} = \binom{4}{2}$, which is exactly the same as saying that the binomial coefficient is the sum of the two above it. This implies that the binomial coefficients are the entries in Pascal's Triangle, since both can be generated in the same way.

We're now ready to prove our main result.

**The Binomial Theorem.** *Let $n$ be a positive integer. Then*

$$(a+b)^n =$$
$$\binom{n}{0}a^n + \binom{n}{1}a^{n-1}b + \binom{n}{2}a^{n-2}b^2 + \cdots + \binom{n}{n-1}ab^{n-1} + \binom{n}{n}b^n.$$

*Proof.* We'll prove this using mathematical induction.

**Step 1:** When $n = 1$,

$$(a+b)^1 = a + b = \binom{1}{0}a + \binom{1}{1}b,$$

so our initial induction step is verified. (From the definition of the binomial coefficient, you can see that $\binom{n}{0} = \binom{n}{n} = 1$ for all positive integers $n$.)

**Step 2:** Assume that the binomial theorem is true for $n = k$, so

$$(a+b)^k =$$
$$\binom{k}{0}a^k + \binom{k}{1}a^{k-1}b + \binom{k}{2}a^{k-2}b^2 + \cdots + \binom{k}{k-1}ab^{k-1} + \binom{k}{k}b^k.$$

We now have to prove that the appropriate formula works for $n = k+1$. We begin by observing that

$$(a+b)^{k+1} = (a+b) \cdot (a+b)^k$$
$$= a(a+b)^k + b(a+b)^k.$$

We use the induction assumption to expand each $(a+b)^k$. This yields

$$a\left(\binom{k}{0}a^k + \binom{k}{1}a^{k-1}b + \binom{k}{2}a^{k-2}b^2 + \cdots + \binom{k}{k-1}ab^{k-1} + \binom{k}{k}b^k\right)$$

$$+ b\left(\binom{k}{0}a^k + \binom{k}{1}a^{k-1}b + \binom{k}{2}a^{k-2}b^2 + \cdots + \binom{k}{k-1}ab^{k-1} + \binom{k}{k}b^k\right)$$

$$= \binom{k}{0}a^{k+1} + \binom{k}{1}a^k b + \binom{k}{2}a^{k-1}b^2 + \cdots + \binom{k}{k-1}a^2 b^{k-1} + \binom{k}{k}ab^k$$

$$+ \binom{k}{0}a^k b + \binom{k}{1}a^{k-1}b^2 + \binom{k}{2}a^{k-2}b^3 + \cdots + \binom{k}{k-1}ab^k + \binom{k}{k}b^{k+1}$$

$$= \binom{k}{0}a^{k+1} + \left(\binom{k}{1}+\binom{k}{0}\right)a^k b + \left(\binom{k}{2}+\binom{k}{1}\right)a^{k-1}b^2$$

$$+ \cdots + \left(\binom{k}{k}+\binom{k}{k-1}\right)ab^k + \binom{k}{k}b^{k+1}.$$

Proposition A.5 allows us to replace the sums of binomial coefficients by a single binomial coefficient. For example,

$$\binom{k}{2} + \binom{k}{1} = \binom{k+1}{2}.$$

Also,

$$\binom{k}{0} = 1 = \binom{k+1}{0} \quad \text{and} \quad \binom{k}{k} = 1 = \binom{k+1}{k+1}.$$

Therefore, we obtain

$$\binom{k+1}{0}a^{k+1} + \binom{k+1}{1}a^k b + \cdots + \binom{k+1}{k}ab^k + \binom{k+1}{k+1}b^{k+1}.$$

We have proved that $(a+b)^{k+1}$ is given by the desired formula. Mathematical induction implies that the formula is true for all $n$.  $\square$

---

**CHECK YOUR UNDERSTANDING**

4. Compute $\binom{7}{3}$.

5. Find the coefficient of $a^2 b^6$ in the expansion of $(a + b)^8$.

---

# A.5   Fibonacci Numbers

Fibonacci numbers occur in many situations in mathematics. In the following, we give a brief treatment of the properties that we need in this book.

**Definition A.6.** *The* **Fibonacci numbers** *are defined recursively by*

$$F_0 = 0, \ F_1 = 1, \ F_2 = 1, \ and \ F_n = F_{n-1} + F_{n-2} \ for \ n \geq 3.$$

As an example, we calculate the first six Fibonacci numbers:

$$F_3 = F_2 + F_1 = 1 + 1 = 2$$
$$F_4 = F_3 + F_2 = 2 + 1 = 3$$
$$F_5 = F_4 + F_3 = 3 + 2 = 5$$
$$F_6 = F_5 + F_4 = 5 = 3 = 8.$$

The notation $F_n$ is also used for Fermat numbers (see Section 2.9). However, the context should always make it clear which type of number is being discussed.

There are many identities involving the Fibonacci numbers. For example, $F_{n+3} + F_n = 2F_{n+2}$. To see why this is true, note that

$$F_{n+3} + F_n = (F_{n+2} + F_{n+1}) + F_n$$
$$= F_{n+2} + (F_{n+1} + F_n)$$
$$= F_{n+2} + F_{n+2} = 2F_{n+2}.$$

Some other identities are in the exercises.

The following result is known as **Binet's Formula**, in honor of Jacques

Philippe Marie Binet (1786-1856), who, among other accomplishments, was the first to describe the rule for multiplying matrices. However, Binet's Formula was probably first discovered by de Moivre around 1730.

The **Golden Ratio** is defined to be

$$\phi = \frac{1 + \sqrt{5}}{2} \approx 1.618034.$$

An easy calculation shows that

$$-\phi^{-1} = \frac{1 - \sqrt{5}}{2} \approx -.618034.$$

**Theorem A.7.** *The nth Fibonacci number $F_n$ is given by the expression*

$$F_n = \frac{1}{\sqrt{5}} \left( \phi^n - (-\phi)^{-n} \right). \tag{A.4}$$

*Proof.* We will use strong induction to prove this.

**Step 1:** Verify that is true for $n = 0$ and $n = 1$.

$n = 0$:

$$\frac{1}{\sqrt{5}} (\phi^0 - (-\phi)^{-0}) = 0 = F_0.$$

$n = 1$:

$$\frac{1}{\sqrt{5}} (\phi^1 - (-\phi)^{-1}) = \frac{1}{\sqrt{5}} \left( \frac{1 + \sqrt{5}}{2} - \frac{1 - \sqrt{5}}{2} \right) = 1 = F_1.$$

**Step 2:** Assume that (A.4) is true for all $n \leq k$ and use this to prove it's true for $n = k + 1$ This means that we assume that

$$F_m = \frac{1}{\sqrt{5}} (\phi^m - (-\phi)^{-m})$$

is true for $m = 0, 1, 2, 3, \ldots, k$ and we want to prove that

$$F_{k+1} = \frac{1}{\sqrt{5}} (\phi^{k+1} - (-\phi)^{k+1}).$$

First, observe that

$$\phi^2 = \frac{6 + 2\sqrt{5}}{4} = \frac{3 + \sqrt{5}}{2} = \phi + 1.$$

Multiply by $\phi^{k-1}$ to get

$$\phi^{k+1} = \phi^k + \phi^{k-1}. \tag{A.5}$$

Similarly,

$$(-\phi^{-1})^2 = \frac{6 - 2\sqrt{5}}{4} = \frac{3 - \sqrt{5}}{2} = (-\phi^{-1}) + 1,$$

which tells us that

$$(-\phi)^{-(k+1)} = (-\phi)^{-k} + (-\phi)^{-(k-1)}. \tag{A.6}$$

Therefore,

$$F_{k+1} = F_k + F_{k-1} \quad \text{(from the definition of the Fibonacci sequence)}$$

$$= \frac{1}{\sqrt{5}}\left((\phi^k - (-\phi)^{-k}) + (\phi^{k-1} - (-\phi)^{-(k-1)})\right)$$

(from the induction hypothesis)

$$= \frac{1}{\sqrt{5}}\left(\phi^k + \phi^{k-1}\right) - \frac{1}{\sqrt{5}}\left((-\phi)^{-k} + (-\phi)^{-(k-1)}\right)$$

$$= \frac{1}{\sqrt{5}}\left(\phi^{k+1} - (-\phi)^{-(k+1)}\right) \quad \text{(from (A.5) and (A.6))}.$$

This is (A.4) for $n = k + 1$. By induction, the formula (A.4) is true for all positive integers. This completes the proof. $\qquad\square$

If we look at the ratios

$$\frac{F_6}{F_5} \approx 1.600000, \quad \frac{F_{11}}{F_{10}} \approx 1.618182, \quad \frac{F_{16}}{F_{15}} \approx 1.618033,$$

we see that the ratios of successive Fibonacci numbers are approximately the Golden Ratio. This is explained by Binet's Formula:

**Corollary A.8.**

$$\lim_{n \to \infty} \frac{F_{n+1}}{F_n} = \phi.$$

*Proof.* From Binet's Formula, we have

$$\lim_{n \to \infty} \frac{F_{n+1}}{F_n} = \lim_{n \to \infty} \frac{\phi^{n+1} - (-\phi)^{-(n+1)}}{\phi^n - (-\phi)^{-(n)}}.$$

Dividing the numerator and denominator by $\phi^n$ yields

$$\lim_{n\to\infty} \frac{\phi - (-\phi)^{-(n+1)}\phi^{-n}}{1 - (-\phi)^{-n}\phi^{-n}}. \tag{A.7}$$

Since $|-\phi| > 1$, we see that $(-\phi)^{-(n+1)} \to 0$ as $n \to \infty$. Similarly, $\phi^{-n} \to 0$ and $(-\phi)^{-n} \to 0$ as $n \to \infty$. Therefore, the limit of the numerator in (A.7) is $\phi$ and the limit of the denominator is 1. Putting everything together, we obtain

$$\lim_{n\to\infty} \frac{F_{n+1}}{F_n} = \phi,$$

as desired. $\qquad\qquad\qquad\qquad\qquad\qquad\qquad\qquad\qquad\qquad\qquad\qquad$ $\square$

If you know enough Fibonacci numbers, the corollary gives you a way to convert between miles and kilometers quickly. For example, 55 miles is approximately 88 kilometers (exactly 88.51392). This is because the successive Fibonacci numbers 55 and 88 have a ratio approximately $\phi \approx 1.618$, while one mile is 1.609344 kilometers, which is almost the same ratio. Similarly, if you know that 6765 and 10946 are successive Fibonacci numbers and you want to convert 110 kilometers to miles, you match the first digits and observe that $10946/6765$ is approximately $110/68$, so the answer is approximately 68 miles.

## A.6    Matrices

A **matrix** is just a rectangular array of numbers. If you've ever used a spreadsheet where entries are specified by their row and column numbers, you've seen a matrix. Most of the matrices we use in this book have two rows and two columns and look like this:

$$\begin{pmatrix} a & b \\ c & d \end{pmatrix},$$

where the entries $a, b, c,$ and $d$ are numbers. In this matrix, $a$ is in the first row and first column, $b$ is in the first row and second column, $c$ is in the second row and first column, and $d$ is in the second row and second column.

We also need to work with **vectors**. The vectors that we'll be using have 1 column and two rows:

$$\begin{pmatrix} e \\ f \end{pmatrix}.$$

For the Hill cipher in Chapter 7, we need to multiply a matrix times a vector and a matrix times a matrix.

To begin, here's the rule for multiplying a matrix by a vector:

$$\begin{pmatrix} a & b \\ c & d \end{pmatrix} \begin{pmatrix} e \\ f \end{pmatrix} = \begin{pmatrix} ae + bf \\ ce + df \end{pmatrix}$$

For example,

$$\begin{pmatrix} 3 & 2 \\ 5 & 8 \end{pmatrix} \begin{pmatrix} 4 \\ 7 \end{pmatrix} = \begin{pmatrix} 3 \cdot 4 + 2 \cdot 7 \\ 5 \cdot 4 + 8 \cdot 7 \end{pmatrix} = \begin{pmatrix} 26 \\ 76 \end{pmatrix}$$

Next, we'll use the method of multiplying a matrix times a vector to multiply a matrix times a matrix. Let's say you want to calculate $M\,N$, where

$$M = \begin{pmatrix} 7 & 10 \\ 3 & 4 \end{pmatrix} \quad \text{and} \quad N = \begin{pmatrix} 3 & 6 \\ 5 & 2 \end{pmatrix}.$$

Think of $N$ as consisting of two vectors, so $N$ is made up of the vectors

$$\begin{pmatrix} 3 \\ 5 \end{pmatrix} \quad \text{and} \quad \begin{pmatrix} 6 \\ 2 \end{pmatrix}.$$

For the product $M\,N$, we get the $2 \times 2$ matrix whose first column is

$$M \begin{pmatrix} 3 \\ 5 \end{pmatrix} = \begin{pmatrix} 7 & 10 \\ 3 & 4 \end{pmatrix} \begin{pmatrix} 3 \\ 5 \end{pmatrix} = \begin{pmatrix} 7 \cdot 3 + 10 \cdot 5 \\ 3 \cdot 3 + 4 \cdot 5 \end{pmatrix} = \begin{pmatrix} 71 \\ 29 \end{pmatrix}$$

and whose second column is

$$M \begin{pmatrix} 6 \\ 2 \end{pmatrix} = \begin{pmatrix} 7 & 10 \\ 3 & 4 \end{pmatrix} \begin{pmatrix} 6 \\ 2 \end{pmatrix} = \begin{pmatrix} 7 \cdot 6 + 10 \cdot 2 \\ 3 \cdot 6 + 4 \cdot 2 \end{pmatrix} = \begin{pmatrix} 62 \\ 26 \end{pmatrix}.$$

Therefore,

$$M\,N = \begin{pmatrix} 7 & 10 \\ 3 & 4 \end{pmatrix} \begin{pmatrix} 3 & 6 \\ 5 & 2 \end{pmatrix} = \begin{pmatrix} 71 & 62 \\ 29 & 26 \end{pmatrix}.$$

In general, if

$$M = \begin{pmatrix} a & b \\ c & d \end{pmatrix} \quad \text{and} \quad N = \begin{pmatrix} e & f \\ g & h \end{pmatrix},$$

then

$$MN = \begin{pmatrix} ae + bg & af + bh \\ ce + dg & cf + dh \end{pmatrix}.$$

Notice that under this multiplication rule, if

$$M = \begin{pmatrix} a & b \\ c & d \end{pmatrix} \quad \text{and} \quad I = \begin{pmatrix} 1 & 0 \\ 0 & 1 \end{pmatrix},$$

then

$$MI = \begin{pmatrix} a \cdot 1 + b \cdot 0 & a \cdot 0 + b \cdot 1 \\ c \cdot 1 + d \cdot 0 & c \cdot 0 + d \cdot 1 \end{pmatrix} = \begin{pmatrix} a & b \\ c & d \end{pmatrix} = M.$$

Similarly, $IM = M$. Because of this, $I$ is called the **identity matrix**. We now use the identity matrix to discuss how (and when) we can form the inverse of a matrix. Recall that if $a$ is a nonzero number, then $a^{-1} \cdot a = 1$. Analogously, we want to compute the inverse of a matrix $M$. That is, given a matrix

$$M = \begin{pmatrix} a & b \\ c & d \end{pmatrix},$$

we want a matrix $M^{-1}$ with the property that

$$M^{-1} M = I.$$

Fortunately, there is a formula: If $ad - bc \neq 0$, then

$$\begin{pmatrix} a & b \\ c & d \end{pmatrix}^{-1} = \frac{1}{ad - bc} \begin{pmatrix} d & -b \\ -c & a \end{pmatrix}.$$

Here, the notation

$$\frac{1}{ad - bc} \begin{pmatrix} d & -b \\ -c & a \end{pmatrix}$$

means that we divide every entry of the matrix by $ad - bc$.

It can be shown that if $ad - bc = 0$, then the matrix does not have an inverse.

**Example.**

$$\begin{pmatrix} 11 & 5 \\ 6 & 3 \end{pmatrix}^{-1} = \frac{1}{3}\begin{pmatrix} 3 & -5 \\ -6 & 11 \end{pmatrix} = \begin{pmatrix} 1 & -5/3 \\ -2 & 11/3 \end{pmatrix}.$$

We can check this:

$$\begin{pmatrix} 1 & -5/3 \\ -2 & 11/3 \end{pmatrix}\begin{pmatrix} 11 & 5 \\ 6 & 3 \end{pmatrix} = \begin{pmatrix} 1 & 0 \\ 0 & 1 \end{pmatrix}.$$

There's an issue that we need to mention. In the Hill cipher, the entries of the matrix are numbers mod $n$ for some integer $n$. The number $ad-bc$ occurs in the denominator in our equation for the inverse of a matrix, and we need $ad - bc$ to have an inverse mod $n$ in order for the inverse of the matrix to exist.

Finally, suppose $v$ and $w$ are vectors and $M$ is a matrix and we have an equation

$$Mv = w,$$

where $M$ is a matrix that has an inverse. Then

$$M^{-1}w = v.$$

For example,

$$\begin{pmatrix} 1 & 2 \\ 3 & 4 \end{pmatrix}\begin{pmatrix} 5 \\ 6 \end{pmatrix} = \begin{pmatrix} 17 \\ 39 \end{pmatrix}.$$

Since

$$\begin{pmatrix} 1 & 2 \\ 3 & 4 \end{pmatrix}^{-1} = \frac{-1}{2}\begin{pmatrix} 4 & -2 \\ -3 & 1 \end{pmatrix} = \begin{pmatrix} -2 & 1 \\ 3/2 & -1/2 \end{pmatrix},$$

we have

$$\begin{pmatrix} -2 & 1 \\ 3/2 & -1/2 \end{pmatrix}\begin{pmatrix} 17 \\ 39 \end{pmatrix} = \begin{pmatrix} 5 \\ 6 \end{pmatrix}.$$

---

**CHECK YOUR UNDERSTANDING:**

6. Compute $\begin{pmatrix} 1 & 2 \\ 3 & 4 \end{pmatrix}\begin{pmatrix} 4 & -2 \\ 1 & 5 \end{pmatrix}$.

7. Find the inverse of $\begin{pmatrix} 5 & 2 \\ 1 & 3 \end{pmatrix}$.

# A.7    Problems

## A.7.1    Exercises

### Section A.1: Geometric Series

1. Evaluate $3 + \frac{3}{4} + \frac{3}{16} + \frac{3}{64} + \cdots + \frac{3}{4^n} + \cdots$.
2. Express $x = .91919191\cdots$ as a geometric series and use the formula for the sum of a geometric series to write $x$ as a rational number.
3. For what value of $r$ does $1 + r + r^2 + r^3 + \cdots = 3$?
4. Why is there no $r$ such that $1 + r + r^2 + r^3 + \cdots = 1/3$?

### Section A.2: Mathematical Induction

5. Use induction to prove that $1 + 2 + 3 + \cdots + n = \frac{1}{2}n(n+1)$ for all $n \geq 1$.
6. Use induction to prove that $n + 2 \leq 3^n$ for all $n \geq 1$.
7. Use induction to prove that $1 + 2^2 + 3^2 + 4^2 + \cdots + n^2 = \frac{1}{6}n(n+1)(2n+1)$ for all $n \geq 1$.
8. Use induction to prove that $1 + 2^3 + 3^3 + 4^3 + \cdots + n^3 = \left(\frac{1}{2}n(n+1)\right)^2$ for all $n \geq 1$.
9. Here is a "proof" due to Pólya that all horses have the same color. We want to show that in every set of $n$ horses, every horse has the same color. This is clearly true for $n = 1$. Assume that the statement is true for $n = k$. Now consider a set of $k + 1$ horses. Remove one horse, call it Ed, from the set. By the induction hypothesis, this set of $k$ horses all have the same color. Now put Ed back into the set and remove another horse. Again, this set of $k$ horses all have the same color, so Ed has the same color as the other horses. We conclude that all $k + 1$ horses have the same color. By induction, all horses have the same color. Where is the flaw in the reasoning?
10. The **Cantor Expansion** of a positive integer $n$ is

    $$n = a_m m! + a_{m-1}(m-1)! + ... + a_2 2! + a_1$$

    where $a_m \neq 0$ and $0 \leq a_k \leq k$ for each $k$. For example, the Cantor Expansion of 15 is $2 \cdot 3! + 1 \cdot 2! + 1$.
    (a) Find the Cantor Expansion of 50, 73, and 533.
    (b) Use strong induction and the fact that for every positive integer $n$ there is an integer $m$ with $m! \leq n < (m+1)!$ to show that every positive integer $n$ has a unique Cantor Expansion.

### Section A.3: Pascal's Triangle and the Binomial Theorem

11. (a) Compute $(a + b)^5$. What can you say about the divisibility of the coefficients by 5?

(b) Compute $(a + b)^7$. What can you say about the divisibility of the coefficients by 7?

(c) Compute $(a+b)^6$. Note that many of the coefficients are not divisible by 6.

(d) Compute $(a+b)^{10}$. Note that many of the coefficients are not divisible by 10.

12.   Show that $\binom{n}{k} = \binom{n}{n-k}$.

13.   (a) Show that

$$\binom{n}{0} + \binom{n}{1} + \binom{n}{2} + \cdots + \binom{n}{n-1} + \binom{n}{n} = 2^n.$$

(*Hint:* Compute $(1 + 1)^n$.)

(b) Show that

$$\binom{n}{0} - \binom{n}{1} + \binom{n}{2} - \binom{n}{3} + \cdots + (-1)^n \binom{n}{n} = 0.$$

(*Hint:* Modify the hint for part (a).)

## Section A.4: Fibonacci Numbers

14.   Use induction to prove that $F_1 + F_2 + \cdots + F_n = F_{n+2} - 1$.

15.   Prove that $F_1 + F_3 + \cdots + F_{2n-1} = F_{2n}$ for all $n \geq 1$
   (a) by induction
   (b) by Binet's Formula plus Proposition A.2.

16.   Cassini's identity says that $F_n^2 - F_{n+1}F_{n-1} = (-1)^{n-1}$ for all $n \geq 2$. Prove this
   (a) by induction.
   (b) by Binet's Formula.

17.   (a) Prove that $\phi^2 - \phi^{-2} = \sqrt{5}$.
   (b) Use Binet's Formula to prove that $F_{2n} = F_{n+1}^2 - F_{n-1}^2$ for all $n \geq 2$.

18.   The **Lucas numbers** 1, 3, 4, 7, 11, ... are defined by

$$L_1 = 1, \quad L_2 = 3, \quad \text{and} \quad L_n = L_{n-1} + L_{n-2} \text{ for } n \geq 3.$$

(a) Show that $L_n = \phi^n + (-\phi)^{-n}$.
(b) Show that $\lim_{n \to \infty} L_{n+1}/L_n = \phi$.
(c) Show that $F_{2n} = F_n L_n$.
(d) Show that $F_n = L_{n-1} + L_{n+1}$.

19.   (Fibonacci) Assume that each pair of rabbits produces a pair of rabbits every month, starting when the pair is 2 months old. Suppose you start with one pair of rabbits. After 1 month, you still have one pair. After 2 months, you have two pairs, the original pair and a pair that was just born. After 3 months, you have three pairs because a new pair is born. After 4 months, you have five pairs since the original pair and the second pair each produce a pair. How many pairs of rabbits are there after 12 months?

**Section A.5: Matrices**

20.  Compute
$$\begin{pmatrix} 3 & 7 \\ 1 & 2 \end{pmatrix} \begin{pmatrix} 2 \\ 3 \end{pmatrix}.$$

21.  Compute
$$\begin{pmatrix} 10 & 1 \\ 3 & 1 \end{pmatrix} \begin{pmatrix} 4 \\ 2 \end{pmatrix}.$$

22.  Compute
$$\begin{pmatrix} 1 & 4 \\ 3 & 1 \end{pmatrix} \begin{pmatrix} 4 & 0 \\ 3 & 2 \end{pmatrix}.$$

23.  Compute
$$\begin{pmatrix} -3 & 4 \\ 2 & 7 \end{pmatrix} \begin{pmatrix} 5 & 1 \\ 1 & 6 \end{pmatrix}.$$

24.  Compute
$$\begin{pmatrix} 1 & 4 \\ 3 & 1 \end{pmatrix}^{-1}.$$

25.  Compute
$$\begin{pmatrix} 7 & 3 \\ 3 & 1 \end{pmatrix}^{-1}.$$

# A.7.2  Answers to "CHECK YOUR UNDERSTANDING"

1.  This is a geometric series with $a = 1$ and $r = 2$. The formula says that the sum equals
$$\frac{2^n - 1}{2 - 1} = 2^n - 1.$$

2.  The decimal can be rewritten as
$$.3 + .3(.1) + .3(.1)^2 + \cdots.$$
This is a geometric series with $a = .3$ and $r = .1$, so the sum is $a/(1-r) = .3/(1 - .1) = 1/3$.

3.  The formula is true for $n = 1$ since $1 = 2^1 - 1$. Assume that it is true for $n = k$, so
$$1 + 2 + 4 + 8 + \cdots + 2^{k-1} = 2^k - 1.$$
Add $2^k$ to both sides of the equation:
$$1 + 2 + 4 + 8 + \cdots + 2^{k-1} + 2^k = 2^k - 1 + 2^k = 2^{k+1} - 1.$$
Therefore, the formula holds for $n = k + 1$. By induction, the formula is true for all positive integers $n$.

4.    Use the formula for binomial coefficients:

$$\binom{7}{3} = \frac{7!}{3!4!} = \frac{7 \cdot 6 \cdot 5 \cdot 4 \cdot 3 \cdot 2 \cdot 1}{3 \cdot 2 \cdot 1 \cdot 4 \cdot 3 \cdot 2 \cdot 1} = \frac{7 \cdot 6 \cdot 5}{3 \cdot 2 \cdot 1} = 35.$$

5.    The coefficient is

$$\binom{8}{6} = \frac{8!}{6!2!} = \frac{8 \cdot 7}{2 \cdot 1} = 28,$$

since 6! cancels from the numerator and denominator.

6.

$$\begin{pmatrix} 1 & 2 \\ 3 & 4 \end{pmatrix} \begin{pmatrix} 4 & -2 \\ 1 & 5 \end{pmatrix} = \begin{pmatrix} 6 & 8 \\ 16 & 14 \end{pmatrix}.$$

7.

$$\begin{pmatrix} 5 & 2 \\ 1 & 3 \end{pmatrix}^{-1} = \frac{1}{13} \begin{pmatrix} 3 & -2 \\ -1 & 5 \end{pmatrix} = \begin{pmatrix} 3/13 & -2/13 \\ -1/13 & 5/13 \end{pmatrix}.$$

# Appendix B

# Answers and Hints for Odd-Numbered Exercises

**Chapter 2**

1. These use the definition of divisibility: $120 = 5 \cdot 24$, $165 = 11 \cdot 15$, $98 = 14 \cdot 7$.

3. These use the definition of divisibility: $21 = 7 \cdot 3$, $21 = 3 \cdot 7$, $42 = 14 \cdot 3$, $42 = 6 \cdot 7$.

5. $15/8$, $20/3$, $72/5$, and $22/4$ are not integers.

7. (a) 1, 2, 4, 5, 10, 20
   (b) 1, 2, 4, 13, 26, 52
   (c) 1, 3, 5, 13, 15, 39, 65, 195
   (d) 1, 7, 29, 203

9. (a) False: $a = 2$, $b = 3$. (b) False: $a = 9$, $b = 4$. (c) False: $a = 1$, $b = 5$.

11. (a) True: If $c \mid a$ then there exists $j$ with $a = cj$. Therefore, $ab = c(jb)$, which says that $c \mid ab$.
    (b) True: If $c \mid a$ and $c \mid b$, there exist $j$ and $k$ with $a = cj$ and $b = ck$. Therefore, $ab = c^2(jk)$, which means that $c^2 \mid ab$.
    (c) False: Let $c = 4$, $a = 1$, $b = 7$. Then $c \nmid a$, $c \nmid b$, but $c \mid a + b$.

13. Note that $n^2 - n = n(n - 1)$. If this is a prime, we must have $n = \pm 1$ or $n - 1 = \pm 1$. These give $n^2 - n = 0, 2, 2, 0$. Therefore, $n = -1$ and $n = 2$ make $n^2 - n$ prime.

15. If $n$ is even, then $n^2$ and $3n$ are even, so $n^2 - 3n$ is even. If $n$ is odd, $n^2$ and $3n$ are odd, and the difference of two odds is even.

17. If $a$ is even, then $a^m$ and $a^n$ are even, so $a^m - a^n$ is even. If $a$ is odd, then $a^m$ and $a^m$ are odd, and the difference of two odds is even.

19. If $a \mid b$, there exists $k$ with $b = ak$. If $b \mid a$, there exists $\ell$ with $a = b\ell$. Therefore, $b = (b\ell)k$. Divide by $b$ to obtain $1 = \ell k$. The only way for the product of two positive integers to be 1 is for each to be 1. Therefore, $k = \ell = 1$, so $a = b$.

21. (a) If $a \mid b$, there exists $k$ with $b = ak$. Therefore, $bc = ack$, which implies that $ac \mid bc$. (b) If $ac \mid bc$, there exists $\ell$ with $ac\ell = bc$. Divide by $c$ to obtain $a \mid b$.

23. $n = 1$

25. (a) All odd primes, (b) None

27. $2 \cdot 3 \cdot 5 \cdot 7 \cdot 11 \cdot 13 + 1 = 59 \cdot 509$,　$2 \cdot 3 \cdot 5 \cdot 7 \cdot 11 \cdot 13 \cdot 17 + 1 = 19 \cdot 97 \cdot 277$

29. (a) $p$ must be odd. If its last digit is 5, it is a multiple of 5. (b) 101, 103, 107, 109

31. $37 = 3 + 17 + 17$, $59 = 3 + 13 + 43$, $2013 = 13 + 7 + 1993$ (there are many other ways)

33. For each prime $p$, there is an $n$ with $n^2 < p < (n+1)^2$. Suppose there is a largest such $n$, call it $N$. Then all primes are less than $(N+1)^2$, which contradicts Euclid's theorem. Therefore, there are infinitely many such $n$.

35. (a) $q = 14, r = 1$, (b) $q = 8, r = 5$, (c) $q = 15, r = 10$, (d) $q = 9, r = 0$

37. (a) quotient $= 1$, remainder $= 0$, (b) quotient $= 0$, remainder $= 9$, (c) quotient $= 24$, remainder $= 6$, (d) quotient $= 6$, remainder $= 0$

39. (a) $q = 4, r = 0$, (b) $q = 0, r = 47$, (c) $q = -29, r = 0$, (d) $q = -18, r = 3$

41. Wednesday

43. (a) 5, (b) 1

45. $n = -3, -2, 0, 1$

47. Yes=5, No=10, Abstain=1

49. (a) 12, (b) 3, (c) 1

51. Let $d = \gcd(n, n+1)$. Then $d \mid (n+1) - n = 1$.

53. Let $d = \gcd(2n - 1, 2n + 1)$. Then $d \mid (2n + 1) - (2n - 1)$.

55. Let $d = \gcd(6m + 5, 7m + 6)$. Then $d \mid 6(7m + 6) - 7(6m + 5) = 1$, so $d = 1$.

57. Let $d = \gcd(k, k - 2)$. Then $d \mid k - (k - 2) = 2$, so $d = 1$ or 2. But $d \mid k$, and $k$ is odd. Therefore, $d \neq 2$, so $d = 1$.

59. $2n^2 - 1 = 2(n+1)(n-1) + 1$. An integer that divides both $2n^2 - 1$ and $n + 1$ must therefore divide 1.

61. Let $d = \gcd(b, c)$. Then $d \mid c$, so $d \mid a$. Also, $d \mid b$.

63. $(n + 1)(n! + 1) - ((n+1)! + 1) = n$, so $\gcd(n! + 1, (n+1)! + 1)$ divides $n$ and $n! + 1$.

65. $(a + b)d - (c + d)b = \pm 1$

67. (a) $2 = 14 \cdot (-7) + 100$, (b) $6 = 6 \cdot 1 + 84 \cdot 0$, (c) $14 = 630 \cdot (-2) + 182 \cdot 7$, (d) $24 = 1848 \cdot 25 + 1776 \cdot (-26)$

69. Let $d = \gcd(n^2, n^2 + n + 1)$. Then $d \mid (n^2 + n + 1) - n^2$, so $d \mid n + 1$ and therefore $d \mid (n + 1)(n - 1) = n^2 - 1$.

71. (a) $13 = 2 \cdot 5 + 3$,　$5 = 1 \cdot 3 + 2$,　$3 = 1 \cdot 2 + 1$,　$2 = 2 \cdot 1 + 0$
(b) $111111111111 = 100001000 \cdot 11111 + 111$,　$11111 = 100 \cdot 111 + 11$,　$111 = 10 \cdot 11 + 1$,　$11 = 11 \cdot 1 + 0$
(c) If $a = bq + r$ with $r > 0$, then $c = dQ + R$, where $Q$ is $q$ ones, each separated by $b - 1$ zeros, and the last 1 is followed by $r$ zeros; and $R$ is $r$ ones.

73. (a) $d' = 42$, (b) $d = 14$, (c) If $e$ is a common divisor, then $e \mid 210$ and $e \mid 294$, so $e \mid \gcd(210, 294) = 42$. Since also $e \mid 490$, we have $e \mid d = \gcd(42, 490) = 14$.

75. (a) 194, (b) 21, (c) 133
77. (a) $234_4$, $1114_4$, $3324_4$, (b) Group the digits into pairs and use $00 = 0, 01 = 1, 10 = 2, 11 = 3$.
79. (a) $2010011_3$, $1102112_3$, $122001_3$, (b) Change each base 9 digit to base 3.
81. $6 = 110_2$, $28 = 11100_2$, $496 = 111110000_2$ (compare with Theorem 16.5).
83. Expand the final number in base 20: $a_0 + a_1 20 + a_2 20^2 + \cdots$. Then $a_i$ is the number of people worth $i$ billion.
85. If $n = rs$ with an odd number $r > 1$, then $a^n + 1$ has $a^s + 1$ as a factor.
87. If $n = rs$ with $r, s > 1$, then $10^n - 1 = (10^r - 1)(10^{r(s-1)} + \cdots + 10^r + 1)$. Therefore, $(10^r - 1)/(10 - 1)$ is a nontrivial factor of $10^n - 1$

## Chapter 3

1. (a) $x = 2 + 4t, y = 1 - 3t$, (b) $x = 6 - 7t, y = 3 - 5t$,
   (c) $x = -5 + 23t, y = 2 - 9t$, (d) No solutions
3. (a) (1,45), (2, 40), (3,35), (4, 30), (5, 25), (6, 20), (7, 15), (8, 10), (9, 5),
   (b) (3,2), (c) $x = 5 + t, y = 1 + 9t$ for $t \geq 0$, (d) None
5. (horses, oxen) $= (51, 9)$, or (30, 40), or (9, 71)
7. Solutions are $x = 88 - 11t, y = 10t$ for $1 \leq t \leq 7$
9. $x = 103$
11. Because $\gcd(a, b) = 1$, there is a solution $x_0, y_0$ with integers that are not necessarily positive. The general solution of $ax - by = c$ is $x = x_0 - bt, y = y_0 - at$. Let $t$ be a large negative number.
13. largest $= 38$, smallest $= 34$.
15. $x = 2 - 14z + 5t, y = -1 + 7z - 3t$, where $t, z$ are integers
17. $x = 4 - 12y + 5t, z = 6y - 2 - 3t$, where $t, y$ are integers
19. The customer.
21. (a) (i) Use the solution for 10 and then add thirty 3-cent stamps
    (ii) Use the solution for 8 and then add eighty-four 3-cent stamps.
    (iii) Use the solution for 8 and then add ninety-eight 3-cent stamps.
    (b) Every number greater than $k + a - 1$ differs from a number on the list by a positive multiple of $a$.
23. Three 12-inch blocks and five 7-inch blocks

## Chapter 4

1. (a) $3^2 \cdot 5^4$, (b) 5625 is a square
3. Use Theorem 4.2 with $a = b$.
5. $a = b = 5$
7. (a) $2a = (a + b) + (a + c) - (b + c)$ is a multiple of 3, so $a = 3a - 2a$ is a multiple of 3.
   (b) Let $p$ be prime and let $p^a$, $p^b$ and $p^c$ be the powers of $p$ in the prime factorization of $r, s, t$. Then $a + b, b + c, a + c$ are multiples of 3.
9. $n = 10^k$ with $k = 0, 1, 2, \ldots$

11.  (a) Write $a = 2^{a_2}3^{a_3}\cdots$ and $b = 2^{b_2}3^{b_3}\cdots$. By Proposition 4.6, $na_p \leq nb_p$ for each $p$. Use Proposition 4.6 again to get $a \mid b$.
     (b) Write $a = 2^{a_2}3^{a_3}\cdots$ and $b = 2^{b_2}3^{b_3}\cdots$. Proposition 4.6 says that $ma_p \leq nb_p$ for each $p$. Use Proposition 4.6 again to get $a \mid b$.
     (c) Let $a = 4, b = 2, m = 1, n = 2$.

13.  $d \mid c$, so $d \mid a + b$. Since $d \mid a$, we have $d \mid b$.

15.  The answer is 3: If we have four consecutive integers, one of them is divisible by 4. An example of three consecutive squarefree integers is 1, 2, 3.

17.  (a) 5, (b) 3, 7, (c) 31, (d) $|p - q| = 2$

19.  Let $n = 2^{n_2}3^{n_3}\cdots$. Then $r = n_2$ and $m = 3^{n_3}\cdots$.

21.  (a) $p^3$, (b) $p$

23.  (a) 84, (b) 84

25.  15, 60

27.  $a = b$ (or $|a| = |b|$ if negative numbers are allowed)

29.  (a) Look at the exponents in the prime factorizations of $a, b$ and use the fact that $\max(x, y) + \min(x, y) = x + y$.
     (b) $a = 2, b = 4, c = 8$.

31.  $a = p^i$ and $b = p^j$ with $0 \leq i \leq 2, 0 \leq j \leq 2$, and $\max(i, j) = 2$.

33.  Let $a = 2^{a_2}3^{a_3}\cdots$ and $b = 2^{b_2}3^{b_3}\cdots$. Then $0 \leq a_p, b_p \leq n$ and $\max(a_p, b_p) = n$ for each $p$. If the max is $a_p$ then there are $n_p + 1$ possibilities for $b_p$. If the max is $b_p$ then there are $n_p + 1$ possibilities for $a_p$. But $(a + p, b_p) = (n_p, n_p)$ was counted twice.

## Chapter 5

1.  (a) $2^3 \cdot 3^4 = 648$
    (b) $2^{15} \cdot 3^{10} \cdot 5^6$

3.  $m^n = m^{ax+by}$ with $x, y \geq 0$.

5.  Write $n = 2^a 3^b 5^c \cdots$. Then $2 + a$ and $1 + b$ are multiples of 3, so $a \geq 1$ and $b \geq 2$.

7.  Write $n = 2^a 3^b 5^c \cdots$. Then $a + 1$ and $c + 1$ are multiples of 3, so $a, c \geq 2$.

9.  $n = 1, 2, 3, 6$

11.  (a) If $x = a/b$ then $x^2 = a^2/b^2$.
     (b) If $\sqrt{2} + \sqrt{3}$ is rational, then $(\sqrt{2} + \sqrt{3})^2 = 5 + 2\sqrt{6}$ is rational. But $\sqrt{6}$ is irrational.

13.  $x = 2$

15.  $x = 1/2$

17.  $x = 5/3$

19.  (a) Use $(\sqrt{2} + \sqrt{3})^2 = 5 + 2\sqrt{6}$ and $(\sqrt{2} + \sqrt{3})^4 = 49 + 20\sqrt{6}$.
     (b) The only possible rational roots of $x^4 - 10x^2 + 1$ are $\pm 1$.

21.  $(37, 684, 685)$

23.  $(63, 16, 65), (33, 56, 65)$

25.  $(24, 18, 30)$ and $(80, 18, 82)$

27. $(85, 132, 157), (85, 3612, 3613), (85, 204, 221), (85, 720, 725)$

29. The case of odd $n$ is given in the hint. If $n = 2^r k$ with $r \geq 2$, use the method preceding Theorem 3.5. If $n = 2k$ with $k$ odd, construct a triple for $k$ and double it.

31. Starting with an odd $z$: $n = (z+1)/2$, $m = (z-1)/2$.
    Starting with $z$ that is a multiple of 4: $n = z/2$, $m = 1$.

33. $(x+y)(x-y) = p$ with $x > y > 0$ implies that $x + y = p$ and $x - y = 1$.

35. $(x, y) = (8, 7), (4, 1)$

37. If $n \geq 3$, then $x^n = (x^{n-1})(x)$, and $x^{n-1}$ and $x$ have the same parity. If $n = 2$: If $x$ is odd, $x^2 = (x^2)(1)$. If $x = 2k$ is even, $x^2 = (2k^2)(2)$.

39. 249

41. $2^{47}$

43. No.

45. The power of $p$ dividing $p^2!$ is $p^{p+1}$, the power of $p$ dividing $(p^2 - p)!$ is $p - 1$, and the power of $p$ dividing $p!$ is $p^1$. Write

$$\binom{p^2}{p} = p^2!/(p!(p^2 - p)!).$$

47. (a) Use the proof of Theorem 3.10, leaving out the factor for $p = 2$ and using $S(3, 5, \ldots, p_m)$ in place of $S(2, 3, 5, \ldots, p_m)$.
    (b) If you multiply the equation in (a) by $(1 - 1/4)^{-1}$, you get $\zeta(2)$.

# Chapter 6

1. (a) 13, (b) 2, (c) 8, (d) 3, (e) 7, (f) 0

3. (a) $96 - 6 = 10 \cdot 9$, (b) $101 - (-9) = 10 \cdot 11$, (c) $-5 - 13 = 9(-2)$

5. (a) 1, (b) 1, (c) 2, (d) 5, (e) 3

7. (a) $95 - 5 = 10 \cdot 9$, (b) $213 - (-7) = 10 \cdot 22$, (c) $4 - (-20) = 6 \cdot 4$

9. 1, 3, 37, 111

11. 1, 7

13. 9, because $12345 \equiv 9 \pmod{12}$

15. (a) Try $n \equiv 0 \pmod 2$ and $n \equiv 1 \pmod 2$. Both give $n(n+1) \equiv 0 \pmod 2$.
    (b) Try $n \equiv 0, 1, 2, 3, 4, 5 \pmod 6$.
    (c) $\binom{n+1}{2} = (n+1)n/2$ and $\binom{n+2}{3} = (n+2)(n+1)n/6$.

17. Try $n \equiv 0, 1, 2, 3, \ldots, 8 \pmod 9$ and compute $n^3 + (n+1)^3 + (n+2)^3 \pmod 9$. A more efficient way is to show that if $x \equiv 1 \pmod 3$ then $x^3 \equiv 1 \pmod 9$, and if $x \equiv -1 \pmod 3$ then $x^3 \equiv -1 \pmod 9$. Among any three consecutive integers, both of these occur and cancel each other. The remaining number is a multiple of 3; when it's cubed it's 0 mod 9.

19. 1 or 5

21. (a) True, (b) False, (c) True, (d) True, (e) True

23. Squares are 0 or 1 mod 3, so $x^2 + 3y^2$ cannot be 2 mod 3.

25. Let $q = p + 2$. (a) We must have $p \equiv 2 \pmod 3$ and $q \equiv 1 \pmod 3$.
    (b) Look at the cases $p \equiv 1 \pmod 4$ and $p \equiv 3 \pmod 4$.
    (c) Use Proposition 4.4.

27. The case $n = 0$ is easy. Assume the statement is true for $n = k$, so $5^k \equiv 1 + 4k \pmod{16}$. Multiply by 5 to obtain $5^{k+1} \equiv 5 + 20k \equiv 5 + 4k \equiv 1 + 4(k+1) \pmod{16}$.

29. (a) $x$ is congruent mod 9 to one of $0, 1, 2, \ldots, 8$. Verify for each of these.
    (b) No combination of $0, \pm 1$ plus $0, \pm 1$ adds up to 3 mod 9.

31. Show that a cube is $0, 1,$ or $-1$ mod 9 (see Exercise 29). The only ways to make a sum that's 0 mod 9 are $0 + 0 + 0$ and $0 + 1 + (-1)$ (in some order).

33. (a) Alice's strategy is to make the number of markers remaining be 0 mod $m + 1$ after she plays. If $n \equiv 0 \pmod{m+1}$, then Bob uses this strategy after Alice's first move and Bob wins.
    (b) Alice's strategy is to make the number of markers remaining be 1 mod $m + 1$ after she plays.

35. $F_{n+2} = F_n + F_{n+1} \equiv F_n \pmod{F_{n+1}}$. Also, $F_{n+3} = F_{n+1} + F_{n+2} \equiv F_{n+2} \pmod{F_{n+1}}$.

37. Suppose there are only finitely many: $p_1, \ldots, p_r$. Form $N$ as in the hint. Then $N$ is a product of odd primes. If all of these primes are 1 mod 4, then their product is 1 mod 4, contradiction. Let $p \equiv 3 \pmod 4$ divide $N$. Then $p$ is not one of the primes already found, contradicting the assumption that $p_1, \ldots, p_r$ are all of the primes that are 3 mod 4.

39. (a) 6 (mod 13), (b) 1 (mod 11), (c) 6 (mod 10)

41. (a) 6 (mod 97), (b) 91 (mod 97), (c) 93 (mod 97), (d) 36 (mod 97)

43. (a) 1 (mod 97), (b) 1 (mod 97), (c) 1 (mod 97), (d) 1 (mod 97) (Look at Section 8.1)

45. (a) 4, (b) 1, (c) 4, (d) 1, (e) 1, (f) 3

47. $a = 6$

49. $ab = 36$

51. $a = 1$

53. $N \equiv z - y + x - w + \cdots - a \pmod{11}$ and $R \equiv a - b + c - d + \cdots - z \pmod{11}$.

55. $x = 7, y = 4$

57. (a) 8 (mod 21), (b) 11 (mod 19), (c) 12 (mod 121)

59. (a) 12 (mod 25), (b) 11 (mod 17), (c) 57 (mod 169)

61. $n = 0, 5, 10, 15, 20, 25$

63. $n = 0, 12, 24$

65. (a) 47 (mod 100), (b) 94 (mod 100)

67. (a) 23 (mod 120), (b) 46 (mod 120)

69. If $5 \nmid y$, divide by $y$ to get $(x/y)^2 \equiv 2 \pmod 5$, which is impossible. If $x, y$ are multiples of 5, then $x^2 - 2y^2$ is a multiple of 25.

71. $(x, y) \equiv (0, 3) \pmod{17}$

73. 37 (mod 231)

75. 119

77. 214, 229, 244, 259, 274, 289, 304, 319, 334, 349, 364, 379, 394

79. 2519

81. 43; 103

83. (a) 2076, (b) 2076

85. Let $p_1, \ldots, p_{100}$ be distinct primes. Solve $x \equiv -i \pmod{p_i^2}$ for $1 \le i \le 100$. Then $p_i^2 \mid x_i$ for $1 \le i \le 100$.

87. Suppose $p$ is a prime dividing $\gcd(b, n)$. Consider the case $p \mid m$ and the case $p \nmid m$, $p \mid n$.

89. $x = -1$ solves the congruences. Add 60 to get $x = 59$.

91. (a) Suppose the gcd is not 1. Let $p$ be a prime dividing the gcd. Then $p \mid M_i = m_1 m_2 \cdots m_i$, so $p$ divides one of the factors $m_j$. Also, $p \mid m_{i+1}$, so $p \mid \gcd(m_j, m_{i+1})$. Contradiction.
(b) By Proposition 2.17, $m_1 m_2 \mid n$. Suppose $M_k \mid n$. We have assumed that $m_{k+1} \mid n$. Since $\gcd(M_k, m_{k+1}) = 1$ from part (a), Proposition 2.17 says that $M_k m_{k+1} \mid n$. This is the same as saying $M_{k+1} \mid n$. By induction, $M_i \mid n$ for all $i$. In particular, $M_r \mid n$.

## Chapter 7

1. ciphertext = NCONDA; decryption function: $3x + 13$

3. plaintext = TODAY

5. plaintext = RAIN

7. encryption function: $3x + 1$, plaintext = DONE

9. A and N encrypt to B

11. $a = 7$, plaintext = SOLVED

13. TIKSJUSFERFV

15. YOUMADEMELAUGH

17. key = SLEEP, plaintext = ZZZZZZZZZ $\cdots$ ZZZZ

19. (a) MATHISFUN, (b) EVERYONEKNOWSTHAT

21. plaintext = ciphertext

23. IFYOUCANREADTHISTHANKACRYPTOGRAPHER

25. 01010101

27. 000100

29. $(c_0, c_1, c_2) = (1, 0, 1)$

31. LADX

33. CASH

35. $B_n = D(C_n) \oplus C_{n-1}$

37. secret = 5

39. forgery = B, secret = 18

## Chapter 8

1. 1 $\pmod{73}$

3.   5 (mod 11)

5.   4 (mod 47)

7.   Use $2222 \equiv 3$ (mod 7) and $5555 \equiv 5$ (mod 6) and Fermat's theorem to get $2222^{5555} \equiv 5$ (mod 7). Use $5555 \equiv 4$ (mod 7) and $2222 \equiv 2$ (mod 6) and Fermat's theorem to get $5555^{2222} \equiv 4^2 \equiv 2$ (mod 7).

9.   9

11.   $11^{16} \equiv 1$ (mod 17) by Fermat, and $16^{11} \equiv (-1)^{11}$ (mod 17).

13.   If $n \equiv 0$ (mod 11), the statement is easy. If $11 \nmid n$, then $n^{10} \equiv 1$ (mod 11), so $n^{23} \equiv n^3$ and $n^{13} \equiv n^3$.

15.   3

17.   75

19.   If $p \mid a$, the statement is easy. Show that if $p \nmid a$ then $a^{(p-1)!} \equiv 1$ (mod $p$).

21.   Fermat's theorem says that $1^n + 2^n + 3^n + 4^n$ (mod 5) depends only on $n$ mod 4. Check the cases $n = 1, 2, 3, 4$ to obtain the result.

23.   $ab \equiv a^{p-1} \equiv 1$ (mod $p$)

25.   (a) 60, (b) 80, (c) 60, (d) 80

27.   1 (mod 21)

29.   18 (mod 35)

31.   11 (mod 35)

33.   0, 1

35.   401

37.   202

39.   2849

41.   (a) 3, 4, 6 , (b) None, (c) 5, 8, 10, 12, (d) 11, 22, (e) None

43.   $n = 4$

45.   $n\phi(n) = n^2 \prod_{p \mid n}(1 - 1/p) = \phi(n^2)$.

47.   $a^{\phi(b)} + b^{\phi(a)} \equiv 1 + 0$ (mod $b$), and similarly for $a$

49.   (a) False, (b) True

51.   (a) $\phi(N) = \phi(2)\phi(3)\phi(5) \cdots \geq 8$.
      (b) There is a prime $p$ dividing $N$, and all primes are assumed to divide $N$.
      (c) Part (b) yields $\phi(N) = 1$, which is impossible by (a).

53.   (a) 869, (b) Work mod 11 and mod 100, then recombine using (a). The answer is 869.

55.   $(p - 1)! \equiv (p - 1) \cdot (p - 2)! \equiv -((p - 2)!$ (mod $p$).

57.   This is easy if $p \mid m$. If $p \nmid m$, then $m^p \equiv m$ and $(p - 1)! \equiv -1$ (mod $p$).

59.   Write the binomial coefficient as $\frac{2p(2p-1)(2p-2)\cdots(2p-p+1)}{p \cdot (p-1)!}$. The $2p/p$ yields 2. The remaining factors in the numerator yield $(-1)^{p-1}(p-1)!$ (mod $p$), which cancels the similar expression in the denominator.

61.   1

63.   (a) The right side is congruent to $(-1^2)(-2^2)(-3^2) \cdots (-((p - 1)/2)^2)$ mod $p$.
      (b) If $p \equiv 1$ (mod 4), then $(p - 1)/2$ is even.

65. Show that if $a + b \geq 2p - 1$ then $a \geq p$ or $b \geq p$.

## Chapter 9

1. (a) No: $\gcd(e, (p-1)(q-1)) \neq 1$, (b) Yes, $d = 103$, (c) No: $\gcd(e, (p-1)(q-1)) \neq 1$, (d) Yes, $d = 275$
3. (a) 28, (b) 4, (c) 11, (d) 54
5. 12
7. $n = 47233$ and $d = 28983$
9. $250519 = $ YES
11. $de \equiv 1 \pmod{p(p-1)}$
13. (a) $c = 87694236463$, (b) $d = 53380876259$, (c) Check that $c^d \equiv 8051216$ $\pmod{n}$.
15. ATTHEMOVIE
17. TOBEORNOTTOBE
19. There exist $x$ and $y$ with $e_1 x + e_2 y = 1$. Then $c_1^x c_2^y \equiv m \pmod{n}$, where $c_1, c_2$ are the two ciphertexts.
21. (a) $r = pq - (p-1)(q-1) + 1$
    (b) The equation is $x^2 - (p+q)x + pq = (x-p)(x-q)$.
    (c) The quadratic equation is $x^2 - 978x + 218957$. The quadratic formula gives the roots 347 and 631.
23. She divides $m_1$ by 123 mod $n$. Note that $(123^e)^d \equiv 123 \pmod{n}$.
25. $(8, 19)$ is the forgery
27. $(921, 636)$ is the forgery.
29. (a) $n = 33$, (b) $d = 3$, (c) $s = 29$
31. (a) $n = 51$, (b) $d = 13$, (c) $s = 5$
33. Yes
35. No
37. $m = 72$
39. $m_0 = 5$, $d = 23$, $m_0^d \equiv 47 \pmod{57}$

## Chapter 10

1. (a) $2, 3 \pmod{19}$, (b) $2, 3 \pmod{19}$, (c) $(x-2)(x-3) \pmod{19}$
3. $10, 39 \pmod{7^2}$
5. $1, 2 \pmod 9$
7. No solutions
9. (a) $1, 8, 10, 17 \pmod{21}$, (b) $1, 31 \pmod{35}$

## Chapter 11

1. 4
3. 2
5. 5

7.  (a) Use Corollary 11.2, (b) 8

9.  $\text{ord}_n(a) = 1$ means that $a^1 \equiv 1 \pmod{n}$.

11. (a) Since $3^{32} \equiv -1 \pmod{p}$, we have $3^{64} \equiv 1$. Theorem 11.1 says that $\text{ord}_p(3)$ divides 64 but does not divide 32.
    (b) Use Corollary 11.2.

13. (a) Imitate the proof of Proposition 11.4, (b) We have $p = 5$, so $q \equiv 1 \pmod 5$. Try dividing 11111 by $q = 11, 31, 41, \ldots$ to find $11111 = 41 \cdot 271$.

15. Since $a^n \equiv 1 \pmod{a^n - 1}$ and no smaller exponent works (because $a^n$ is too small), $\text{ord}_{a^n-1}(a) = n$. Now use Corollary 11.2.

17. Let $\ell = \text{lcm}(b, c)$ and $z = \text{ord}_{mn}(a)$. Then $a^\ell \equiv 1 \pmod m$ and $a^\ell \equiv 1 \pmod n$, so $a^\ell \equiv 1 \pmod{mn}$, so $z \mid \ell$. Since $a^z \equiv 1 \pmod{mn}$, we have $a^z \equiv 1 \pmod m$, so $b \mid z$. Similarly, $c \mid z$, so $\ell \mid z$.

19. $b^x \equiv b^y \pmod n \Leftrightarrow b^{x-y} \equiv 1 \pmod n \Leftrightarrow \text{ord}_n(b) \mid x - y$

21. (a) Show that $2^j \not\equiv 1 \pmod{13}$ for $1 \le j \le 11$. Alternatively, since $\text{ord}_{13}(2) \mid 12$, show that $2^j \not\equiv 1 \pmod{11}$ for $j = 1, 2, 3, 4, 6$.
    (b) 2, 6, 7, 11.

23. 3, 5, 6, 7, 10, 11, 12, 14

25. $(y^2)^{(p-1)/2} \equiv y^{p-1} \equiv 1 \pmod p$

27. If $g$ is a primitive root, so is $g^{-1}$, by Exercise 12. Pair $g$ with $g^{-1}$. Since $p > 3$, $g$ does not pair with itself.

29. (a) The sum is $(p-2)(p-1)/2 = (p-1)/2 + (p-1)(p-1)/2$.
    (b) Use (a) and Proposition 7.6.
    (c) The numbers $g^0, g^1, g^2, \ldots, g^{p-2} \pmod p$ are a rearrangement of $1, 2, \ldots, p-1$.

31. (a) Use Proposition 7.6.
    (b) $x^2 \equiv y^2 + 2 + y^{-2} \equiv y^{-2}(y^4 + 1) + 2 \equiv 2 \pmod p$.

33. (a) If $a$ is a square then $x^2 \equiv (g^j)^2 \equiv a \equiv g^i \pmod p$. Conversely, if $i$ is even then $a \equiv (g^{i/2})^2$.
    (b) There are $(p-1)/2$ even numbers $i$ and $(p-1)/2$ odd numbers $i$ with $0 \le i \le p - 2$.

35. (a) $.\overline{027}$, (b) 3

37. (a) $3/7 = .\overline{428571}$ and $428 + 571 = 999$, (b) $5/7 = .\overline{714285}$ and $714 + 285 = 999$, (c) $58823529 + 41176470 = 99999999$, (d) $29411764 + 70588235 = 99999999$, (e) $692 + 307 = 999$

39. Here are three random examples:

$$1/47 = 0.\overline{0212765957446808510638297872340425531914893617}$$

$$1/19 = 0.\overline{052631578947368421}$$

$$1/11 = 0.\overline{09}$$

41. Use Proposition 11.11. Let $\ell = \text{lcm}(a, b)$ and $z = \text{ord}_{mn}(10)$. Then $10^\ell \equiv 1 \pmod m$ and $10^\ell \equiv 1 \pmod n$, so $10^\ell \equiv 1 \pmod{mn}$, so $z \mid \ell$. Since $10^z \equiv 1 \pmod{mn}$, we have $10^z \equiv 1 \pmod m$, so $a \mid z$. Similarly, $b \mid z$, so $\ell \mid z$.

43. (a) 12, (b) 6

45. 5
47. 14
49. Raise both sides of $g^x \equiv h \pmod{p}$ to the power $(p-1)/2$ and obtain $(-1)^x \equiv h^{(p-1)/2}$.
51. 16

## Chapter 12

1. The key is 2.
3. The key is 18.
5. If Alice chooses $p - 1$ as the secret exponent, then she sends $1 \equiv g^{p-1}$ (mod $p$) to Bob. When Bob raises this to a power, he still gets 1. Moreover, Eve sees Alice send 1 to Bob, so Eve knows that $g^{ab}$ will still be 1.

   If Alice uses 1 as her secret exponent, then she sends $g$ to Bob. Eve sees this and deduces that the exponent is 1. Therefore, when Bob sends $g^b$ (mod $p$) to Alice, Eve knows that this is $g^{ab}$ and therefore obtains the key.
7. (a) $a = 1$ and Alice computes that $g_2^n \equiv a \pmod{p}$.
   (b) $(g_i^n)^{(p-1)/d} = (g_i^{p-1})^{n/d} \equiv 1 \pmod{p}$ (note that $n/d$ is an integer, so $1^{n/d} \pmod{p}$ is defined).
   (c) Let $a \equiv g^z \pmod{p}$. Then $a^{(p-1)/d} \equiv 1 \pmod{p} \Leftrightarrow z(p-1)/d \equiv 0 \pmod{p-1} \Leftrightarrow d \mid z$.
   (d) Bob calculates $g_2^{di} \pmod{p}$ for $0 \le i < (p-1)/d$ until he obtains $a$.
9. $(r, c) = (4, 6)$
11. $(r, c) = (22475007, 45118009)$
13. 14
15. 1948
17. $m_1/m_2 \equiv c_1/c_2$. This yields $m_2 = 5$.

## Chapter 13

1. $x \equiv 5$ or $6 \pmod{11}$
3. $-1$
5. Yes, it has solutions.
7. (a) No solutions because 2 is not a square mod 11.
   (b) Yes. There are solutions mod 11 and solutions mod 19. They can be combined via the Chinese Remainder Theorem.
9. (a) Note that $2^e \equiv 1 \pmod{3}$ for every even integer $e$.
   (b) Since $p \equiv 1 \pmod{4}$, $\left(\frac{3}{p}\right) = \left(\frac{p}{3}\right)$.
   (c) The order divides $p - 1 = 2^{2^m}$, so it must be a power of 2.
   (d) Use Euler's Criterion to show that

   $$-1 \equiv 3^{2^{2^m - 1}} \pmod{p},$$

   so $\mathrm{ord}_p(3) \nmid 2^{2^m - 1}$.
   (e) Change 3 to $a$ in (d).

11.  Because $5 \equiv 1 \pmod 4$, we have

$$\left(\frac{5}{p}\right) = \left(\frac{p}{5}\right) \quad \text{and} \quad \left(\frac{p}{5}\right) = \left(\frac{5}{q}\right).$$

Now use $p \equiv q \pmod 5$.

13.  Let $g$ be a primitive root mod $p$. Write $b \equiv g^i \pmod p$ for some $i$. Show that

$$b^{(p-1)/3} \equiv 1 \pmod p \Leftrightarrow i \equiv 0 \pmod 3,$$

and $i \equiv 0 \pmod 3 \Leftrightarrow b$ is a cube mod $p$. (This proof needs $p \equiv 1 \pmod 3$ so that the exponent $(p-1)/3$ is an integer.)

15.  $p \equiv 1, 3, 7, 9 \pmod{20}$.

17.  (a) First suppose that $n \geq 4$. Then $q \equiv 1 \pmod 4$. Since $n \geq p$, we have $q \equiv 1 \pmod p$. Quadratic Reciprocity yields

$$\left(\frac{p}{q}\right) = 1.$$

If $n < 4$ then we must have $p = 3$ and $n = 3$.
(b) The values of $n$ are 11, 27, 37, 41, 73, 77, 116, 154. The values of $q$ are $n! + 1$. For $p$, take any odd prime $p \leq n$. For example, we could always use $p = 3$ or $p = 11$.

19.  (a) Note that $2^p \equiv 1 \pmod q$.
(b) Use the Supplementary Law for 2, plus part (a).

21.  $\left(\frac{ab}{p}\right) = 1$ and $\left(\frac{b}{p}\right) = \left(\frac{b}{p}\right)^{-1}$.

23.  Use Wilson's Theorem and the Supplementary Law for $-1$.

25.  Let $p \mid N$. Then 2 is a square mod $p$. Now use the Supplementary Law for 2 to get $p \equiv 7 \pmod 8$.

27.  (a) If $n \equiv 0$ or $1 \pmod 2$, then $n^2 + n + 1 \equiv 1 \pmod 2$, so $p$ must be odd.
(b) Rewrite $4(n^2 + n + 1)$.
(c) Part (b) says that $-3$ is a square mod $p$.

29.  $y \equiv 5, 18 \pmod{23}$.

31.  $y \equiv 6, 23 \pmod{29}$.

33.  No.

35.  $x \equiv 12, 18 \pmod{23}$.

37.  $-1$

39.  $+1$

41.  (a) $-1$, (b) Squares make the Jacobi symbol equal $+1$, (c) Compute $\left(\frac{35}{11}\right) = -1$.

43.  (a) $+1$, (b) No, it does not imply that 3 is a square mod 35, (c) Compute $\left(\frac{3}{5}\right) = -1$.

45.  For $\left(\frac{3}{7}\right)$, the set of residues is $\{3, 6, 2\}$. One of them, namely 6, is above $p/2$, so $n = 1$ and $\left(\frac{3}{7}\right) = (-1)^1 = -1$.
For $\left(\frac{3}{13}\right)$, the set of residues is $\{3, 6, 9, 12, 2, 5\}$. Two of them, namely 9 and 12, are above $p/2$, so $n = 2$ and $\left(\frac{3}{13}\right) = (-1)^2 = +1$.

47. Show that $n = (p-1)/2$.

49. (a) $\{10, 12, 14, 16\}$, $n = 4$, $(2/17) = +1$, (b) $\{8, 10, 12\}$, $n = 3$, $(2/13) = -1$, (c) $\{10, 12, 14, 16, 18\}$, $n = 5$, $(2/19) = -1$, (d) $\{16, 18, 20, 22, 24, 26, 28, 30\}$, $n = 8$, $(2/31) = +1$, $x \equiv 8, 23 \pmod{31}$

51. $p = 11$: $n = 2$, $(5/11) = +1$
    $p = 13$: $n = 3$, $(5/13) = -1$
    $p = 17$: $n = 3$, $(5/17) = -1$
    $p = 19$: $n = 4$, $(5/19) = +1$

## Chapter 14

1. $2880 = 2^6 \cdot 3^2 \cdot 5$

3. $1152 = 2^7 \cdot 3^2$

5. $391 = 17 \cdot 23$

7. (a) $621 = 27 \cdot 23 = 3^3 \cdot 23$, (b) $621 = 3^3 \cdot 23$

9. Compute $2^{14} \equiv 4 \pmod{15}$.

11. (a) $2^{11} - 1 = 2047$, (b) 1023 is a multiple of 11, (c) $2047 = 23 \cdot 89$ so 2047 is not prime. Now use the definition of strong pseudoprime.

13. Use Fermat's theorem for $p = 7, 11, 13, 41$ to show if $\gcd(a, 41041) = 1$ then $a^{41040} \equiv 1 \pmod{p}$ for each of these $p$.

15. Write $ed = 1 + k(p-1)(r-1)$. Use Fermat for $p$ and the Carmichael property of $r$ to show that $m^{ed} \equiv m \pmod{p}$ and $\pmod{r}$.

17. The primes $q$ dividing $F$ are 2 and 3. Show that $3^{30} \equiv 1 \pmod{31}$, that $3^{15} \equiv 30 \pmod{31}$, and that $3^{10} \equiv 25 \pmod{31}$.

19. Use the second half of the proof of Pépin's Test with $F = 2^n$ and $N = k$.

21. $\gcd(427 - 333, 893) = 47$ and $893 = 47 \cdot 19$

23. $(937 \cdot 1666)^2 \equiv (2 \cdot 3 \cdot 7)^2 \pmod{n}$ and $\gcd(937 \cdot 1666 - 2 \cdot 3 \cdot 7, 28321) = 223$. Therefore, $28321 = 127 \cdot 223$.

25. (a) Show that $4^{14} \equiv 1 \pmod{15}$, (b) Show that $4^7 \not\equiv \pm 1 \pmod{15}$, (c) Compute $\gcd(4^7 - 1, 15) = 3$ and obtain $15 = 3 \cdot 5$

27. Compute $\gcd(2^6 - 1, 77) = 7$ to get $77 = 7 \cdot 11$.

29. Compute $2^{24} \equiv 1 \pmod{91}$, so $\gcd(2^{24} - 1, 91) = 91$. The method of Subsection 14.3.2: write $24 = 2^3 \cdot 3$, compute $b_0 \equiv 2^3 \equiv 8 \pmod{91}$, $b_1 \equiv 64$, $b_2 \equiv 1$. Then $\gcd(64 - 1, 91) = 7$ and $91 = 7 \cdot 13$.

## Chapter 15

1. Let $N$ be large. Take a rectangle with corners at $(.1, -N)$, $(.9, -N)$, $(.1, N)$, $(.9, N)$. Its area is $1.6N$. If $N = 1000$, the area is larger than 1000.

3. For each side of the polygon, there is a corresponding side that is symmetric across the origin. Therefore, the sides can be paired up. Clearly, no side gets paired with itself.

5. $41 = 5^2 + 4^2$

7. $157 = 11^2 + 6^2$

9.  (a) Yes, (b) No, (c) No

11.  $221 = 5^2 + 14^2$

13.  $1237 = 34^2 + 9^2$

15.  (a) Use the matrix $A = \begin{pmatrix} p & u \\ 0 & \sqrt{2} \end{pmatrix}$.

(b) $\pi\sqrt{2} > 4$

(c) Use the fact that $u^2 + 2 \equiv 0 \pmod{p}$.

(d) Imitate the end of the proof of Theorem 15.4.

17.  The prime factorization of $n$ contains a prime $p \equiv 3 \pmod 4$ to an odd exponent. Suppose $n = x^2 + y^2$, where $x, y$ are rational. Express $x, y$ with a common denominator: $x = a/d$, $y = b/d$. Then $d^2 n = a^2 + b^2$. But $d^2 n$ still has $p$ to an odd exponent, contradiction.

19.  $123 = 7^2 + 7^2 + 5^2 + 0^2 = 11^2 + 1^2 + 1^2 + 0^2 = 8^2 + 7^2 + 3^2 + 1^2$.

21.  (a) Just multiply it out. (b) Again, multiply it out. (c) Take norms of everything in the formula in part (a).

23.  Cube each number from 0 to 8 mod 9. Note that a sum of three numbers that are $0, \pm 1$ mod 9 cannot be 4 or 5 mod 9.

25.  79

27.  $(x, y) = (8, 3)$

29.  (a) $x^2 - 11y^2 = 1 \Rightarrow x^2 \equiv 1 \pmod{11}$.

(b) $(x, y) = (10, 3)$

31.  (a) If both $x$ and $y$ are odd, then $1 - d \cdot 1 \equiv 4 \pmod 8$ (note that $-4 \equiv 4 \pmod 8$). This implies that $d \equiv 5 \pmod 8$, contradiction. Therefore, one, and hence both, of $x, y$ are even.

(b) See the solution to (a). Note that if $dy^2 \equiv 0 \pmod 4$ then $y$ must be even, since $d$ is squarefree.

(c) Combine (a) and (b).

33.  (a) $39^2 - 61 \cdot 5^2 = -4$

(b) $29718^2 - 61 \cdot 3805^2 = -1$

(c) $1766319049^2 - 61 \cdot 226153980^2 = 1$

## Chapter 16

1.  (a) for 2: 1, 2;

for 4: 1, 2, 4;

for 15: 1, 3, 5, 15;

for 30: 1, 2, 3, 5, 6, 10, 15, 30;

for 60: 1, 2, 3, 4, 5, 6, 10, 12, 15, 20, 30, 60.

(b) $\sigma(2) = 3, \sigma(4) = 7, \sigma(15) = 24, \sigma(30) = 72, \sigma(60) = 168$

(c) $\sigma(4)\sigma(15) = 7 \cdot 24 = 168 = \sigma(60)$ but $\sigma(2)\sigma(30) = 3 \cdot 72 \neq 168 = \sigma(60)$.

3.  $\sigma(20) = 42$

5.  The case $n = 1$ is easy, so assume that $n \geq 2$. If $\phi(n) = n - 1$ then every positive integer less than $n$ is relatively prime to $n$.

7.  $\sigma(2^p - 1) = 1 + (2^p - 1) = 2^p$, so $\sigma(\sigma(2^p - 1)) = 1 + 2 + \cdots + 2^p = 2^{p+1} - 1$.

9.  $\phi(5) = \phi(8) = 4$, $\phi(7) = \phi(9) = 6$, $\phi(5) = \phi(10) = 4$, $\phi(13) = \phi(21) = 12$.

11. Let $n = p^a q^b r^c \cdots$ be the prime factorization of $n$. If $\phi(n) = 14$ then there can be only one odd prime power factor (since each supplies $p-1$, which is even). Therefore, $n = 2^a p^b$. We cannot have $2^a > 4$ nor $(p-1)p^{b-1} > 14$. If $2^a > 2$ then there is no odd prime factor, so $n = 2^a$, which cannot yield $\phi(n) = 14$. Therefore, $n = 2^a p^b$ with $a = 0, 1$ and $(p-1)p^{b-1} \le 14$. Trying all cases yields no example with $\phi(n) = 14$.

13. Take the relation $2n = \sum_{d|n} d$ and divide by $n$. Note that is $d \mid n$ then $n/d$ is a divisor of $n$.

15. The squares are open, the rest are closed.

17. $\sigma(pq) = (p+1)(q+1)$ and $\phi(pq) = (p-1)(q-1)$. Add them together.

19. (a) $\sigma(p^a) = 1 + p + p^2 + \cdots + p^a$ and each summand is odd.
    (b) $\sigma(2^a) = 1 + 2 + 2^2 + \cdots + 2^a$ and each term except the first is even.
    (c) Use (a) and (b), plus the multiplicativity of $\sigma$.
    (d) If $n$ is perfect then $\sigma(n) = 2n$ is even.

21. (a) $\mu$ is multiplicative, so $\mu^2$ is multiplicative.
    (b) If $\gcd(m, n) = 1$ then the prime factors of $m$ and $n$ are distinct, so $\omega(mn) = \omega(m) + \omega(n)$.
    (c) Prove the equation for prime powers (it becomes $1 = 1$) and then use multiplicativity.

23. Write $n = 6k$ with $k \ge 2$. Then $1, k, 2k, 3k, 6k$ are divisors of $n$ (there might be more).

25. There are $\binom{k}{k-j} = \binom{k}{j}$ divisors that have exactly $j$ prime factors. Such $d$ have $\mu(d) = (-1)^j$.

27. Because $\sigma$ is multiplicative, if $\sigma(m)$ is prime then $m$ is a prime power. Use the hint plus $\tau(p^a) = a + 1$.

29. $\sigma(n) = \sigma(3^{k-1})\sigma(q) = \left((3^k - 1)/2\right)(1 + q) = \frac{1}{2}(3^k - 1)\frac{1}{2}(3^k + 1)$. If $q$ is prime, then $k \le 3$. Now use the fact that $\frac{1}{2}(3^k + 1) \le \frac{1}{2}(\cdot 3^k + 3^{k-3}) < 2 \cdot 3^{k-1}$.

## Chapter 17

1. $p_0/q_0 = 1/1$, $p_1/q_1 = 5/4$, $p_2/q_2 = 16/13$, $p_3/q_3 = 21/17$, and $q_4 = 64$.
3. $a/b = 4/17$
5. $a/b = 6/11$
7. $43/10$
9. (a) $[2, \overline{1, 1, 1, 4}]$, (b) $(x, y) = (8, 3)$
11. (a) $[3, \overline{1, 6}]$, (b) $(x, y) = (4, 1)$
13. (a) $[n - 1, \overline{1, 2n - 2}]$, (b) $(x, y) = (n, 1)$
15. $(4 + \sqrt{37})/7$
17. $\sqrt{2}$
19. Obtain the equation $x^2 - 2y^2 = -1$ with $x = 2a + 1$ and $y = c$. The odd powers of $1 + \sqrt{2}$ yield $x_n + y_n\sqrt{2}$ that are solutions of $x_n^2 - 2y_n^2 = -1$. Note that $x_n$ is always odd.

## Chapter 18

1.  (a) Expand both sides. (b) Take the square of the absolute value of the two sides of the equation $(x_1 + y_1 i)(x_2 + y_2 i) = (x_1 y_1 - x_2 y_2) + (x_1 y_2 + x_2 y_1)i$ to get an expression for $|z_1 z_2|^2$, and then use part (a) to relate this to $|z_1|^2 |z_2|^2$.

3.  $1 + i, 1 - i, -1 + i, -1 - i$

5.  $(2 + i)(4 + i)$ (or $(-1 + 2i)(1 - 4i)$)

7.  False: Let $\alpha = 2 + i$ and $\beta = 2 - i$.

9.  (a) $85 = (2 + i)(2 - i)(4 + i)(4 - i)$, (b) Use $(2 + i)(4 + i) = 7 + 6i$ to get $85 = 7^2 + 6^2$. Use $(2 + i)(4 - i) = 9 + 2i$ to get $85 = 9^2 + 2^2$.

11. (a) $a + bi$ is irreducible and divides $(c + di)(c - di)$.
    (b) $N(\gamma) = 1$.
    (c) The four possibilities correspond to $\gamma = 1, -1, i, -i$, respectively.

## Chapter 19

1.  The multiplicative inverse of $u_1 u_2$ is $u_1^{-1} u_2^{-1}$, which is a product of elements of $R$, hence is in $R$.

3.  (a) If $a$ is odd, then $1 + b - b^2 \equiv 0 \pmod 2$, which has no solutions. Similarly, $b$ cannot be odd.
    (b) If both $a$ and $b$ are even, then the left side is a multiple of 4.
    (c) If $\alpha\beta = 2$ with $\alpha$, $\beta$ non-units, then $N(\alpha) = \pm 2$. Now use part (b).

5.  (a) $\phi^{-1} = -1 + \phi$, $(1 + \phi)^{-1} = 2 - \phi$, $(1 + 2\phi)^{-1} = \sqrt{5} - 2 = -3 + 2\phi$, $(2 + 3\phi)^{-1} = (7 - 3\sqrt{5})/2 = 5 - 3\phi$.
    (b) $(\phi)(F_{n-1} + F_n \phi) = \phi F_{n-1} + F_n(\phi + 1) = F_n + \phi F_{n+1}$.

7.  (a) $(\sqrt{2} - 1)^{-1} = \sqrt{2} + 1$.
    (b) Multiply it out.
    (c) The norm of the left side is $-(a^2 - 2b^2)$.
    (d) $a = \sqrt{3}b < 2b$, and $a = \sqrt{3}b > b$.
    (e) Take norms of the relation $(2 - \sqrt{3})(a + b\sqrt{3}) = (2a - 3b) + (2b - a)\sqrt{3}$ and use the assumptions that $a^2 = 3b^2$ and that $b$ is smallest.

9.  If $\pi \mid \alpha\beta$ then $\pi\gamma = \alpha\beta$ for some $\gamma$. Since $\pi$ is in the factorization into irreducibles on the left, it (or an associate) must occur in the factorization of $\alpha\beta$ into irreducibles, which comes from the factorization of $\alpha$ and $\beta$.

## Chapter 20

1.  There is a prime $p$ with $n/2 < p \le n$. Since $2p > n$, the prime $p$ occurs exactly once in $n!$.

3.  There is a prime $p$ with $n^2/2 < p \le n^2$. This $p$ occurs in only one column.

5.  (a) $\pi(2x) > .921(2x)/\ln(2x)$ and $\pi(x) < 1.106x/\ln(x)$. Therefore,

$$\pi(2x) - \pi(x) > \frac{1.842x}{\ln x + \ln 2} - \frac{1.106x}{\ln x} > \frac{9x}{(\ln x + \ln 2)(\ln x)} > 0$$

when $x \ge 10^6$.
    (b) The calculation in (a) shows that $\pi(2x) - \pi(x) \to \infty$ as $x \to \infty$.

**Appendix A**

1. 4

3. $r = 2/3$

5. If it's true for $n = k$ then $1 + 2 + \cdots + k + (k+1) = \frac{1}{2}k(k+1) + (k+1) = \frac{1}{2}(k+1)(k+2)$.

7. If it's true for $n = k$ then $1^2 + 2^2 + \cdots + k^2 + (k+1)^2 = \frac{1}{6}k(k+1)(2k+1) + (k+1)^2 = \frac{1}{6}(k+1)(k+2)(2k+3)$.

9. The "proof" fails for the passage from $n = 1$ to $n = 2$.

11. (a) $(a+b)^5 = a^5 + 5a^4b + 10a^3b^2 + 10a^2b^3 + 5ab^4 + b^5$. All of the middle coefficients are multiples of 5.
    (b) $(a+b)^7 = a^7 + 7a^6b + 21a^5b^2 + 35a^4b^3 + 35a^3b^4 + 21a^2b^5 + 7ab^6 + b^7$. All of the middle coefficients are multiples of 7.
    (c) $(a+b)^6 = a^6 + 6a^5b + 15a^4b^2 + 20a^3b^3 + 15a^2b^4 + 6ab^5 + b^6$.
    (d) Coefficients are $1, 10, 45, 120, 210, 252, 210, 120, 45, 10, 1$

13. (a) $(1+1)^n = \sum_{j=0}^{n} \binom{n}{j} 1^j 1^{n-j}$.
    (b) $(1-1)^n = \sum_{j=0}^{n} \binom{n}{j} 1^j (-1)^{n-j}$.

15. (a) Use the identity $F_{2n} + F_{2n+1} = F_{2(n+1)}$.
    (b) $F_1 + F_3 + \cdots + F_{2n}$ can be written as

$$(1/\sqrt{5}) \left( \phi + \phi^3 + \cdots + \phi^{2n-1} \right)$$
$$- (1/\sqrt{5}) \left( (-\phi)^{-1} + \cdots + (-\phi)^{-(2n-1)} \right).$$

Sum the two geometric series, use the fact that $\phi^2 - 1 = \phi$, and simplify the result to get Binet's Formula for $F_{2n}$.

17. (a) $\phi^2 = (3 + \sqrt{5})/2$ and $\phi^{-2} = (3 - \sqrt{5})/2$, (b) Use Binet's Formula to rewrite $F_{n+1}^2 - F_{n-1}^2$ and then use (a) to simplify the resulting expression.

19. 233 pairs

21. $\begin{pmatrix} 42 \\ 14 \end{pmatrix}$

23. $\begin{pmatrix} -11 & 21 \\ 17 & 44 \end{pmatrix}$

25. $\frac{1}{-2} \begin{pmatrix} 1 & -3 \\ -3 & 7 \end{pmatrix} = \begin{pmatrix} -1/2 & 3/2 \\ 3/2 & -7/2 \end{pmatrix}$

# Index